QA21 .A73 2006
The architecture of
 modern mathematics :
Northeast Lakeview Colleg
33784000121509

THE ARCHITECTURE OF MODERN MATHEMATICS

The Architecture of Modern Mathematics
Essays in History and Philosophy

Edited by

J. FERREIRÓS

and

J. J. GRAY

UNIVERSITY PRESS

OXFORD
UNIVERSITY PRESS

Great Clarendon Street, Oxford OX2 6DP

Oxford University Press is a department of the University of Oxford.
It furthers the University's objective of excellence in research, scholarship,
and education by publishing worldwide in

Oxford New York

Auckland Cape Town Dar es Salaam Hong Kong Karachi
Kuala Lumpur Madrid Melbourne Mexico City Nairobi
New Delhi Shanghai Taipei Toronto

With offices in

Argentina Austria Brazil Chile Czech Republic France Greece
Guatemala Hungary Italy Japan Poland Portugal Singapore
South Korea Switzerland Thailand Turkey Ukraine Vietnam

Oxford is a registered trade mark of Oxford University Press
in the UK and in certain other countries

Published in the United States
by Oxford University Press Inc., New York

© Oxford University Press, 2006

The moral rights of the authors have been asserted
Database right Oxford University Press (maker)

First published 2006

All rights reserved. No part of this publication may be reproduced,
stored in a retrieval system, or transmitted, in any form or by any means,
without the prior permission in writing of Oxford University Press,
or as expressly permitted by law, or under terms agreed with the appropriate
reprographics rights organization. Enquiries concerning reproduction
outside the scope of the above should be sent to the Rights Department,
Oxford University Press, at the address above

You must not circulate this book in any other binding or cover
and you must impose the same condition on any acquirer

British Library Cataloguing in Publication Data

Data available

Library of Congress Cataloging in Publication Data

Data available

Typeset by Newgen Imaging Systems (P) Ltd., Chennai, India
Printed in Great Britain
on acid-free paper by
Biddles Ltd., King's Lynn, Norfolk

ISBN 0-19-856793-6   978-0-19-856793-6

1 3 5 7 9 10 8 6 4 2

Contents

Acknowledgements	vii
List of Contributors	ix

Introduction	1
J. Ferreirós and J.J. Gray	

REINTERPRETATIONS IN THE HISTORY AND PHILOSOPHY OF FOUNDATIONS — 45

Frege and the Role of Historical Elucidation: Methodology and the Foundations of Mathematics Michael Beaney	47
Riemann's Habilitationsvortrag at the Crossroads of Mathematics, Physics, and Philosophy José Ferreirós	67
The Riemannian Background to Frege's Philosophy Jamie Tappenden	97
Axiomatics, Empiricism, and *Anschauung* in Hilbert's Conception of Geometry: Between Arithmetic and General Relativity Leo Corry	133

EXPLORATIONS INTO THE EMERGENCE OF MODERN MATHEMATICS — 157

Methodology and metaphysics in the development of Dedekind's theory of ideals Jeremy Avigad	159
Emmy Noether's 'Set Theoretic' Topology: From Dedekind to the Rise of Functors Colin McLarty	187
Tarski on models and logical consequence Paolo Mancosu	209
A Path to the Epistemology of Mathematics: Homotopy Theory Jean-Pierre Marquis	239

ALTERNATIVE VIEWS AND PROGRAMS IN THE PHILOSOPHY OF MATHEMATICS 261

Felix Hausdorff's Considered Empiricism 263
Moritz Epple

Practice-related symbolic realism in H. Weyl's mature view of mathematical knowledge 291
Erhard Scholz

From Kant to Hilbert: French philosophy of concepts in the beginning of the twentieth century 311
Hourya Benis Sinaceur

Relative Consistency and Accessible Domains 339
Wilfried Sieg

CODA 369

Modern mathematics as a cultural phenomenon 371
Jeremy Gray

References 397
Name Index 433
Subject Index 439

Acknowledgements

We wish to thank the contributors to this volume for their comments on a previous version of our introduction. This collective volume was arranged and prepared in the course of the last four years, especially at two events that served us to discuss common viewpoints, choose contributors, and select topics – a conference on the history and philosophy of modern mathematics (May 2002) at the Open University (*Centre for the History of Mathematical Sciences*) and University College London; and an international meeting on the same topic (September 2003) at the Universidad de Sevilla (Spain). The Seville meeting was financed thanks to an 'acción especial' of the Spanish government (BFF2002-10381-E) and complementary financial means offered by the Junta de Andalucia and Vicerrectorado de Investigación (Universidad de Sevilla).

List of Contributors

Jeremy Avigad is an Associate Professor of Philosophy at Carnegie Mellon University, with research interests in mathematical logic, proof theory, automated deduction, and the history and philosophy of mathematics. He was supported by a New Directions Fellowship from the Andrew W. Mellon Foundation in 2003–2004.

Michael Beaney is Reader in Philosophy at the University of York, with research interests in the history of analytic philosophy, methodology and the foundations of reasoning. He is author of *Frege: Making Sense* (Duckworth, 1996) and *Analysis* (Acumen, forthcoming), and editor of *The Frege Reader* (Blackwell, 1997) and (with Erich Reck) of *Gottlob Frege: Critical Assessments of Leading Philosophers* (4 vols., Routledge, 2005).

Hourya Benis Sinaceur has taught Philosophy of Science at the University Paris I-Sorbonne. She is at present Director of Research at the CNRS (France), working at the Institut d'Histoire et de Philosophie des Sciences et des Techniques in Paris. She is the author of *Corps et Modèles. Essai sur l'histoire de l'algèbre réelle*, Paris, Vrin, 1991 (second edition : 1999) and *Jean Cavaillès. Philosophie Mathématique*, Paris, Presses Universitaires de France, 1994. She edited A. Tarski's 'Address at the Princeton University Bicentennial Conference on Problems of Mathematics' (The Bulletin of Symbolic Logic, March 2000) and she translated into French P. Bernays' 'Abhandlungen zur Philosophie der Mathematik', Paris, Vrin, 2003. She has published many papers, in particular on Bolzano, Cauchy, Sturm, Poincaré, Hilbert, and Tarski. She is presently preparing a new French edition of R. Dedekind's 'Was sind und was sollen die Zahlen?' and 'Stetigkeit und irrationale Zahlen'.

Leo Corry is Director of the Cohn Institute for History and Philosophy of Science and Ideas, Tel-Aviv University, where he has taught history and philosophy of mathematics for more than twenty years. He is coeditor of the international journal *Science in Context* published by Cambridge University Press. His recently published books are: *Modern Algebra and The Rise of Mathematical Structures* (Basel: Birkhauser – 2nd, softcover edition: 2003) and *David Hilbert and the Axiomatization of Physics* (2004).

Moritz Epple is Professor of the History of Science at the University of Frankfurt am Main. His interests include the history of topology, the history of mathematical physics, philosophical aspects of modern mathematics, and the relations between the mathematical sciences and society. Recent publications include: *Die*

List of Contributors

Entstehung der Knotentheorie: Kontexte und Konstruktionen einer modernen mathematischen Theorie, Wiesbaden: Vieweg, 1999; 'Topology, matter, and space I: Topological notions in 19th Century natural philosophy', *Archive for History of Exact Sciences* 52 (1998) pp. 297–392; *The end of the science of quantity: Foundations of analysis, 1860–1910*, Chapter 10 in N. Jahnke (ed.): *A history of analysis*, Oxford University Press, 2003.

José Ferreirós is *profesor titular* for Logic and Philosophy of Science at the University of Seville. The author *of Labyrinth of Thought: A history of set theory and its role in modern mathematics* (Basel, Birkhäuser, 1999), he has edited in Spanish works of Riemann, Dedekind, and Cantor, and has published papers on the history and philosophy of logic, mathematics, and general science in journals like *The Bulletin of Symbolic Logic, Historia Mathematica, Science in Context,* and *Crítica*. Among his forthcoming papers is 'Ο $\theta\varepsilon o\varsigma$ $\alpha\rho\iota\theta\mu\eta\tau\iota\zeta\varepsilon\iota$: The rise of pure mathematics as arithmetic with Gauss', in Goldstein, Schappacher & Schwermer, eds., *The Shaping of Arithmetic: Number theory after Carl Friedrich Gauss's Disquisitiones Arithmeticae* (Springer).

Jeremy Gray is Professor of the History of Mathematics and Director of the Centre for the History of the Mathematical Sciences at the Open University, and an Honorary Professor at the University of Warwick. He works on the history of mathematics in the nineteenth and twentieth centuries, and also on issues in the philosophy and social significance of mathematics. He wrote *The Hilbert Challenge*, Oxford University Press, 2000, and he was the editor of and a contributor to *The symbolic universe: geometry and physics, 1890–1930*, Oxford University Press, 1999. His book *Janos Bolyai, non-Euclidean geometry, and Nature of Space*, a Burndy Library publication, MIT, came out in 2004.

Paolo Mancosu is Associate Professor of Philosophy at University of California at Berkeley and Chair of the Group in Logic and the Methodology of Science. His main interests are in mathematical logic and the history and philosophy of mathematics. He is the author of *Philosophy of Mathematics and Mathematical Practice in the Seventeenth Century*, Oxford University Press 1996) and *From Brouwer to Hilbert* Oxford University Press 1998, and a coeditor of *Visualization, Explanation and Reasoning Styles in Mathematics* (2005).

Jean-Pierre Marquis is professor of logic, philosophy of science and epistemology at the Université de Montréal, where he has taught for 12 years. He is particularly interested in philosophy of mathematics and is presently finishing a book on the philosophical aspects of category theory and categorical logic, entitled *From a geometrical point of view: the categorical perspective on mathematics and its foundations*. He is the author of the article on category theory in the *Stanford Encyclopedia of Philosophy*, and has published in *Philosophia Mathematica, Synthese, The Notre Dame Journal of Formal Logic,* and *The Journal of Philosophical Logic*. Recent contributions include "Categories, Sets and the Nature of Mathematical Entities", to appear in *The Age of Alternative Logics. Assessing philosophy of logic and mathematics today*, J. van Benthem, G. Heinzmann, Ph. Nabonnand, M. Rebuschi, H.Visser, eds., Kluwer, and a paper in collaboration with Elaine Landry, "Categories in Contexts: historical, foundational and philosophical", *Philosophia Mathematica*, 2005 vol 13, no 1, 1–43.

Colin McLarty is an Associate Professor of Philosophy, and of Mathematics, at Case Western Reserve University where he has taught for 18 years. He is interested in mathematical and philosophic aspects of category theory, and the history of twentieth-century mathematics. He is the author of *Elementary Categories, Elementary Toposes*, Oxford University Press (1992, third printing 1999); 'Richard Courant in the German Revolution' in *Mathematical Intelligencer* (2001) and 'Exploring Categorical Structuralism' in *Philosophia Mathematica* (2004).

Erhard Scholz is Professor of the History of Mathematics at the University of Wuppertal, where he has been teaching since 1980. He is author of a book on the history of the concept of a manifold (*Geschichte des Mannigfaltigkeitsbegriffs von Riemann bis Poincaré*. Basel – Birkhäuser 1980) and the history of the relationship between theoretical and applied mathematics in the nineteenth century (*Symmetrie – Gruppe – Dualität. Zur Beziehung zwischen theoretischer Mathematik und Anwendungen in Kristallographie und Baustatik des 19. Jahrhunderts*. Basel – Birkhäuser 1989). He has edited and contributed to a book on the history of algebra (*Geschichte der Algebra. Eine Einführung*.) Mannheim, Bibliographisches Institut 1990) and more recently *Hermann Weyl's Raum – Zeit – Materie and a General Introduction to His Scientific Work*. Basel–Birkhäuser 2002.

Wilfried Sieg is Professor of Philosophy and Mathematical Logic at Carnegie Mellon University. He was Head of the Philosophy Department from 1994 to 2005 and is Co-Director of the Laboratory of Symbolic and Educational Computing. He works in proof theory and automated proof search, history of mathematics of the 19th and 20th Centuries, and in the philosophy of mathematics – on issues related to Hilbert's Program and to the conceptual analysis of computability. With W. Buchholz, S. Feferman, and W. Pohlers he wrote the volume Iterated Inductive Definitions and Subsystems of Analysis: Recent proof-theoretical studies (Springer Verlag, 1981) and edited a number of volumes among them Logic and Computation (Compositio Mathematica, 1990), Acting and Reflecting (Synthese Library, 1990), Reflections on the Foundations of Mathematics with Sommer and Talcott (Lecture Notes in Logic, 2002). He has published widely in Studia Logica, Journal of Symbolic Logic, Synthese, Bulletin of Symbolic Logic, Philosophia Mathematica, and Archive of Mathematical Logic. During the last few years he has been involved in editing works by Gödel, Hilbert, and Bernays, but also in constructing a fully web-based introduction to logic, Logic & Proofs.

Jamie Tappenden is Associate Professor in the Department of Philosophy at the University of Michigan. His recent work in Frege scholarship and the history of mathematics has addressed explanation and 'fruitfulness' as objectives motivating mathematical investigation, as sketched in 'Extending Knowledge and "Fruitful Concepts": Fregean Themes in the Philosophy of Mathematics' (*Noûs* December 1995). More recent developments of these themes include 'Frege on Axioms, Indirect Proof and Independence Arguments in Geometry' (*Notre Dame Journal of Formal Logic*, Spring 2000), 'Proof Style and Understanding in Mathematics I: Visualization, Unification and Axiom Choice' (forthcoming in *Visualization, Explanation and Reasoning Styles in Mathematics*, Paolo Mancosu and Klaus Jørgensen (eds.) Kluwer) and 'Recent Work in the Philosophy of Mathematics' (*Journal of Philosophy*, September 2001). He is currently finishing a book on Frege and the mathematics of the nineteenth century.

Introduction

J. Ferreirós and J.J. Gray

The aim of this book of essays is to advance contemporary work in creating stronger links between the history and philosophy of mathematics. It has become clear through several conferences and publications that the present situation at the beginning of the twenty-first century is congenial to this kind of historico-philosophical enterprise. The editors have brought together an important international group of scholars whose contributions focus on the history and philosophy of modern mathematics, roughly from 1800 to 1970.

For most of the twentieth century philosophers of mathematics tended to study their subject from an ahistorical systematic perspective. In the Anglo-American tradition, despite the presence of attractive proposals such as those of Lakatos or Kitcher, historical approaches have remained in the minority. The once powerful German trend of historico-philosophical studies died out, largely as a result of the Nazi period and the Second World War. Philosophers in the Latin traditions (e.g. Cavaillès in France, Gonseth in Switzerland, or De Lorenzo in Spain) have long been groping toward historical perspectives, but here also we lack satisfactory overall pictures and an adequate integration of such proposals with systematic knowledge.

Two complementary developments in recent years have been crucial for the present state of affairs. Within the English-speaking sphere, analytical philosophy has ceased to be regarded as a monolithic, obviously correct approach, and its historical origins have begun to be subjected to careful scrutiny. Through the intermediary of symbolic logic, a core element in that approach, the history of analytical philosophy led into the history of mathematics and its foundations. Recent historians have consciously followed this lead, and our book stresses this connection. Meanwhile, the emergence of a community of professional historians of mathematics led to more sophisticated inquiries into the subject, raising questions about the nature of mathematical knowledge and its origins in mathematical practices.

And yet there are significant problems in researching and writing the history of mathematics that are bringing historians and philosophers together. The history of modern mathematics provides opportunity for realizing that many pivotal mathematical contributions (especially in the period 1870–1930) were not philosophically neutral. Great mathematicians such as Gauss, Riemann, Kronecker, Dedekind, Hilbert, Poincaré, Brouwer or Weyl (to name but a few) have concerned themselves with eminently philosophical questions in the course of their research.

Thus, an adequate historical understanding of modern mathematics calls for the consideration of philosophical issues both in the mathematics itself and in the writing of its history.

As one can see, three different ways of inquiry (the philosophy of mathematics, the historical understanding of recent philosophy, and the history of mathematics) have begun to converge. The result of this process should be a more sophisticated and adequate understanding of history, philosophy, and mathematics. We are confident that the essays included in this collection will help to advance and consolidate such an understanding.

1 The architectural metaphor

There was a time when architecture (civil or military) was regarded as one of the branches of mathematics. This happened not so long ago, still being the case in the eighteenth century when the separation between mathematics and engineering had not yet been clearly established. To be an architect, and not a mere artisan, it was essential to know geometry and to calculate strengths – mathematics was the keystone for turning the art into science. But, reciprocally, architecture has often been the source of analogies applied to science in general and mathematics in particular: the whole world has been seen as an edifice, but so have been each of the sciences.

The architectural metaphor is an old one. If ancient mythologies talk about the sky being supported by columns, in the seventeenth century at least it becomes easy to find talk of the sciences as buildings. We shall give only two examples, taken from Francis Bacon and René Descartes. In the preface to his *Instauratio magna* or *Great Restoration* [1620], Bacon says that previous knowledge of nature is not 'well constructed and built,' but rather a 'splendid bulk lacking foundations.' By contrast, he offers a 'total Restoration, from adequate foundations, of the sciences, arts and all of human knowledge.' As regards Descartes, in his first *Meditation* he states that the point of the famous methodical process of doubt that he once undertook, was in 'commencing anew the work of building from the foundation, if I desired to establish a firm and abiding superstructure in the sciences.' As is well known, the best examples of sound foundations he found in the basic truths of arithmetic and geometry, before he arrived at the basic metaphysical truths of his system.[1] It would be easy to multiply such examples.

In thinking about mathematics and its philosophy, architectural metaphors have played an important role. The big question here has been about the *foundations* of mathematics: the totality of this science is thereby compared to

---

[1] Based on 'demonstrations' that he thought 'to be equal or even superior to the geometrical in certitude and evidence.'

an edifice, with its higher and lower stages, and of course sound foundations to support its weight.

Until the late nineteenth century it was customary to accept that studying the soil into which those foundations penetrate was not itself a mathematical question, but rather a philosophical one, and the diagnosis about where the pillars end and the soil begins was very different from today. In 1854 Riemann begged his listeners pardon when he proceeded to analyse the concept of a manifold or class,[2] this being work 'of a philosophical character.' In his 1887 paper on the number concept, Kronecker would still declare that the study of that concept belonged in 'the open field' of philosophical discussion, not in the 'fenced' domain of a particular science.

By contrast, a very important ingredient in twentieth-century images of mathematics (though not omnipresent) has been the idea of a self-contained discipline, one that would account for its own foundations. This move was begun by authors like Dedekind, Cantor, Frege and Peano, who turned the elementary theory of numbers and sets into a purely mathematical topic, but it was Hilbert who played a decisive role in bringing that image centre stage.[3] Strange image of an edifice whose pillars and beams would be essentially of the same material as the terrain on which it has been built. But this was indeed the hope, to show that the stony edifice of mathematics was built on nothing but stone – and not of course on sand.[4]

Two conscious and sophisticated modern uses of the architectural metaphor are worth mentioning. In a lecture course of 1905, 'The Logical Principles of Mathematical Thinking,' Hilbert turned the image away from its naïve implications, and accommodated it to the actual historical development of mathematics:

> The edifice of science is not raised like a dwelling, in which the foundations are first firmly laid and only then one proceeds to construct and to enlarge the rooms. Science prefers to secure as soon as possible comfortable spaces to wander around and only subsequently, when signs appear here and there that the loose foundations are not able to sustain the expansion of the rooms, it sets about supporting and fortifying them. This is not a weakness, but rather the right and healthy path of development.

As we see, Hilbert took great liberty in adapting the basic metaphor, to the extent of risking to lose sight of its key meaning. Recall that, according to Descartes, the work of the sciences can only begin when the philosophical foundations have been solidly established, 'as the removal from below of the foundation necessarily involves the downfall of the whole edifice' (which is of course the case in real architecture).

---

[2] Riemann Über die Hypothesen, etc., in Riemann [1990]. For the identification of manifolds as classes, see Torretti 1978, Ferreirós 1999.

[3] Hilbert himself should not be fully blamed, as he was far from the dull idea that metamathematics could solve all epistemological questions about mathematics (Peckhaus, 1900).

[4] As a radical Hermann Weyl declared in the heated post-war atmosphere of 1921.

But even such freedom could be insufficient to make room for the complexity of mathematical knowledge. More faithful to the complex panorama of twentieth century mathematics, Bourbaki shifted the metaphor and talked about mathematics as a *polis*:

> It is like a big city, whose outlying districts and suburbs encroach incessantly, and in a somewhat chaotic manner, on the surrounding country, while the centre is rebuilt from time to time, each time in accordance with a more clearly conceived plan and a more majestic order, tearing down the old sections with their labyrinths and alleys, and projecting towards the periphery new avenues, more direct, broader, and more commodious. (Bourbaki 1948, in Ewald 1996, 1275)

Logicians and experts in foundations would here be seen as engaged in developing some particular new quarters, and perhaps in refurbishing and restoring old edifices and parts of the town.

In spite of the problems encountered by important modern architects such as Hilbert and Bourbaki, many scientists and philosophers still think of mathematics as a finished building, whose foundations are certainly solid. And, since philosophers have a sound and understandable tendency to concentrate on the basics, they normally focus their attention upon logic and set theory, identified as the foundations of the whole edifice. Meanwhile, historians have an equally understandable tendency to look at the rooms and floors, trying to understand how they were built and took their present shape. In the process, they often focus their attention on the build*ers* and not so much the build*ing*.

Needless to say, both attitudes involve a large measure of simplification. If the metaphor has some truth in it, we should keep in mind that any building has been erected by builders – beginning with the main architects, men like the ones studied in the chapters that follow, but not ending with them (one also needs contractors, technicians, masons, and so on). In mathematics, as in science generally, the lineaments of the finished building are never the same as the main architects initially envisioned, which reminds us of the collective nature of knowledge production. On the other hand, the metaphor as elaborated by Bourbaki is certainly more adequate than its original version. The contributions in this volume represent attempts to explore the implications of these observations.

2   The received picture in philosophy of mathematics

The received picture has not changed much during the last 50 years. The Frege–Russell–Hilbert strategy of turning philosophical questions about mathematics into logico-mathematical questions was highly successful in the twentieth century, although there is general agreement that it has not led to the definitive solution of those questions. That strategy has come to *define* the phrase 'philosophy of mathematics' for many, in spite of the efforts of authors such as Lakatos or Kitcher.

Many philosophers and mathematicians simply equate philosophy of mathematics with the study of the pros and cons of three foundational positions, called logicism, intuitionism and formalism. Those who have a closer knowledge of the game will either be aware of more foundational alternatives (say, predicativism) or have a preference for subtler philosophical positions about mathematics (platonism, varieties of structuralism, naturalism). This more sophisticated understanding is well represented in the recent *Oxford Handbook of Philosophy of Mathematics and Logic* edited by Stewart Shapiro. Apart from sections dealing with the history of the discipline, the book has 7 sections devoted to the already mentioned 'big three,' and 15 sections discussing topics that have recently been of relevance. These topics are the positions of naturalism, nominalism, structuralism; logical questions about predicativity, relevance, higher-order logic; and basic questions about the application of mathematics and the concept of logical consequence.

Upon reflection, it is not difficult to realize how much that received picture owes to early twentieth-century studies in logic and foundations, on the one hand, and to the analytic tradition in philosophy, on the other. Significantly absent from that broad picture are a number of topics and positions that have been present for 30 years or more, constituting what Aspray and Kitcher 16 years ago called a 'maverick tradition'. No doubt, recent developments of Anglo-Saxon philosophy of mathematics have absorbed some of the sensitivities of that rebel tradition. In this connection, one can mention especially the cases of naturalism and structuralism, say the work of P. Maddy and S. Shapiro during the 1990s, where one finds attention to issues of methodology and the intention to stay close to mathematical practice. But we shall come back to Kitcher's 'mavericks' in the next section, reserving the present one to more traditional philosophical questions.

Today one may say that we are in a phase of transition. After a time in which foundationalist and anti-foundationalist understandings fought to fill the space, we have come to understand that the dichotomy was too restrictive and ultimately misplaced: to go *beyond* foundational studies is by no means the same as to go *against* them. At the same time, the issue of mathematical practice, which at some point represented the main divide between alternative views, is now present in all philosophical agendas.

2.1 Foundational studies

If we are allowed an attempt at fixing terminology, we might borrow two traditional expressions and use them to mark a distinction. *Mathematical philosophy*[5] will denote the Frege–Russell–Hilbert strategy of solving by logic or mathematics

---

[5] Historically a French phrase, a heritage of nineteenth century positivism, famously adopted by Russell [1919].

Introduction

philosophical questions about mathematics. *Philosophy of mathematics*, by contrast, will designate the broader space in which questions about mathematical knowledge, its underpinnings, and its development are asked – including questions about mathematical practices. Foundational studies belong squarely to mathematical philosophy in this sense, indeed they represent the core of this orientation, its main technical problems. By tradition we consider logicism, intuitionism, Hilbert's program, Gödel's incompleteness results, and the status of set theory as the main issues to be discussed here.[6]

Once there was a general belief that foundational studies began in earnest with the path-breaking work of Frege and Hilbert, and that the famous triad of Russellian logicism, Brouwerian intuitionism, and Hilbertian metamathematics represented the main philosophies of mathematics as of 1930 – 'the big three.' Leaving aside the historical accuracy of this view (for which see below), the fact is that the much-discussed 'foundational crisis' of the 1920s was mainly about two viewpoints, namely the *constructivist* and the '*postulational*' varieties of mathematical philosophy.[7] We prefer to use these terms, instead of contrasting – as usual – intuitionism with formalism, because the latter labels (imposed by Brouwer, 1912b) do not adequately represent the key traits of the mathematical traditions then in dispute.[8]

The key issue in dispute was whether mathematical *definitions and proofs* have to be *constructed* from a restricted set of basic objects and operations, or can they be based on free (but consistent) *postulates* such as the axioms of Infinity and Choice. Brouwer had an elaborate idealistic/intuitionistic philosophy on which he justified his choice of basic objects and operations, but it is and was perfectly possible to offer other accounts, independent of that particular worldview (see, e.g., Weyl 1918, Skolem 1923, Bishop 1985). Thus, more than the emphasis on intuition, what seems of lasting significance in these proposals is (i) what they reject, and (ii) how they aim to proceed:

(I) The constructivists reject full-blown set theory, and with it some of the most characteristic methods of modern mathematics – purely existential proofs are not allowed, mathematical analysis cannot rely on the existence of a supremum for a bounded infinite set of real numbers, etc.

(II) On the basis of some fundamental epistemological and ontological positions, constructivists place at the basis of mathematics a restricted set of basic objects and operations, say the natural numbers and their arithmetic

---

[6] Two very good texts presenting the tradition of foundational studies to a philosophical audience are George and Velleman, *Philosophies of Mathematics* (2001), and Giaquinto, *The Search for Certainty* (2002).

[7] See, e.g., the recent account in Hesseling, *Gnomes in the Fog* (2003), which employs a statistical survey of more than 1,000 primary sources from the 1920s.

[8] It is now well known that 'formalism' is far from a good description of the constellation of philosophical beliefs about mathematics that was characteristic of Hilbert (see especially Corry's chapter in this volume).

(to which different versions of constructivism may add new elements). Everything else must be explicitly defined and proved on that basis, by appropriate logical means (Brouwerians rely on intuitionistic logic, others may accept first-order logic or some predicative system).

By so restricting their theoretical means, constructivists resist some of the driving tendencies of modern mathematics. The whole development of modern geometry, and axiomatics more generally, tended to suppress old claims of evidence and absolute truth for mathematical knowledge. In a sense, the constructivists attempt to preserve evidence and truth.

Characteristic of 'classical' postulational mathematics is the rejection of restrictions such as mentioned under (ii) above, and wholehearted acceptance of the modern objects and methods indicated in (i). During the first third of the twentieth century, the most noteworthy example was the Axiom of Choice (AC), essential in the context of modern analysis and algebra. The atmosphere of insecurity created by the foundational debate led many mathematicians to treat AC as a conjecture rather than as a basic postulate; it became customary to be explicit about the conditional dependence of theorems on its assumption. Authors such as Hilbert and Zermelo were fierce defenders of AC, but a mathematician of first rank like Weyl is said to have *never* published a proposition that depended on AC. Van der Waerden, who in Göttingen spirit had freely employed the axiom in the first edition of his famous *Moderne Algebra* (1930), backed down for the second 1937 edition. But AC is not the only example: from a constructivist perspective, the set-theoretical axioms of Infinity and Power Set are no less characteristic of the postulational approach and its free adoption of 'dubious' postulates.

Although Hilbert and Bernays famously tried to establish the admissibility of the postulates of 'classical mathematics' by metamathematics, it is and was possible to offer alternative accounts for them, such as the platonism of Cantor and Gödel, or the 'inductivism' that Russell felt forced to adopt. All of these represent very different philosophies of mathematics that aim to respect the whole edifice erected on the classical postulates, and account for it (note that, among them, only metamathematics is purely a mathematical philosophy in our sense).

The divergence of two mathematical styles, constructive and postulational, started around 1870, consolidated in the 1920s, and has continued to the present date (in a very unbalanced distribution of powers, to be sure). To summarize, the divide between 'the big two' lies in at least the following points:

- attitudes towards infinity – potential infinity (Brouwer) or a sentential approach to infinity leading to predicative restrictions on set theory, vs. a quasicombinatorial approach, leading to set-theoretical views of actual infinity;
- the acceptance or not of rather speculative postulates, resp. the will or not to abandon strong claims of truth in mathematics;
- and concomitant, all-important changes of mathematical methodology, which can even affect the basic logical framework.

Introduction

The contrast is extremely significant, especially for what it reveals about mathematics as usually pursued (a classic study is Bernays, 1935a). But it has been explored many times, and it figures prominently in philosophy of mathematics texts; thus it will not be our goal to offer yet another exploration.

As will become clear to the reader from a look at the table of contents, we have intentionally restricted our attention to aspects of what we have called postulational mathematics. Indeed, we have tended to simplify and equate 'modern mathematics' with the mainstream postulational trend. It is our belief, however, that much is still to be done by way of reinterpreting the history and philosophy of foundational studies. Thus, Part I of this book gathers together a number of essays that shed new light on the development of reflections about foundations. These essays, and several others that the reader will find cited there, offer a starting point for a deeper understanding; also relevant are the studies reunited in Part III.

It would be very interesting to provide a detailed analysis of alternatives in mathematical philosophy, but obviously this is not the place. Such a detailed analysis would have to discuss subvarieties of 'the big two.' On the constructivist side, alongside intuitionism, one can mention the work of Bishop, the predicativism of Weyl and Feferman, and reverse mathematics; much work in the general area of proof theory has implications here. On the side of postulational mathematics, alongside formal axiomatic systems like ZFC and their metatheoretical study one should mention divergent systems for set theory, and developments in category theory including topos theory.[9]

2.2 Systematic philosophy of mathematics

A crucial part of the received picture in the English-speaking countries, and by implication at the international level, consists in problems and proposals properly belonging to the *philosophy of mathematics* (in our sense) that arise from the systematic perspectives of the analytic tradition. Our comments in the present section will focus on this tradition, but they are meant to apply more generally to systematic approaches that somehow subordinate the agenda in philosophy of mathematics to the main problems of general philosophy.

A key belief in the analytic tradition, though not shared by everyone in its purest form, is that the method of formulating and studying philosophical questions by formal logic is a quintessential ingredient, without which no crisp posing of problems (let alone attempt at solution) is even possible. In this sense, one often finds authors who seem to believe that we are living in the year 126 a. F.[10] But, whatever the novelty of the method, the basic questions that one intends to

---

[9] The book of Feferman (1998) provides a very good point of entry into the first realm. On alternative systems for set theory, Fraenkel *et al.* (1973) is still commendable. As concerns category theory, see McLarty (1992) for introduction, and the recent Lawvere and Rosebrugh (2003).

[10] That is, after Frege's *Begriffsschrift*, the locus classicus where the Method was born; for an example of that pure belief, haphazardly chosen among the many available, see the remark by J. Floyd in *Phil. Math.* 10 (2002) 1: 70.

Introduction

answer are rather classical:

- to analyse the system of mathematics and fix its basic fundamental principles,
- to lay out the ontology of mathematics, clarifying the status of its objects – if any – and otherwise, to show that there are no objects,
- to analyse the epistemology of mathematics and explain how we attain knowledge of it,
- and, related to all of the above, to justify mathematical principles and mathematical knowledge, clarifying in the process the nature of mathematical rationality.

Although we are still far from a definitive solution of these basic questions, many hope that all that is needed is a new turn in the development of logical formalism: a deeper logical system, say some version of relevant logic, or context logic; perhaps a heavier use of weapons from category theory, or some new development in topos theory. Once this is available and its proper application to the questions above has been found, the key to the secrets of knowledge will be in our hands. The set of beliefs that we have just caricatured is perhaps on the retreat, and may have been so for some decades, but it agrees rather well with the venerable attitudes of figures like Carnap, to name but one leader.

Mainstream philosophy of mathematics is strongly entrenched in preferred views of the analytic tradition, in particular the celebrated (but controversial) linguistic turn, and the so-called Tarskian theory of truth understood as a restriction on any acceptable ontology and epistemology. Philosophy of mathematics here

> appears to be a microcosm for the most general and central issues in [analytic] philosophy – issues in epistemology, metaphysics, and philosophy of language – and the study of those parts of mathematics to which philosophers most often attend (logic, set theory, arithmetic) seems designed to test the merits of large philosophical views about the existence of abstract entities or the tenability of a certain picture of human knowledge. (Aspray and Kitcher 1988, 17)

Among the issues that have figured prominently, one finds many of the preoccupations and theses of Quine, starting with his celebrated 'compromise Platonism,' based on arguments about the indispensability of mathematics in scientific theorizing. Leading authors such as P. Maddy and S. Shapiro have been in favour of realism on the basis of such arguments, also (again like Quine) because attempts to develop alternative positions like nominalism and empiricism face great difficulties. No less prominent have been the difficulties posed by Benacerraf and Putnam, and very especially Benacerraf's dilemma (1973) about the impossibility of accounting for knowledge of abstract objects on the basis of a causal theory of knowledge.[11]

We certainly agree that the philosophy of mathematics is a wonderful proof-field for general epistemological ideas, as it shows that most of our doctrines founder

---

[11] There is abundant literature on all of these questions. See, e.g., Maddy 1997, Shapiro 2000.

Introduction

one way or another.[12] Indeed, we believe that the epistemology of mathematics can have such a great effect that eventually it may lead to the abandonment of entrenched philosophical views. In this respect, one can mention the usual understanding of knowledge in terms of truth *simpliciter*, as elaborated in correspondence theories. As we have already remarked (and develop further in Section 3), the emergence of modern mathematics was linked with the abandonment of traditional claims of truth about its axioms and body of results. Naïvely considered – but perhaps also from a sophisticated standpoint – what we find here is a form of knowledge that cannot be reduced to knowledge of truths. This suggests that one should be careful in employing pet arguments, like the ones previously mentioned, to dismiss either historical accounts or attempts to develop alternative epistemologies (examples will be found in Part III of this book; see, e.g., Scholz's chapter).

Systematic philosophical approaches such as the ones just evoked tend to alienate both historians of mathematics and practicing mathematicians. They simply don't seem to appeal to their interests. This contrasts strongly with the situation that the reader will find abundantly represented in this book: the history of modern mathematics presents us with many interesting instances in which the practitioners faced philosophical problems, and therefore the historian has to make sense, some way or another, of those problems in order to provide a reasonable historical account. Is it not possible to find common ground where professional philosophy meets the problems faced by practitioners and historians?

All authors in this volume share a strong conviction that the answer is – Yes. Perhaps the difference in reasoning style that we are emphasizing might be presented as a difference in conceptions of philosophy, systematic philosophy vs. philosophical reflection. The first conception takes one particular, historically important version of 'the' mathematical building, and employs it as a playground and testing ground for key issues in systematic philosophy (as conceived at time t). The latter gives much room for philosophy understood as a reflection on conceptual and methodological problems that *arise from* mathematical practice. This kind of philosophy is, quite obviously, practised by mathematicians themselves (better or worse), and for that same reason it tends to be perceived as relevant and interesting by mathematicians and historians.

Going a little further, we have emphasized how systematic philosophy tries to focus on 'the' system of mathematics, 'the' edifice of this *Wissenschaft*. But the conceptual and methodological problems encountered by mathematicians in practice arise, more often than not, from efforts to present mathematical knowledge in new ways, or to provide novel solutions to mathematical problems. The interesting questions here have to do with conceptual dynamics and methodological dynamics – and these are topics that systematic philosophy tends to avoid

---

[12] See Putnam 1994, originally published in 1979.

Introduction

systematically. Thus the main criticism that we may raise against traditional perspectives is that they cannot help analyse the dynamics of the discipline.

We are not advocating here the view that philosophy of mathematics should be entirely freed from general philosophy. This would be very naïve, as there is no way of reaching a kind of philosophical limbo. Any attempt to reflect on the conceptual and methodological problems of mathematics will, consciously or not, be informed by theoretical and philosophical presuppositions. What we call for is more modest, but crucial in practice. We ask philosophers that enter this area to avoid looking at it as a field for 'application' of what they 'know' from general philosophy. We ask them to engage in a real exchange with the complexities of mathematical practice, past or present, and to be open to the questions that worry practising mathematicians about the nature of their subject, especially in times of change. Such engagement, after all, was characteristic of investigations into the philosophy of mathematics precisely in the period when the main currents referred to above began.

To make our discussion more concrete, and in order to stress some salient features of traditional approaches, take for instance the interesting and successful trend of structuralism.[13] This trend tends to go along the lines of our previous characterization of systematic philosophy of mathematics, inspired by the analytic tradition. Structuralism is by no means the only currently important trend in the philosophy of mathematics, not even if we include the existence of different varieties of structuralism (modal structuralism, 'ante rem' structuralism, categorical structuralism)[14] and explore them. One has to mention, at the very least, nominalism, the view that there are no mathematical objects, and naturalism, of which radically different versions exist.[15] For our present purposes, however, it is more important to reflect on the limitations of these traditions in the philosophy of mathematics, rather than try to describe all the relevant brands. One finds here a tendency to proceed as if one could fix the true basic principles of mathematical knowledge, normally presented as some form of axiomatic set theory; understand them in terms of a correct ontology (that seems to be valid eternally), namely an ontology of abstract *structures*, variously interpreted;[16] and study, on the basis of general epistemology or even with the help of some cognitive science, the way in which our minds have access to those structures and gain knowledge of them. Obviously this is a reflection on a supposedly fixed building of mathematics, and an attempt to analyse it by deploying the general tools of systematic philosophy.

---

[13] For which see the relevant chapters in Shapiro (2005).

[14] Modal structuralism is the version proposed by Hellman, 'ante rem' is that of Shapiro. See the references given in previous footnotes.

[15] Compare Kitcher (1984), Maddy (1990) and Maddy (1997). For nominalism, see especially Burgess and Rosen (1997). See also the *Oxford Handbook of Philosophy of Mathematics and Logic* edited by Shapiro (2005).

[16] And thus giving rise, *malgré tout*, to a variety of philosophies; compare Hellman (1989) and Shapiro (1997).

In practice, this project can be quite different from the attempt to understand how structuralism is part and parcel of the mathematical practices associated with the triumphant form of mathematical philosophy (Section 2.1). What we have called 'postulational' mathematics won the day in twentieth-century mathematical practice, and is still the main form of mathematics in higher education. One of the reasons why set theory was important is that sets and mappings were actually used to characterize and study mathematical structures (like groups or topological spaces; see the chapter by McLarty). This way of looking at mathematical topics led to new kinds of definitions, problems, and proof methods, generally referred to by the name 'structuralism.'

Structuralism can thus be studied from the viewpoint of a philosophy of mathematical practice, and such studies need not be oversimplistic interpretations. By this we mean interpretations that identify this mathematical practice with *the* system of mathematics; reification of the structures as elements of basic ontology; an understanding of structuralism as the supreme form of mathematical rationality; or the posing of a complex philosophical problem by postulating that our preferred epistemological views should be able to account for knowledge of reified structures.

3 Prospects for a philosophy of mathematical practice

For centuries, the Western tradition tended to conceive of mathematics as an idealized collection of theories, living perhaps in an ideal space and waiting to be pinpointed by human beings. The perspective change in philosophy of science from the 1960s involved ceasing to think about science primarily as a body of theory, to focus attention on science understood as a human activity. Axiomatic theories are one of the products of mathematical practice, but only one among others that are equally important for the practice: problems, methods, conjectures. As products of human activity, theories are subject to change; they are historical and in principle contingent, at least to some extent (Section 3.2).

Today it is easy to find approaches that focus on *mathematical practice*, indeed (as we have remarked) practice is to a greater or lesser extent on the agenda of basically all active philosophers of mathematics. We shall now discuss some related issues in connection with twentieth-century developments in foundational studies and their philosophical implications, namely what we call the hypothetical conception and its consequences for ideas of certainty and objectivity. Then we shall turn to those who presented mathematical practice as the keystone of their attempt to overturn traditional philosophies of mathematics, what Aspray and Kitcher (1988) called the 'maverick tradition.' Finally, the reader will find some reflections on the variety of practices that contribute to the formation of mathematical knowledge.

Introduction

3.1 Framing hypotheses

Around 1900, at least in the context of geometry, it became common to argue that mathematical axioms are hypotheses. This viewpoint goes back to Riemann, but it can be reinforced and generalized on the basis of results obtained in foundational research. It is equivalent to what some call, following Lakatos, the 'quasiempirical' nature of mathematics. As this adjective is obscure and even misleading, we propose to speak instead of a *hypothetical conception* of mathematics. After all, 'quasi-empirical' aims to mean that the methodology of mathematics is not so different from that of the empirical sciences, but this is because mathematical knowledge, far from being a priori, builds on postulates, hypotheses.[17] This position, however, is not forced to assert that *all* mathematics is hypothetical (see below).

Key authors in all the main tendencies of mathematical philosophy came to agree on the hypothetical conception of the 'classical' edifice. We shall just give four examples. Brouwer and other constructivists proposed to abandon 'classical' mathematics because it depends on ungrounded assumptions; by refusing to accept postulates such as the Axiom of Choice and even the Axiom of Infinity, they underscored the hypothetical nature of mainspring mathematics. Hilbert accepted this diagnosis and came to speak of 'ideal elements' (Hilbert 1926) as an inescapable ingredient in mathematics. Even a logicist like Russell would speak of axioms accepted on the basis of 'inductive evidence' (logical axioms!), namely because seemingly evident propositions can be derived from them, no good alternative route to these is in sight, and no contradiction can be derived.[18] And the platonist Gödel was also open about the hypothetical methods at play in mathematics (see Gödel 1964 and the elaborations in Maddy 1990, 1997).

By considering great debates in modern mathematics, in particular but not only those between constructivists and postulationists, it is easy to isolate hypothetical elements in the classical edifice. These include:

(I) the Euclidean axiom of the parallels (see Gray 1989),
(II) the completeness axiom for the real number system (see Mancosu 1998, Ferreirós 1999),
(III) the axiom of infinity and associated issues in the logic of quantifiers (Mancosu 1998, Feferman 1998), and of course
(IV) that quintessential example of hypothesis, the Axiom of Choice.[19]

But the hypothetical conception is difficult for philosophers and mathematicians alike to accept, because it runs against time-honoured views of mathematics as a body of certain, evident truths. Such views were hegemonic from the time of

---

[17] For a sharp critique of mathematical apriorism, see Kitcher (1984).
[18] A. N. Whitehead and B. Russell, *Principia Mathematica*, vol. 1 (1910).
[19] See Moore (1982). This history of the axiom of choice is a wonderful display of mathematical methods employed in the study of proposed hypotheses.

Plato and Euclid to the days of Kant and Gauss. However, the period 1830–1930 led to their abandonment, in spite of the contrary wishes of influential authors such as Kronecker and Frege.

When we speak of a hypothetical conception, 'hypothesis' is used in its etymological sense, without the connotation that mathematical hypotheses may represent aspects of physical reality.[20] A *hypothesis* is etymologically a supposition, 'what is placed at the basis,' just like the axiom of choice is put at the basis of set theory and classical analysis. Indeed, Aristotle used the Greek 'hypothesis' for what we call axioms.

A possible way to emphasize the peculiar nature of mathematical hypotheses, the absence of that connotation, is to say that these hypotheses are *constitutive* and not representational – they constitute or lay out a mathematical domain. But it is necessary to clarify further what the claim of a hypothetical nature of modern mathematics entails.

Obviously, there is no need to take this as an all-or-nothing approach: one can, and in our view must, accept that parts of mathematics are *not* hypothetical. Recall that Bernays and Hilbert endorsed that conception (and talked about 'ideal elements') while emphasizing the special epistemological status of finitary mathematics. Thus the hypothetical conception does not entail that all of mathematics is of the same character. One could certainly argue that some parts of the theoretical body (say elementary arithmetic, or even parts of elementary geometry) enjoy a special non-hypothetical status. A rather obvious example is the theory of natural numbers, as embodied in the first-order Peano–Dedekind axioms – what is not obvious is *how* to justify the certainty we achieve in this domain.[21] The open question is, further, how exactly to circumscribe the non-hypothetical domains, and how to argue for their epistemological character (for a seemingly related problem, see Sieg's chapter).

Thus mathematical hypotheses can also be called 'idealizations' or even 'fictions,' provided that we make room for talk of *well-founded fictions*, to adapt an expression of Leibniz. One must emphasize that we talk about assumptions that are not completely arbitrary, as they arise 'naturally' in connection with previously known phenomena and results, and they come in good measure *conditioned* by this previous context. Consider the assumption of an infinite set $\mathbb{N}$ of natural numbers, i.e. the axiom of infinity: in our view this is a very clear example of a hypothesis. Note also that some consequences of a hypothesis may be *forced* by its linkage with previous knowledge: our knowledge of the theory of natural numbers forces us to accept that $\mathbb{N}$ has subsets $\mathbb{S}$ such that $\mathbb{N}$ is equipollent to $\mathbb{S}$ (examples being the even numbers, the square numbers, or the primes).

---

[20] This comes only when mathematical hypotheses are employed to formulate some scientific model, say when semi-Riemannian geometries are used as a basis for relativistic models of the Universe.

[21] One can be against *apriorism* even here, and adopt a line of argument (reminiscent of Kitcher, 1984, and thus of Piaget and Mill) which might be called naturalistic. An alternative tradition is reviewed in Potter 2000.

Introduction

Another obvious example is Cantor's theorem about the impossibility of establishing a bijective mapping between $\mathbb{N}$ and $\mathbb{R}$; once the set-theoretical approach has been adopted, once the sets $\mathbb{N}$ and $\mathbb{R}$ are assumed to exist, this is an inevitable consequence of previously known properties of the natural and real numbers. One could also study more interesting cases from this angle, like the Riemann surfaces in function theory, which 'inherit' properties of the functions they describe, or Dedekind's ideals in algebraic number theory, which generalize the properties of principal ideals.

3.2 On certainty and objectivity

Although it may seem so at first sight, the above does not entail that we abandon talk of logic and certainty. First, to the extent that some portions of mathematics (e.g., the theory of natural numbers) are not hypothetical, we have certainty in our knowledge of them (for whatever epistemological reasons). But second, one can achieve certainty in knowing *what follows from what*, that is, *conditional* certainty concerning what follows deductively from given axioms. The proofs of mathematics can be timeless even if some axioms are hypothetical. Indeed we would argue, in disagreement with Lakatos and many historically inclined philosophers, that the methods developed in twentieth-century mathematical logic have provided a kind of absolute rigour in that sense.

Needless to say, we are not unaware of problems in the philosophy of logic, particularly those related to alternative logical systems such as intuitionistic logic. In this respect we would offer a line of argument justifying the logic of modern 'postulational' mathematics on the basis of the goals and values linked with the mathematical practices it embodies – in particular, those practices that go *beyond* the area of foundations (see Section 3.4). But this introduction is not the place to elaborate on that topic.

Can one make compatible this kind of hypothetical conception with the peculiar objectivity of mathematical knowledge? Weyl for instance wrote:

> The constructs of the mathematical mind are at the same time free and necessary. The individual mathematician feels free to define his notions and set up his axioms as he pleases. But the question is will he get his fellow mathematicians interested in the constructs of his imagination. We cannot help the feeling that certain mathematical structures which have evolved through the combined efforts of the mathematical community bear the stamp of a necessity not affected by the accidents of their historical birth. Everybody who looks at the spectacle of modern algebra will be struck by this complementarity of freedom and necessity.[22]

That compatibility can reasonably be attained at the price of defining objectivity to be intersubjectivity, and accepting that, if the development of mathematics seems to be endowed with necessity, to force some results upon us, on closer

---

[22] See also Cantor 1883, Dedekind 1888, etc.

examination this comes down to a (moderate) lack of arbitrariness – more precisely, the existence of strong constraints on arbitrariness. We have seen an elementary example above, with ℕ and the properties of its subsets: there is no necessity to accept the set ℕ, but *once* that is done, some consequences are inevitable given its connections with previous theory.

Adoption or rejection of a certain hypothesis is the outcome of consensus within the mathematics community. It is perfectly conceivable that the Axiom of Choice, or the Completeness Axiom for ℝ, might have been rejected by the vast majority of twentieth-century mathematicians (the present statistical distribution of 'classical' vs. constructivist mathematicians would be exactly reversed from what we observe). As already remarked, that is not to say that such hypotheses are 'purely conventional,' that no restriction operates on the free choices that lead to general consensus. Indeed, we would defend the existence of different kinds of restrictions or linkages – from those related with our constitution as human beings and linked with the activities of counting, measuring and constructing, to others deriving from the role of mathematics in scientific modelling of phenomena.[23]

We hasten to add, however, that such *conceivable* scenarios (rejection of Completeness for the reals, or of the Axiom of Choice) are highly unlikely in view of the established practices of mathematics in the nineteenth century, especially the practices linked with mathematical physics and with analysis. Even so, philosophically the hypothetical conception entails as an inescapable consequence the contingent nature of (parts of) mathematical knowledge.

If one could define the function of mathematicians as bringing to light some pre-existing structures, given in advance of the development of their activities, there would be no reason to talk about an essential historicity. It is precisely because mathematical hypotheses do not (at least not necessarily) represent an independent reality, that they could always have been rejected by the community of mathematicians. Herein lies the historicity of mathematical knowledge, which is only one (but also the deepest) of the sources of that diversity of mathematical practices that historians make their business to explore. Contingency is inevitably installed at the heart of modern mathematics, although the highly interconnected web of mathematical and scientific activity (past and present) acts in such a way that it becomes disguised and can even seem invisible.

3.3 The 'maverick' and us

In the last three decades there have been a number of attempts to elaborate a philosophy of mathematical practice, exemplified in the work of authors such as Lakatos, Kitcher, and Maddy. It could be said, however, that these attempts remain isolated and have not been able to attain the same level of attention and open debate as more classical issues in the philosophy of mathematics. This may be changing in recent times, especially in the case of P. Maddy. But more important

---

[23] All of this would require very detailed treatment; an overview can be found in Ferreirós 2005.

Introduction

than judging the success of these proposals, is to elaborate on the similarities and differences with views such as the reader will find in this volume. For the differences are indeed noteworthy.

We begin by offering the reader some orientation about this alternative tradition, but first we remark that there have always been alternatives to what was perceived as the mainstream in philosophy of mathematics. In fact, the reader will find comments on some interesting proposals (by Hausdorff, Weyl, the French tradition from Cavaillès, and by Sieg recently) in Part III of this book. Restricting our attention to the English-speaking world, the 1960s marked the arrival of some influential and powerfully argued alternatives to the reigning views in foundational studies (formalism) and systematic philosophy (logical empiricism). Quine proposed his well-known holistic view of mathematics as continuous with empirical knowledge, and his ontology of compromise platonism. Putnam argued that mathematics does not need foundations, and suggested a modal view of mathematical structures deviating from the usual, ontology-laden, set-theoretical view. These ideas have since become essential ingredients in systematic philosophy of mathematics (Section 2.2).

Contemporaneously with Quine and Putnam, Imre Lakatos elaborated ideas of Pólya and his own in his celebrated essay *Proofs and Refutations*, which has unquestionably been one of the most successful philosophical proposals among mathematicians. One may say that he proposed a particular model of the *microevolution* of mathematics. His model focused on the 'dialectics' of conjectures, attempted proofs, counterexamples, refined concepts, and new proofs. It was an interesting beginning for a methodology of mathematical practice, although it could only aspire to be partial; also, the peculiar nature of the case study on which Lakatos based the proposal (Euler's conjecture for polyhedra) did not speak much in its favour. But these ideas have not been absorbed into the mainstream, partly (perhaps) because of party allegiances, and partly (no doubt) because of Lakatos' overheated polemics against formalism in particular and foundational work in general.[24] Being a good polemicist, and delighting in it, is not always the best way of promoting one's ideas.

In fact, one of the most important transformations in the philosophy of mathematics over the last 20 years concerns the apparent incompatibility between the sensitivities of the maverick and attention to foundational issues. It seemed in the 1980s that the rebel tradition was forced to subsist alongside the hegemonic tradition, emphasizing historical development and methodological issues while opposing the assumption that logic and foundational studies may shed real light on mathematical knowledge. Today a number of important philosophers of mathematics no longer share that vision. They ask about a whole range of issues concerning mathematical practice, they know how to employ historical material in a methodologically sophisticated way, and at the same time they are

---

[24] See the introduction to his booklet. There is, however, a small tradition of followers of Lakatos, including interesting authors such as Tymoczko and Corfield, among others.

well grounded in logic and take advantage of the results of foundational studies. The contributors to this volume clearly demonstrate that we no longer see an opposition today between foundational studies and attention to mathematical practice. But let us go back to the maverick.

Lakatos's attempt was followed two decades later by Philip Kitcher's *The Nature of Mathematical Knowledge*. Besides offering a detailed critique of apriorism, this book elaborated ideas about the naturalistic origins of key mathematical concepts, and proposed particular models of mathematical practices and of 'transitions' between practices. Kitcher's model aims to capture the *macroevolution* of mathematics using a conception of practices formed by the *language* in use at some time, the accepted *statements* (propositions, theorems, axioms), sanctioned modes of *reasoning*, unsolved *questions*, and general *images* about the discipline. To speak in the parlance of philosophers of science, it is easy to realize the proximity of Kitcher's idea of mathematical practice to the 'statement view' characteristic of logical empiricism. It seems likely that this approach will have to be replaced by a richer analysis of the ingredients of mathematical practice, taking into account a wider spread of cognitive and cultural ingredients.

It was Kitcher's aim to show that mathematical knowledge is much more similar to general scientific knowledge than traditional views suggested. The key principle of his philosophy is that all the main transitions between mathematical practices are rational. As usual among philosophers of science, he hoped to distil an idea of rationality that can be recognized as such, while being wide enough to make possible a reconstruction of actual historical changes. According to Kitcher, changes or transitions happen because of instability induced in an older practice by the introduction of new language, new statements, new modes of reasoning, and the like. He has typified several important kinds of 'rational transition,' applying his scheme in a rather detailed study of the development of mathematical analysis.

Concern with the progress and rationality of mathematics was a common element of Lakatos and Kitcher. Sensitivity to those questions was obviously inherited from mainstream philosophy of science at the time, and here lies one of the differences between the 'maverick' and us. It may strike the reader that no single chapter in the present volume addresses the grand problems of progress and rationality directly. This is not necessarily a sign that we reject such problems, indeed the different contributors to this volume may be of different opinions. We all agree, however, that excessive attention to all-inclusive claims has diverted the attention of philosophers of mathematics away from a long list of interesting questions waiting for careful scrutiny. In this respect, once again, our agendas are guided less by systematic philosophy, and more by the questions that worry mathematicians themselves and that call for philosophical reflection.

In a number of recent contributions, new questions are being asked about the role played by visualization (and not formal deduction or conceptual understanding) in mathematical practice; about the explanatory role often played by new mathematical developments; about the epistemological roles played by proofs other than simply warranting a result (e.g., diverse proofs for a single result); or about the criteria and values at work in mathematical practice. Some of the contributors to the present work have in fact worked along those lines.[25] Questions like these call naturally for making extensive use of history, even if their aim is not to develop a historical philosophy of mathematics, and their analysis also calls for deploying the tools of logic and other disciplines. Furthermore, those questions are often found while doing historical work; to give a few examples, one cannot study the work of Klein without encountering the question of visualization, or the works of Riemann, Dedekind, and others without meeting the problem of purity of method. Thus we find here natural meeting points for historians and philosophers.

3.4 Foundational and other practices

A successful recent example of a philosophical study of practice is offered by the work of Maddy. She has developed a program of study centred on the introduction of new axioms into mathematics and how they are justified, culminating in her *Naturalism in mathematics* (1997). Her focus has been on the development of set theory, past and present, and thus she has drawn on diverse examples from the practices of a community of experts in mathematical logic and foundations. In this way, she has remained close to the usual concerns of more traditional philosophers of mathematics, while moving toward fresher questions, questions that have (at least at times) been important for practising mathematicians.

For obvious reasons, many mathematicians and historians tend to become impatient with philosophers' obsession with logic and set theory. Could this simply be the outcome of a biased, very incomplete knowledge of mathematics? Certainly there are large fields of study in geometry, in algebra, in combinatorics, in topology, and so on, that await philosophical scrutiny. Conceptual and methodological issues do arise in these domains, no less than they do in set theory. And we should not be too quick in assuming that there are no significant philosophical differences between the practices in different branches of mathematics. Work in logic and foundations has had several particular traits that distance it from 'more mainstream' work, as many a mathematician would say. Just think of the kinds of questions mentioned in the last paragraph of the previous section. Visualization plays very different roles in the history of geometry and set theory; the meaning of explanation may change in significant ways from one branch of mathematics to another; multiple proofs are more often found in number theory than they are in geometry; the values and criteria that guide an expert in differential equations are not at all the same as in abstract algebra.

As a cursory overview of the chapters in this volume will show, our contributors have made it their concern to explore the richness of concepts, knowledge and methods that mathematics in all its extensions displays before us. Some branches

---

[25] See, e.g., Tappenden 1995b on fruitfulness, Mancosu 2001 on explanation, and the collection Mancosu, Jorgensen and Pedersen (eds.).

are of course particularly well represented: geometry, algebra, and logic are the main examples. We hope the reader will believe that, if set theory is not more strongly represented, this is not simply because of lack of knowledge on the part of the authors.

Also well represented is the question of 'applied mathematics' or the 'applicability' of mathematics – terms that, in themselves, may already beg some questions. It is represented here under the guise of the relation between mathematical knowledge and physics, a topic discussed in connection with Riemann, Hilbert, Hausdorff, and Weyl – no dearth of 'pure' mathematicians, representative of modern mathematics, in this selection. This deserves particular emphasis, because historians and philosophers alike have tended to incorporate the mathematicians' own set of values and images of their discipline in the twentieth century. The role of the pure in modern mathematics has created an abundance of 'purist' histories and philosophies of mathematics.

One of the last examples is the already-mentioned work of Maddy. In the late 1990s she arrived at what she calls a 'naturalistic' perspective, meaning recognition of the autonomy of mathematical activities, a practical autonomy reflected in the autonomy of their evaluation criteria. Recall that Maddy worries especially about criteria for accepting new axioms: methodological rules having to do with the goals of mathematical practice explain the adoption of axioms, judged mainly in terms of their fruitfulness. Maddy's naturalism is a discipline-oriented viewpoint, guided mainly by the self-image of mathematical logicians in the twentieth century. But one may worry whether in this case, when we are asking about very deep and basic questions concerning the key lineaments of a whole discipline, it may be unwise to follow too closely the present views of some of its practitioners.

Such discipline-oriented approaches can usefully be contrasted with sophisticated epistemological ideas at work in recent histories of mathematics. A case in point are the ideas about 'epistemic configurations' employed by one of our authors, Moritz Epple, in some previous contributions.[26] While the more presentistic accounts may, quite naturally, tend to be anachronistic in their judgements of the past, Epple's viewpoint focuses directly on 'local' configurations in all their complexity. Thus it underscores the need for complex analyses in which the work of mathematical practitioners may be guided by the interaction of different disciplines, by the views of particular research schools or by wider traditions, and where particular institutional and sociocultural factors may influence the development of mathematics. Notice that, as just presented, this is simply a point concerning the methodology for a historico-philosophical analysis of practices, which does not pre-empt questions about the interplay of epistemological and sociological factors in the development of mathematics.

---

[26] See Epple 1999 and 2004.

Introduction

―※―

4 Challenges in the history of modern mathematics

Alongside the changes in the philosophy of mathematics sketched above, there has been a comparable sophistication in the history of mathematics that will be worth sketching. The task facing historians of mathematics concerned with the last two centuries is a large one. The period since 1800 saw the emergence of the current university mathematical syllabus, almost in its entirety, and much of what is taught to today's mathematics majors is at least a hundred years old – the rigorous calculus is but one example. This syllabus is the international syllabus. It is taught from, if not the same, then at least very similar books everywhere it is taught. Autonomous mathematical traditions, as for example the Japanese *wasan*, went the way of the buffalo. Modern mathematics is taught in more-or-less identical settings, the modern university, everywhere in the world. Research-minded professionals belong to an active international group of perhaps 10 000 people that read and publish in the same journals, have strikingly similar standards and priorities, and feel they belong to the same group. The institutions that propagate modern mathematics and the journals that publish it are likewise almost all the creations of the last two hundred years.

4.1 Contextualizing the history of mathematics

All of this forms a legitimate topic of historical research yet, paradoxically, a good way in to understanding the present state of the history of mathematics is to ask how mathematics can have a history at all. There is a perfectly satisfactory answer, which is to document the creation or discovery (according to the author's taste) of new ideas and new theorems, alongside the lives of the mathematicians and their institutions. If we are not concerned with this style of history in this book, it is not from any lack of sympathy with it, but because it does not bear closely on the convergence of history and philosophy of mathematics. But it should be noted that there are now rich historical accounts of many topics in the nineteenth century and a number of topics in the twentieth century, and their growing sophistication illuminates our theme. To meet the philosophers half-way, we must go elsewhere, taking the opportunity on our journey to understand better why, from the historians' side, the meeting did not come about long ago.

The work being done in history of mathematics around 1970 was for the most part routine, dominated as it was in the English-speaking world by the production of compendious textbooks. Whatever merits these may have possessed (the one by Morris Kline from 1972 still retains its place as a convenient reference work) they also shared the disadvantages of the genre: little scope for originality in the questions asked and the topics pursued. The standard textbook takes much for granted, and simplifies readily to be understood. In the American context, where history of mathematics was a means to expose students of subjects other than mathematics to a little of the subject, the valuable aim of increasing mathematical awareness dictated that the modern period was little discussed, and that the calculus loom large in volumes bold enough to go that far. The main reason, other than market forces, for this poor state of affairs was the collapse of German scholarship after the First World War and through the Nazi period. Where there had once been a community of professional mathematicians with a genuine sensitivity to history, an acquaintance with sources, and an audience of their peers, there was now a diaspora, and a tradition was lost.

The practice of history also declined among mathematicians. Insofar as the history of mathematics was taken to mean the history of mathematical ideas, it was only the mathematically trained, those with sufficient understanding of the ideas, who were competent to write such a history, and only they, it seemed, who had an audience. An almost exclusive focus on the mathematics, bulked out by an anecdotal interest in the mathematicians themselves (most egregiously on display in E.T. Bell's only too well known *Men of Mathematics*) held writer and reader together. The process of research was truncated because of the easy consensus about what needed to be said, which was the acknowledged highlights – the famous names, the great theorems, the exceptional biographies. Books of this kind tend to confirm what their authors set out to discover. They may turn up unexpected details, usually to do with priorities, but they go looking for the key developments that 'must' have been there in the past or mathematics could not be as it is today. Earlier historical work was taken over uncritically, because it told the author what to look for.

All this meant that the new generation of historians of mathematics that came along in the 1970s did not have mountains to climb in order to get started. To take but one example, the outlines of Cauchy's rigorization of the calculus were well known, but the prevailing style of the history of mathematics had not thought to ask how it caught on. It was simply assumed that it had to. The rigorous calculus spread, like all good things should, from one great mind to another until eventually it was the common property of every educated mathematician, and had, in the process, been tidied up. Outright errors were removed, the arguments were improved and extended, justice was done and a substantial part of the modern mathematical syllabus brought into being. Unsurprisingly, historians found that the actual story was very different: Cauchy's methods were not, in fact, immediately accepted at the École Polytechnique where he taught and he was ordered to tone them down, but when he went into self-imposed exile some years later he left no-one behind to promote them. And if Cauchy's new ideas were adequate for the tasks for which they used them, mathematicians found that they needed considerable refinement when solving differential equations, for example.

The prevailing style of the history of mathematics, insofar as it noticed these complexities, dismissed them as if they were noise. 'Of course', such books would imply, 'the full story is more complicated, but not in any significant way'. The newer generation of historians of mathematics found that the complexity was essential. The questions 'Why and how was it done?' began to acquire a new historical dimension, concerned not just with the life of the ideas in their day but with the course of their journey down to the present day.

4.2 Mathematics in various social contexts

Without leaving the framework of the history of ideas, what had been a history of results gradually became a history of problems, methods, and results. It became possible to ask: why was this result wanted? What were people looking for when it turned up? Why was it done in that way, and not in some other way? But it became clear, once these questions were asked, that there were obvious answers that took one out of that framework. It emerged, for example, that there were significant ways in which French and German mathematicians did not read each other, and within Germany there were significant differences between the two main centres for mathematics, Berlin and Göttingen. In this modest way, certain social and contextual questions entered the subject, as part of the 'historical background', and this became a flourishing, and increasingly sophisticated, approach.

The German case was the first to be studied in any detail. In Berlin, a university created at the start of the nineteenth century in the spirit of neo-humanism and where mathematics was run, from roughly 1850 to 1890 by Kummer, Kronecker, and Weierstrass, the ethos was that of analysis, number theory, and algebra. These were subjects with strong intellectual connections to Gauss, they fitted the neo-humanist ethos because they set no great store by applications, and they established deep roots in the German mathematical psyche as a result. Echoing Gauss's misgivings, and consonant with the personal preferences of these powerful individuals, geometry was regarded as less than rigorous and not entirely pure. On the other hand, when in the 1890s Felix Klein set about creating an alternative vision, one he was able to build up in Göttingen, he emphasized the importance of geometry both as a subject in its own right and as a means to enhance intuition and discovery in mathematics. He sought explicitly to slow the push towards pure analysis for its own sake, and as the years went by he also pushed strongly for applied mathematics and mathematics done with applications in mind. In all this he was remarkably successful, and no one disputes that Göttingen overtook Berlin as the centre of the mathematical world by about 1900. Thus a tidy dichotomy was created in the sphere of history of mathematics that has proved illuminating in a number of ways. It helps articulate different approaches to what is modern in modern mathematics, it plays a part in the story of Einstein's route to the general theory of relativity, and it usefully reminds us that indeed different mathematics gets done in different places.

A somewhat similar story proved to be available in the history of mathematics in France (Gispert, [1991]). The École Polytechnique began the nineteenth century as the sole producer of research mathematicians, while the École Normale Superieure was intended only to produce teachers. But the École Polytechnique soon became a military school, and its primary function was to produce candidates for the engineering schools, and by the 1870s, after the debacle of the Franco-Prussian war, the École Normale Superieure became steadily more attractive and the composition of the French mathematical profession changed. The result was that after 1880 there was a generational split between those who adhered to the French tradition of analysis that inclined towards differential equations and the interests of mathematical physics, and the younger men who were interested in exploring the implications of set theory and topology for the study of functions in their own right.

As these capsule accounts of the state of mathematics in France and Germany indicate, there was also a perceptible difference in the priorities of mathematicians in France and Germany: French mathematicians had a tradition of running mathematics and physics together that German neo-humanism weakened. Historians of science documented a parallel development: in France by 1900 physics was taken in the main to be experimental physics, and there were very few notable theoretical physicists; in Germany there arose a powerful discipline of theoretical physics.

There have been similar accounts of the development of mathematics in other countries, the USA, Britain, and Italy to name the three most significant. The American story (see Parshall and Rowe, 1994) shows a strong, direct influence of Germany, much of it exerted by Klein on the many Americans who came to study with him and returned to America to create mathematics departments very much on the German model. The major figures here all inclined to pure mathematics, often in geometry and algebra. The British story could not be more different: parochial, ill inclined to learn from any Continental model, but exemplary in applied mathematics. And where the British situation was dominated by one institution (Cambridge) the Italian situation was much more heterogeneous, and dominated by the many changes put in train after the Unification of Italy in 1870–71. However, partly because mathematicians had done their bit in the struggle for unification, and partly because they seized their opportunities afterwards, mathematics blossomed in Italy as it had not before in a number of traditional and new fields from algebraic geometry and analysis to logic.

National studies such as these typically detected a useful distinction in the type of mathematics being done, and grounded it in institutional differences. They did not analyse the way these differences operated, and simply accepted them as facts. A more local analysis has recently grown up that takes an institution and seeks to analyse its workings, or, at the very least, to describe them in some detail. In the literature in history of mathematics such work is most commonly found in biographical books and essays and papers focused on a single individual. Jesper Lützen's *Joseph Liouville* is among the most successful large-scale realisations of this approach, but there have been others.

A fruitful topic here has been the study of research schools. It has recently been argued (Parshall 2004) that the concept of a research school was developed by historians of science to deal with a particular situation in the laboratory-based experimental sciences, and that it has largely been used by historians of mathematics in an *ad hoc* way. There is a good deal of justice in this charge, although historians of mathematics often study places that were in fact so small that they have adopted *de facto* the approach to a school that Parshall suggests. This emphasizes the importance of a leading figure whose research is inspiring (the individual may not be), a research ethic focused on a particular area and a set of evolving techniques, the production of students who go on to successful careers in somewhat the same area, and external agreement on the importance of the work done. The thrust of Parshall's critique may be to highlight one specific way in which research is fostered, as opposed to the rather hit-or-miss, one-on-one style of the French in the nineteenth century, or the apprenticeship model often found. It belongs to a formulation of the history of mathematics that studies the professionalization of the mathematician.

Here, mention should be made of a book that approaches these issues from the side of history of science. Andrew Warwick's *Masters of Theory* [2003] might well be described as a microlocal study. It takes one place, admittedly over a long period of time, and studies it in considerable detail, drawing lessons about how teaching and training shaped the practice of generations of graduates. The place is Cambridge, between the end of the eighteenth century and the First World War, and Warwick argues that such detailed studies are the best way to trace the evolution of mathematical practice, the articulation of paradigms (here he has new points to make about the Kuhnian idea of normal science) and the way mathematical physicists read and respond to others – those in the school and those outside.

What these historical accounts have highlighted is the importance of difference in the life of mathematics. They document different priorities within mathematics, different estimations of the value of mathematics and of this or that branch of mathematics, a growing appreciation of the difficulty of mathematics – over questions of rigour, for example – an ever-shifting balance between discovery and proof. Cumulatively, they raise a great deal of evidence for the importance of mathematical practice in any account of mathematics.

These socially oriented accounts are enriched by the long tradition in the history of mathematics of studying the mathematics close up. Done sensitively it has been one of the strengths of the subject, although it can produce a literature some historians of modern science find difficult to read and even affect to disdain (for its presumed lack of methodological sophistication). Regrettably, it is true that some, even many, of these books and articles have something of a survey character about them, they are guide books and route maps, and as much to be criticized for finding only what they set out to look for as for their lack of analytical depth. The reasons for this deficiency, if such it be, may well derive from an inadequate appreciation of the nature of mathematics as a historical enquiry. It is surely unfortunate that there has never been, in the history of mathematics, the equivalent of a Thomas Kuhn. For better or worse, with all the turmoil generated in the wake of events like *The Structure of Scientific Revolutions*, history of science acquired a book that directly challenged a model of the linear accumulation of scientific knowledge and contested the philosophy of science that saw science as a steadily growing approximation to the truth. The history of mathematics has never been shaken up in that fashion, and it has few small-scale attempts in that direction.

4.3 Meeting the philosophy of mathematics

This is where new directions in the philosophy of mathematics enter the discussion. It has always been true that some mathematicians are so evidently philosophers, or so philosophically minded, that they require a philosophical treatment. Leibniz and Descartes are obvious examples, but in the more modern period Bolzano and Grassmann spring to mind. On many of these occasions, it was once the case that historical analysis lagged behind the philosophical accounts, and much valuable work has been done redressing the picture (see, for example, Russ's work on Bolzano, [2004]). The period around 1900 famously sees major transformations in the philosophy of mathematics brought about by mathematicians with philosophical agendas, as well as by philosophers themselves. Hilbert and Brouwer are the obvious examples among mathematicians, and Frege among philosophers (for such he is today taken to have been). Inevitably the philosophical literature crystallises out the philosophical kernels of their work, at some cost to the historical complexity. Two essays in this book seek to show how our understanding of their work is in fact deepened by a reimmersion in history. Corry shows that Hilbert's views move significantly away from Kant as a result of his continuing interest in many branches of physics; Tappenden that Frege's ideas about concepts reflect his commitment to what Frege saw as a pro-Riemannian, anti-Weierstrassian position on complex analysis. Among the leading mathematicians of the next generation Hermann Weyl, as is well known, was another with a philosophical agenda, but it was much less well known what that agenda was, and how and why it shifted over the years. It is scarcely known at all that Felix Hausdorff was another mathematician with significant philosophical interests. Indeed, it was in pursuing these that he came to dive deeply into foundational questions. Essays in this book by Epple, Scholz, and Ferreirós, respectively, are devoted to these themes.

The present situation in history and philosophy of mathematics may be said to be poised to unite the historians' work of the last twenty to thirty years with philosophical analyses of mathematical practice that will integrate the philosophical approaches of the well-remembered names into a fuller study of modern mathematics as a whole. The many detailed studies we now have, for the first time, of mathematical change in the modern period call out for a detailed, insightful philosophical analysis. Now that we can see such richness in the historical development, it is inadequate to dismiss it as froth and continue to discuss only those topics in the philosophy of mathematics that permit the reduction of mathematics to issues in set theory.

Others before us have called for a coming together of historians and philosophers of mathematics. Perhaps the most famous in recent years has been Imré

Introduction

Lakatos, whose death at the age of 51 in 1974 deprived the community of a proper chance to see what his initiative might have produced. To those who find mathematics a dry, if not indeed lifeless, accumulation of detail, Lakatos's book *Proofs and Refutations* seemingly comes as a revelation. Mathematics, it emerges from these pages, is full of disagreement, putative theorems jostle with apparent counterexamples as mathematicians argue over what is relevant and what is not. No better indictment of the poor state of the history of mathematics when Lakatos wrote (and of mathematics teaching in general, one must presume) can be given than the lasting popularity of this book, because as a work of history it is disappointing. Every research mathematician knows the to and fro of proof and refutation, which is only left out of the histories of the subject because it tends to be missing from the final, published papers and books in mathematics itself. Moreover, this aspect of mathematics was much better described in Polyá's books of the period. But if Lakatos had discovered a trade secret, the example with which he displayed it was too thin to nourish research. The rhetorical structure of the book is that a pleasant but minor example of mathematical research in the early nineteenth century – Euler's formula for polyhedra – is made to stand in for mathematics as a whole. What is one aspect of the way mathematics is done becomes the central revelation of the book, and because mathematics in fact develops in many ways historical complexity is replaced by personal recognition. Mathematics is humanized, but its history was trivialized.

Nor did it help that among the few examples immediately offered of more substantial topics one was irredeemably flawed. Cauchy had claimed that a convergent sum of continuous functions converges to a continuous function, and his successors tiptoed round this issue until, oddly late in the day, Cauchy suddenly leapt up to correct his former account. Could it be, it was asked, that this was because he had had another idea of continuity all along, one only made visible to historians with the advent of non-standard analysis? True, there was a way of using non-standard analysis to spin what Cauchy had originally said that made it correct. But the defects of this historical analysis are patent. Cauchy would have had to have had some feeling for the non-standard version, of which he gave no hint at all while generally being rather up front in his attachment to rigour in mathematics. Rigorous non-standard analysis is a large departure from nineteenth-century mathematics, not a route it was possible to take in the 1820s. On the other hand, a certain naivety about continuity, otherwise apparent in Cauchy's use of language, is all that is required for Cauchy to have made a mistake. The complexities of the historical record are not illuminated at all by a strange desire to make Cauchy correct even if he becomes the only mathematician of his day to have a hint of Robinson's work in the 1950s.

On the other hand, Lakatos's directing attention to the role of examples in mathematics, be they putative counterexamples or invitations to extend known results, was welcome, and should have had a richer effect on historians, and philosophers, of mathematics than it did. Among historians, at least, the response was more to note historical topics that did not fit Lakatos's framework (e.g. Bos, [1984]

Fisher [1981]) than those that did. It might be that the more philosophically sensitive approach to history of mathematics that he advocated has more chance of success now than it did 30 years ago. In this connection, the generally positive remarks by Feferman are interesting.[27] His criticisms of Lakatos's approach do not dispute the validity and importance of most of the insights Lakatos offered, rather they add to the list of manoeuvres mathematicians indulge in. For example, he points out that Lakatos does not discuss how mathematical theories can be reorganized around new fundamental concepts, how criticisms of mathematics arguments can arise without the production of counterexamples, how proofs can be improved, or the full range of types of conjectures that can be made. He also noted that Lakatos made no attempt to say what was distinctive about mathematics, as opposed to any other area of rational discussion, and this can perhaps be read as another way in which Lakatos's work is incomplete.

Lakatos's attempt was followed two decades later by the more substantial, and philosophically lasting, work of Kitcher, most notably his *The Nature of Mathematical Knowledge* and the set of essays edited with William Aspray, *History and Philosophy of Modern Mathematics*, as we discussed above. Whether Kitcher dented the view of mathematics as a source of a priori knowledge or not, the view these days seems to be that mathematicians have done a good job looking after mathematics over the years, and something valuable might be gained by looking more carefully at how they have done it. This complements the view of those historians of mathematics who seek to enrich the accounts of mathematical discoveries, however well grounded in a social context, with some greater intellectual coherence.

4.4 History of mathematics in the philosophy of mathematics

Some words should be said here in support of the claim that mathematics, in its core, does permit a history capable of sustaining philosophy. Feferman [1998] noted that the natural numbers are as good an example of a crystal clear concept as any we possess, and he could think of no significant confusion on this score in the entire history of number theory, which stretches back, he observed, at least as far as Euclid's *Elements*. He noted that this clarity threatened the gradualist philosophy advocated by Lakatos, according to which things could only be said to be progressing. Mathematicians do indeed habitually speak of theorems in number theory as being true, whether they are trivial (13 is a prime number) or very deep indeed (Fermat's Last Theorem). The still-elusive Riemann hypothesis will turn out, they say, to be true or false. The central difficulty, if such it be, presented by the proofs of many theorems in pure mathematics is that, according to most philosophies of mathematics, such proofs are timelessly valid and intellectually accessible across cultures to anyone with sufficient ability and time, and their conclusions are therefore timelessly true. Accordingly, a theorem and its proof are

---

[27] Feferman [1998].

Introduction

paradigmatic ahistorical objects, and the defining activity of the mathematician admits historical enquiry only in ways devoid of philosophical substance.

To this challenge, the historian can reply first that it is well to insist on the distinction between what it is for something to be proved, and for something to be true. It has been clear since the acceptance of non-Euclidean geometry that mathematics is capable of building up distinct theories, each of them consistent on their own but which are mutually inconsistent. Likewise, there are distinct systems of analysis, such as constructive and non-constructive analysis, that disagree about results. As we discussed above (section 3.1) when describing the role of hypothetical reasoning in mathematics, since assumptions and hypotheses cannot be taken to be necessarily true, it follows that mathematics based upon hypotheses cannot be taken to be true either in any naïve sense (such as 'true in the world we live in'). Useful mathematics therefore requires choices to be made, and it is clear that large bodies of hypothetical mathematics, together with the existence of mutually incompatible systems of validly proved results, at once admit the possibility of historical analysis. Mathematics, like the sciences, is immersed in a human world of aims, purposes, and decisions.

It must be admitted, however, that most, if not all, of the mathematics needed for science can be done with the Zermelo–Fraenkel axioms without the need for either the full axiom of choice or the continuum hypothesis, which seem to be necessary only for the higher reaches of set theory, and intermittently and often unexpectedly in other branches of mathematics. There are also other possibilities. Much can be done with the Zermelo–Fraenkel axioms and only the denumerable version of the Axiom of Choice, or, more radically, by adopting an extension of arithmetic such as the predicative system studied by Feferman. The full axiom of choice, the continuum hypothesis, and other stronger axioms about large cardinals and so on, seem to be necessary only for the higher reaches of set theory. A mathematical conclusion derived from Zermelo–Fraenkel Choice axioms is as good as it gets. What then is the challenge on such occasions posed by insisting that it is timelessly true that 6 700 417 is prime? In what way is this a different challenge from the claim that this piece of rock contains gold?

One answer is, of course, that there is no essential difference. The historian may say that a theorem is like a nugget of gold. Like the nugget, it lies where it does, and when dug up it has the properties it has, which are not those of the discoverer or of society to alter. Gold is a malleable metal that does not corrode, Fermat's Last Theorem simply says what it says and, once proved, is known with complete certainty. But why society sets the value it does on gold, why it does these and not those things with it, gives gold certain metaphorical meanings, and so forth, these are all historical questions much discussed in the literature. In the same way, one can discuss why society values Fermat's Last Theorem, why mathematicians sought proofs for it in the ways they did, what value they, and society, attaches to work of this kind – all this likewise has a historical dimension.

Such historians may go on to add that some philosophers since Quine have doubted that the analytic-synthetic distinction is sufficiently clear to make the truths of mathematics different in kind from those of science, but this erosion of a once fundamental dichotomy should not give comfort to the historian of mathematics unless and until the history of mathematics can match the history of science in areas that interact with philosophy. There are schools of thought that seek to compare the abstract entities of mathematics with those of theoretical physics, and that may well be a propitious analogy. Uncomfortable though it is to admit, however, there is still much history of mathematics done with too much deference to naïve ideas about mathematical truth.

Other historians of mathematics sympathetic to the philosophical enterprise might begin by taking comfort in another of Feferman's observations, that by defining the concept of a natural number in terms of the notion of a set, one reduces 'a completely clear concept to one that is quite unclear'.[28] From Benacerraf to Kitcher and since, it is indeed contested if the traditional philosophies of mathematics permit an adequate analysis of mathematical knowledge. But it is beyond dispute that most practising mathematicians, as they establish what they see as the truth of this or that result, do not find solace in the philosophy of set theory. The practice of set theory, theorems, even recondite theorems, in mathematical logic, these do find their ways into contemporary mathematics, but that is another matter. Mathematicians value proofs. But a philosophy of mathematics that focuses so exclusively on set theory is not the one that promises a fruitful union with the history of mathematics.

The analysis of proofs, and the elaboration of a clear, useful distinction between syntactic and semantic analyses of proofs is a twentieth-century development, with a rich history of its own (see the essay by Mancosu). Mathematicians often discuss if this or that analysis of a problem is satisfactory or unsatisfactory. Bourbaki, for example, believed that a good explanation of a mathematical topic involved drawing out its underlying structural features, and coined the phrase 'a calculation is not an explanation.' The historian of mathematics must follow these debates. An answer in the manner of Bourbaki might be anachronistic, but the question remains, because in various ways mathematicians have often asked for more from a proof than that it be correct. Poincaré, for example, regarded a theorem as good if it organized one's ideas, and therefore if it is enabling, and opens the way to further discoveries. The purely logical aspect of mathematical work is not, for a mathematician, the whole story, and so it cannot suffice for the historian of mathematics.

Mathematics is concerned with concepts, and these concepts can be unclear and subject to refinement. A sound logical elucidation of a mathematical concept may well not help one to analyse how that concept came about and how it changed, and therefore not help the philosopher interested in how mathematics develops. Even if this or that piece of mathematics can be distilled into set theory, it is likely that its imperfectly formed antecedents cannot be, and not only

---

[28] See for instance, Feferman's *In the Light of Logic*, p. 90, but similar thoughts may be found in Quine's writings and elsewhere.

do large swathes of remarkable work become invisible to the philosophical set theorist, but the implications are that only those parts of mathematics that are so reducible count as mathematics at all. This is a view so remote from the actual beliefs of mathematicians – despite the formalism they are said to adhere to when cornered – that one might wish to distinguish between the philosophy of set theory and the philosophy of mathematics. This is not a distinction we would wish to insist upon, but the fact remains that the large body of excellent work on the philosophy of set theory in the last 80 years has little to say about what most mathematicians do, or ever have done.

The long tradition in mathematical epistemology since the 1950s has focused on the elusive nature of mathematical objects, conceived of as sets, and discussed how and what we might know about them. Here, a useful analogy with the history and philosophy of science can be drawn. Scientists and philosophers of science readily admit that the fundamental laws of science – whatever they might be – can only be studied through a web of auxiliary hypotheses governing such things as the behaviour of the equipment needed to make the requisite observations. Historians of science have been active in the last 15 to 20 years explaining how old experiments work and how they contributed to the acquisition of scientific knowledge. They claim that the way people come to know scientific facts or laws is inseparable from the laws, and that a proper understanding of science – such as a scientist or philosopher of science would wish to have – does not cream off the discoveries. Or rather, and more precisely, that the proper philosophical picture of science would display the discoveries (call them facts or laws, or whatever you may wish) together with their place in theory and with an account of the relevant experimental methods.

The analogous claim for mathematics has much to commend it. The long, demanding technical arguments that fill up the pages of research journals are the mathematicians' means of knowing about the objects the theorems describe. Mathematicians savour, perhaps in different ways, the perception of an original and valuable idea and the proof of its correctness. Philosophers of mathematics who have discussed proof can meet profitably with similarly minded historians of mathematics. For such analyses we may also turn to the thoughts of the mathematicians themselves, and especially the research mathematicians. These people necessarily construct conceptual narratives that guide them in regions where there are no proofs, and spelling out what these systems have been is a part of the history of mathematics that is perhaps neglected and where the analytic skills of a philosopher may well assist.

An analysis of proofs as a means of acquiring knowledge of mathematical concepts will bring other features of mathematics to the surface. Quite obviously, mathematics does not advance anything like as fast as it would if it was just a matter of logical deduction. There ought to be ways of explaining why this theorem follows from that one but it took twenty years for anyone to notice. It may be because the technicalities were forbidding, albeit logical and in that sense routine. Or it may be that new ideas were needed. It may be because existing ideas were imperfectly understood, Some people understand some issues better than others, have different skills, sensitivities, and indeed knowledge. All of these factors are invisible when mathematics is taken to be purely logical.

One way forward, we believe, is the elucidation of mathematical narratives. The construction of conceptual systems (if 'system' is not too grand a word for what may be little more than an organized collection of hunches or preferences) is built up in the training of a mathematician, doubtless much as Warwick described in the case of nineteenth-century Cambridge. These systems reflect the skills and knowledge of the mathematician, they may shape his or her priorities, hint or even circumscribe where they look for answers and how they go about finding them. They also generate the tools to evaluate another mathematician's work.

4.5 The example of Desargues' theorem

To give one example, with the rediscovery of projective geometry in the nineteenth century Desargues' theorem first became important, and then contentious. In 1847 von Staudt published his *Geometrie der Lage*, a book that might well have languished in obscurity had not Otto Stolz urged Klein to read it in 1870/71. The effort of thinking through what von Staudt was doing in that book seems to have been a major source for Klein to place the emphasis he did on the role of transformations of figures in geometry in his *Erlangen Program* of 1872, not because the *Geometrie der Lage* does so clearly and forcefully, but because the book is quite obscure. Von Staudt took the primitive figures of geometry to be points, lines, and planes, which are somehow moved around in a way that sends points to points, lines to lines and planes to planes. When this happens the principal invariant of projective geometry is preserved, the cross-ratio of four points (or of four lines). But whether a projective transformation is any transformation of this kind, or whether it is some combination of simpler transformations (such as central projections) that happen, as the result of a theorem, to preserve cross-ratio – all that is not discussed.

To prove Desargues' theorem, von Staudt used arguments involving three-dimensional plane geometry, and showed that the proof follows from the fundamental assumptions of three-dimensional projective geometry. But over the next few years, as mathematicians began to work over the jumble of ideas bequeathed to them in the first half of the nineteenth-century in projective geometry, doubts began to set in. They began to sort out what the fundamental features of projective geometry were: what figures did it discuss, what transformations were allowed, what properties of those transformations go into the definition, which into the theorems. With growing clarity about this, they began to doubt if Desargues' theorem followed from any plausible starting point for plane projective geometry, even though it did so as a theorem about figures in planes in any three-dimensional projective geometry.

As is well known, it was finally discovered that it does not. In his *Grundlagen der Geometrie* of 1899 Hilbert gave a complicated example replaced in later editions by a delightfully simple argument due to the American astronomer R.F. Moulton. It follows that there are projective planes that cannot be embedded in projective three-dimensional spaces. There ensued a series of investigations, of which Hessenberg's theorem of 1905 is a positive highlight: Pappus's Theorem implies Desargues's. On what might be called the negative side (which is not to make it less attractive) there must be projective geometries in which various familiar properties are lacking, and these too were sought out. What transformations does a geometry possess if Desargues' theorem fails in it? What sorts of projective planes do not permit coordinates? What is the role of the axiom of Archimedes?

This is not the place for a full historical account, but what has been said already highlights a number of issues for the philosopher and the historian of mathematics. There can be a lack of clarity about concepts that is not apparent to a mathematician but is apparent to his or her successors, but not for want of a good exposition. I have in mind the entirely elementary ideas of two- and three-dimensional geometry. Von Staudt was unaware of the implications of slipping between two and three dimensions because he thought it was entirely natural to study geometry in that setting – it was, after all, the setting given to it in Euclid's *Elements*.

Von Staudt was of the view that two- and three-dimensional projective geometry – and by this he meant the inherited family of objects, transformations, and results – needed tidying up. He sketched out a basis free of Euclidean concepts, but with its own muddles. Among these muddles one might decide that the lack of clarity about what constitutes a projective transformation is a failure of exposition (because it would have been easy enough for von Staudt to put it right on his own terms) but the sliding between two and three dimensions is not. One lesson Klein drew from it was that projective geometry should be the fundamental geometry, with Euclidean geometry as an offshoot or special case, but that work still needed to be done to make this happen.

In particular, it is somehow inadequate if the principal results of projective geometry are proved using theorems in Euclidean geometry. This is not because it risks a vicious circle, rather that it imperils any chance of establishing projective geometry on a sufficiently strong base. It would be possible that purely projective proofs of those results were artificial and complicated by comparison with proofs that use Euclidean concepts, and preferable to show that that is not the case. To keep Euclidean geometry at bay is quite a task, because many situations involving cross-ratio tempt one into arguing about four lengths, the use of coordinates is drilled into mathematicians, the role of the Archimedean axiom was eventually uncovered, and so on.

In the course of assigning some concepts to the list of the truly fundamental, and others to the list of derived concepts, it became clear that the objects under discussion had unexpected properties. When the list of fundamental aspects of projective geometry was subjected to scrutiny, what had been allowed to grow up naturally was sorted through in ways that culminated in many different axiomatizations of the subject in the period from 1890 to 1906 (and indeed beyond). A handful of results had been taken to be the high points of projective geometry (Pappus' Theorem, Desargues' Theorem, and so on). Now these results were seen as key points in the elaboration of a theory, their presence or absence exerting a governing role over what else could be said.

In terms of the jointly historical and philosophical tasks sketched out earlier, the example of Desargues' theorem suggests something like the following. The narrative, in this context, is a story about what two- and three-dimensional projective geometry really is. Von Staudt embarked upon this enterprise, which he saw as a worthwhile task that could be carried out in a certain way. Certain concepts are to be highlighted, such as points, lines, and planes in space, concurrence of lines, collinearity of points, and cross-ratio. Other concepts, the metrical geometric concepts, are to be avoided. To establish, for example, Desargues' theorem, von Staudt suggested certain procedures, in the form of constructions and arguments, which lead to the result. But there was what later mathematicians saw as an anomaly in his reasoning, which became apparent – and here we oversimplify, but not destructively – as clarity grew about the fundamental concepts and techniques, until finally it emerged that lines in the plane do not necessarily have the properties that ensure the validity of Desargues' theorem, and there is a remarkable twist to the relationship between two- and three-dimensional geometry.

It seems to the editors at least that just as historians and philosophers of science agree that an analysis of experiments contributes to understanding how scientists come to know what they do, and what it is that they do know, so too an analysis of narratives, and particularly of purported proofs will contribute to our understanding of the nature of mathematical knowledge.

4.6 Correctness, conjecture, and error in mathematics

There are many ways mathematicians find arguments convincing. It depends, for example, on whether an argument is written out and purports to be a proof, or is a sketch suggesting a profitable line of enquiry. Of course there are times when there is nothing for it but a long technical argument, and the standards for such arguments are high. Mistakes in such matters are real, and they can be anything from trivial to fatal. Generating such a proof for the first time may be much harder than following it, but there are rules such proofs ought to obey, and in principle they can be checked. However, confronted with such a proof, mathematicians do not necessarily check it before forming an expert opinion. They may instead consider if the result fits well with everything else they know, and if the method of proof seems appropriate. Of course the proof will be rejected if it leads immediately to a contradiction with a known, solid result. The more interesting case is what happens if the result is merely a bad fit – and what does that mean, in any case? If the result is thought to be deep, but the proof seems to

Introduction

be routine, mathematicians may be very doubtful about it. Surely, they argue, if the result was this easy to obtain, we'd have proved it long since. It is more likely that a new idea is needed. All these judgements are provisional, of course. What is important is that they are routine.

They matter even more when all that is offered is a potential line of enquiry. On these occasions, mathematicians often work by analogy. A very common way of working for mathematicians is to take an unsolved problem, and find one (or, better) a graded series of simpler problems like it that have been or can be solved, and work back towards the unsolved one, progressively clearing the techniques of peculiarities that do not generalize. This is not to suggest that the daily life of a mathematician is unimaginative and routine, but to articulate one of the entirely reasonable strategies they employ. The individual steps may be very difficult, the sequence of them exceptionally arduous and difficult to hold in one's head.

In short, and as they themselves often say, mathematicians tell themselves stories, summary explanations of why the theory goes the way it does so far, and they use these stories to evaluate ideas. It is the stories that generate the idea of a good or bad fit: this idea is consistent with what is already known; this one illuminates what is already known; this one, however, in some way jars. It follows that one task facing the philosophically inclined historian of mathematics is to make sense of the stories that mathematicians tell themselves.

Recent work by Tappenden on fruitfulness of concepts, and Mancosu on explanation, are indications of what can be done by philosophers deeply immersed in the history of mathematics, and more is going on. There needs to be more work from the side of the historians too. Apart from proofs, two published sources are available for the historian: conjecture and error. The frequent use of conjecture may well be a feature of twentieth-century mathematics[29], and some have decried it as a cheap way to present an idea, but at its best it is a major focus of ideas. The most famous of these conjectures is the one known as Fermat's Last Theorem, and it is well known that the turning point in the study of the celebrated problem was the discovery of an intimate connection with a well-developed theory of modular curves. At this point it was predicted, accurately, that Fermat's Last Theorem would fall within ten years. To be more precise, the discovery of the connection put Fermat's Last Theorem within the network of ideas thought to be governed by the Taniyama–Shimura conjecture, which is another story about how a substantial body of ideas should be organized. The famous four Weil conjectures offer another dramatic example of how a theory developed in one setting should supposedly be generalized and adapted to fit a much larger setting. It called for quite explicit tools to be created, and, interestingly, from that standpoint the first of the Weil conjectures to be solved was solved the 'wrong' way, thus offering the historian

---

[29] On this topic, see Mazur [1997].

Introduction

of mathematics further evidence of the organizing nature of mathematical narratives.

Error in mathematics is another rich field for the historian, and the historian's use of error needs to be sharpened. Plausible but inadequate arguments offer another way of examining the grounds for confidence in the persuasive powers of mathematics. It is in any case open to historical question how rigorous proofs in mathematics have been. Just as the working masses failed to conform to the theoretical prescriptions of Marxist political parties and real existing socialism didn't look like 'Socialism', real existing proofs often fell far short of 'proof' and can be analysed accordingly. Not to put them right, but to open up the question of how they (and not some idealized, but non-existent object) persuaded and convinced.

Historians of science have shown that there is much more to be said about mistaken experiments than that they were wrong. Buchwald's account of how Hertz could conclude that cathode rays do not carry an electric charge when ten years later J.J. Thomson showed that they do is a case in point. His essay is provocatively entitled 'Why Hertz was right about cathode rays' and he shows that Hertz was – given the theoretical beliefs he had at the time and the constraints on his equipment (in particular, the relatively poor vacuums he could create by comparison with Thomson's). Buchwald shows how these beliefs guided Hertz's design of his experiment and shaped his conclusions.

Historians of mathematics could do more of the same. Mathematicians make mistakes that derive from an imperfect grasp of the terms involved, inadequate abilities to calculate and estimate quantities, and in various other ways. Their investigations into the parallel postulate and what became non-Euclidean geometry invoked beliefs about the nature of the straight line. These beliefs turned out to be poorly founded. However, the resolution was not the discovery of a definition that, though perhaps obscure and difficult to grasp, could be seen to be defining the object that had been under discussion all along. Rather, it turned out that the term could have two, or even three mutually incompatible meanings, all unified as geodesics on two-dimensional surfaces of different constant curvature. Nonetheless, the satisfactory resolution of the matter as a piece of pure mathematics did seem to have implications for applied mathematics, as the geodesics and their accompanying spaces afforded various interpretations.

Cauchy's famous mistake noted above is harder to analyse, but we can agree that to elucidate it is, in part, to elucidate what he meant by continuity of a function. This turns out to be something about which one can be vague. Indeed, the history of a large amount of nineteenth-century mathematics can be made to hang on the way the precise concept of continuity departs in major but unpredicted ways from the intuition mathematicians had of it. Whatever problems there are in pointing to real-world candidates for straight lines (and there are many, starting, in the opinion of some people, with the absence of good candidates) one can imagine that geometers were aware of the real world. Understanding the role of adjectives in mathematics, such as 'continuous', is not the same as understanding nouns (such as 'line').

5 The chapters: mathematics, history, and philosophy in interaction

Although the foundational debate and related issues have frequently been discussed, it is our contention that many of its episodes stand in need of philosophical and historical reinterpretation. The paper by Mike Beaney that opens this volume takes issue with the 'Opinionated Introduction' of Aspray and Kitcher. There, they had noted two general programmes in the philosophy of mathematics as they saw it at the time. One, originating in Frege's work, sought to centre the philosophy of mathematics on the problem of the foundations of mathematics, while a newer, less-developed programme focused on articulating the methodology of mathematics through detailed historical investigation. Beaney argues that while these two programmes can still be found, possibly in sharper forms, the interesting question, raised by Aspray and Kitcher, is whether the two programs are consistent. It might seem, he admits, that they are not, and certainly Frege himself repudiated all historical and empirical approaches. But if there is not a timelessly true foundation of mathematics then the philosophy of mathematics might well benefit from historical investigation. In fact, Beaney goes further, and argues that the solution of the problem of the foundations of mathematics *requires* an historical approach. Beaney analyses Frege's own methods of investigation and finds that, contrary to what might be expected, he indeed made use of historically grounded 'elucidation.' His analysis of Frege's 'historical elucidation' takes us on a tour through related issues in Kant, Dedekind, Russell and Hilbert. Philosophical and mathematical methodologies are compared, and different conceptions of philosophical analysis (regressive, decompositional, and interpretive) are discussed. According to Beaney, even the famous debate between Frege and Hilbert on the axiomatic method and geometry took place to a large extent at the elucidatory level. Beaney's conclusion is that 'historical elucidation is called for at the very deepest levels,' and that a careful study of the analytic tradition in philosophy of mathematics suffices to show that there is no incompatibility between concern with the foundations of mathematics and historical investigations of mathematical methodology. The same conclusion could have been reached starting from the situation in mainstream foundational studies during the second half of the twentieth century: hopes for a priori knowledge have largely been abandoned, and this has led (as we remarked above) to recent fruitful cross-breedings of foundational and historical work.

Introduction

Ferreirós has previously argued (1999) that Riemann's work constituted a historically crucial background for the emergence of set-theoretical mathematics. In his contribution to the present volume, he places great emphasis on the figure of Riemann as a philosopher, showing how his famous contribution to differential geometry and spaces of variable 'curvature' emerged within the context of his epistemological studies. As Riemann's philosophical fragments have not been easily available in languages other than German, the reader will find here a discussion of little-known material. Riemann's inaugural lecture is an extremely rich document, owing also a great deal to Riemann's deep involvement with physics, a connection that is again emphasized. Interestingly, the three strands of physics, philosophy and mathematics also converged in Riemann's vision of the nature and origins of mathematical knowledge. Riemann set himself decidedly apart from apriorism, and conceived of mathematical concepts as evolving from attempts to understand real physical phenomena. In Ferreirós's view he was perhaps the first to think of mathematics as a free conceptual exploration of possible models of real-world phenomena. This aspect of Riemann's mind set has clear links with later work of Poincaré and Weyl, and it is taken up in Scholz's essay on Weyl, included in Part III.

In his thought-provoking study of 'The Riemannian background to Frege's philosophy,' Tappenden considers not only Riemann and Frege, but what he argues has been a distortion of the debate about the philosophy of mathematics that follows from a failure to appreciate the importance of complex analysis in nineteenth-century mathematics. The theory of complex functions was one of the greatest creations of nineteenth-century mathematics, and the central branch of the discipline according to the most influential German mathematicians at the time. By not appreciating this, and Frege's professional interest in it, philosophers and historians of mathematics have failed to see Frege as an essentially pro-Riemannian, anti-Weierstrassian philosopher. If this failure derives from a stunted appreciation of mathematical practice, its consequences are nonetheless a diminished understanding of what Frege was trying to do. Tappenden pursues the Fregean themes of fruitful concepts and centrality of a foundational concept, concentrating on the concept of function. He argues that Riemann's work presented these methodological issues (which were there connected to both the role of geometry and intuition and applications) in an especially urgent way, and that Frege's non-foundational work has live roots in this new style of mathematics.

The last chapter in this Part is devoted to Hilbert's conception of geometry. The once prevalent view of Hilbert as the champion of an exclusively formalist philosophy of mathematics is now superseded. However important finitism was to his celebrated metamathematical programme, plenty of evidence attests to the fact that his overall conception of the discipline went far beyond formalism. Corry presents Hilbert as a practicing mathematician who often changed his views, depending on his current scientific interests, and who was hardly willing to reflect on those changes or even acknowledge their existence. This is shown by Corry through the example of changes in Hilbert's views of geometry, from the 1890s to about 1920. Starting from considerations about the roles of empirical knowledge and intuition as the sources of geometry, Hilbert continued to shift emphasis from one to the other throughout his career (even after having erected the famous axiomatic structure contained in the *Grundlagen*). He finally came to favour empiricism as a result of his involvement with General Relativity Theory. One should perhaps emphasize that there is no inconsistency between a generally empiricist epistemology of geometry, a mathematical preference for a structural-axiomatic presentation of the theory, and a deep interest in the Hilbert programme as a promising avenue for foundational research. The historical Hilbert thus defies simplistic reductions and presents us with an interesting and multifacetted perspective on mathematics.

The expression 'modern mathematics' has become intimately associated with the study of mathematical structures, the heavy use of morphisms, of topological methods, and related themes. This image of mathematics was convincingly presented for algebra by van der Waerden in his (1930), expanded to the world of topology and associated subdisciplines, and became essential to the Bourbakist project. Part II of the book is devoted to an exploration of the emergence and consolidation of this modern framework, over a span of almost a century from Dedekind to Quillen, from set-theoretical structures to modern logic and category theory.

Avigad discusses the role of methodological criteria and more general philosophical principles ('metaphysics') in Dedekind's successive attempts to perfect his theory of ideals. The ideal theory of (1871) marked the beginning of modern number theory, which Dedekind redefined by turning away from the Gaussian study of binary forms and reciprocity laws to focus on rings of algebraic integers analysed by means of set-theoretical structures such as his ideals (infinite sets of algebraic integers in the ring under study, closed under certain operations). But that initial version of the theory was still too marked by pre-set-theoretical methods, and for more than 20 years Dedekind kept making substantial revisions in order to achieve his desired goals of generality, uniformity, invariant definition, direct deployment of sets and morphisms, and so on. Avigad subjects these goals and values to closer scrutiny and in this way advances towards an analysis of key elements of mathematical practice that have not been studied by traditional philosophy of mathematics. This is not to say that older studies of proof and axiomatic structure are deficient, but Avigad emphasizes the need to complement them with a broader account of the conceptual resources at play in the dynamics of mathematics as a discipline.

While several number theorists have found disadvantages and loss of information in Dedekind's preferred methodology, modern algebraists and very especially Emmy Noether were delighted to see how those methods lent themselves to application far beyond the realm in which Dedekind presented them. McLarty calls attention to the ways in which Noether promoted structural mathematics by going beyond Dedekind with her original vision of a 'set-theoretic foundation' for algebra. Noether's image of algebra gave centre stage to homomorphism theorems between structures and the related isomorphisms between substructures. Furthermore, and this makes McLarty's chapter particularly noteworthy, Noether persuaded topologists such as Alexandroff and Vietoris to extend this structural viewpoint to their discipline. In this way, the two most characteristic branches of modern mathematics became infected with the structuralist virus. The correlations between maps (among topological spaces) and homomorphisms (between homology groups) that came to characterize algebraic topology were, in turn, the model for functors in category theory.

Marquis takes up a related topic by scrutinizing the history of homotopy theory, and fibrations in particular, from their introduction by Hopf in the 1930s to Quillen's work in the 1960s. Apart from being a very significant further development in algebraic topology, the story of fibrations is of interest to Marquis for what it reveals about distinctive patterns of development of mathematics in the twentieth century. Once the importance of homotopy groups for the classification of spaces became clear, and their simple definition was laid out, it became apparent that it was extraordinarily difficult to compute these groups. Homotopy theory studies systematically the different methods for the computation of homotopy groups, among which fibrations figure prominently. The philosophical lesson that Marquis extracts and emphasizes throughout his detailed study is that mathematicians do not simply deal with *objects* (whatever their ontological status), but also and very often with *tools*: homotopy theory is not primarily a 'scientific' theory aimed at gaining knowledge of some mathematical objects, but rather can be seen as a 'technological' theory aimed at developing necessary tools; it is not only a theory of, say, electrons, but also a theory of electron microscopes. This pattern of development can be traced back at least to the days when groups emerged in nineteenth-century mathematics. Marquis offers further reflections on the contexts in which these new tools are configured, and the functions that serve to define them axiomatically, in an attempt to provide general concepts that can be generally applied to such developments.

A third key branch in the modern configuration of mathematics is mathematical logic, a subject that is particularly well known to many of the contributors to the present volume. Beginning in the 1930s, mathematical logic subdivided into different branches, such as higher set theory, model theory, and proof theory. The concept of a model was an elaboration of early notions of mathematical structures, based on set theory and not yet category theory. Mancosu's contribution to this volume is a detailed analysis of a celebrated and philosophically very influential contribution, Tarski's theory of logical consequence, on the basis of previously unknown manuscripts. Since the publication of Etchemendy's work in the late 1980s, there has been considerable debate over the interpretation of Tarski's theory and the reasons for its adequacy. One of the problems under discussion is the notion of model that Tarski upheld back in 1936. In recent years a number of authors expressed the belief that this issue had been conclusively solved to show, contrary to Etchemendy's claims, that Tarski's notion

Introduction

was the current one. However, Mancosu provides strong evidence to show that as late as 1940 Tarski was still employing a fixed-domain conception of model. This chapter will thus significantly alter the ongoing debate on logical consequence.

In and around the heyday of logical empiricism, a limited list of topics and viewpoints seemed to define the philosophy of mathematics (see, for example, the very good compilation Benacerraf and Putnam, 1983), but it should come as no surprise that there have been noteworthy alternative programs throughout the twentieth century. Part III of this volume, which returns us to a largely German setting around 1900, discusses a sample of such programs that partly connect with topics broached in Part I, and are also well worth reconsidering today.

Moritz Epple considers the epistemological views of Felix Hausdorff, an important but distinctly unconventional mathematician, philosopher, and writer of the period around 1900. Despite his lasting and profound contributions to mathematics, which include his book on set theory, his analysis of the concept of dimension, and his celebrated paradox in measure theory, Hausdorff was probably better known even in the 1930s under the pseudonym Paul Mongré. Using that name he wrote philosophy, cultural criticism, poetry and a successful play. In his essay, Epple considers the threads that join the two authors, Mongré and Hausdorff. He shows that a genuine epistemological interest brought Mongré, the Nietzschean, to study Georg Cantor's theory of sets in 1897/1898, and more importantly that Mongré and Hausdorff shared an epistemological position that drew on both Nietzschean philosophy, set theory, and modern mathematical axiomatics (which Hausdorff learned from Hilbert's *Grundlagen der Geometrie* of 1899). What Hausdorff called his 'considered empiricism', is the claim that beyond mathematics, no scientific knowledge can claim to be more than a more-or-less plausible, economical, or complex system of beliefs compatible with current empirical information. No empirical information completely determines any system of scientific beliefs. This position is more precise than some present-day variants of the idea that theory is underdetermined by experiment, for Hausdorff attributes a special role to mathematics. In Hausdorff's view mathematics is an autonomous creation of disciplined thinking. It is not an empirical science, and all empirical phenomena are capable of being described in alternative mathematical terms. Consequently, any mathematization relies on conceptual decisions, not on absolute truths. This viewpoint preceeded that of the Vienna circle by some 15 years, and Epple ends his paper by comparing Hausdorff's position with some views of Henri Poincaré and Moritz Schlick.

Weyl's mature view of mathematical knowledge is the topic of Scholz's essay. Far from the mid-twentieth-century tendency to think of mathematics as a perfectly autonomous intellectual domain, Weyl regarded mathematics as indissolubly linked with the rest of scientific knowledge (physics in particular) and as influenced by practical and cultural life. This came partly as a result of his strong involvement with the foundations of mathematics and with mathematical physics (both the general theory of relativity and quantum mechanics) during the 1920s. It also reflected the influence of several philosophical tendencies, and of Riemann's and Hilbert's ways of thinking, upon this multifaceted mathematician and intellectual. Weyl emphasized that scientific theories can no longer be regarded as faithful representations of the 'outside world,' but must be seen as 'symbolic constructions' of mankind. Because he also underscored the links to the empirical sciences and technology, as well as to cultural practices, Scholz calls Weyl's viewpoint a 'practice-related symbolic realism' (it must be clear from the previous remarks that it was in no way a naïve realism). Furthermore, Scholz suggests that such a viewpoint, properly understood and complemented, is appropriate to our present historical situation and may serve as a guide to reorienting social and scientific practices.

French philosophy of mathematics is seldom discussed in the Anglo-Saxon tradition of analytic philosophy, and the essay by Hourya Benis Sinaceur shows some of what we have been missing. She describes an evolution from what might be called French neo-Kantianism in the writings of Brunschvicg to Cavaillès and his philosophy of concept, and beyond, to the work of Desanti and Granger. While Cavaillès accepted that mathematics had a central role to play in any theory of knowledge, he also wanted the philosophy of mathematics to give an account of mathematical practice and how mathematicians are led to productive new results. Such an account was to be both pragmatic and abstract, rather than sociological or descriptive, and Cavaillès was led to it by an analysis of recent work in the foundations of mathematics. This in turn generated a structuralist account of mathematics, which, even before Bourbaki took up the term, he called architectural. Cavaillès, Lautman, and others added to this a somewhat phenomenological perspective taken from the later writings of Husserl, an attention to the semiotics of mathematics, and a philosophical perspective on the historical character taken from Brunschvicg's dispute with Couturat a generation earlier.

Cavaillès and others sought a dynamic objectivity for mathematics in which each constellation of mathematical concepts produces, in a rational way, the next one. They saw an objective and autonomous content to mathematics, as well as a formal aspect, to which they added a Husserlian dimension of human, individual action. The content of mathematics is, they argued, rationally organized and the constraints of the situation at any time are a force for change. Benis Sinaceur discusses some of the problems this evolving philosophy of mathematics encountered, including the tendency to see mathematics as producing itself without human actors, so objectivity is produced at the cost of removing the thinking subject (the individual mathematicians). She notes that this is an old conundrum for Kantians in a new guise, and that the French adopted a solution that took them some steps away from a Husserlian approach. This involved the distinctively French take on what Sinaceur calls the false question of how to give a rational foundation for knowledge, which replaces this false perspective with a historicized account of mathematics. This account invokes a dialectic of concepts that, it was claimed, was superior to a logical account because it was not so

Introduction

rigid and permitted the possibility of progress. The essay concludes with a brief consideration of the views of French philosophers in the years 1940 to 1970.

Needless to say, philosophical reflections and proposals, stimulated by developments in foundational studies, keep cropping up in recent times. Among them, the editors of this volume thought adequate to rescue, so to say, a recent paper due to Sieg and originally published in *Synthese*. Even though its title, 'Relative consistency and accessible domains,' does not suggest this, Sieg's paper is based to a large extent on reflections about the mathematical developments that led historically to modern axiomatics, structural mathematics, and proof theory. Departing from the failure of Hilbert's Program to attain its main goal of a consistency proof, Sieg makes a call for philosophical studies that may clarify the 'objective underpinnings' of the theories that have been employed in recent proof theory. Proof theory is often regarded as the branch of mathematical logic that stays closer to epistemological concerns. Interesting results about relative consistency have been proved for large portions of modern mathematics, relative to 'quasiconstructive' theories. Sieg proposes to look at these theories through the feature of accessibility in a well-defined sense (hence his talk of accessible domains), and he thus highlights a fruitful area where gains are to be expected from a collaboration between logicians and epistemologists. In this way, he makes an explicit proposal to go beyond the exclusive dichotomy between platonism and constructivism, and to bridge the gulf between experts in foundations and philosophers interested in mathematical practice.

The coda, written by Gray, proposes a particular and explicit way of responding to the challenge the study of mathematics poses to historians: how can mathematics have a history, and how can this history most adequately be described? He moves beyond discipline-oriented writing of history, to explore one of the possible venues for bringing the history of mathematics in line with history of science and intellectual history. Beyond its links with particular theories, methods, and foundational assumptions, the arrival of 'modern' mathematics happened precisely at the time of Modernism, a well-known phenomenon in intellectual and cultural history, of which the cubism of Picasso or the phenomenology of Husserl may serve as examples. Gray highlights the gains that our understanding of history can obtain by adding this novel perspective to our exploration of twentieth-century mathematics. His main examples, beyond mathematical theories, are taken from contemporaneous philosophical movements.

Reinterpretations in the History and Philosophy of Foundations

Frege and the Role of Historical Elucidation: Methodology and the Foundations of Mathematics

Michael Beaney

> Without the concepts, methods and results found and developed by previous generations right down to Greek antiquity one cannot understand either the aims or achievements of mathematics in the last 50 years. (Hermann Weyl, 1951.)

1 Introduction

In their 'Opinionated Introduction' to the collection of papers published in 1988, *History and Philosophy of Modern Mathematics*, the editors William Aspray and Philip Kitcher suggested that there were two general programs in the philosophy of mathematics as they saw it at the time. The first, originating in Frege's work, 'conceives of the philosophy of mathematics as centered on the problem of the foundations of mathematics', while the second, newer and less developed, 'takes the central problem to be that of articulating the methodology of mathematics', pursued through detailed historical investigation (1988, 19). Fifteen years later, the same two general programs can be discerned – if anything, in sharper forms. Recent work by neo-Fregeans such as Bob Hale and Crispin Wright has restored logicism to the agenda; and there have been a number of fine historical studies over the last few years. Aspray and Kitcher also raised the question of whether the two programs were consistent. At first sight they seem clearly inconsistent. For if mathematics has foundations, particularly if those foundations are knowable a priori, then historical investigations appear irrelevant. This certainly seems true, if Fregean views lie at the core of the program, for in arguing that arithmetic had logical foundations, Frege himself repudiated all historical and empirical approaches. However, one can take the problem of the foundations of mathematics as the central issue in the philosophy of mathematics without necessarily believing that there *is* a foundation; and one can argue that a proper understanding of the problem itself requires historical investigation. Indeed, as I will argue in this chapter, one can go further. It is not just that historical investigation can *help* in understanding the problem, but that its solution *requires* an historical approach.

The obvious way to show this, and to bring together the two programs that Aspray and Kitcher distinguished, is to consider Frege's own methodology in his work on the foundations of arithmetic. Frege may have repudiated historical approaches himself, but as the history of any discipline shows, the practitioner of a method is not necessarily the best person to reflect upon it or explain its motivation and operation. Frege came to recognize the need for what he called 'elucidation', but he failed to appreciate the historical understanding that this presupposes. In clarifying the role of what I shall call 'historical elucidation' in Frege's work on the foundations of mathematics, I will compare his views with some of Dedekind's, Russell's and Hilbert's views. Before looking at Frege's methodology, however, I will say something briefly about Kant, whose philosophy of mathematics forms an essential part of the background to Frege's work and to the debate in which Frege staked out such a clear position.

2 Kant, Methodology and Analyticity

Kant's central concern in his *Critique of Pure Reason* with the question, 'How are synthetic a priori judgements possible?', and his view that mathematical propositions are synthetic a priori, are well known. What is less well known is that it was his reflections on the differences between mathematical and philosophical methodology that motivated his critical project, though this was admittedly obscured by Kant himself in placing his discussion of methodology at the back of the *Critique*, in the 'Transcendental Doctrine of Method'. Kant opens the first section of Chapter 1 with a claim that Frege too would have endorsed: 'Mathematics gives the most resplendent example of pure reason happily expanding itself without assistance from experience' (A712/B740). Where Kant and Frege differ, however, is in their views of the role played by 'intuition' ('*Anschauung*'). According to Kant, mathematical cognition involves the construction of concepts, and to construct a concept, Kant writes, 'means to exhibit *a priori* the intuition corresponding to it' (A713/B741). For Frege, on the other hand, in the case of arithmetic, 'intuitions' are not required.

Kant takes the example of a simple geometrical problem to illustrate his conception of the difference between philosophical and mathematical method:

> Give a philosopher the concept of a triangle, and let him try to find out in his way how the sum of its angles might be related to a right angle. He has nothing but the concept of a figure enclosed by three straight lines,

and in it the concept of equally many angles. Now he may reflect on this concept as long as he wants, yet he will never produce anything new. He can analyze and make distinct the concept of a straight line, or of an angle, or of the number three, but he will not come upon any other properties that do not already lie in these concepts. But now let the geometer take up this question. He begins at once to construct a triangle. Since he knows that two right angles together are exactly equal to all of the adjacent angles that can be drawn at one point on a straight line, he extends one side of his triangle, and obtains two adjacent angles that together are equal to two right ones. Now he divides the external one of these angles by drawing a line parallel to the opposite side of the triangle, and sees that here there arises an external adjacent angle which is equal to an internal one, etc. In such a way, through a chain of inferences that is always guided by intuition, he arrives at a fully illuminating and at the same time general solution of the question. (*CPR*, A716–7/B744–5.)

The problem that Kant considers here is in fact taken from Euclid's *Elements* (I, 32), and it is his reflections on Euclidean methodology that lead him to emphasize the need for 'intuitions' in geometry. By 'intuition' he means a particular spatio-temporal representation. Without such representations, according to Kant, I would be unable to solve geometrical problems. 'I construct a triangle', he writes, 'by exhibiting an object corresponding to this concept, either through mere imagination, in pure intuition, or on paper, in empirical intuition, but in both cases completely a priori, without having had to borrow the pattern for it from any experience' (A713/B741). For the purposes of understanding or finding a proof concerning the properties of a triangle, it does not matter what particular triangle we take (imagine or draw), but there must be *some* representation of a triangle. It is in this way that geometrical judgements can be both a priori (since they are not essentially dependent on any particular empirical experience) and synthetic (since they still require an 'intuition' – a 'pure intuition' where there is no actual object). Similar considerations apply in the case of arithmetic, on Kant's view, where 'symbolic constructions', involving the manipulation of symbols in accordance with rules, rather than geometrical constructions are required (cf. A717/B745).

How does Kant see philosophical methodology? The 'philosopher' to whom Kant is referring in this discussion is the Leibnizian, who believes that advances can be made by 'conceptual analysis'. Clearly, what Kant has in mind here is a particular model of analysis, in which complex concepts are regarded as 'composed' of simpler concepts, whose extraction it is the task of the philosopher to undertake. This conception is reflected in his official criterion for the 'analyticity' of a judgement, which can be formulated as follows:

(AN_K)   A true judgement of the form '*A* is *B*' is *analytic* iff the predicate *B* is contained in the subject *A*.

According to Kant, however, merely showing what concepts are 'contained' in a given concept cannot genuinely advance our knowledge. Given that mathematics does advance our knowledge, it cannot simply consist in conceptual analysis. Now we may well regard Kant's view of conceptual analysis as too crude, being legitimate only where the metaphor of 'containment' can be cashed out.[1] But this does not necessarily undermine his point about the need for some kind of spatio-temporal representation or 'intuition' in mathematics.

3 Frege, Analyticity and Elucidation

Frege's project is often characterized as the attempt to show that arithmetic – *pace* Kant – is a system of analytic a priori truths. Frege suggests as much himself in his *Grundlagen* of 1884. But it is not how he described it in his major work, the *Grundgesetze der Arithmetik* (1893, 1903), where instead he talks simply of 'reducing' arithmetic to logic. By the time of the *Grundgesetze*, the introduction of his distinction between *Sinn* (sense) and *Bedeutung* (reference or meaning) had complicated matters, and, so it seems, Frege had come to have doubts about the idea of analyticity.[2] However, if we do characterize Frege's project in a Kantian context, and bring the *Grundlagen* and *Grundgesetze* together, then the following can be offered as Frege's criterion for analyticity:

(AN_F)   A truth is *analytic* if its proof depends only on general logical laws and definitions. (Cf. *GL*, §3.)

This looks quite different from Kant's official criterion. So has Frege simply changed the terms of the debate? If so, then Kant and Frege cannot be considered to be offering different answers to the same question ('Are arithmetical judgements analytic or synthetic?'), since they interpret the question in different ways. Indeed, the conflict between their positions appears to disintegrate, since Frege would agree with Kant that arithmetical propositions are not analytic in the sense captured in (AN_K). Frege is well aware of the methodological issue here. In explaining his own distinction between analytic and synthetic truths in the *Grundlagen*, he comments: 'By this I do not, of course, wish to introduce new senses, but only to capture what earlier writers, in particular *Kant*, have meant' (§3, fn. E; *FR*, 92). He returns to the issue in the concluding part of the *Grundlagen*. The key passage is §88:

> Kant obviously underestimated the value of analytic judgements – no doubt as a result of defining the concept too narrowly, although the

---

[1] Cf. Beaney 2003, especially §§4 and 6.

[2] For an account that elaborates on this, see Beaney 2005, especially §3. I return to Frege's doubts about analyticity below. 'Bedeutung', as Frege uses the term after 1891, is generally translated as either 'reference' or 'meaning', though I prefer to leave the term untranslated, for the reasons given in Beaney 1997, §4.

broader concept used here does appear to have been in his mind. On the basis of his definition, the division into analytic and synthetic judgements is not exhaustive. He is thinking of the case of the universal affirmative judgement. Here one can speak of a subject concept and ask – according to the definition – whether the predicate concept is contained in it. But what if the subject is an individual object? What if the question concerns an existential judgement? Here there can be no talk at all of a subject concept in Kant's sense. Kant seems to think of a concept as defined by a conjunction of marks; but this is one of the least fruitful ways of forming concepts. Looking back over the definitions given above, there is scarcely one of this kind to be found. The same holds too of the really fruitful definitions in mathematics, for example, of the continuity of a function. We do not have here a series of conjunctions of marks, but rather a more intimate, I would say more organic, connection of defining elements. The distinction can be clarified by means of a geometrical analogy. If the concepts (or their extensions) are represented by areas on a plane, then the concept defined by a conjunction of marks corresponds to the area that is common to all the areas representing the marks; it is enclosed by sections of their boundaries. With such a definition it is thus a matter – in terms of the analogy – of using the lines already given to demarcate an area in a new way. But nothing essentially new comes out of this. The more fruitful definitions of concepts draw boundary lines that were not there at all. What can be inferred from them cannot be seen from the start; what was put into the box is not simply being taken out again. These inferences extend our knowledge, and should therefore be taken as synthetic, according to Kant; yet they can be proved purely logically and are thus analytic. They are, in fact, contained in the definitions, but like a plant in a seed, not like a beam in a house. Often several definitions are needed for the proof of a proposition, which is not therefore contained in any single one and yet does follow purely logically from all of them together. (*GL*, §88; *FR*, 122.)[3]

In a footnote to the first sentence of this section, Frege refers to a passage in the *Critique* in which Kant alludes to a different criterion, which might be formulated as follows:

(AN$_L$) A true judgement of the form '*A* is *B*' is *analytic* iff its negation '*A* is not *B*' is self-contradictory. (Cf. Kant, *CPR*, B14; A150–1/B189–91.)

This offers a broader criterion, which can be readily extended to judgements not obviously of subject-predicate form, and which is closer to Frege's own criterion. But Frege is right that Kant does think primarily of universal affirmative judgements, and that Kant's official criterion is too narrow for application in the case of mathematics. There is therefore some justification in Frege's earlier claim too that what he is doing is not introducing an entirely new meaning, but merely making explicit what Kant really had in mind – in the sense that, had Kant considered it properly, he would have endorsed Frege's reformulated criterion.

---

[3] For detailed discussion of this key passage, see Tappenden 1995b.

However, whether or not one agrees with Frege's claim, the crucial point for present purposes is that Frege recognizes the need to address the methodological issue. It is not enough for him simply to stipulate what he means by 'analytic', and then proceed to prove that arithmetical judgements are analytic in his sense. In calling them 'analytic', he is using a term with an established history, and some comment is necessary on the relationship between his use of the term and previous uses. Furthermore, Frege realizes that he cannot depart too radically from earlier uses, on pain of being misunderstood and of making no significant contribution to the traditional debate. In motivating his own project, Frege himself has to do history of philosophy. He may do so in a crude form, and may rely (as he does in the *Grundlagen*) on second-hand sources, but he cannot avoid saying something about previous philosophers, and this requires interpreting them. If he offered incorrect interpretations of Kant, say, then he would be confusing rather than clarifying the debate, and hence not showing the value of his own project.

But would it not have been better to have dispensed with the concept of analyticity altogether? As noted above, this is just what Frege does from the *Grundgesetze* onwards. After the *Grundlagen*, he never again characterizes his project by talking of arithmetical truths as 'analytic'. Perhaps he came to the view that he was not in fact using the term in the same sense as Kant – or at least, that there were problems in claiming that he was. However, if we characterize his project simply as showing that arithmetic is *reducible to logic*, then the problem only re-emerges in a different form. For what is now required is explanation of the relevant conceptions of reducibility and logic.[4] According to Kant, arithmetic is not reducible to Aristotelian logic, and on this Kant is quite right, as Frege would have been the first to agree. But what Frege did, of course, was to broaden the scope of formal logic, by means of which the reducibility of arithmetic to logic became a genuinely feasible project. But this too is something that needs to be spelled out, as Frege both recognized and undertook. He was at least as much concerned with expounding and justifying his 'Begriffsschrift' – his 'concept-script', as Frege called his logical system – as he was in using it to demonstrate logicism. And he was well aware that this also required comparing his system with those of others, as he does, for example, in his two papers on Boole's logic (*BLC* and *BLF*), and criticizing the various accounts of arithmetic and logical conceptions of other mathematicians and philosophers, as he does in the first three parts of the *Grundlagen* and at various points in the two volumes of the *Grundgesetze*. Once again, in other words, Frege

---

[4] See MacFarlane 2002 for discussion of whether Kant and Frege can be said to disagree about logicism, understood as the view that arithmetic is reducible to logic, if they have different conceptions of logic. MacFarlane comes to the conclusion that there is enough similarity in their conceptions to substantiate the claim that there is genuine disagreement. But the issue is controversial, and even if we were convinced by MacFarlane, his account only confirms the point that detailed argumentation is required to establish that Frege is genuinely offering a refutation of Kant – as opposed to just 'changing the subject'.

recognized the need to do a certain amount of history of logic and philosophy to motivate his project.

This is not to say that Frege saw himself as doing history of logic and philosophy as such. Nor is it to say that he did it well. The caustic tone he increasingly adopted in his criticisms of other writers reflects a lack of the charity and sensitivity that is essential in the proper interpretation of other points of view.[5] But Frege did realize that clarification of the most basic concepts of a system or theory was a necessary complement to its formal construction, and that this clarification had a different methodological status. This comes out most explicitly in his late work, 'Logic in Mathematics', written in 1914 (and only published posthumously). What he draws attention to here is the role of what he calls 'elucidation' ('*Erläuterung*'), which makes clear the sense of an expression by example – by using it in context.[6] Since not everything can be defined, we must rely on something else – elucidation – to explain the meaning of the most basic terms of all. Since there is always the risk of misunderstanding in attempts at elucidation, Frege writes, 'we have to be able to count on a meeting of minds, on others' guessing what we have in mind' (*LM*, 224; *FR*, 313). But, Frege goes on, 'all this precedes the construction of a system and does not belong within a system' (ibid.).

Clearly, Frege sees elucidation as a non-scientific or pre-theoretical activity. But, one might suggest, the most effective method of elucidation will involve history of philosophy and science. We may still have to count on a meeting of minds, but drawing on, making explicit, and to some extent refining, what is our common conceptual heritage may be the best way of achieving this. Frege relegates elucidation to mere propaedeutics, but if our concern is genuinely with the foundations of a system, then elucidation is just as important a part of the foundationalist enterprise as the formal construction itself. Significantly, Frege himself uses the term 'logical analysis' ('*logische Zerlegung*') in describing the work of elucidation (*LM*, 225–8; *FR*, 314–8), suggesting an admission of its fundamental role. So although Frege's own emphasis is much more on systematic construction, even Frege, who is often regarded as the archetypical denigrator of historical considerations, in fact makes room for them.

To make explicit that the elucidation required here has an historical dimension, let us call it 'historical elucidation'. On Frege's account, while historical

---

[5] For an account of this lack of charity and sensitivity, with particular reference to Cantor and Dedekind, see Tait 1997.

[6] The importance of Frege's conception of elucidation is unfortunately obscured in the English translation of Frege's *Nachgelassene Schriften* by the rendering of 'Erläuterung' as 'illustrative example' (*PW*, 207–8). In *The Frege Reader*, I corrected this to 'elucidation' (*FR*, 313). There are three places in Frege's writings where he talks about elucidation. The passage in 'Logic in Mathematics' might be taken as his definitive statement (*FR*, 313–4). The two other places are in his letter to Hilbert of 27 December 1899 (*PMC*, 36–7) and in his published critique of Hilbert in the second series of his essays 'On the Foundations of Geometry', which appeared in 1906 (*FGII*, 300–1). It is worth noting, then, that Frege's appeal to elucidation originated in seeking to respond to Hilbert's work on the foundations of geometry. I return to their dispute in Section 6 below.

elucidation serves a preliminary pedagogic purpose, it plays no further role once the concepts are fixed. To use Wittgenstein's famous metaphor of the *Tractatus*, elucidation is merely a ladder to be kicked away once the conceptual framework is in place. But fixing concepts is itself an essential part of the foundational enterprise, and if fixing concepts requires historical understanding, then historical considerations enter at a much deeper level than Frege was willing to admit. Indeed, if the aim of a foundational enterprise is to reveal and articulate the underlying conceptual framework (and the ontological commitments, and so on) of one of our actual practices, then does this not presuppose understanding of that practice as it has been manifested up to now? And how can we assess whether a proffered set of concepts is foundationally adequate without reference to our use of such concepts – and related concepts – in the past?

By 'elucidation' Frege meant explaining the sense of the most basic terms by using them in context. But clarity about their senses is achieved not just by giving examples illustrating those senses but also by saying how those terms are *not* to be understood.[7] To find good examples, we need to draw on our past and present practice. But since that practice will also involve the use of related terms, and those same terms used in slightly different senses, and indeed, those same terms used in confusing ways, we also need to distinguish 'correct' from 'incorrect' uses. So critique of past and present practice will also be required. Properly understood, in other words, elucidation requires critical awareness of our practice to date, and hence possesses an historical dimension. It is this that I want to reflect by talking of 'historical elucidation'.[8]

4 Frege, Dedekind and the Definition of Number

The importance of historical elucidation can be illustrated by considering Frege's central concern, the analysis of number, a concern that he shared with his contemporary, Richard Dedekind, whose own work, *Was sind und was sollen die Zahlen?* (1888), appeared between the publication of Frege's *Grundlagen* (1884) and the publication of the first volume of Frege's *Grundgesetze* (1893), and was influential in the development of Frege's ideas. Frege's *Grundlagen* opens with the complaint that no one, up to then, has properly understood what numbers are, and as noted above, the first three parts of the book consist in a criticism of various earlier views of number. Frege discusses the views of, among others, Euclid, Kant,

---

[7] Cf. Frege *PMC*, 36–7; *FGII*, 300–1 (mentioned in the previous fn.).

[8] For further discussion of Frege's conception of elucidation, with particular reference to his distinction between concept and object (as articulated in Frege *CO*), see Weiner 1990, ch. 6; 2001; 2005. I stress more than Weiner does the historical dimension of elucidation, but we agree on the important – but often overlooked – role that elucidation plays in Frege's philosophy.

Leibniz, Mill, Hankel, H. Grassmann, Newton, Schröder, Locke, Berkeley, Hobbes, Hume, Thomae, Descartes, and Jevons. His source for many of their works is the two-volume collection edited by Johann Julius Baumann, *Die Lehren von Zeit, Raum und Mathematik*, published in 1868, which might make one reluctant to regard Frege as a genuine historian of philosophy, searching out previous views for himself. He is also largely critical of those views, and generally insensitive to what might have motivated them. But even if his source is second-hand, and he shows little sign of interpretive charity, historical elucidation is what he is supplying, deliberately preparing the way for his own positive account. A number statement, we eventually learn in §46, more than half-way through the book, is an assertion about a concept. But without the preceding critique, we would not really understand what this conception amounts to, and how it is an advance on earlier views.

This can be illustrated in the particular case of existential statements, which, on Frege's account, are number statements involving the number 0 (cf. *GL*, §53). To say that unicorns exist, for example, is to assert that the concept *unicorn* is instantiated, i.e. to deny that the number 0 applies to the concept *unicorn*. Frege's view is often encapsulated in the claim that 'existence is not a predicate'; but unless we had some understanding of the difficulties that arise in treating existence as a predicate – most notoriously, in the ontological argument for the existence of God – then we would not appreciate Frege's claim.[9]

In Part IV of the *Grundlagen*, he then develops his positive account, defining the numbers in terms of extensions of concepts. His definitions of the numbers 0 and 1, for example, can be formulated as follows:

(E0) The number 0 is the extension of the concept 'equinumerous to the concept *not identical with itself*'.
(E1) The number 1 is the extension of the concept 'equinumerous to the concept *identical with* 0'. (Cf. FR, 118–19.)

But these definitions are far from immediately compelling. Could anyone even understand them, let alone judge that they are true and capture the essence of the natural numbers, without prior elucidation? Frege admitted himself that his definitions 'will hardly, perhaps, be clear at first' (*GL*, §69), and would strike people as 'unnatural' (*GL*, xi). A similar admission was made four years later by Dedekind, who also sought to define the numbers in terms of what he called 'systems', corresponding to Frege's extensions of concepts: 'in the shadowy forms which I bring before him, many a reader will scarcely recognize his numbers which all his life long have accompanied him as faithful and familiar friends'

---

[9] That existence is not a predicate was also argued by Kant, in his criticism of the ontological argument in the *Critique of Pure Reason* (A592–602/B620–30). And Frege's conception of existential and number statements as statements about concepts was also influenced by Johann Friedrich Herbart; see Gabriel 2001. But it is only in Frege's work that we find the explicit claim that existential statements are to be analysed in terms of the second-level concept 'is instantiated'.

(1888, 791). Clearly, there are other considerations that need to be understood in order to judge the legitimacy of the definitions. As Frege put it, 'To those who might want to declare my definitions unnatural, I would suggest that the question here is not whether they are natural, but whether they go to the heart of the matter and are logically unobjectionable' (*GL*, xi). We might agree; but whether they go to the heart of the matter is not something that can be appreciated without historical elucidation.

In the *Grundlagen*, Frege had himself expressed doubts about the appeal to extensions of concepts (cf. *GL*, §68, fn.; §107). By the time of the *Grundgesetze*, however, he had convinced himself that the appeal was unavoidable, a conviction he encapsulated in his infamous Axiom V, which asserts the equivalence between the following two propositions:

(Va) The function F has the same value for each argument as the function G.
(Vb) The value-range [*Werthverlauf*] of the function F is identical with the value-range of the function G. (Cf. *GG*, I, §§3, 9.)

Since concepts, according to Frege, are functions whose values are truth-values, Axiom V entails the equivalence between the following instantiations of (Va) and (Vb):

(Ca) The concept F applies to the same objects as the concept G.
(Cb) The extension of the concept F is identical with the extension of the concept G.

Admittedly, Frege still expressed unease about whether Axiom V could really be taken as a logical law, but he saw no other way of providing arithmetic with foundations (*GG*, I, vii; cf. *GG*, II, §147; *PMC*, 213). The appeal to extensions of concepts was justified precisely to the extent that Axiom V was legitimate.

What convinced Frege? One reason is almost certainly the publication of Dedekind's *Was sind und was sollen die Zahlen?* in the intervening period. We know that Frege took it as the text for discussion in the mathematical seminar that he organized at Jena during 1889–90 (cf. Kreiser 2001, 295–6). Although published four years after Frege's *Grundlagen*, Dedekind's monograph was written independently of it. As Dedekind mentions in his 1893 preface to the second edition, he only became aware of Frege's book a year after the publication of his own. Although there are problems in Dedekind's conception of a 'system',[10] his introduction of

---

[10] Dedekind (1888, §1) talks of a 'system' as 'an aggregate, a manifold, a totality', but he also sees a system as a 'thing', a 'thing' being whatever is an 'object of our thought'. He writes that a system S 'is completely determined when, for every thing, it is determined whether it is an element of S or not'. This suggests that for every system, it must be determined whether it is itself an element of itself or not. So an analogue of Russell's paradox threatens: is the system of systems that are not elements of themselves itself an element of itself or not? Dedekind also writes that 'we intend here for certain reasons wholly to exclude the empty system which contains no elements at all, although for other investigations it may be appropriate to imagine such a system'. If we can 'imagine' it, then this suggests

what are essentially extensions of concepts would surely have reinforced Frege's own thinking.[11]

However, the differences between the two projects are no less important, and Frege's recognition of these differences was also influential in the development of his work. The key methodological difference is Dedekind's appeal to abstraction in deriving the natural numbers from an arbitrary 'simply infinite system', as Dedekind called it. According to Dedekind, if, in considering such a system, 'ordered by a mapping ϕ we entirely neglect the special character of the elements, simply retaining their distinguishability and taking into account only the relations to one another in which they are placed by the ordering mapping ϕ, then these elements are called *natural numbers* or *ordinal numbers* or simply *numbers*' (cf. 1888, §6, no. 73). Frege had already criticized, in the *Grundlagen*, the idea of abstraction as somehow generating featureless 'units' (*GL*, §§34–9), and he was to heap further scorn on abstraction in the *Grundgesetze* and elsewhere (cf., e.g., *GG*, II, §99; *RT*). But, one might say, what Frege did was not so much to outlaw abstraction as to make it logically respectable, respectability that was to be achieved through Axiom V. This comes out most clearly in §147 of Volume II of the *Grundgesetze*, in the context of Frege's attempt to distinguish his own method of 'constructing' value-ranges, and hence numbers, from the methods adopted by, among others, Dedekind.

> People have indeed clearly already made use of the possibility of transformation that I have mentioned; only they have ascribed coincidence to functions themselves rather than value-ranges. When one function has in general the same value as another function for the same argument, it is usual to say: 'the first function is the same as the second' or 'the two functions coincide'. The expression is different from ours, but all the same here too we have an equality holding generally transformed into an equation (identity).
>
> Logicians have long since spoken of the extension of a concept, and mathematicians have used the terms set, class, manifold; what lies behind this is a similar transformation; for we may well suppose that what mathematicians call a set (etc.) is nothing other than an extension of a concept, even if they have not always been clearly aware of this.
>
> What we are doing by means of our transformation is thus not really anything novel; but we do it with full awareness, appealing to a fundamental law of logic. And what we thus do is quite different from the lawless, arbitrary construction of numbers by many mathematicians.

---

that it is an 'object of thought' and hence a thing. But if not a 'system', then what kind of thing is it? Frege, on the other hand, was quite clear that the empty set was just as legitimate as any other set. (Cf. *GG*, I, Introd., 1–3/*FR*, 208–10.)

[11] Here it should be noted that in his later *Grundgesetze*, Frege defines numbers in terms not of equivalence classes of concepts but of equivalence classes of classes. See Reck 2005, §2.1. Dedekind may have influenced this change. On the mathematical background to Frege's appeal to extensions of concepts, see also Wilson 2005; and for a detailed history of set theory, see Ferreirós 1999.

> If there are logical objects at all – and the objects of arithmetic are such objects – then there must also be a means of apprehending, of recognizing, them. This service is performed for us by the fundamental law of logic that permits the transformation of an equality holding generally into an equation. Without such a means a scientific foundation for arithmetic would be impossible. For us this serves towards the ends that other mathematicians intend to attain by constructing new numbers. ... Can our procedure be termed construction? Discussion of this question may easily degenerate to a quarrel over words. In any case our construction (if you like to call it that) is not unrestricted and arbitrary; the mode of performing it, and its legitimacy, are established once for all. And thus here the difficulties and objections vanish that in other cases make it questionable whether the construction is a logical possibility; and we may hope that by means of our value-ranges we shall attain what has been missed by following any other way. (*GG*, II, §147; *FR*, 278–9.)

Dedekind is not the only person Frege has in mind here, and Frege's target is also the 'construction' of the real numbers through Dedekind cuts, as set out in Dedekind's earlier *Stetigkeit und irrationale Zahlen* (1872), as well as the 'creative definitions' of Hermann Hankel and Otto Stolz (see *GG*, II, §§138–46). But it clearly shows just how deeply Frege's own work on the foundations of arithmetic presupposed and implicated the work of others. Even though Frege treated Axiom V as a logical law – indeed, as one of his basic logical laws – he still felt the need to offer historical elucidation.

Axiom V was intended to legitimize Frege's appeal to extensions of concepts in defining the natural numbers. Although, as noted above, Frege's conception of extensions of concepts and Dedekind's conception of systems are similar, Dedekind did not himself choose any particular set-theoretic construction by means of which to define the natural numbers; rather, 'the' natural numbers were identified with the elements of a kind of 'abstracted' set. So why did Dedekind offer a different account? According to Dedekind, sets or classes have different properties to numbers themselves. We do not talk ordinarily, for example, of numbers 'containing' other numbers, or of having 'members', although we do talk of every natural number having a successor. Identifying the numbers with sets or classes or extensions of concepts accords them properties that they do not in fact have; and what is required of a definition is that it captures what we take as their essential properties.[12]

---

[12] Cf. Dedekind 1932, 490 (tr. and quoted in Reck 2003, 386): 'One will say many things about a class (e.g., that it is a system of infinitely many elements, namely all similar ones) that one would surely attach only very unwillingly to the numbers ...Does anyone think about the fact, or does one not quickly and willingly forget, that the number four is a system containing infinitely many elements? (However, that the number 4 is the child of the number 3 and the mother of the number 5 will always and for everyone remain present.)' Cf. also Frege *GL*, §69, where he admits such worries but suggests that we need only insist on the logical equivalence of number statements and corresponding statements about extensions of concepts, rather than on any stronger equivalence. The issue is raised in Beaney 1996, §4.5, and discussed in detail in the chapters that follow.

Now there are various ways in which one might go in clarifying, defending or developing Dedekind's conception.[13] But what this shows, in the case of Dedekind too, is the role that historical elucidation plays in the foundational enterprise – in the articulation and explanation of the most basic conceptions and methodologies. Dedekind is clearly sensitive to existing practice – to our use and talk of numbers, from the natural numbers right up to the complex numbers – and critically aware of alternative attempts to ground that practice. The foundational dispute between Frege and Dedekind is as much a battle for the soul of our past and present arithmetical practice as it is the establishment of a new order – or more accurately, that battle is an essential part of the establishment.

5 Frege, Russell and Analysis

In their preface to *Principia Mathematica*, Russell and Whitehead wrote that 'In all questions of logical analysis, our chief debt is to Frege' (*PM*, viii). Frege's analysis of number was clearly what they had in mind; and from his first knowledge of Frege's work, Russell had seen 'analysis' as the method to adopt. Russell's *Principles of Mathematics* (1903) opens with the claim that 'Our method will ... be one of analysis' (*POM*, 3; cf. *IMP*, 1–2); and in 1918 he wrote that 'the chief thesis that I have to maintain is the legitimacy of analysis' (*PLA*, 189). The last chapter of his *History of Western Philosophy* (1945) is entitled 'The Philosophy of Logical Analysis', referring to what he obviously saw as the correct form of philosophy, in which the historical process had culminated. But if we look at the work of Frege and Russell, there is nothing that we can identify as *the* method of analysis. Rather, there are a number of different conceptions in play, which are far from clearly distinguished.

For the purposes of this chapter, we can distinguish three main conceptions of (philosophical) analysis, all of which play a role in Frege's and Russell's work – the regressive conception, the decompositional (or resolutive) conception, and the interpretive conception.[14] On the regressive conception, which has its origins in ancient Greek geometry, analysis involves the working back to first principles by means of which something can then be deduced or explained (through a corresponding process of 'synthesis'). There is a clear sense in which Frege's and Russell's logicist projects involve regressive analysis – working back to the fundamental

---

[13] For an excellent discussion of the options, in the context of exploring Dedekind's 'structuralism', see Reck 2003.

[14] For further details of these conceptions, see Beaney 2003b, and for more on their role in Frege's and Russell's work, in particular, see Beaney 2003a, §6; Levine 2002. I make no attempt here to relate philosophical conceptions of analysis to what is generally understood today as 'mathematical analysis' – involving, in particular, the differential and integral calculus.

axioms, definitions and principles from which, on their view, arithmetic can be derived. This sense can be found in the preface to Frege's *Grundgesetze*, for example, where Frege discusses his Euclidean motivation (*GG*, I, vi; cf. *GL*, §§1–2), and it is also explicit in Russell's paper, 'The Regressive Method of Discovering the Premises of Mathematics' (1907). In this paper, Russell suggests that Peano's five axioms are not as 'ultimate' as Frege's logical definitions, since they can be derived from Frege's definitions (*RMDP*, 276–7).[15] On Russell's view, Frege had provided a deeper analysis – analysis being understood here in the regressive sense.

However, it is not the regressive but the decompositional conception of analysis that was dominant at the time that Frege and Russell were writing, and that is arguably still seen as the main conception today. This has its roots too in ancient thought, but came to prominence in Descartes' work. On the decompositional conception, analysis involves the decomposition of something, such as a geometrical figure, problem, proposition or concept, into its constituents, taken as including its form or structure. This is certainly Russell's official conception. As he defined it in his chapter on 'Analysis and Synthesis' in his 1913 manuscript, *Theory of Knowledge*, 'analysis' is 'the discovery of the constituents and the manner of combination of a given complex' (*TK*, 119). The conception can also be found in Frege's work, for example, in 'Logic in Mathematics' (*LM*, 225; *FR*, 314), although, in comparison with Russell, Frege places much more emphasis on function-argument rather than whole-part analysis. (I use 'resolutive analysis' to cover both function-argument and whole-part analysis, reserving 'decompositional analysis' for the latter only. The differences need not concern us here.[16])

Frege and Russell (together perhaps with Moore and Wittgenstein) are generally seen as the founders of 'analytic' philosophy, whose very name suggests the importance of analysis. But neither of the two conceptions of analysis just identified can be what distinguishes 'analytic' philosophy, for the simple reason that they have always been a part of philosophical methodology, from Plato and Aristotle onwards. What does, I think, provide the distinctive core of analytic philosophy is not regressive or decompositional analysis, although these are also involved, but interpretive analysis. On the interpretive conception, with the particular case of the analysis of propositions in mind, analysis involves rephrasing the proposition into its 'correct' logical form, with the aim of avoiding the problems that may be generated by its surface grammatical form. It was the idea that many of our ordinary propositions may be systematically misleading in one way

---

[15] That the Dedekind–Peano axioms can be derived – within second-order logic, from the Cantor–Hume Principle – was shown by Frege in the *Grundlagen*, a thesis that has since been called 'Frege's Theorem'. The Cantor–Hume Principle states that the number of *F*s is equal to the number of *G*s iff *F* and *G* can be correlated one-to-one. The principle is more usually called 'Hume's Principle', but this fails to do justice to Cantor's role in the story of its formulation and use: Hume only considered finite numbers and it is more of a 'principle' in Cantor's work. Cf. Beaney 2005, §1; Reck and Beaney 2005.

[16] Again, for details, in relation to Frege and Russell, see Beaney 2003a, §6; and cf. Levine 2002.

Frege and the Role of Historical Elucidation

or another, and hence required some kind of 'translation', that has been particularly characteristic of analytic philosophy since its origins in the work of Frege and Russell.

Consider the example used above to illustrate Frege's analysis of number statements (involving the number 0):

(0a) Unicorns do not exist.

Negative existential statements such as (0a) have caused problems throughout the history of philosophy. If (0a) is meaningful, then must unicorns not exist in some form (e.g., 'subsist') to be attributed the property of non-existence? On Frege's view, however, (0a) is to be analysed or 'interpreted' as (0b), which can then be formalized in his logic, using modern notation, as (0c):

(0b) The concept *unicorn* is not instantiated.
(0c) $\neg(\exists x) Fx. \,[(\forall x)\neg Fx.]$

If (0a) is understood or interpreted as (0b), then there is no need to invoke any mysterious objects (such as 'subsistent' unicorns) to make sense of it. The statement is not about any objects, in particular, but about a concept, of which we are merely asserting that it is not instantiated. This can be formalized logically, but the crucial point is that before we do so, or *in* doing so, we must *interpret* the statement.

So too, in the case of Russell's theory of descriptions (*OD*; *IMP*, ch. 16), which Frank P. Ramsey famously called a 'paradigm of philosophy' (1931, 263), (Ka) is interpreted as (Kb), which can then be formalized as (Kc):

(Ka) The present King of France is bald.
(Kb) There is one and only one King of France, and whatever is King of France is bald.
(Kc) $(\exists x)(Kx \,\&\, (\forall y)(Ky \rightarrow y = x) \,\&\, Bx).$

Again, we do not need to suppose that the definite description 'the present King of France' must have some kind of reference for the sentence as a whole to have meaning: the definite description is 'analysed away' through Russellian interpretation.

Of course, in both cases, we must have some understanding of the conceptual resources drawn upon or developed in offering the 'interpretation', and of the point of doing so. But this is precisely where *elucidation* is required – in the Fregean sense that I have argued includes historical elucidation. To understand that (0a) is really about a concept, and not a particular type of object, and that it can be rephrased as (0b) and formalized as (0c), we must understand the distinction between object and concept, the problems that negative existential statements have caused in the history of philosophy, the claim that 'existence is not a predicate', the role of the quantifiers, and so on, all of which require elucidation.

At the very heart of analytic methodology, then, is a process of 'interpretation' that reveals just how deeply elucidation – with its historical dimension – is required in analysis. The naïve decompositional conception might suggest that there is something 'out there' just waiting to be decomposed into its elements, which we can all immediately recognize. But analysis in fact involves new forms of conceptualization,[17] the nature of which cannot be understood or explained independently of the historical context. What Frege's analysis of number statements and Russell's theory of descriptions opened up is the eliminativist strategy of 'analysing away' problematic propositions, which Russell was to employ further himself in arguing that classes are 'logical fictions', which was part of his response to what we now know as Russell's paradox.[18] If analysis involves 'analysing away', then it presupposes that there is some problem or conceptual confusion to which the analysis is an answer. 'Analysing away' would make no sense without that background: such analysis is intrinsically critical, and the associated elucidation intrinsically historical.

6   Frege, Hilbert and Axiomatization

The dispute between Frege and Hilbert over the nature of axioms and definitions, and the associated issues of consistency and formalism, is of fundamental methodological significance. According to Frege, axioms were to be seen as true propositions, and their consistency follows from their truth. For Hilbert, on the other hand, there was nothing more to the truth of a set of axioms, and the existence of whatever is referred to in those axioms, than their consistency, and it was the demonstration of consistency that was the main task. As Hilbert summed up their disagreement in a letter to Frege:

> I was very much interested in your sentence: 'From the truth of the axioms it follows that they do not contradict one another', because for as long as

---

[17] Hallett (2003) has distinguished two approaches in work on the foundations of mathematics – generalization and reductionism. Generalization, he writes, 'is characterised by the attempt to reveal that the central laws governing an established area are restricted instances of general laws which govern some new or different theory, which can thus be assimilated to the first', while reductionism involves 'the attempt to explain the central concepts of a theory by characterising it as using the conceptual apparatus of another (usually better-established) theory' (2003, 130). Projects that are typically taken as reductive, he then argues, often turn out to involve generalization. In discussing Frege's appeal to extensions of concepts, for example, Hallett comments that 'Frege's system is not the reduction of arithmetic to anything well known at all, but rather a form of generalisation' (2003, 133). Dedekind's analyses too, he writes, are 'classic examples of conceptual generalisation and unification' (2003, 137). I am in complete agreement with Hallett here: as I would put it, 'reductive' analysis is indeed often subservient to interpretive analysis, understood as involving the embedding of a given set of statements in a richer conceptual system.

[18] This is not to suggest that Frege adopted an eliminativist strategy himself (about, e.g., extensions of concepts), though he arguably *should* have done so to solve his own problems with the so-called paradox of the concept *horse*. Cf. Beaney 1996, pp. 11, 185–92. I discuss Russell's paradox in Beaney 1996, §7.2; 2003a, §5.

> I have been thinking, writing, lecturing about these things, I have been saying the exact reverse: If the arbitrarily given axioms do not contradict one another, then they are true, and the things defined by the axioms exist. This for me is the criterion of truth and existence. (Letter to Frege, 29 December 1899; in Frege *PMC*, 42.)

For Hilbert, axioms were to be understood as implicitly defining the concepts they contained. As he put it in a later letter to Frege, 'In my opinion, a concept can be fixed logically only by its relations to other concepts. These relations, formulated in certain statements, I call axioms, thus arriving at the view that axioms (perhaps together with propositions assigning names to concepts) are the definitions of the concepts.' (Letter to Frege, 22 September 1900; in Frege *PMC*, 51.)

There is much to be said about this dispute. Here I just want to highlight the connection between this dispute and what I have said about elucidation.[19] On what might be thought of as the traditional view of the axiomatic method, in developing a theoretical system, we first find our axioms and principles (perhaps through 'analysis' in the regressive sense), and having grasped and accepted them as true, we then proceed to deduce our theorems. Presupposed here is that the axioms do have a determinate meaning, which can be fully grasped before we derive anything else. These are our solid foundations, on which everything else can be built. But what is really involved in understanding the axioms? Can we really be said to understand them independently of appreciating their purpose and role in the system? As we have seen, both Frege's definitions of the natural numbers and Axiom V hardly seem immediately intuitable as true. It was in recognition of the problems involved here that Frege appealed to elucidation. Elucidation was seen as providing the understanding that is presupposed in the construction and use of the system.

Hilbert comes at the issue from the other direction. The meaning of the axioms is not established by elucidation, understood as a logically prior activity, but by their use or application. Until interpreted, the axioms are mere schemata, encapsulating the formal relations between the concepts contained but leaving room for alternative instantiations, governed only by the requirement of consistency. For Frege, this was formalism, and in places Hilbert does seem to be offering the kind of formalism to which Frege rightly objected. Hilbert writes, for example, that on his standpoint,

> the objects of number theory are for me – in direct contrast to Dedekind and Frege – the signs themselves, whose shape can be generally and certainly recognized by us – independently of space and time, of the special conditions of the production of the sign, and of insignificant differences in the finished product. The solid philosophical attitude that I think is required for the grounding of pure mathematics – as well as for all scientific thought,

---

[19] As remarked in fn. 6 above, Frege's appeal to elucidation was first made in responding to Hilbert's work on the foundations of geometry.

understanding, and communication – is this: *In the beginning was the sign.* (1922, 1121–2.)

As Frege pointed out, however, treating the objects of number theory as mere signs explains neither the applicability of arithmetic nor what may be in common to different notational systems.[20] But Hilbert is only suggesting here that this is the starting point, and he goes on to argue that what grounds arithmetic are the finite combinations of the number-signs. It is these combinations that constitute the content of what he calls 'real arithmetic', the finitary core on the basis of which 'ideal arithmetic' (covering higher mathematics) is to be established through proofs of consistency. (It would have been better, therefore, to have said: *In the beginning was the use of signs.* Such a view might be described as Kantian rather than formalist. As noted above (§2), what is fundamental in arithmetic, according to Kant, is the manipulation of symbols in accordance with rules.)

Now the details of Hilbert's programme are not our concern here. What is of interest is Hilbert's conception of axiomatization, and the idea, in particular, that axioms can encode a formal structure of concepts, which can then be interpreted in different ways. Frege found this idea deeply problematic, and took issue with Hilbert's conception both in correspondence with him from 1895 to 1903 (Frege *PMC*, 31–52) and in a series of articles published in 1903 and 1906 (*FGI*, *FGII*). In his letter of 27 December 1899, prompted by the publication of Hilbert's *Grundlagen der Geometrie*, Frege writes: 'The first thing that seems to me necessary is to come to an understanding about the expressions 'explanation', 'definition', and 'axiom', where you strongly diverge from what is familiar to me as well as from what is customary' (*PMC*, 38). According to Frege, in seeing axioms as implicitly defining concepts, Hilbert is confusing axioms with definitions. Hilbert may wish to leave open the precise meanings of 'point', 'line', and so on, in his axiomatization, but Frege accuses him of presupposing their everyday meanings in other explanations he gives (*PMC*, 35). In his reply, Hilbert picks up on just this accusation.

> This is apparently where the cardinal point of the misunderstanding lies. I do not want to assume anything as known in advance; I regard my explanation in sect. 1 [of Hilbert 1899] as the definition of the concepts point, line, plane – if one adds again all the axioms of groups I to V as characteristic marks. If one is looking for other definitions of a 'point', e.g., through paraphrase in terms of extensionless, etc., then I must indeed oppose such attempts in the most decisive way; one is looking for something one can never find because there is nothing there; and everything gets lost and becomes vague and tangled and degenerates into a game of hide-and-seek. If you prefer to call my axioms characteristic marks of the concepts which are given and hence contained in the 'explanations', I would have no objection to it at all, except perhaps that it conflicts with the customary

---

[20] For Frege's critique of formalism, see especially *GG*, II, §§86–137, translated in *TPW*.

practice of mathematicians and physicists; and I must of course be free to
do as I please in giving characteristic marks. (Letter to Frege, 29 December
1899; in Frege *PMC*, 39.)

The dispute between Frege and Hilbert is clearly happening at the elucidatory level, with both taking note of, and to some extent appealing to, existing scientific practice and our ordinary understanding of key terms. Frege accuses Hilbert of disregarding the usual meaning of 'axiom', 'point', and so on, while surreptitiously presupposing them in places. Hilbert regards the ordinary meanings of 'point' and 'line' as permitting implicit definition, and takes his use of 'axiom' to be actually in line with its use by mathematicians and physicists.

Hilbert's elucidatory justification of his conception of axiomatization comes out clearly in his later paper 'Axiomatic Thought' (1918), where he gives a host of examples from the history of science to show how there has been a gradual '*deepening of the foundations* of the individual domains of knowledge—a deepening that is necessary for every edifice that one wishes to expand and to build higher while preserving its stability' (1918, 1109). The 'crowning achievement', he writes, was the axiomatization of logic. But even this is incomplete, he goes on, until answers have been provided to 'difficult epistemological questions which have a specifically mathematical tint', such as those concerning decidability, and the relationship between content and formalism (1918, 1113). Moving from axioms in the traditional sense to axiomatic schemata was thus seen by Hilbert as just a further 'deepening' of the foundationalist project.[21]

Hilbert too, then, uses historical elucidation in explaining his fundamental ideas and methodology. Frege may have been right that axioms have traditionally been seen as true propositions, but as the case of his own Axiom V showed, even what we take as a basic logical law can turn out not only not to be fundamental but also not even to be true. So deeper questions arise about axiomatization itself. But Frege is on stronger ground when he points out how Hilbert surreptitiously appeals to our ordinary understanding of 'point', 'line', and so on, in his axiomatization of geometry. For even if we allow that axioms lay down a 'framework of concepts' (cf. Hilbert 1918, 1107), we still need to instantiate those concepts to grasp the structural relationship (cf. Frege *FGI*, 278–9) – to represent it in 'intuition', in Kantian terms. It is our everyday understanding of Euclidean geometry that provides this, as a precondition for the abstraction required for subsequently seeing Euclidean geometry as just one kind of geometry. Hilbert is famously reported to have said: 'It must be possible to replace in all geometric statements the words *point, line, plane,* by *table, chair, mug.*'[22] But even if the *words* 'point', 'line', and so on, might be variably interpreted, the structure remains the same, and elucidation is required here – implicitly taken for granted by Hilbert in choosing the words 'point', 'line', and so on, in laying down his axioms.

---

[21] On the significance of Hilbert's development of the axiomatic method for philosophy of mathematics, cf. Bernays 1922b.

[22] Cf. Ewald 1996, 1089.

The dispute between Frege and Hilbert, then, highlights the important role played by elucidation – with its historical dimension. Unlike Frege, Hilbert may not have talked of the need for elucidation, but he certainly offered and presupposed it in motivating, articulating and defending his ideas and methodology.

7 Conclusion

We began with the apparent opposition between two programs in the philosophy of mathematics – concern with the foundations of mathematics and historical investigation of mathematical methodology. But that opposition, we can now see, is spurious. Frege's logicist project might be taken as a paradigm example of a foundationalist project pursued ahistorically, but Frege himself recognized the important role played by elucidation. And elucidation, I have argued, has an essentially historical dimension, strikingly manifested in Frege's own work.

Historical elucidation is called for at the very deepest levels. In foundationalist projects, our most basic concepts and methodologies are themselves in dispute – and new concepts and methodologies emerge. By developing quantificational logic, Frege made genuinely feasible a positive answer to the Kantian question of whether arithmetical statements are analytic. But the concept of analyticity was arguably transformed in the process and eventually abandoned as problematic. Frege and Dedekind may have shared a concern with the logical definition of number, but their accounts nevertheless differed, reflected in different conceptions of abstraction and definition. Taking Frege and Russell as joint founders of 'analytic' philosophy might suggest that there is some underlying agreement on a method of analysis that transcends historical context. But here too there are many conceptions in play, and even if we take what I suggested is most characteristic of analytic philosophy, namely, interpretive analysis, historical elucidation is implicated. The dispute between Frege and Hilbert over axiomatization was pursued at the elucidatory level, with both making historical appeals in arguing for their views. Indeed, what all these comparisons suggest is that the very idea of foundations has meant different things to different people, in their different conceptual and methodological contexts. Any fruitful work on the foundations of mathematics involves new and richer conceptualizations and methodologies, which can only be understood in relation to earlier conceptualizations and methodologies. Historical elucidation lies at the heart of foundational projects.[23]

---

[23] A predecessor of this chapter, under the title 'Philosophical and mathematical methodology: philosophical conceptions from Kant to Hilbert, and the role of historical elucidation', was given at the conference on the history and philosophy of mathematics held in Seville from 17 to 19 September 2003. I am grateful to the audience for discussion, and to Jeremy Gray and José Ferreirós, in particular, for written comments on the penultimate version.

Riemann's Habilitationsvortrag at the Crossroads of Mathematics, Physics, and Philosophy

José Ferreirós

> *Et his principiis via sternitur ad majora.* And by these principles the road is open to higher things. (Newton, quoted by Riemann in 1861)[1]
>
> With Dirichlet and Riemann, Göttingen has remained the plantation of the most profoundly philosophical orientation in mathematical research that it became with Gauss. (Wilhelm Weber)[2]

There is no doubt that Bernhard Riemann was one of the main architects of modern mathematics, a visionary planner who delineated new outlines for quarters like complex analysis or abstract geometry, and designed magnificent modern avenues to link mathematics with physics. Riemann's work emerged from a most noteworthy interaction between the three disciplines of physics, mathematics, and philosophy – the 'magic triangle,' as Sánchez Ron has aptly put it in a paper about Einstein.[3] In fact, the evolution of Riemann's ideas affords a better example of the magic triangle at work than Einstein's in 1905–1916. We shall examine this in the relatively localized domain of mathematical and physical geometry, but also at the more global level of Riemann's epistemology of mathematics.

I have found it opportune to enter into a somewhat detailed discussion of the complex of ideas from which the celebrated *Habilitationsvortrag* emerged, but my real interest here is in presenting Riemann as a philosopher. The details are meant to clarify and emphasize the impact of Riemann's philosophical understanding of

---

[1] Riemann (1892), 391, taken from I. Newton's 'Tractatus de quadratura curvarum,' first published 1704.

[2] Weber to von Warnstedt, 26.12.1866, discussing candidates for replacing the late Riemann (quoted in (Dugac 1976, 166), from *Akten Alfred Clebsch*, Archiv der Universität Göttingen): 'Die Pflanzstätte der tiefer philosophischen Richtung im mathematischen Forschen, die Göttingen durch Gauss geworden, ist es unter Dirichlet und Riemann geblieben; es ware betrübt, wenn es von jetzt an diesen Ruhm verlieren sollte.'

[3] The need to consider the role of physics and philosophy in Riemann's *Vortrag* has been emphasized in recent years by Nowak (1989) and Laugwitz (1999, Chap. 3), apart from myself (1996) (1999) (2000).

mathematics, with its conceptual and 'genetic' turn and its vision of a continuity between mathematics and physics, upon his actual scientific practice. Indeed, I believe that Riemann's views still have something to offer us today, in our search for an adequate understanding of mathematical knowledge and its bearing on science. Here, too, Riemann was a pioneer, whose contributions inaugurated a small but powerful tradition that includes Poincaré and Weyl, a tradition in which a developmental, genetic philosophy of mathematics was elaborated.

1 Riemann and Grundlagen

Having entered Göttingen University in 1846 with the aim of studying theology, Riemann was drawn to mathematics and physics, but he kept a most active interest in theological and philosophical matters. Although Riemann led a secluded life, his immediate world in Berlin and Göttingen was so rich, judged by contemporary standards, that he found there the most advanced stimuli available. It is, I believe, the richness of this environment and of Riemann's intellectual engagement that explains some key traits of his work.

Riemann played a decisive role in the conformation and delineation of the conceptual and structural orientation of mathematics, later understood as modern. His name is closely associated with the *conceptual approach* that emerged by the mid-nineteenth century in Germany, more precisely with what I have termed a radically *abstract* reorientation of this conceptual approach.[4] 'Abstract' signifies here that the (concrete) means of study traditionally employed by mathematicians are avoided; e.g., in function theory Riemann seeks to study the functions 'independently of [analytical] expressions for them.'[5] This is one of the reasons why Laugwitz (1999, Chap. 4) considered Riemann's work as a crucial 'turning point' along the road to modern mathematics.

One way to substantiate the novel orientation of Riemann's mathematics is to reflect on the characteristic emphasis on *foundations* (*Grundlagen*) that we find in his early work. This is of particular interest here, as we shall find links between new mathematical practices, philosophical issues, and foundational problems. It will prove instructive to consider the question more closely, and examine how the term changed from Riemann to Hilbert.

The word 'foundations' appears of course in the title of his epoch-making lecture of 1854, *On the hypotheses which lie at the foundations of geometry*, which

---

[4] See Ferreirós (1999), especially pp. 10, 26–38. On the conceptual approach, see Klein (1926), Laugwitz (1999), Wussing (1984), 270, note 81, and older articles such as Klein (1897), Minkowski (1905), Hilbert (1921).

[5] See the 1851 dissertation, art. 19 (Riemann 1892, 35), and the program sketched in art. 20.

we shall discuss below.⁶ But there is more. His 1854 work on the representability of functions by trigonometric series was justified by the fact that the topic 'stands in the closest connection with the principles of the infinitesimal calculus' and can contribute to make them 'clearer and more precise.'⁷ Here, a refinement of the 'concept of definite integral' (resulting in the Riemann integral) was the basis for a very general study of trigonometric series and discontinuous functions. But, above all, we have Riemann's dissertation (1851), *Grundlagen für eine allgemeine Theorie der Functionen einer veränderlichen complexen Grösse* [*Foundations for a general theory of functions of one complex variable*]. This provided a new basis on which to establish a general theory, a new *conceptual* approach to complex functions – elaborating from the concept of analytic function as defined by the Cauchy–Riemann differential equations, emphasizing that such functions are conformal mappings, and introducing independent but sufficient conditions for the characterization of a function (Riemann surfaces included).⁸

Let us now consider more closely what is meant by *Grundlagen* in Riemann's geometric work. He did not mean the establishment of an axiom system for a previously given body of knowledge, nor the investigation of properties like independence and consistency. Rather he was interested in a *deepening of the foundations* (to borrow Hilbert's apt phrase, (1918, 417)), in the development of new, deeper and broader geometric theories, with the aim of opening up new conceptual possibilities for mathematical *and physical* theory. Again, the reference to foundations in his title indicates that he wanted to offer a completely new basis on which to establish a general theory, a new *conceptual* approach – elaborating this time on the basis of the concept of n-dimensional manifold, and introducing metrics and a concept of curvature by differential-geometric means.

Incidentally, it is worth noting that the first paragraph of the 1854 lecture did refer to questions that we presently associate with logic and foundations. Riemann remarked that Euclid's geometry offers only 'nominal definitions' for the basic concepts, while the essential specifications are presented as axioms.

> The relations between these assumptions [the axioms, JF] thus remain in the dark; one does not see whether and to what extent their connection is necessary, nor *a priori* whether it is possible. (Riemann 1868, 272)

As far as I know, this is the first time the problem of the consistency of geometry is posed in the mathematical literature: the question whether, a priori, the system formed by the assumptions or axioms 'is possible,' can only be understood as asking whether the system may not lead to contradictions. Riemann thinks that the answer to this question will come as an offshoot of a deeper analysis of the basic geometric concepts and assumptions, the kind of analysis he proceeds to

---

⁶ Riemann's title says literally '*zu Grunde liegen*,' that is just a linguistic variant of *Grundlagen*.

⁷ Riemann (1892), 238. Riemann offers also a second justification, the applicability of more general Fourier series in questions of analytic number theory.

⁸ On this topic, see Bottazzini (1986), Laugwitz (1999).

offer. It is not logic alone, nor of course symbolic or syntactic considerations (as in later proof theory) that will answer the problem of consistency, but rather a deeper conceptual understanding of space and geometrical stipulations.

During the nineteenth century, the word 'foundations' does not seem to have been used often in languages other than German, at least not in mathematical titles. To give two outstanding examples, the investigations of both Peano in Italy and Russell in England, be they on geometry or arithmetic, were presented as a study of the 'principles' of mathematics.[9] Even more interesting, a cursory review suggests that, when the word 'foundations' appeared in a German title published before 1900, there was some connection with Riemann. Let me briefly mention four examples.

In 1881, the young Hurwitz published in *Mathematische Annalen* 'Grundlagen einer independenten Theorie der elliptischen Modulfunctionen,' a paper based on his dissertation, dealing with a topic studied by Dedekind and linking back to Riemann. It is well known that Hurwitz's mentor, Felix Klein, was then trying to establish himself as the heir and continuator of Riemann's tradition in German mathematics. Klein saw this tradition as *intuitive-conceptual*, but arguably his emphasis on intuition does not reflect Riemann's standpoint, which was more abstract except for purposes of exposition.[10] In his obituary, Hilbert remarks:

> The contact with Klein at Leipzig (1881–1882) brought Hurwitz in particular *one* scientific gain, which was of decisive influence for his whole development and is always recognizable in his publications, namely the familiarity with Riemannian ideas. By then these were not yet general property, as today, and in a sense their knowledge signified promotion to a higher class of mathematicians. In Leipzig Hurwitz became acquainted not only with Riemannian methods in general, but with their so fruitful application to the theory of automorphic functions, which Klein was at the time exploiting with the greatest success. (Hilbert (1921), 161–62)[11]

Two years later, Cantor published a booklet entitled *Grundlagen einer allgemeinen Mannichfaltigkeitslehre* (1883). Here the link to Riemann is explicit in the title itself by the word 'manifold' understood as set, but there are also connections at the level of natural philosophy and Cantor's analysis of the continuum.[12] Like Riemann, Cantor was not trying to produce a foundational study (in the sense of twentieth-century logic) of a previously given discipline or body of theory, but rather offering new conceptual foundations for a nascent theory that went far beyond established knowledge.

---

[9] Consider, e.g., Peano's *Arithmetices principia* (1889), and Russell's *Principles of mathematics* (1903).

[10] A case in point is the geometric exposition of topological ideas in his famous paper on 'Abelian functions' (Riemann 1892, 91–95) – compare the much more abstract manuscript on n-dimensional topology, written four years earlier (Riemann 1892, 479–82).

[11] See also Klein (1897).

[12] On Cantor's manifold concept, see (Ferreirós 1999); on the broader question of natural philosophy, (Ferreirós 2004).

But many readers must have been thinking, naturally enough, of a case that I have not yet mentioned, Frege's *Grundlagen der Arithmetik* (1884), where the link to modern foundations seems much more obvious. Here the argument for the influence of Riemann becomes most interesting, proceeding through the realm of complex analysis: see the chapter by Tappenden in this volume. And so I come to my fourth example, which is none other than Hilbert's *Grundlagen der Geometrie* (1899). Hilbert's links with Klein and especially with Hurwitz are very well known, and his affinity to the views of Riemann's friend and foundational thinker Dedekind are becoming more and more clear (see Sieg's chapter). His adherence to Riemann-style conceptual mathematics was also explicit, especially in the preface to the *Zahlbericht* (Hilbert 1897, 67). In fact, I believe it is instructive and quite adequate to read Hilbert's *Grundlagen* as a work still far from twentieth-century logico-mathematical studies of foundations, and, for all its deep differences with Riemann's treatment of geometry, closer to his understanding of 'foundations.'

Indeed, Hilbert was proposing new conceptual tools for dealing with geometry, beginning first and foremost with a radical turn away from the empirical-intuitive roots of the theory, to formulate it instead as an abstract branch of modern mathematics. His model of theory building came from the new algebra and number theory, his main fields up to that time, and the work led him to developing new geometrical ideas, like non-Archimedean geometries. In a word, the first edition of Hilbert's *Grundlagen* was not oriented towards a foundational study in the logical spirit of the twentieth century, but rather towards a new conceptual approach to geometry, imparting the topic a characteristically modern turn.[13] The later crucial question of consistency was, quite obviously, a subsidiary concern for Hilbert, as it had been for Riemann 45 years before. It was Cantor's discovery of the set-theoretical antinomies, known to Hilbert especially after the publication of *Grundlagen*, that led him to emphasizing the problem of consistency.[14] Thus, both Hilbert's 1900 paper on the real numbers and the second problem in his celebrated Paris talk represent already a different stage from the first edition of his geometry book. The story of how this constellation of questions led to modern foundational research is well known, and cannot be recounted here (see Sieg's chapter and the references therein).

Coming back to Riemann, his emphasis on reconception of mathematical theories was a reflection of his philosophical and epistemological views. As Scholz (1982) (2001a) has convincingly argued, on the basis of manuscripts found in

---

[13] To be sure, Hilbert's axioms certainly are of a greater level of finesse (logical decomposition) than Riemann's geometrical hypotheses (see Section 3.2), and his way of dealing with axioms incorporated very interesting novelties. As Jeremy Gray has emphasized while discussing this issue, Hurwitz remarked immediately that Hilbert had 'opened up an immeasurable field of mathematical investigation ... the "mathematics of axioms" ' (Gray 2000b).

[14] As I argue in a forthcoming paper on 'Hilbert, Logicism, and Mathematical Existence.' The letters between Cantor and Hilbert were published in Purkert and Ilgauds (1987), and translated in (Ewald 1996, vol. 2).

Riemann's *Nachlass*, those views were decisively shaped by his reading of the philosophy of Johann Friedrich Herbart. This author, active at Königsberg and Göttingen, emphasized that philosophy ought to develop '*with* and *inside*' the other disciplines, in an immanent relation to them. He also remarked that, of all the sciences, mathematics stands closer to philosophy, and properly treated may be seen as a part of philosophy. This 'philosophical study' of mathematics would entail focusing directly on its concepts; more concretely, Herbart recommended reconstructing each discipline around one single central concept, acting like an axle around which the theoretical development would revolve.[15] The expositions of Riemann, and curiously enough also of his close friend (and to some extent disciple) Dedekind, complied to this recommendation. Thus, Riemann's function theory revolves around his abstract definition of an analytic function, and Dedekind thought of algebra as erected around the axial concept of a field [*Körper*].[16]

2 On Riemann's epistemology

It has been said that discovery and revolution in the realm of thought were a German specialty during the nineteenth century: despite living in a reactionary society, the Germans operated their philosophical revolution from the time of the shaking political events in France. Riemann continued this tradition, engaging in a daring but lonely journey of mental discovery, under the guidance of mathematicians like Euler and Gauss, physicists such as Newton and Weber, and philosophers like Leibniz, Kant and Herbart. Herbart is mentioned at key places in Riemann's work of the early 1850s, including the *Vortrag*, and it is worth emphasizing that there was no need to mention him (there is no evidence that Gauss and Herbart got along particularly well,[17] and the man who succeeded Herbart at Göttingen, H. R. Lotze, was no orthodox Herbartian).

2.1 Beyond apriorism

As Erhard Scholz has remarked, what Riemann took from Herbart amounts to a theory of knowledge and its associated methodology.[18] In his youth, Herbart

---

[15] Herbart's quotations are taken from Scholz (1982), 424–25, 436–37. They were excerpted by Riemann himself from the early work *Über philosophisches Studium* published in 1807.

[16] See Riemann (1851) and Dedekind (1873, 409), where algebra is presented as 'die Wissenschaft von der Verwandtschaft der Körper.' Cf. Avigad in this volume.

[17] Unlike with Kant or Fries, Gauss never expressed interest in his colleague Herbart; indeed he opposed his hiring.

[18] On Herbart's philosophy and his influence, see especially the work of Scholz (1982) (2001a) but also Torretti (1984, 107–108), and of course Riemann's philosophical fragments. I have made some remarks myself in (1999, 45–47) and (2000, *xxviii–l*).

studied some sciences carefully, especially the calculus and mechanics, and he later started the development of a scientific (though still speculative) psychology. This attention to science led him to interesting epistemological ideas, which explains why several German scientists became his followers. These epistemological ideas, as developed by Riemann, are in my judgement very modern, and sometimes strongly reminiscent of famous views of twentieth-century philosophers.[19]

An example that I would like to mention briefly is the pictorial theory of representations, which in all likelihood influenced Hertz, and through him Wittgenstein. In the epistemological fragment there is a short but pregnant discussion of the correspondence theory of truth, which begins as follows:

> The elements of our image of the world are totally different from the corresponding elements of the real represented. They are something in us; the elements of the real are something outside us. But the connections between the elements of the image and of the represented should coincide, if the image is to be true. The truth of the image is independent from its degree of thinness; it does not depend on whether the elements of the image represent greater or lesser amounts of the real. But the connections should correspond with each other ... (Riemann 1876, 522)

Riemann goes on to discuss what happens when there is a refinement of the image by introduction of more elements, of a greater 'degree of thinness.' There is reason to think that Riemann's conception of function theory, as expressed already in 1851, was important in the emergence of this pictorial theory of representations and truth. The mathematical idea was further refined and generalized by Dedekind in his theory of mappings [Abbildungen] and morphisms; Dedekind did not forget that such ideas apply not only to mathematics, but also to 'other sciences,' to human thinking in general.[20] All of this belongs in the context within which Hertz elaborated his oft-quoted ideas.

Herbart regarded himself as a follower of Kant to some extent, but he rejected his most famous epistemological ideas: the view that there are some innate concepts (the categories) in human understanding, and the idea of space and time as pure forms of intuition.[21] Gauss himself, though more sympathetic to Kant's epistemology of geometry than Herbart, published in a famous paper of 1831 a 'decisive refutation' of it. The argument is most interesting, as it depends only on some characteristically Kantian thoughts – in particular the famous argument of

---

[19] It must be said, too, that some of his ideas – like the Fechnerian hypothesis of the earth's soul – may seem extremely speculative and foreign to a modern reader. For commentary and a charitative interpretation, see my (2000, l–lvi).

[20] See also Dedekind's text of 1879 introducing the idea of mapping, quoted in (Ferreirós 1999, 89), and Dedekind (1888, 335–36). The reference for Hertz is his *Principles of Mechanics presented in a new form* (New York, Dover, 2004), p. 1.

[21] See the heated rejection of the latter in *Psychologie als Wissenschaft* (Herbart 1824, 428), quoted in (Scholz 1982, 422) or (Laugwitz 1999, 290).

the incongruous counterparts (left and right hand), what we call the orientability of space.[22]

Herbart's departure from Kant amounted, in a nutshell, to a complete rejection of *apriorism*, which Riemann simply inherited. In both thinkers' view, all knowledge originates in experience, but their position was not naïve empiricism: experience displays a manifold of relations between sensible elements, and thus not only rough contents, but also forms given with them (these being as objective as the former). They also departed from classical empiricism by considering that knowledge is not just based on experience and mental association, but rests on a faculty of *reflection* [*Nachdenken*] (perhaps one could translate 'theorization') which is the source of coherent conceptions of reality. By the time Riemann composed his *Vortrag* he had long superseded the idea that geometry may be based on an a priori intuition. Those who have read his work as a critique of Kant simply impose a pre-determined interpretation that has no internal connection with his statements and arguments.[23]

'Natural science,' wrote Riemann, 'is the attempt to conceive Nature by precise *concepts*.' These concepts are the outcome of *Nachdenken*, in deliberate attempts to reorganize and interpret our experience, which introduce *hypotheses* (his term) going beyond the immediately given. One might criticize Riemann from a logical standpoint, saying that his emphasis on concepts is slightly misplaced, as hypotheses must be more complex logically (interrelations of concepts). On his behalf it can be said that at different points he speaks rather of *conceptual systems*, which bring him closer to a modern, theory-centred analysis of science. In any event, according to Riemann and Herbart, the human understanding has an active, creative role to play in scientific knowledge.

One could say that experience and reflection amount to the two fundamental aspects of the hypothetico-deductive methodology, which was beginning to be accepted by scientists precisely at that time (largely as a result of the revolutionary overturn of Newtonian optics). Note that the physicist Wilhelm Weber, whose assistant Riemann was, adopted a form of this methodology.[24] Experience would be the initial source of knowledge of phenomena, for which *Nachdenken* provides a hypothetical explanation; then deduction from the hypotheses leads to predictions

---

[22] Gauss (1831), explicitly referred to by Riemann (1868, 273). Gauss calls his argument a 'decisive refutation of [Kant's] figuration [Einbildung]' in a letter to Schumacher, 8.02.1846 (Gauss 1973, vol. VIII, 247); I discussed this topic and argued that the refutation is correct in (Ferreirós 2003). Let me add that, in a letter probably not known to Riemann, Gauss stated that Kant's analytic/synthetic distinction is either trivial or false (Gauss to Schumacher, 1.11.1844 (Gauss 1973, vol. XII, 63)): 'seine Distinction zwischen analytischen und synthetischen Sätzen ist meines Erachtens eine solche, die entweder nur auf eine Trivialität hinausläuft oder falsch ist.' This obviously pre-figures the twentieth-century criticism of authors like Quine and Putnam, but unfortunately we know nothing about the details of Gauss's reasoning.

[23] An example is Nowak (1989), on which I agree with Laugwitz (1999, 291), although Nowak's paper is an interesting source for the reception of Riemann's work among philosophers.

[24] See Dedekind's interesting comments, reproduced in (Ferreirós 1999, 25).

that can be tested by precise observations (new experiences). This viewpoint was put forward by Riemann in his epistemological writings (1876, 521); interestingly, he saw notions of possibility, necessity, and probability in light of the consequences of established hypotheses.

2.2 Spatial ideas

For all those reasons, and because he understood the word 'axiom' in the traditional sense, Riemann entitled his lecture 'On the *hypotheses* which lie at the foundations of geometry.' Recall that, traditionally, axioms had been conceived as evident, *a priori* truths; for Kant (1787, 760), they were 'immediately certain' synthetic *a priori* principles. Riemann speaks of hypotheses and not axioms precisely because he wants to speak of *axioms* in the twentieth-century sense. He wrote that 'today' one understands by hypothesis anything that is added in thought to the phenomena [*Alles zu den Erscheinungen Hinzugedachte*].[25] The properties that distinguish space from other conceivable 3-manifolds, he wrote, are only to be established from experience. To the best of his knowledge, experience confirmed that physical space is Euclidean, but 'these matters of fact – like all matters of fact – are not necessary, but only of empirical certainty; they are hypotheses' (1868, 273).

Riemann and Herbart tended to prefer a Leibnizian view on space, seeing space not as a basic entity, but as an expression of the relations between physical entities (see end of Section 3.1). It is particularly interesting that this viewpoint led Herbart to think geometrically about all kinds of subjects, which again was noticed and further developed by Riemann. This is a feature of Herbart's work that makes it vaguely reminiscent of modern mathematics, and perhaps it has not been sufficiently appreciated. By that time, it was customary (especially among philosophers) to believe that geometrical thinking only had sense as applied to physical space, the framework where objects and phenomena take place. Herbart liberated himself from this. In his view, most concepts give rise to continua,[26] as their associated representations [*Vorstellungen*] order themselves in such a way that 'unnoticeably' they generate a 'qualitative continuum' (Scholz 1982, 422–23). Each object is thus a 'bundle [*Complexion*] of properties', and each of these lies in a continuum; some are merely linear continua, as in the 'line of tones,' some can be 2-dimensional, as the 'triangle of colours' (Herbart 1825, 192–93). His freedom went to the point of postulating, at least in a work of 1806, an 'intelligible space' for the intelligible beings of metaphysics, a kind of Leibnizian monads 'with windows' that he would later call '*Realen*' (see Scholz (2001a), 175).

---

[25] Riemann (1876), 523 (notice the critique of Newton's distinction between 'laws of movement' and 'hypotheses').

[26] Riemann disagreed because common life concepts normally give rise to discrete manifolds, and among them only 'the positions of perceived objects and the colours' make up continuous manifolds (1868, 274).

This freely 'geometrizing' standpoint was further developed by Riemann, both in his function theory and in the broad general view that the *Vortrag* proposed. His starting point here was of extreme generality, suggesting a new foundation for pure mathematics in direct relation to questions of epistemology and logic. His definition of manifolds made implicit (but clear) reference to usual notions of formal logic at the time,[27] concretely to the distinction between comprehension (*Inhalt*) and extension (*Umfang*) of a concept. Whenever there is a 'general concept that admits of different instances [*Bestimmungsweisen*],' the corresponding extension (the class of all instances) constitutes a manifold (Riemann 1868, 273). With this, Riemann stretched the usual extensional approach to concepts, to the point of formulating a *principle of comprehension* in explicit reference to natural language and implicit to traditional logical theory.[28] A typically philosophical emphasis on logic was again an important element in the emergence of Riemann's set-theoretically inclined understanding of the basic language of mathematics.

As a side remark, it is interesting to note that the concept of manifold, a most fundamental idea in mathematics, seems to have connections with Riemann's psychological ideas too. His psychological fragments begin as follows:

> With every simple mental act [*Denkact*] something permanent, substantial enters our soul. This substantial element appears to us like a unity, but it seems to contain an inner manifoldness [*Mannigfaltigkeit*] (insofar as it is the expression of something extended spatially and temporally); therefore I shall call it a 'mental mass' [*Geistesmasse*]. (Riemann 1876, 509)

The soul is then defined as a compact *Geistesmasse*, interwoven in the most intimate and manifold way, and Riemann thinks that the assumption of the soul as 'a unitary support' may be perfectly 'superfluous' (1876, 511), since the intimate interconnection of mental masses suffices. His psychology is not spiritualistic, it takes the physiological basis quite seriously and looks rather like an intermediate step towards a non-dualistic understanding of the mind. As one can see, the topic of unity in multiplicity, which was to become so important in set theory, is here linked with the analysis of 'perceptions' or 'mental acts.'

Remember that knowledge was regarded by Riemann as the outcome of an interplay between the observations provided by experience and the assumptions or theories offered by reflection. In his view, both the logical analysis of *concepts*, basic elements in the elaboration of theories, and the psychological analysis of *perceptions*, the core ingredients of any possible empirical test, led to this idea of a manifold, of something that 'appears to us like a unity, but . . . contain[s] an inner manifoldness.'

---

[27] In Germany, the expression 'formale Logik' was then associated with the names of Kant and Herbart; see (Ueberweg 1882, 47–53).

[28] There has been considerable misunderstanding of Riemann's definition of manifolds among twentieth-century historians and philosophers. The above interpretation was first offered and justified in (Ferreirós 1996); see also Ferreirós (1999), 47–53, 63.

2.3 Genesis of science, and antinomies

Another important element that Riemann inherited from Herbart was a developmental, genetic understanding of science. Far from the usual idea that there exists (in some Platonic realm) a ready-made theory of everything, in his view all concepts of natural science, and of mathematics in particular, have evolved gradually from older explanatory systems. Scientific theories are for Riemann the outcomes of a process of gradual transformation of concepts, starting from the basic ideas of object, causality, and continuity. Development takes place under the pressure of contradictions or else implausibilities [*Unwahrscheinlichkeiten*] revealed by unexpected observations – unexpected in light of the hypotheses proposed by reflection at some particular stage.

The three basic concepts of object, causality, and continuity are the ones that concern Riemann in his short epistemological writings. In his opinion, Herbart had satisfactorily

> proved that also *those* concepts which are employed in conceiving the world, but whose formation we cannot follow, either in history or in our own development, since they are inadvertently transmitted to us with language – all of them, in so far as they are something else than mere forms of connection of simple sensorial representations, can be derived from that source[29] and therefore need not be derived from a special constitution of the human soul, prior to all experience (as the categories according to Kant).
>
> This proof of their origins in the conception of the given by sensorial observation is important for us, *because only thus can their significance be established in a manner that becomes satisfactory for natural science*. (Riemann 1876, 522)

Riemann means that his and Herbart's arguments show, that the basic concepts of object, causality, and continuity have true bearing for the attempt to conceive the world. He is thus arguing for a realistic (albeit critical rather than naïve) interpretation of scientific concepts.

With this reference to Herbart, he meant the study of the formation of a concept of 'things existing in themselves,' objects that subsist in time, in *Psychologie als Wissenschaft*. This highly non-Quinean understanding of objects had important resonances in twentieth-century psychology, especially in the genetic epistemology of Jean Piaget (well known to Quine), but also in more recent work. As Riemann was satisfied with Herbart's explanations, he felt no need to discuss the notion of an object further, and thus he concentrated on the concepts of causality and continuity, which of course were generally regarded in the nineteenth century as essential for scientific knowledge. The text in question bears the title, 'Attempt [at an] Outline [of a] theory of the basic concepts of mathematics and physics, as foundations for the explanation of Nature' (1876, 521).

---

[29] Namely, the gradual transformation of concepts by confrontation with experience.

Interestingly enough, Riemann thinks that causality and continuity arise from a common source. The notion of a permanent object is contradicted by the experience of change, and so the conceptual system must be gradually revised. The notion of permanent object, which after all had been previously 'assayed' [bewährten], must be partly preserved. But we need something else, and this is the origin of the concepts of continuous variation and of causality:

> We only observe the transition of an object from one state into another, or to speak more generally from one determination [*Bestimmungsweise*] into another, without any jump being observed in the process. While complementing [*Ergänzung*] the observations one can either assume that the transition takes place by a very large but still finite number of jumps which are unnoticeable for our senses, or else that the object is transformed from one state into the other through all the intermediate degrees. The strongest reason for this last conception lies in the requirement, that one should preserve as far as possible the previously assayed concept of the subsistence of objects [*für sich Bestehens der Dinge*]. However it is not possible to really conceive [*vorstellen*] a transition through *all* intermediate steps, but this applies to all concepts, as we have already remarked.[30]
>
> At the same time, according to the previously formed concept of the subsistence of objects, which has been tested in experience, one will conclude that the object would remain what it is, without something else being added. Here lies the stimulus [*Antrieb*] for searching a cause to each variation. (Riemann 1876, 522; see also 524)

Knowledge is always a matter of complementing the observed, of adding hypothetical elements to the data of experience, and even such basic concepts as continuity and cause are regarded by Riemann as hypothetical—open to revision and even substitution, as may be the case with our continuous conception of space.

The reader must have noticed the way Riemann, in the quotation above, makes room for a discrete conception of transitions; the same happens at key places in his *Vortrag* (Riemann 1868, 284–86). All of this is further clarified by a text that the editors of Riemann's *Werke* classified as metaphysical, but that in fact bears close relation to his epistemological writings (and to which the words 'as already remarked' above seem to refer). This is Riemann's noteworthy table of *Antinomien*, obviously modelled upon Kant's legendary piece of work.[31] Recall that an antinomy, in Kant's sense, is an irreducible conflict in our thinking, and Riemann too seems to be very serious about this connotation. Like Kant, Riemann presents four antinomies, and there are interesting parallels between the topics broached by both authors (the 2nd discusses freedom vs. determinism, the 3rd a world-governing vs. a transcendent God, and the 4th 'mere' immortality (which in

---

[30] Riemann's use of his technical term *Bestimmungsweise* suggests that this text was written during roughly the same period as the *Vortrag*.

[31] Riemann (1876), 518–20. Here I must disagree with Scholz's provisional suggestion (1982, 418) that the model could be Fichte and Schelling. For comments on the relations between this table and Kant's, see (Ferreirós 2000, *xliii–xlv*).

Riemann's view does not require a substantive soul, see above) vs. a transcendent soul). However, the general commentary offered by Riemann (1876, 519–20) makes it clear that, in his case, there is not just a sharp contrast, but also an internal relation between thesis and antithesis in each antinomy: the model for this relation is given by the limit methods of mathematical analysis.

The general heading of Riemann's antinomies, the first one, and the final commentary are all very interesting to read from the point of view of the foundations of mathematics. He subsumes his four theses under the heading 'finite, representable,' while the antitheses come under 'infinite, conceptual systems that lie on the limits of the representable' (1876, 518). Riemann remarks that the concepts in the antithesis are well defined but not positively representable [*vorstellbar*], applying to all of them (infinity included) a feature that theologians traditionally applied to God. The first antinomy is between 'finite elements of space and time' and the 'continuum.' Riemann sees here an irreducible choice, and with hindsight one can only wonder at the wisdom of his vision, as it can be linked to current debates in physical theory, and also (if we focus on the conceptual aspect and eliminate reference to space–time) to the divide between constructivist and postulational mathematics. Considering a long view of the whole foundational debate, from the time of Riemann to the present, it seems fair to say that the question of the continuum has been the key problem, 'the mother of all battles.'

Riemann was not only an expert analyst and geometer, but also a mathematical physicist who devoted much time to theoretical physics. In all of this work, and even in his deep contribution to number theory, the continuity hypothesis is a firm and constant foundation. It is thus noteworthy how he emphasized the idea of an irresoluble tension in our conceptual systems (and our attempts to model physical reality) between the usual continuous representation and an alternative, discrete one. In the *Habilitationsvortrag* he mentions the possibility of a discrete spatial world, thus expressing his doubts about the adequacy of a continuous representation of physical reality. This is a most clear example of the magic triangle at play, another instance in which his work resonates with 'cryptic messages for the future.'[32]

In Riemann's view, mathematics appears as part and parcel of this attempt to conceive nature by precise concepts that we call natural science. Mathematics is not a priori and belongs to the conceptual side of the effort to obtain scientific understanding. At the end of his *Vortrag*, Riemann remarks that his investigation has started 'merely from general concepts,' stopping short of the consideration of experimental results; such investigations

> can only be useful in preventing this work [of changing and improving upon previous conceptual systems of science, especially Newtonian mechanics, JF] from being hampered by too narrow concepts, and progress

---

[32] The quote is from Fields medallist Lars V. Ahlfors in 1953, see his *Collected Papers*, vol. 1 (Basel, Birkhäuser, 1982), 493.

in knowledge of the interdependence of things from being checked by traditional prejudices. (Riemann 1868, 286)

Mathematical research is seen as a propaedeutics for physics or other natural sciences, or better, as a necessary accompaniment and counterpart to empirical researches. It depends on hypotheses expressed in its concepts and axioms, which are immersed in a historical process of change, to which Riemann himself contributed forcefully.

3 Physics, mathematics and philosophy in the Vortrag

Riemann's *Habilitationsvortrag* of 1854 is one of the most celebrated achievements in the history of mathematics. The context from which the lecture grew was extremely rich, ranging from epistemology to physics, from logic to function theory. It seems implausible that Riemann could have reached the breakthrough represented by his geometrical ideas without this complex mixture of ideas and motivations.

The lecture was very neatly structured into three parts, the first more philosophical, the second mathematical, the third turning to physics. Here is a brief summary, for which I use slightly modernized language:[33]

I. *Concept of n-dimensional manifold.* 1) general ideas about manifolds, distinction between topology and metric geometry; 2) topological notion of the dimension of a manifold; 3) parametrization, requiring n co-ordinates for n-dimensional manifolds.

II. *General differential geometry for Riemannian manifolds.* 1) line element given by positive definite quadratic differential form; 2) concept of curvature generalizing Gauss, manifolds of variable curvature; 3) manifolds of constant curvature, with geometric examples.

III. *Applications to the space problem.* 1) 'simplest matters of fact from which the metrics of space may be determined' empirically; 2) properties of physical space in the extremely large; 3) properties of physical space in the extremely small.

The work had great impact from the beginning, starting with the well-known reaction of Gauss, then in the last year of his life: Riemann's lecture 'superseded all of his expectations and left him most astonished.'[34] A comparable response would come from wider circles after Dedekind published the lecture in 1868; a

---

[33] Compare his own summary in (Riemann 1868, 286–87).

[34] Dedekind (1876a, 547), who knew the story first-hand from Wilhelm Weber.

witness, Felix Klein, reminisced:
> This lecture caused a tremendous sensation upon being published... For Riemann had not just embarked in extremely profound mathematical researches... but had also considered, throughout, the question of the inner nature of our idea of space, and had touched upon the topic of the applicability of his ideas to the explanation of nature. (Klein 1926, 173)

But the legend of Riemann's contribution arose after the emergence of relativity theory, beginning with Minkowski's application of Riemann-style geometry to reformulate special relativity. Although it took Einstein several years to realize that Minkowski's work was not merely a mathematical game, in 1922 he remarked that Riemann had 'foreseen the physical meaning of this generalization of Euclid's geometry with prophetic vision.'[35]

That is obviously an overstatement, as Riemann had no inkling of the kind of connection between space–time metrics and gravitation that would form the key to Einstein's theory. We shall discuss his ideas concerning the explanation of gravity below. Yet it is true that, in the last section of his lecture, he proposed directly the idea of a linkage between spatial metrics and physical forces, using very suggestive aphoristic words. Perhaps one day, after new developments in physics (along the lines of quantum gravity or some other grand unification), we shall read new meanings into those words, but this of course does not allow mythic interpretations of the kind that the author himself 'had seen it all.'

3.1 The road of physics

Riemann was very seriously involved with physics in the early 1850s. Already on the occasion of his PhD, the theses he proposed for open disputation were mainly of a physical nature: only one dealt with mathematics from a didactical standpoint. For instance, Riemann denied both the existence of magnetic fluids and Faraday's notion of 'induction along curved lines;' he threw doubts on the empirical basis of the novel doctrine of energy conservation, and defended the need for a better definition of the tension concept.[36] We have mentioned that he entered the newly established Mathematico-Physical Seminar in 1850, being Weber's laboratory assistant in the crucial years 1853–1854.

In that position he assisted the famous precision experiment done by Weber and Kohlrausch, which established that a certain constant c in electromagnetic theory equalled, within the limits of observational error, the speed of light. A few years later he presented for publication a paper using that result to argue for a

---

[35] On page 42 of *Vier Vorlesungen über Relativitätstheorie* (1923) Einstein states: 'Zuerst dehnte Riemann den Gaussschen Gedankengang auf Kontinua beliebiger Dimensionszahl aus; er hat die physikalische Bedeutung dieser Verallgemeinerung der Geometrie Euklids mit prophetischem Blick vorausgesehen' (See the new edition in the *Collected Papers*, vol. 7: Einstein 2002). Weyl had already commented along these lines in his 1919 edition of Riemann's *Vortrag* (Berlin, Springer).

[36] See Riemann (1892), *Nachträge*, p. 112. His doubts may sound strange to the modern reader who ignores the history of physics, but they were all quite sound at the time.

close connection between light and electromagnetism, in which he employed, apparently for the first time in history, a retarded potential.[37] Discussions with Kohlrausch also led to a paper on residual electric charge in the Leyden bottle, which, however, was left unpublished.[38] And yet the very first paper that Riemann published was on a topic of physics and appeared in Poggendorff's *Annalen der Physik und Chemie* (1855), under the title 'On the Theory of Nobili's Coloured Rings' (see the detailed analysis in (Archibald 1991)).

Electromagnetism, energy, electrochemistry, potential theory, precision experiment... all of this shows that Riemann was very much up to date in physical knowledge. He worked on experimental physics, mathematical physics,[39] and theoretical physics. Indeed on many occasions during those days he emphasized that his main research topic was not function theory (the basis for his enormous fame in life) or partial differential equations, but rather a new unified theory of the physical forces:

> My main occupation concerns a new conception of the known natural laws – their expression by means of different basic concepts – which makes it possible to employ experimental data on the interaction between heat, light, magnetism, and electricity, in order to investigate their interrelations. To this I was led essentially by the study of the works of Newton, of Euler, and (from a different angle) of Herbart. (Riemann 1876, 507)

This is nothing short of astonishing, but in fact these endeavours became deeply entangled with the new geometrical ideas that he created in 1853–1854.

This, however, has been the topic of some debate. No lesser figure than Weyl claimed in 1919 that Riemann's natural philosophical ideas had no objective relation to the content of his geometry lecture. Eight years later, his colleague Speiser (1927, 108) stated that the *Neue mathematische Principien* were 'the departing point for the investigations which he made public one year later in his *Habilitationsvortrag*.'[40] More recently, Bottazzini and Tazzioli (1995, 27) have sided with Speiser, finding that Riemann's differential geometry generalizes some of the ideas stated in his 1853 manuscript. A similar standpoint has been adopted by Laugwitz (1999, 222). The work of Bottazzini and Tazzioli contains a detailed presentation of the mathematical details, to which I refer the reader, as I shall try to present the main point without too much mathematical machinery.

The first expression of those ideas was a manuscript of 1853 bearing the remarkable title, *New mathematical principles of natural philosophy*, to emphasize – the shy

---

[37] 'Ein Beitrag zur Elektrodynamik' (Riemann 1892, 288).

[38] The reason seems to have been that Riemann did not want to introduce some changes proposed by Poggendorff or by Kohlrausch himself (Dedekind 1876a, 548–49).

[39] Partial differential equations was the topic of his first university course, the basis for a long and very successful tradition of textbooks, starting with Hattendorff 1869 and continuing through H. Weber to Frank–von Mises (see Laugwitz (1999), 257–68).

[40] Andreas Speiser is best known as an editor of Euler's work. He was a historian of mathematics and science, but also a mathematician who did his PhD with Hilbert.

Riemann's Habilitationsvortrag at the Crossroads of Mathematics, Physics, and Philosophy

Riemann was beyond doubt a bold thinker – the ambitious project of overshadowing Newton. Riemann's letters to his family make it clear that, from December 1853 to April 1854, he was engrossed in natural philosophy, 'so fully immersed' that he postponed the preparation of his lecture long after Gauss had chosen the topic (Dedekind 1876a, 547–48).[41]

Riemann's attempt in theoretical physics was based on the assumption of a *plenum*, a 'continuous material filling of space' by some physical 'substance' or 'ether.' Although the extant manuscripts date from the early 1850s, we know through Schering (1866) and Dedekind that he entertained the same view until his death. This is how Dedekind presents the matter in one of his earliest letters to Weber, his collaborator in the edition of *Werke*:

> As far as I am concerned, I'm totally for the continuous material filling of space and the [Riemannian] explanation of the phenomena of gravitation and light, and from the years 1856, 1857, when I stood in close contact with Riemann, I know that he valued these thoughts very much. During a common stay in Harzburg he showed me the letter of Newton (from his biography [by Brewster]) in which, without being blinded by the superb success of his hypothesis, he talks about the philosophical impossibility of immediate action at a distance.[42] Riemann adopted these thoughts very early, not in his last years ... No doubt, his efforts were directed towards basing the most general principles of mechanics, which he did not want to revoke at all, upon a novel conception, more natural for the explanation of Nature. The effort of *self*-preservation and the dependence – expressed in partial differential equations – of state variations upon the states immediately surrounding in space and time, should be seen as the primitive and not the derived. At least, this is how I understand his plans.... Unfortunately, it's all so fragmentary![43]
>
> (Was mich betrifft, so bin ich für die stetige materielle Erfüllung des Raumes und die Erklärung der Gravitations- und Lichterscheinungen im höchsten Grade eingenommen, und ich weiß aus den Jahren 1856, 1857, wo ich eifrigen Verkehr mit Riemann hatte, daß er auf diese Gedanken großen Werth legte; er zeigte mir bei einem gemeinschaftlichen Aufenthalte in Harzburg den Brief von Newton (deßen Biographie), in welchem er sich, ungeblendet durch die großartigen Erfolge seiner Hypothese, über die philosophische Unmöglichkeit der unmittelbaren Fernwirkung ausspricht. Diese Gedanken hat Riemann sehr früh, nicht erst in seiner letzten Zeit, ergriffen,

---

[41] The geometry lecture was prepared in just five weeks, starting about the first day of May, and read the 10th of June.

[42] Third letter to Bentley, in I. B. Cohen (ed.), *Newton's Papers & Letters on Natural Philosophy* (Harvard Univ. Press, 1958), p. 302–03: 'innate gravity' acting through a vacuum is 'so great an absurdity that I believe no man who has in philosophical matters any competent faculty of thinking can ever fall into it.' Riemann quotes it in a footnote to his Gravitation und Licht (1876, 534). Ideas about the possible role of a 'subtle spirit' were suggested in the *Scholium generale* to Newton's *Principia*.

[43] Dedekind to Weber, 14.03.1875, *Nachlass* Riemann in Göttingen (Cod. Ms. Riemann 1, 2, 24).

und wenn ich nicht ire, so findet sich auf einen übel aussehenden Folioblatte die Bemerkung: 'Gefunden am 1 März 1853'. Sein Streben ging ohne Zweifel dahin, den allgemeinsten Principien der Mechanik, die er keineswegs umstoßen wollte, bei der Naturerklärung eine neue, natürlichere Auffassung unterzulegen; das Bestreben der *Selbst*erhaltung und die in den partiellen Differentialgleichungen ausgesprochene Abhängigkeit der Zustandsveränderungen von den nach Zeit und Raum unmittelbar benachbarten Zuständen sollte er als das Ursprüngliche, nicht Abgeleitete angesehen werden. So denke ich mir wenigstens seinen Plan. Unter den Fragmenten findet sich doch auch eines, wo die allgemeinen Bewegungsgleichungen im Fall der Existenz einer Kräftefunktion besprochen werden? Leider ist Alles so lückenhaft!)

Following in the footsteps of Newton and Euler,[44] in 1853 Riemann employed a geometrically conceived system of dynamic processes in the ether, a kind of *ether field*, to propose a unified explanation of gravitation, electromagnetism, heat, and light. It was an attempt to revise and modify the classic physical theories in the gradualist spirit of his epistemology (sect. 2). The historian of physics Norton Wise wrote (1981, 289) that Riemann's was the first attempt to establish a unified field theory on a mathematical basis, much in the spirit of Einstein's late work. One can thus apply to Riemann what Maxwell said of Faraday, that he saw a medium where 'mathematicians' had only seen centres of force and distance, and real actions propagated through that medium where they postulated instantaneous attraction.[45]

The key new insight of March 1853 was a bold psycho-physical hypothesis that solved a deep riddle in Euler's theory, establishing at the same time a connection between nature and mind. This kind of approach, in which physics and psychology are intertwined, is typical of German authors of the nineteenth century, be they idealistic or post-idealistic. In general, they had a clear tendency to eliminate mind dualism, either in the sense of spiritualism or materialism.[46] Riemann's metaphysics was a geometrical worldview, assuming a single reality of which the mental and the physical are different but strongly intertwined expressions. Multidimensionality seems to have played a key role in his later development of this worldview.

---

[44] Euler made a 'magnificent attempt' to explain gravity by means of an ether (see Speiser (1927); quotation on p. 106). Even Gauss was against action-at-a-distance, at least in electromagnetism, as Riemann must have known through Weber (see Gauss's letter to Weber, 19.03.1845, in (Gauss 1973, vol. V, 627–29)).

[45] Maxwell (1873), *ix*. See also Klein (1897). Incidentally, Maxwell's *Treatise* has been the source of an erroneous view, that Riemann was a fierce defender of Weber's action-at-a-distance approach in electromagnetism. At the time of his *Treatise*, Maxwell did not know Riemann's manuscripts on physics, published three years later, but only the (posthumously published) paper of 1857 where he in fact had employed action at a distance, in all likelihood for expository purposes.

[46] For further details see (Ferreirós 2000), (Laugwitz 1999). The same intertwining is prominent in Cantor (Ferreirós 2004).

The new idea was simply that 'in all ponderable atoms, [ether] substance enters continually from the corporeal world into the mental,' forming 'mental substance' (Riemann 1876, 529). In subsequent work, perhaps early in 1854, Riemann made an attempt to elaborate on the physico-mathematical aspects of his explanation of gravity and light, leaving the 'metaphysical' question aside (1876, 534). Here the main idea is presented as follows:

> The cause, determined in magnitude and direction (accelerating force of gravity), which according to 3. is given at each point in space, I shall search for under the form of the movement of a substance which extends continually throughout the whole of infinite space, and I shall assume that its direction of movement is equal to the direction of the force it must explain, its velocity proportional to the magnitude of this force. The substance can thus be imagined as a physical space, whose points move within the geometrical. (Riemann 1876, 533)

Riemann's 'substance,' which he called ether in lectures of 1861, was conceived as an elastic, homogeneous, isotropic medium. The stream of ether, flowing towards the atoms, provided a satisfactory, intelligible explanation of gravitational forces. The search for intelligibility puts Riemann on a par with scientists like Newton, Einstein and even Kepler, far from the mere desire for empirical adequacy characteristic of positivistically inclined physicists (on this general topic, one can see (Torretti 1990)).

Already in Riemann's first attempt to develop these ideas, the study of the variation in time of an element of ether led to one of the characteristic problems of his differential geometry. "According to a known theorem," he wrote (1876, 530), one can determine ds_i as linear expressions on dx, dy, dz such that $dx'^2 + dy'^2 + dz'^2 = G_1^2 ds_1^2 + G_2^2 ds_2^2 + G_3^2 ds_3^2$ (where the G_i are functions of time *and* of the spatial coordinates), and so that the line element ds is expressed by a positive definite quadratic form:

$$ds^2 = dx^2 + dy^2 + dz^2 = ds_1^2 + ds_2^2 + ds_3^2.$$

This gives a family of Riemannian metrics in 3-dimensional Euclidean space, and it thus seems to be the germ for the extension and generalization one year later. It posed the analytical problem of studying the transformation of differential forms, which Riemann discussed (without formulas) in the *Vortrag*, and subjected to a deeper study in 1861.

Riemann assumed a minimality principle accounting for the 'resistance' of the ether to deformation in time; this 'effort of self-preservation' (cf. Dedekind above) and the resulting moment of forces became the basis for his explanation of the propagation of gravitational and other forces. The mathematical development of his physical hypotheses led him to conclude that electromagnetic forces are an expression of the alteration of the physical line element, which is clearly related to thoughts expressed in Part III of the *Vortrag* (see below). At the end of *Neue mathematische Principien*, he derived a differential equation in two members, one depending on dV, and one on ds, which he then analysed as follows (Riemann

1876, 531–532). The first member would express 'the resistance of a particle of [ether] substance to a change of volume' and would explain 'gravity and electrostatic attraction and repulsion'. The second, expressing 'the resistance with which a physical line element opposes a change of length,' would be linked to 'the propagation of light and heat, and electrodynamic or magnetic attraction and repulsion'. This, however, was only a temporary conclusion. In later work, he concentrated solely on the explanation of gravity and the propagation of light, as he had come to conclude that the explanation of the remaining phenomena involved more than just the ether dynamics:

> these two phenomena, gravitation and light propagation through open spaces, are the only ones that should be explained *merely* by the movements of this substance. (Riemann 1876, 533)

There is one further physico-philosophical aspect of these investigations that we should comment upon. As we have seen, the manuscripts of 1853/54 are intended to explain the main physical phenomena on the basis of the assumption of a geometrical 'infinite space,' of 'ponderable atoms,' and of 'a physical space' of ethereal 'substance' (see Riemann (1876), 532–33). The dynamics of this ether medium, the variations in the metrics (the line element ds) of the 'physical space,' explain the transmission of forces throughout space. At some point during the year that passed between *Neue mathematische Principien* and the *Vortrag*, Riemann realized that a more sophisticated conception of physical space was possible, a conception on which geometry and physics would become most intimately intertwined. The former separation of an ambient space and the physical phenomena happening in it would disappear, thus bringing to fruition Leibniz's insight that space may be just the relational 'order of coexistence' of things.

The *Habilitationsvortrag* seems to propose eliminating the duplicity of geometrical space and ether medium in the somewhat obscure but often-quoted passage:

> The question of the validity of the assumptions of geometry in the infinitely small is bound up with the question of the inner ground of the metric relations in space. In this last question, which can still be counted among those pertaining to the theory of space, is found the application of the remark that was made above; that in a discrete manifold the principle of its metric relations is already given in the concept of this manifold, while in a continuous one, the ground must come from outside. Either therefore the reality which underlies space is constituted by a discrete manifold, or we must seek the ground of its metric relations outside it, in binding forces which act upon it. (Riemann 1868, 285–86)

In my view, this has to be interpreted as an attempt to fully implement a Leibnizian viewpoint, informed and updated by developments in nineteenth-century physics. Physical space is not an ultimate reality, but rather the geometrical expression of a 'reality that underlies' it, the metric relations among physical elements determined by their 'binding forces.' Riemann comes to view spatial geometry as an expression of the metric properties of the continuous *plenum*, getting close to replacing the ether by a space of varying curvature endowed with resistance to variation.

Nevertheless, the Leibnizian conception is a philosophical viewpoint that becomes difficult to implement in mathematical physics. There is reason to think that the analytical standpoint, essential to mathematical physics, makes it hardly possible to avoid a duality between a geometry and the properties of (at least some) physical elements. Thus, it has been possible to argue that Einstein's GRT does not dispense with absolute space, although it certainly presents space as more intimately linked with aspects of physical reality. The metrics of geometry *is* the gravitational field, space becomes as physical as matter, but it remains different from matter and the theory presents it as interacting with matter.[47] The resulting viewpoint, characteristic of GRT but also of Riemann's work on natural philosophy, is a compromise position between the desired Leibnizianism and the necessary mathematical treatment. It deserves a proper name, alongside the Newtonian and the Leibnizian perspectives on space, and I propose to name it the Riemannian viewpoint – one can then say that Einstein's GRT adopted the Riemannian view on space.

I mentioned before that multidimensionality seems to have played a key role in the later development of Riemann's worldview. Schering (1866, 377) reported about conversations in which he would have regarded the atoms as 'infinitely dense spots of the [ether] medium,' or else, 'more intuitively, as places at which the medium goes from the definite three-dimensional space to the multiply extended space that surrounds it everywhere.'[48] This would have been an important further step towards unification and the realization of Leibniz's vision, but as far as we know it remained sketchy. With that, even the first hypothesis (see Section 3.2) in part I of the *Vortrag*, that space is a 3-manifold, is substituted by the hypothesis of n-dimensionality. Everything suggests that these later speculations would represent a materialistic development of the views elaborated in the early 1850s: mental life would take place in the $n-3$ remaining dimensions, and there would be no further reason to differentiate 'mental stuff' from the physical medium.

3.2 Mathematical roads

The new view of geometry and physics intertwined was the heart and soul of Riemann's geometry lecture, but it required extremely sophisticated mathematics: an intrinsic geometry of n-manifolds, generalizing Gauss's intrinsic theory of surfaces (1828). As he was running for the Habilitation as mathematics professor, Riemann concentrated on this aspect of the topic. What Riemann inherited from

---

[47] For arguments for and against absolute space after Einstein, see (Friedman 1983) and (Earman 1989).

[48] The idea is of course reminiscent of Clifford's, who must have been influenced by reading Schering's paper. For further details on Schering's interesting report, see (Laugwitz 1999, 284–85).

Gauss was not only intrinsic geometry, but also the terminology of 'manifolds,'[49] and two basic convictions: that it was necessary to create a theory of topology (*analysis situs*), and that 3-dimensionality may be merely due to the limitation of our minds.[50]

The first part of the lecture presented a new philosophical conception of pure mathematics, with the concept of manifold seen as its key fundamental notion, and went on to develop some aspects of the topological theory of manifolds that were essential for the topic at hand. Riemann's starting point was of extreme generality and related directly to questions of epistemology and logic (see Sections 2.2 and 3.3). One should emphasize that the key realization making possible Riemann's novel approach was that a single topological structure may support widely divergent metric structures. Thus, e.g., we may entirely respect the topological structure of the 'manifold' of real numbers $\mathbf{R}$ (or of Euclidean space $\mathbf{R}^n$) while defining a deviant notion of distance. This aspect of the lecture posed significant difficulties for the contemporary reader, as Riemann was charting entirely new land in a very concise, almost aphoristic manner. For instance, many readers may have thought that, by introducing a parametrization of the manifold (in part I.3), Riemann was already giving metric structure to it.

The *leitfaden* that guided Riemann's brilliant exposition was to establish a series of more and more restrictive hypotheses (postulates in our terminology), leading from pure topology to the concretion of Euclidean space. The main hypotheses were:

1. Space is a continuous (and differentiable)[51] manifold of 3 dimensions.
2. Lines are measurable and comparable, so that their length does not depend upon position in the manifold.
3. The length of a line element can be expressed by a positive definite quadratic differential form.
4. Solids can move freely without metric deformation ('stretching').
5. Angle sum in one triangle is 180°.

It was in the second part of the *Vortrag* that Riemann introduced the last three hypotheses, and with them concepts that made it possible to develop an analytic theory of an immensely wide class of spaces with metrics, his new differential geometry. The core notion for this purpose was what we now call the *curvature tensor*, generalizing the concept of curvature developed by Gauss in his theory of surfaces. (On the occasion of his 1861 prize essay for the Académie des Sciences

---

[49] See Scholz (1980, 16–17), Ferreirós (1999, 43–44).

[50] Topology was a key topic in the fourth proof of the fundamental theorem of algebra (Gauss 1849), one of the papers mentioned explicitly by Riemann. He must have known of Gauss' obsession with n-dimensionality orally, especially through Weber (see S. von Waltershausen, *Gauss zum Gedächtniss* (Leipzig 1856), quoted in Gauss (1973), vol. VIII, 268).

[51] It is well known that Riemann, as then usual, was careless about the conditions for a continuous manifold to be differentiable. His manuscripts show awareness of the difficulties involved here, but probably this came after 1854.

(concerning conditions for constant distribution of isotherms in a homogeneous body), Riemann introduced a quantity (ij, kl) that coincides, but for a factor of 2, with the Riemann curvature tensor R_{ijkl}.[52])

One of the keys to Riemann's approach, and particularly to the kind of conceptual generality that he was so interested in obtaining, was to be found in his very starting point. He made the delicate choice (hypothesis 2) to start with the metrics of the expression for *line elements*, so that 'the length of the lines is independent from position, i.e. that any line is measurable by any other' (1868, 277). Lengths were to be locally Euclidean, defining what we call a 'Riemannian' manifold (although Riemann himself was open to more complicated options). This meant that it was possible to define functions g_{ik} varying continuously with the co-ordinates $x_1, \ldots, x_n$, such that ds is expressed by the *fundamental quadratic form*:

$$ds^2 = \sum_{i,k=1}^{n} g_{ik} dx_i dx_k \quad \text{(with } ds^2 \geq 0\text{)}.$$

But as Riemann acknowledged, many other choices could have been made, and not only in the expression for ds (say semi-Riemannian or 'non-Riemannian' manifolds).

One could also have started fixing the metrics through a completely different kind of stipulation. Indeed, from a naïve or an empiricistic standpoint, it would have been natural to introduce hypothesis 4 right away, without the detour through hypotheses 2 and 3. Actually this was the case with Helmholtz and later authors working on the *space problem*, as Helmholtz was successful in convincing his contemporaries that the conditions for physical measurement made it *necessary* to narrow the terms of the problem. This had the effect that Riemann's manifolds of variable curvature appeared to be a figment of the imagination, useless for the business of natural science. We are then left with one of the very few formulas that Riemann actually wrote in the *Vortrag* (1868, 282), given here for the particular case of 3 dimensions:

$$ds = \frac{1}{1 + \frac{\alpha}{4}(x_1^2 + x_2^2 + x_3^2)} \sqrt{(dx_1^2 + dx_2^2 + dx_3^2)},$$

where α represents the constant curvature. The Helmholtzian move marred much of the conceptual freedom that Riemann was precisely intent on creating, but even Poincaré, a genius comparable to Riemann in so many respects, accepted that formulation of the space problem.

Riemann was no positivist, and he regarded geometry as possibly linked with the most subtle laws of electromagnetism and other physical forces (Section 3.1). He was convinced that all gross phenomena are extremely complex, and this includes Helmholtz's physical processes of measurement. Simple laws of nature

---

[52] For further details, see (Bottazzini and Tazzioli 1995).

will only be valid for infinitesimal points at infinitesimal time intervals, as he remarked in his first lecture course at Göttingen (Winter 1854/55):

> Truly elementary laws can only occur in the infinitely small, only for points in space and time.[53]

The same idea can be found in an early essay of 1850 (Dedekind 1876a, 543) and, as Klein and Weyl liked to emphasize, constitutes a veritable leitmotiv in Riemann's work. The effort to analyse global phenomena from a *local* point of view is not only characteristic of his physical speculations, but also of his approach to function theory, of his thoughts on differential geometry and physical space, and it even finds expression in his metaphysical and psychological ideas. In the *Vortrag* he opined that considerations about the immeasurably great (cosmology!) are 'idle questions' for science, while considerations about the immeasurably small are not at all idle—they open the road to 'knowledge of the causal dependence' of phenomena, to the truly elementary laws (Riemann 1868, 285).

These were some of his reasons for preferring to start from the weakest assumption that would guarantee a smooth transition to metric geometry, while at the same time not compromising the scope of possible hypotheses about the local behaviour of space, matter, and the ether. But in his lecture Riemann gave no detailed explanation, presenting a most subtle option in a rather careless way, no doubt because of the limits imposed by an oral exposition.

Interestingly enough, the relatively unorthodox formulation of the space problem by Riemann (unorthodox as compared with the classical Helmholtz–Lie version) liberates the philosophy of geometry to the point required by Einstein's later work on Relativity. Unaware of this,[54] Einstein had to move on the verge of inconsistency while dealing with the problem of the rotating disk around 1915 (see Friedman 2002).

Finally, part three of the lecture discussed the application of the new geometrical ideas to the understanding of physical space. Here, Riemann introduced a whole series of remarks that made it plain how much more general and ambitious his approach to the matter was, compared to his predecessors (Gauss, Lobatchevskii, Bolyai). Even so, Riemann considered this part of his lecture as the least satisfactory (1868, 287 footnote), so we are certain that it would have suffered great changes if the author himself had prepared it for publication.

3.3 Foundational questions

We may take it that the links between mathematics and physics have been sufficiently explored in the foregoing, but the role of philosophical elements

---

[53] Hattendorff (1869), 4; this text was reproduced word by word from a manuscript written by Riemann in 1854, see *op. cit.*, p. vi).

[54] Einstein had read Riemann's *Vortrag* as a student (see his *Collected Papers*, vol. 2 (Princeton University Press, 1989), p. xxv), but it seems that only the new edition by Weyl (1919a) made him aware of its deeper implications.

(conceptual or foundational problems) deserves further treatment. We shall here touch on two aspects: the foundational role of manifolds, and n-dimensionality.

After their definition in the *Vortrag* (see Section 2.2 above), manifolds are classified into discrete and continuous ones. This establishes a direct link to old conceptions of mathematics – from Aristotle to Euler it was customary to define mathematics as the science of magnitudes, and to distinguish two main branches, the theories of discrete and continuous magnitudes. Simplifying, the mathematics of the discrete gives rise to arithmetic and algebra, that of the continuous to geometry and analysis. The link was further stressed by Riemann's use of the words 'manifold' and 'magnitude' as quasi synonyms. All of this emphasized the foundational import of Riemann's ideas, and it must have constituted a most intriguing reading during this transitional period, when the foundations of the discipline were deeply in question.[55] The general conception of pure mathematics as having for objects discrete and continous classes (manifolds, sets) was taken over by Dedekind, Cantor, Peano and Hilbert, leading to the development of set-theoretical mathematics. The more particular study of continuous manifolds led to the topological and differential manifolds of modern mathematics.[56]

The other issue, how the problem of n-dimensionality rose in Riemann's mind, sheds much light on the path towards his novel geometry, complementing in an essential way what we have seen before. Constantin Schmalfuss, who had been Riemann's devoted mathematics teacher at the *Gymnasium* and kept contact with him, stated that it was already during the first year of university studies that Riemann began developing 'his abstractions concerning spatial dimensions,' i.e. thoughts about the possibility of n-dimensional geometry.[57] If true, this is noteworthy because of the early date, but Gauss could have played a decisive role here. In the Winter semester 1846–47 Riemann attended his course on least squares, and in this context (in 1850/51) Gauss talked about multidimensional manifolds.[58] In this way, Riemann became acquainted with early attempts to develop geometric language applied to the study of analytic relations between systems of n variables and functions thereof. In his Göttingen dissertation of 1853 another student of Gauss, August Ritter, wrote about 'analytic spaces of n dimensions' (Scholz 1980, 17). But, as Scholz remarks, this was still far from a systematic development of n-dimensional geometry.

What led Riemann to a truly geometrical approach was a foundational question, the problem of understanding the nature of Riemann surfaces as employed in the dissertation of 1851. This conclusion has been put forward by Erhard Scholz (1982b) on the basis of manuscripts from the period 1851/53, which show

---

[55] Resonances of Riemann's views can be found at unexpected places. An example is the article 'Mathematics' in the *Encyclopaedia Britannica* for 1883, written by George Chrystal.

[56] For the connection with set theory, see (Ferreirós 1999); the emergence of the modern theory of manifolds has been carefully studied by Scholz (1980) and (1999).

[57] Schmalfuss to Schering, letter of 27.11.1866, in Riemann (1892), 852.

[58] Scholz (1980), 16; Gauss (1973), vol. X.1, 477ff.

clearly the links between manifolds and Riemann's previous work on function theory (links suggested also in Riemann (1868), 274). These manuscripts deal with continuous n-dimensional manifolds, n-dimensional topology, and the relation between manifolds and geometry – in this last topic, Riemann was still a long way from the vision that he presented in the 1854 lecture. On the basis of the manuscript published by Scholz as Appendix 4 (1982b, 228–29), I find reason to believe that, more specifically, the *foundational* problems raised by the Riemann surfaces had a crucial role to play in the creation of manifolds.[59]

Riemann's approach to function theory had the great advantage of making possible an eagle's view of the topic, but it posed all kinds of difficulties. The abstract approach that he wished to see applied required much clarification and refinement, the methods that he employed had to be criticized and replaced, or considerably polished. It is well known that it took some 50 years to set in place all the essential elements: differentiable manifolds, Dirichlet principle, and so on. Was Riemann blind to these problems? The evidence suggests that he was not; in papers and manuscripts he showed he was aware of the critical problems of analysis.

After all, he was a student of Gauss and of Dirichlet, the greatest specialist in mathematical rigor in the generation before Weierstrass, and indeed to somebody trained by Dirichlet, recourse to geometrical constructs like the Riemann surfaces in function theory had to appear puzzling. The tendency in contemporary analysis was to avoid resorting to geometry and intuition; on the contrary, Cauchy, Dirichlet and their followers reformulated previous geometrical ideas (function, continuous function, etc.) in abstract terms. As Dedekind (1888, 338) made clear, Dirichlet was fond of emphasizing the standpoint known as the *arithmetization* of analysis, a view that can also be traced back to Gauss. Riemann had been educated in this tradition, and he expressed allegiance to it both in the introduction to his dissertation and in his second dissertation on trigonometric series.[60] The questions were, did his approach to function theory make it dependent upon geometry? Was the use of Riemann surfaces an *ad hoc* intuitive means for understanding multivalued functions? Did the general theory outlined in his dissertation thus lack rigor? In fact, the manuscripts unearthed by Scholz provide very interesting answers to these questions.

Riemann surfaces did not fit into traditional geometry, since they had to be either conceived in a piecemeal fashion (as the covering surfaces of the plane actually described in his writings on function theory), or regarded as objects in higher-dimensional space. Was there a satisfactory foundation for n-dimensional geometry? Riemann had analysed the behaviour of Riemann surfaces from the viewpoint of *analysis situs*, but this again lacked a satisfactory foundation at the time. Could a new approach to topology and geometry be sketched, that answered all of the above problems? All these questions called for a 'philosophical study' of

---

[59] This is argued at length in Ferreirós (1999), 57–60.

[60] In both cases he named Dirichlet explicitly; see Riemann (1892), 3–5 and 239–44.

this area of mathematics, in the Herbartian spirit. They required a deeper analysis of the foundations of function theory, the search for new and more adequate basic concepts. Riemann was perfectly sensitive to such questions.

The manuscripts edited by Scholz offer a definition of continuous manifolds, presenting them as a satisfactory starting point for (a yet to come rigorous development of) the Riemann surfaces and even for an abstract development of geometry independent from intuition; they also explore the rudiments of an abstract topology of n-dimensional manifolds (only this part appeared in the 1876 edition of Riemann's *Werke*). In its abstraction and boldness, this material goes far beyond what was known to contemporaries from Riemann's publications before his death. It provides very interesting complements to the ideas surveyed so quickly in Section I of the lecture on foundations of geometry. Let us discuss a few details.

The definition of manifolds in these fragments is less general than the one in the lecture, but clearer for a normal reader. (The new definition in the lecture was obviously aimed to embrace all of pure mathematics, not just the mathematics of continuous domains and functions; see Ferreirós (1999), chap. 2.) Riemann considers a 'variable object' that admits of different 'determinations' [Bestimmungsweisen], i.e. that can be in different states; the states or 'determinations' constitute the points of the manifold, defined as the totality of all these points.[61] In Appendix 4 he offers concrete illustrations by considering experiments in which two or more physical magnitudes are measured; one reaches manifolds of two or in general n dimensions (Scholz 1982b, 229). Here the precedent of Herbart was important, as this approach to the properties of physical objects was inspired by his conception of objects as 'complexions' of properties, where each property is represented by a continuum (Scholz 1982, 419) (Ferreirós 1999, 45).

Riemann emphasized that the notion of a manifold, so defined, is entirely independent of geometrical intuition, *Anschauung* (Scholz 1982b, 228). Space, plane and line are only the simplest intuitive examples of three-, two- and one-dimensional manifolds. Here, Riemann was improving upon Gauss's views on the difference between an abstract theory of magnitudes and manifolds, and its intuitive exemplification by spatial notions. However, by the time he wrote this manuscript, Riemann regarded such an abstract approach to geometry only as a theoretical possibility, thinking that it would be 'extremely unfruitful' as it would not yield a single new theorem (Scholz 1982, 229). It would make complex and obscure what appears simple and clear in intuitive spatial language. Little did he foresee the new vistas that he would reach within one or two years.

While such an abstract approach to geometry appeared unfruitful, the opposite held for the use of geometrical imagery as a tool in understanding multidimensional manifolds. Thus, when mathematicians 'stumbled upon manifolds of many dimensions . . ., as in the doctrine of definite integrals' within function theory, they

---

[61] See, e.g., Appendix 1 to Scholz (1982, 222), where incidentally Riemann uses the word 'Menge' to define his 'Mannigfaltigkeiten.'

'had recourse to spatial intuition:'

> It is well known how one thus wins a true overview of the matter, and how only in that way the essential points become evident. (Scholz 1982, 229)

Riemann regarded the use of geometrical imagery in function theory as a paradigmatic example. The ideas elaborated in this manuscript offer an explanation of how the Riemannian approach to function theory does not make it dependent upon geometry. The strictest presentation of the topic would start with an abstract theory of manifolds, but in expository papers such as his celebrated work on 'Abelian functions' (1857) Riemann made concessions to readers and relied on geometrical imagery. All of this substantiates my opinion, expressed above, that Riemann's own views were not so *anschaulich* as Felix Klein thought; they were more in the spirit of twentieth-century developments. Surprisingly enough, the manuscripts of 1851/53 call for an abstract account like the one Weyl adopted in his 1913 book on Riemann surfaces.

4   Concluding remarks

According to Riemann, mathematics belongs to the conceptual side of the scientific enterprise and is not a priori. His research on differential geometry was a propaedeutics for physics, or better, a necessary accompaniment and counterpart to empirical researches. In contrast to many of his German contemporaries, Riemann had no place for a philosophy of mathematical knowledge that might assume its absolute autonomy,[62] and even less for any form of Platonism. In the view he presented so forcefully in the *Vortrag*, far from being a priori, mathematics depends on hypothetical postulates. Furthermore, mathematical concepts and axioms are immersed in a historical process of change. Hopes for an all-embracing system, that may codify once and for all the foundation of mathematics, can hardly blossom here.

Using very anachronistic terms, we could say that Riemann's conception of mathematical knowledge is not only post-Kantian from the beginning, but also post-positivistic. A long-standing tradition in philosophy has tried to understand mathematics as a completely autonomous source of knowledge, whose roots are independent from Nature and the study of natural phenomena. This applies in very different forms to Kant, to the logical positivists, with their implausible view of mathematics as tautological, and to authors like Cavaillès, who viewed the

---

[62] Epistemologically speaking, I mean. Riemann certainly granted the possibility of an autonomous development of the conceptual problems generated by pure mathematics, but this is more of a disciplinary issue.

development of mathematics as an autonomous dialectics of concepts.[63] Not so for Riemann, who underscored that mathematics has emerged as part and parcel of the larger enterprise of natural science.

Abandoning apriorism and emphasizing the role of hypotheses naturally leads to an emphasis on historical development, obvious in Riemann's epistemological manuscripts and in the *Vortrag*. Thus Riemann became a pioneer in yet another field, starting a tradition that would count names such as Henri Poincaré and Hermann Weyl, and that can be linked with the philosophical ideas of Lakatos and Kitcher, among others. Although this tradition did not carry on to the logical empiricists, it remained alive among a select group of European authors, including Bernays, Gonseth, Piaget, and some Marxist mathematicians and philosophers.

There is evidence that the nineteenth century was not prepared for that kind of approach, especially not in the German environment. The most influential authors of the time, Gauss, Weierstrass, Cantor, Dedekind, were decidedly for aprioristic conceptions (with different rationales, logicistic in some cases, Platonistic in others). This line was of course closer to tradition, as the received view from Euclid to Gauss, from Plato to Kant, was to understand mathematics as a body of truths *simpliciter*.

The hypothetical conception advanced by Riemann was in obvious conflict with this tradition, but it won more and more adherents. The reconception of geometry in the wake of the non-Euclidean revolution played a key role in this process, a well-known representative being the Italian Mario Pieri (1900). The new conception was less easy to find in connection with arithmetic and analysis, but it emerged gradually with constructivist criticism of analysis and the foundational problems of logic and set theory. Russell's epistemology of axioms, developed in the 1900s, Hilbert's ideal elements of the 1920s, Weyl's views (1927) (1949) on classical mathematics, and Quine's famous arguments (1951) belong here.

Indeed, one of the outcomes of the foundational debate, transversal to the different basic viewpoints, emerging among constructivists and postulationists alike, was the rise of hypothetical conceptions of mathematics. To give an extreme example, even Kurt Gödel, despite the basic Platonism in his understanding of mathematics, accepted the presence of hypothetical elements in mathematical theories, adopting a sophisticated, 'critical' version of Platonism (Gödel 1964). The emergence of the hypothetical conception was emphasized by Lakatos long ago (1967), and since then it has been customary to call it 'quasi-empiricism,' although this label is in my view poorly chosen and misleading. One of the reasons why I find it objectionable is that it is based on the dubious distinction between formal and empirical sciences – the problem being not so much with the idea of a demarcation line between disciplines, but with the ill-chosen adjectives 'formal' and 'empirical,' which suggest a wrong way of labelling the areas.

---

[63] See the chapter by Benis-Sinaceur.

The emergence of this hypothetical conception can be seen as yet another testimony of Riemann's far-reaching vision. Let me finish by pointing to another such testimony, which however, cannot be discussed here in detail. In Section 2.3 we saw that Riemann had the courage to emend Kant and elaborate his own table of antinomies. The first antinomy poses the conflict between the discrete ('finite elements of space and time') and the continuum. In retrospect this antinomy appears, to this writer at least, as extraordinarily insightful and profound. Its formulation is all the more surprising, as at that time no physicist or mathematician seriously doubted the correctness of the traditional continuous representation of spatio-temporal reality; the topic gained a new urgency after the arrival of quantum theories and is still very much alive today. Looking at it from the viewpoint of the foundational debates, it reveals its deep nature again, as it can be argued that the debates centred essentially (and still do) around the problem how to understand the continuum. The cleavage between a mathematical perspective firmly grounded in the discrete (intuitionism, constructivism), and another that insists on an infinitary understanding of the continuum as a point-set, can be seen as the main source of the profound methodological divergences that separated streams of mathematical thought in the twentieth century.[64]

---

[64] This is a rich topic, to be argued at length elsewhere. Some further indications can be found in the Introduction and in Ferreirós (forthcoming).

The Riemannian Background to Frege's Philosophy[1]

JAMIE TAPPENDEN

There was a methodological revolution in the mathematics of the nineteenth century, and philosophers have, for the most part, failed to notice.[2] My objective in this chapter is to convince you of this, and further to convince you of the following points. The philosophy of mathematics has been informed by an inaccurately narrow picture of the emergence of rigour and logical foundations in the nineteenth century. This blinkered vision encourages a picture of philosophical and logical foundations as essentially disengaged from ongoing mathematical practice. Frege is a telling example: we have misunderstood much of what Frege was trying to do, and missed the intended significance of much of what he wrote, because our received stories underestimate the complexity of nineteenth-century mathematics and mislocate Frege's work within that context. Given Frege's perceived status as a paradigmatic analytic philosopher, this mislocation translates into an unduly narrow vision of the relation between mathematics and philosophy.

This chapter surveys one part of a larger project that takes Frege as a benchmark to fix some of the broader interest and philosophical significance of nineteenth-century developments. To keep this contribution to a manageable

---

[1] I am grateful to many people for input, including Colin McLarty, Jim Joyce, Juha Heinonen, Karen Smith, Ian Proops, Rich Thomason, Howard Stein, Abe Shenitzer and Göran Sundholm. An early version of this chapter was read at UC Irvine, Notre Dame, the Berkeley Logic and Methodology seminar, and at a conference at the Open University. I am grateful to those audiences, especially Pen Maddy, David Malament, Robert May, Aldo Antonelli, Terri Merrick, Mic Detlefsen, Mike Beaney, Marcus Giaquinto, Ed Zalta, Paolo Mancosu for comments and conversation. Jay Wallace and his students in Berlin provided invaluable help in obtaining photocopies of archival materials. Heinrich Wansing and Lothar Kreiser were also generous with photocopies from the archives. Input at several stages from José Ferrierós led to substantial changes. I should especially acknowledge a debt to Jeremy Gray for a range of things that have influenced this chapter including his conversation, writings, and patience.

[2] One exception is Howard Stein [1988], who discusses the importance of Dirichlet, rather than Riemann, to this revolution, but both exemplify the reorientation in method of which he writes: 'Mathematics underwent, in the nineteenth century, a transformation so profound that it is not too much to call it a second birth of the subject – its first having occurred among the ancient Greeks...' (Stein [1988] p. 238)

length, the chapter will emphasize the philosophical interest and methodological underpinnings of the mathematical history, with the connection to Frege given in outline, to be fleshed out elsewhere.³

I'll concentrate on issues intersecting one specific theme: the importance of fruitfulness in applications as a criterion for the importance and 'centralness' of a foundational concept, with special attention to the concept of function. We'll find a connected family of Fregean passages mirrored in methodological views and research strategies explicitly enunciated by Riemann and his successors. In the abstract, of course, this sort of parallel could be just a coincidence: perhaps Frege was unaware of these developments or out of sympathy. The role of the biographical detail about Frege's mundane life as a mathematics professor will make it clear that Frege was both aware and in crucial respects sympathetic.

I will recall some points about Frege; I can be brief since I've explored them extensively elsewhere. (Tappenden [1995b]) It is well known that a major Fregean innovation was the choice of function/argument rather than (say) subject/predicate as the basis of his logic. Frege is also explicit about why he chooses the framework he does:

> All these concepts have been developed in science and have proved their fruitfulness. For this reason what we may discover in them has a far higher claim on our attention than anything that our everyday trains of thought might offer. For fruitfulness is the acid test of concepts, and the scientific workshop is logic's genuine field of observation. ([BLC] p. 33)

This stance resonates with other facets of Frege's epistemology, notably his view that it is because of 'fruitful concepts' that logic can extend knowledge, and his view that only through applications to physical reality is mathematics elevated above the status of a game. Moreover, it fits with a Fregean aspiration I have called (in Tappenden [1995b]) the 'further hope'. In answering the rhetorical question of why one should prove the obvious, Frege cites several reasons, among them:

> ... there may be justification for a further hope: if, by examining the simplest cases, we can bring to light what mankind has there done by instinct, and can extract from such procedures what is universally valid in them, may we not thus arrive at general methods for forming concepts and establishing principles which will be applicable also in more complicated cases? ([FA] p. 2)

How did Frege expect that his many interconnected remarks on such themes would be received? Would he expect his audience to hop through them as *obiter dicta* or to see in them charged declarations of methodological principle? The rest of this chapter will bring out that Frege could not have failed to realize that these

---

³ I develop the story in detail in Tappenden [----].

remarks would be read as a kind of declaration of allegiance, or at least affinity with one of two competing schools.

One rhetorical obstacle to the recognition of the facts about Frege's mathematical context and the interest they bear should be noted at the outset. There is a story about Frege's background that is so generally accepted as to deserve descriptions such as 'universally acknowledged dogma'. The story has it that the mathematics of the nineteenth century, so far as it is relevant to Frege's career, can be summed up as the process of the 'arithmetization of analysis' exemplified by Weierstrass. I will argue that this folk legend is not just wrong, but *wildly* wrong. It is not just wrong in that it excludes a wide range of facts and events but wrong about the *spirit* of what was happening at the most revolutionary turns, and wrong about what philosophical issues that events rendered salient. First, there was much more going on than the arithmetization of analysis. Second, the other things going on were much more relevant to Frege's training and research. Third, to the extent that Frege had any stance concerning the Weierstrass 'arithmetization of analysis' program it was as a *critic*.[4]

The story that will unfold here will run as follows:

a) In Germany from the 1850s on, there was a clash of mathematical styles in complex analysis and neighbouring fields in which the issue of the fruitfulness of concepts was of paramount importance; this was bound up in intricate and sometimes surprising ways with the development of geometry and geometric interpretations of analysis.

b) The coarsest division separates 'followers of Riemann' and 'followers of Weierstrass'. Riemann's mathematics was revolutionary, exhibiting for the first time a variety of the styles of reasoning that we associate with contemporary mathematics, while Weierstrass' was a continuation of a broadly computational mathematics that continued what had gone on before. The Riemann material was foundationally in flux, and it gave rise to several different ways of rendering the basic results tractable. In addition, the work presented certain methodological issues – clarity and fruitfulness of concepts, the role of geometry and intuition, connection to applications – in an especially urgent way.

---

4 The idea of 'the development of rigour in the nineteenth century' has been so thoroughly identified with 'the arithmetization of analysis' in the style of Weierstrass that I should take a moment to make explicit that by describing Frege as a critic of the programme I am in no way making the preposterous suggestion that he opposed rigour. The arithmetization of analysis (to the extent that this means specifically Weierstrass' broadly computational program of extending the techniques of algebraic analysis by exploiting power series) was one among many programs for imparting rigour. Dirichlet and Dedekind, to name just two of the nineteenth-century mathematicians committed to rigour, had no part of the procrustean restrictions that were bound up with the Weierstrass techniques.

Also, the issues I will be discussing have little to do with 'δ-ε' definitions of continuity, differentiability, etc. In the mid-nineteenth century this style of definition was common coin. Riemann's definition of (what we now call) the Riemann integral was also in the 'δ-ε' mould.

c) Frege's non-foundational work and intellectual context locate him in reference to these live issues of mathematical method. He is securely in the *Riemann* stream.[5]

d) These methodological issues were reflected in Frege's foundations – notably in his assessment of the importance of the concept of function, in his emphasis on applications, the theory of negative and real numbers in *Grundgesetze II*, his principle that objects should be presented independently of particular modes of representation, his opposition to piecemeal definition, his view of the relation of arithmetic to geometry, his stance on 'fruitful concepts' and its connection to extending knowledge, and his consistently critical stand on Weierstrass and his acolytes.

I Myth and Countermyth: Frege's position in nineteenth-century mathematics or: the 'arithmetization of analysis' in the style of Weierstrass is a red herring

Frege's foundational work was developed in a context in which the issues it addressed were still volatile and disputed. Before spelling out this history, it will be useful to indicate two commonly accepted positions that I'll be arguing against. (I'll call them 'the myth' and 'the countermyth' for reasons that will become apparent.)

a) The myth (Russell, folklore): Weierstrass to Frege – a natural succession?

The thesis part of this thesis–antithesis pair is familiar. I expect that most people with even a glancing interest in mathematics or philosophy have heard it. For those of my generation I think it fair to describe it as the 'received view' against which the subsequent counter-myth struck a powerful blow. I don't know the precise origins of the myth, but I expect that a genealogy would assign a large role to Russell's popular writings like *Introduction to Mathematical Philosophy* and their echoes in writers like Quine. In this familiar tale, the foundations of mathematics develops as a series of reductions of clearly delineated and well-understood fragments of real (not complex) analysis. Derivatives are reduced to limits of reals,

---

[5] To keep some focus to this already extensive discussion I am leaving out several topics that would be relatively distracting here, though of course they would have to be included in a complete account of Riemann's influence on our conception of geometry and complex analysis. In particular, I'm leaving out Riemann's role in the early development of topology ('*analysis situs*') and non-Euclidean geometry. Some of these are addressed in José Ferrierós' contribution to this volume. This omission is not to be taken to suggest that these developments are uninteresting, or unrevealing either in themselves or as background to Frege's work. It is just an editorial decision about how to present this piece of the story.

reals and limits of reals are reduced to sets of rationals, rationals are reduced to sets of pairs of integers, and finally integers are reduced to sets via a (so-called) 'Frege–Russell' definition of number.

One objective is taken to be the introduction of rigour through the elimination of geometry in favour of the 'arithmetization of analysis' represented by Weierstrass. Frege and Russell are viewed as the culmination of a mathematical trend that begins with Cauchy and winds through the 'δ-ε' definition of continuity and other concepts, and the approach to functions in terms of convergent infinite series. The payoff of the reduction is taken to be the possibility of knowing mathematical truths for certain. Another desideratum, on this picture, is reduction in the number of primitive logical and ontological categories. From this point of view, the paradigms of foundational work are the reduction of ordered pairs to sets (Quine), and the Sheffer stroke (Russell).[6]

Of course, if the nineteenth-century foundational enterprise could be summed up in the myth, that enterprise would be *mathematically* idle. Reduction motivated purely by 'bare philosophical certainty' and ontological economy – whatever its philosophical merits – is in practice of little interest to the working mathematician.[7] Consequently the efforts of Frege and others also appears mathematically idle, a point upon which the countermyth fastens.

A related, similarly misleading picture locates the momentum for logical foundations of mathematics in a formless 'loss of certainty' arising from the realization that Euclid's parallel postulate is not a certain truth about space and reinforced by the discovery of the set-theoretic paradoxes.[8] On this picture, the purpose of logical foundations is to restore the unshakable confidence in mathematics that had been lost.

As far as Frege and German work in foundations is concerned, this thumbnail picture is so drastically off the mark that it is hard to find anything right about it, but I want to concentrate on one particular cluster of shortcomings. The picture completely neglects the diagnostic need for a more adequate understanding of key

---

[6] For Quine's views on the reduction of ordered pairs to sets as a 'philosophical paradigm' see (for example) Quine [1960]. The jaw-dropping Russell assessment of the importance of the Sheffer stroke appears in Whitehead and Russell ([1925] p. xiv–xv).

[7] I do not mean to deny that reductions of one theory to another are sometimes seen as important successes in mathematical practice. The point is just that such reductions are rarely, if ever, motivated by 'ontological economy'. Unifications of diverse theories that reveal crucial structure as achieved through the Klein program in geometry, the unification of the theories of algebraic functions and of algebraic numbers in Dedekind–Weber [1882], or more recently the unification of algebraic geometry and number theory via the concept of 'scheme' in algebraic geometry are valued, to cite just three examples. But reductions like those given by the Sheffer stroke or ordered pair, which have as their sole advertisement a reduction in the number or kind of basic entities or expressions seem to be regarded with indifference in practice.

[8] I don't know if the view that a foundational crisis arose from the 'loss of certainty' inspired by non-Euclidean geometry and the paradoxes has ever been given a serious scholarly defense, but I have often encountered it proposed or apparently assumed in conversation and lectures and in writing. Certainly it informs much writing on Frege.

topics. It assumes that 'mainstream' mathematical practice was more or less in order as it stood, when in fact key areas were, and were recognized to be, in wild disarray. Though it is hard to tell from Russell's bloodless reconstruction, the early foundational researchers were responding to direct and pressing mathematical needs. This is also, as it happens, the direction taken by what I will call the 'countermyth', which takes over the basic historical picture of the myth and argues that Frege should be relocated outside the mathematical stream.

b) The Countermyth (Kitcher, recent consensus) Weierstrass and Frege: the Mathematics/Philosophy boundary?[9]

The countermyth holds Frege to be crucially *different* from Weierstrass and, by extension, from nineteenth-century mathematics generally, in that Frege was purportedly *not* moved by mathematical considerations.[10] According to the countermyth Frege was proposing a foundational program that makes sense only according to *philosophical* desiderata while Weierstrass exemplified the *mathematical* tradition. He proposed his rigourous definitions to solve well-defined mathematical problems, such as: 'If a sequence of continuous functions converges to a function, is the result continuous?' Proofs demanded distinctions like that between uniform and pointwise convergence, requiring in turn more rigorous definitions of the ingredient concepts. This conclusion – perhaps best articulated by Philip Kitcher ([1981], [1984] (ch.10) and elsewhere) – seems to have been widely accepted in philosophical circles.[11] Indeed, I have the impression that it has displaced its predecessor as 'reigning conventional wisdom'.

Kitcher draws attention to some of the delicate ways that problem-solving efforts and increases in rigour have historically interacted, with specific attention to the events presented in the myth: the foundational treatments of sequences,

---

[9] Lest my use of the terminology of 'myth' and 'countermyth' leave a misimpression, I should stress that despite my disagreements on points of detail, I regard Kitcher's work as important and as genuinely driven by an honest and diligent engagement with the historical record. I speak of the 'countermyth' as itself having the character of a myth not because of Kitcher's original writings but rather because of a subsequent uncritical ratification. The subsequent wide acceptance seems to me to be based more on conformity to philosophers' prejudices about mathematical activity than on the historical evidence.

In this connection it is worth mentioning a particularly astute point that Kitcher has emphasized in his writing on Dedekind: We gain only a meager part of Dedekind's method if we restrict attention to his explicit methodological *dicta*. We have to recognize how much methodology is implicit in the specific decisions that inform the presentation of his actual mathematics. Part of my objective here is to bring out how much unstated methodology is implicit in Frege's mathematical choices as well.

[10] I am concentrating on Weierstrass for the sake of focus, but of course the countermyth also separates Frege from Dedekind, Cantor, etc., along the 'mathematical motive'/'philosophical motive' boundary.

[11] In addition to the writings of Weiner to be discussed in a moment, the Kitcher version of the countermyth is endorsed by Wagner [1992] p. 95–96 and Currie [1982], to cite just two. Furthermore, Dummett ([1991] p. 12) responds to Kitcher's point in a way that seems to reflect a grudging acceptance of the core historical thesis, though he quarrels with Kitcher's 'spin'.

derivative, limits and continuity for functions of real numbers. Kitcher's conclusion is that Frege stands outside the mathematical mainstream, and was just mistaken to think himself in its midst:

> If we disentangle the factors which led to the Weierstrassian rigorization of analysis, we find a sequence of local responses to mathematical problems. That sequence ends with a situation in which there were, temporarily, no further such problems to spur foundational work. Frege was wrong to think of himself as continuing the nineteenth-century tradition... (Kitcher [1984] p. 269–270)
>
> [Frege advanced] an explicitly philosophical call for rigour. From 1884 to 1903 Frege campaigned for major modifications in the language of mathematics and for research into the foundations of arithmetic. The mathematicians did not listen... [because] none of the techniques of elementary arithmetic cause any trouble akin to the problems generated by the theory of series or the results about the existence of limits. Instead of continuing a line of foundational research, Frege contended for a new program of rigour at a time when the chain of difficulties that had motivated the nineteenth-century tradition had, temporarily, come to an end. (Kitcher [1984] p. 268)

To the extent that Kitcher showed the need to set aside the Russell picture, his work represents a clear advance over previous research: it is absolutely correct that *the myth* is inaccurate. There *were* important differences between Weierstrass and Frege (and even greater differences between Weierstrass and Russell) in the way they were engaged with mathematical practice. Furthermore, if all that had been going on were the problems of taming infinite series and continuity of real functions, Frege would have indeed come on the scene decades after the shouting had died down. But Kitcher has only looked at parts of the story. The picture looks quite different if we extend our view from real analysis to include the areas of mathematics that Frege was professionally engaged with. Besides foundations, these were principally geometry and *complex* analysis (especially elliptic and more generally Abelian integrals and functions). In these areas there were new central questions that wouldn't be satisfactorily addressed until into the twentieth century.

A problem shared by *both* myth and countermyth is that the motives driving the nineteenth-century evolution to more explicit foundational grounding are presented in an emaciated form that leaves it hard to understand why that evolution should have gripped many of the greatest minds of that century. It is important to get the history right, both because it is rich and interesting in its own right and for more specific metaphilosophical reasons. The specific reasons arise from the fact that Kitcher's picture has come to be used rather uncritically in the Frege literature, for example in the writing of Joan Weiner.[12] This is unfortunate not just because the picture is historically inaccurate. It also nourishes an unduly meager conception of the relations of mathematical and philosophical

---

[12] See Weiner [1984] and [1990], especially Chapters I and II.

investigation. If we take Frege to be a paradigmatic analytic philosopher, these presumptions can support a quietism about philosophy that sees it as rightly disengaged from mathematical practice. To the contrary, I'll argue that both in our reading of Frege, and in our attitude to the relations of philosophy and mathematics, we should view the intellectual streams as essentially interwoven.

It is easy to see how the countermyth could gain a foothold in Frege studies. The myth induces narrow horizons around real analysis. This in turn narrows our vision when reading Frege's words. Here is one example, a passage both Kitcher and Weiner cite, that appears in the motivational opening pages of *Grundlagen*:

> Proof is now demanded of many things that formerly passed as self-evident. Again and again the limits to the validity of a proposition have been established for the first time. The concepts of function, of continuity, of limit, and of infinity have been shown to stand in need of a sharper definition....
>
> In all directions the same ideals can be seen at work – rigor of proof, precise delimitation of extent of validity, and as a means to this, sharp definition of concepts. [FA] p. 1

If we take Frege to be speaking exclusively of functions of a real variable the countermyth might seem reasonable. Central questions in the foundations of *real* analysis had been settled for decades, and Frege's work could be argued to be importantly different.[13] But Frege says nothing to prompt that restriction. In what was called *Funktionentheorie* – functions of a *complex* variable – the foundations were in wild disarray. Questions of limit and boundary behaviour, behaviour at infinity and the nature of infinity, the proper construal of functions and legitimate patterns of function-existence arguments, and the natural definitions of continuity and differentiability *remained* up in the air.

That Frege has *complex* analysis in mind is clear throughout his writings; for example, in the essay [FC] he describes what he means when he speaks of the widening of the concept of function. As in *Grundlagen* ([FA] p. 1) he describes the expansion as resulting from the development of 'higher analysis' ([FC] p. 138) where, as Frege puts it, 'for the first time it was a matter of setting forth laws holding for functions in general.'[14]

---

[13] Since it is worthwhile to concede points to bring out how far off target the myth is, and I don't intend to contest the point here, I'll temporarily concede that the foundations of real analysis were settled. But I should register a footnote demurral. Even with this tunnel vision Frege's motives don't turn out as starkly different from those of the other mathematicians around him as Weiner takes them to be. (For example, even with the restriction to real analysis it is hard to explain Weiner's neglect of the definition of integral, which even for functions of a real variable remained an unsettled topic of research throughout the second half of the nineteenth century and (with the introduction of the Lebesgue integral) into the twentieth century.)

[14] That 'higher analysis' was understood to include complex analysis is easy enough to show by consulting any of the many textbooks devoted to it. For concreteness, consider a text Frege took out of the Jena mathematics library in the 1870s: Schlömilch's *Übungsbuch der Höheren Analysis* of 1868. About a seventh of the book is devoted exclusively to functions of a complex variable.

> Now how has the word 'function' been extended by the progress of science? We can distinguish two directions in which this has happened.
>
> In the first place, the field of mathematical operations that serve for constructing functions. Besides addition, multiplication, exponentiation and their converses, the various means of transition to the limit have been introduced...
>
> Secondly, the field of possible arguments and values for functions has been extended by the admission of complex numbers... ([FC] p. 144)

Weiner suggests in a recent book that so far as the relation between geometry and analysis was concerned:

> The attempt to clarify these notions (of limit, and continuity) involved arithmetizing analysis, that is, showing that its truths could be proved from truths of arithmetic. By the time Frege began his work, most proofs of analysis had been separated from geometry and the notion of magnitude. It is not surprising, then, that it would have seemed less evident to Frege that the truths of analysis are synthetic a priori. (Weiner [2000] p. 19–20)

This opinion completely *inverts* the historical situation in a way that crucially misrepresents the relation between Frege's foundations and ongoing non-foundational research. *Prior* to Riemann's work (which drew on Gauss' geometric representation of complex numbers) complex analysis in Germany had been done almost exclusively computationally, as 'algebraic analysis' with no connection to intuition.[15] It was in the years *after* Riemann's revolution in the 1850s that complex analysis became something different. During this period, it seemed *less* obvious that complex analysis could be carried out independently of appeals to geometric facts than it had before. In this area, Weierstrass' 'arithmetization' program conservatively clung to old certainties in the face of a revolutionary novel style of mathematical reasoning.

One terminological refinement: there was a great deal of talk about 'arithmetizing mathematics' even by people like Dedekind and Felix Klein who were

---

[15] In connection with Jena in particular, the textbooks of Oskar Schlömilch flesh out the context, providing a glimpse into the computational flavour of the 'algebraic analysis' that underwrote most pre-Riemannian research in complex analysis. See Schlömilch [1862] and [1868]. Frege checked the latter book out of the library; it is safe to assume that Frege consulted either Schlömilch [1862] or some other text with essentially the same content when preparing for and teaching courses that are explicitly devoted to 'algebraic analysis'. In these textbooks the treatment of complex analysis is relentlessly computational, without a hint of any appeal to intuition or geometric fact.

Also illuminating in this connection is Enneper [1876], which Frege borrowed from the library when preparing to teach courses on Abelian/Elliptic functions/integrals. Here again the presentation is relentlessly computational, with geometry appearing only for illustration, never in arguments. The only reference to Riemann is to rule his work outside the scope of the treatment. (Enneper [1876] p. 82).

ideologically and stylistically in sympathy with Riemann (though in different respects) and opposed to much of Weierstrass' method.[16] So for example, Dedekind wrote 'From just this point of view it appears as something self-evident and not new that every theorem of algebra and higher analysis, no matter how remote, can be expressed as a theorem about natural numbers – a declaration I have heard repeatedly from the lips of Dirichlet.' (Dedekind [1888/1901] p. 35) Everyone knows Kronecker's 'God created the natural numbers, all else is the work of man.' However, such pronouncements often meant very different things and pointed in very different directions.[17] In this chapter when I write of 'the arithmetization of analysis' I mean specifically *Weierstrass'* style. The features of Weierstrass' program that are typically at issue when philosophers discuss 'arithmetization of analysis' in connection with Frege (such as the use of power series and the broadly computational, series-based perspective) were not shared by Dedekind or Klein.[18] Indeed, the continuation of Riemann-follower Dedekind's quote represents the throwing down of a methodological gauntlet in an anti-Weierstrass direction:

> But I see nothing meritorious – and this was just as far from Dirichlet's thought – in actually performing this wearisome circumlocution and insisting on the use and recognition of no other than rational numbers. On the contrary, the greatest and most fruitful advances in mathematics and other sciences have invariably been made by the creation and introduction of new concepts, rendered necessary by the frequent recurrence of complex phenomena which could be controlled by the old notions only with difficulty. On this subject I gave a lecture (Dedekind [1854])... but this is not the place to go into further detail. Dedekind [1888/1901] p35–36

Weierstrass differed from Frege in many deep ways: in his opposition to mixing metaphysical reflection and mathematics, in his conditions on an adequate foundation for mathematics, in his view of the connection between pure mathematics and applications, and in many other ways as well. But this does not reflect a divide between Frege and 'mathematicians'. Weierstrass differed from *many* mathematicians of the nineteenth century in these ways. To argue that Frege was different from Weierstrass in some respects isn't yet to argue that Frege is outside

---

[16] For Klein see his [1895]. It is worth mentioning in passing – since Klein's writings are sometimes described as presenting Riemann's point of view and I myself had long understood them that way – that in fact the extent (if any) that Klein's presentation reflects Riemann's own conception is quite unclear. It was indeed a matter of considerable controversy among Riemann's students just how much of the story was Riemann's own understanding and how much was Klein's elaboration. I'm grateful to Jose Ferrierós for helping me appreciate the force of this point. Conversations with him, and his contribution to this volume, have left me convinced that Klein departed significantly from Riemann's understanding.

[17] The fragmented character of nineteenth-century 'arithmetization' is dissected in Schappacher [2007].

[18] Kronecker was also opposed, though from the 'right wing', so to speak.

the 'mainstream' of mathematics. The members of a broad and internally varied cluster of schools inspired by Riemann were different from Weierstrass and his followers, and Frege was in that Riemannian tradition.

The next two sections will be devoted to explaining the key features of the Riemannian tradition (Section II) and Frege's immersion within it (Section III). Here are the features of the Riemann style and the contrast with Weierstrass to be developed:[19]

— Weierstrass and Riemann had different definitions of the object of study in complex analysis, and the difference had significant ramifications in practice.
— Riemann's approach involved *apparent* appeals to geometric intuition (for example, in the construction of Riemann surfaces) and intuitions about physical situations (for example, in the Dirichlet principle). Distinguishing geometry and analysis was therefore a crucial unresolved topic. The Weierstrass approach, restricting itself to algorithmic techniques for working with power series, faced no such problems.
— Riemann's approach treated functions as given *independently* of their modes of representation. Riemann's techniques systematically exploited indirect function—existence arguments that need not correspond to any formula. Weierstrass dealt with explicitly given representations of functions. (Weierstrass: 'The whole point is the representation of a function'.)
— Riemann's methods were directly bound up with applications in electromagnetism, hydrodynamics and elsewhere. Weierstrass' work was relatively 'uncontaminated' with applications in physics.
— Overall, Riemann was introducing an entirely new style of mathematics that presented a different family of methodological problems. Weierstrass confronted the problems involved with managing intricate algorithms (finding more efficient procedures, discovering convenient and simple normal forms to reduce complicated expressions to, . . .) Riemann and his successors addressed problems in a way that made even the formulations of problems and *the choice of fundamental concepts* up for grabs. This made it especially important to identify the most *fruitful* ways to set problems up, as well as the proper contexts in which to address them.

Frege, though well versed in the prior tradition of algebraic analysis, was also (with Dedekind) the first significant philosopher to be immersed in mathematics of this recognizably modern form. In this light, Frege's reflections on mathematical method (for example on fruitful definitions or the nature of mathematical reasoning) take on a special force as early confrontations with a new style of mathematical reasoning.

---

[19] The technical terms in this list of points will be explained in the coming pages.

II Beyond the Myth and Countermyth: Riemann versus Weierstrass on Complex Analysis

> One of the most profound and imaginative mathematicians of all time, [Riemann] had a strong inclination to philosophy; indeed, was a great philosopher. Had he lived and worked longer, philosophers would acknowledge him as one of them. His style was conceptual rather than algorithmic – and to a higher degree than that of any mathematician before him. He never tried to conceal his thought in a thicket of formulas. After more than a century his papers are still so modern that any mathematician can read them without historical comment, and with intense pleasure. (Freudenthal [1975] p. 448b)

If we look at the record – taking into account Frege's lectures and seminars, his education, his teachers, mentors and colleagues, his library borrowings and *Nachlass* fragments, and the other benchmarks that give us what reference points we have on Frege's life as a mathematics professor, a regular theme from his undergraduate studies to his late lectures is the theory of functions of a complex variable, in the distinctive style of Riemann.[20] The rest of this section will explain what the Riemann style is and why we should regard it as significant that Frege worked in this intellectual environment. To set the stage, here is a broad observation: it is oversimplified and admits qualifications and exceptions, but for preliminary orientation the simplifications should be harmless.[21] Berlin and Göttingen were distinct centres of activity with profoundly different senses of what counted as core subjects, as acceptable and preferred methods, as central problems, and even as favoured journals for publication.[22] They diverged on

---

[20] This represents a large fraction of Frege's non-foundational research. The other component – algebraic projective geometry – is treated in Tappenden [----].

[21] Of course there was a great deal of variation and internal differentiation within the streams as well. In particular, both Dedekind and Klein devoted themselves to developing Riemann's results as well as his conception of mathematics, but this took them in strikingly different directions. A particularly stark contrast is the purely algebraic reconstruction of Riemann's theory of algebraic functions in Dedekind–Weber [1882] set beside to the presentation of essentially the same theory structured around geometric intuition and physical examples in Klein [1884/1893]. For two more recent presentations, Chevalley [1951] is faithfully in the style of Dedekind, while the early chapters of Springer [1957] are in large part a clear and readable rigorous reworking of Klein [1884/1893], with some good pictures.

[22] There is some useful discussion of some of the institutional and individual differences separating late nineteenth-century Berlin and Göttingen mathematics in Rowe [1989] and [----a]. A good introduction to some of this general background is Laugwitz [1999], especially Chapter 4, which surveys some of the broader philosophical themes animating Riemann's work and some of the mathematicians he influenced. This chapter also contains some useful insights into some of the 'conceptual' and more broadly metaphysical elements dividing the Riemann stream in Göttingen with the 'anti-metaphysical' and 'computational' attitudes of Berlin.

the role of geometry in analysis and on the importance of physical applications to pure research. There were different paradigmatic figures. Riemann's methods (and, after 1866, his memory) dominated Göttingen in the 1860s and 1870s in a way that found no echo in the Berlin dominated by Weierstrass, Kronecker and Kummer. The circle around Clebsch at Göttingen at the time of his death in 1872 (Klein, Lie, Brill, M. Noether, Lindemann, Voss...) carried on lines of investigation growing out of the earlier Göttingen ones, retaining a principled independence of developments in Berlin. Writing a little over twenty years later, Lie reflects on the differences in style in one of the rare books that we can document that Frege read.[23]

> Riemann... knew how to apply geometric tools to analysis magnificently. Even though his astonishing mathematical instinct let him see immediately, what his time didn't allow him to prove definitively by purely logical considerations, nonetheless these brilliant results are the best testimony to the fruitfulness of his methods.
>
> I regard Weierstrass, Riemann's contemporary, as also a successor of Abel, not only because of the direction of his investigations, but even more because of his purely analytical method, in which the appeal to geometric intuition is strongly avoided. However outstanding Weierstrass' accomplishments may be for the foundations and supreme fields of analysis, I nevertheless think that his one-sided emphasis on analysis has not had an entirely favorable effect on some of his students. I believe I share this opinion with Klein, who like Riemann has understood so well how to take from geometric intuition fruitful stimuli for analysis.[24] (Lie and Scheffers [1896] p. v–vi)

The basic datum is the just-mentioned divide in approaches to functions of a complex variable between a Riemann–Göttingen axis and a (then relatively dominant) Weierstrass–Berlin axis.[25] A sad accident of history shaped the events that followed. After transforming complex analysis (and other fields) with work that

---

[23] The *Nachlass* catalogue records 41 pages of notes, calculations and diagrams on 'Two remarkable proofs in *Geometry of Contact-Transformations* by Lie and Scheffers'. (Cf. Veraart [1976] p. 101)

[24] Unless otherwise indicated, translations from German are my own. To save space I haven't reproduced the German original text in footnotes, though it will be available in the book version.

[25] By narrowing the scope I am omitting many interesting things, such as a contrast with a more algorithmic approach worked out by Eisenstein and Kronecker. Here too there is a contrast with Riemann:

> Riemann later said that [he and Eisenstein] had discussed with each other the introduction of complex magnitudes into the theory of functions, but that they had been of completely different opinions as to what the fundamental principles should be. Eisenstein stood by the formal calculus, while [Riemann] saw the essential definition of a function of a complex variable in the partial differential equation. [i.e. The Cauchy–Riemann equations] (Dedekind, cited in Bottazini [1986] p. 221)

was stunningly novel but also full of compressed and often opaque arguments, Riemann died in 1866. The mathematical community, especially in Göttingen, where Frege was to attend graduate school four years later, was left to pore over and decode what Ahlfors [1953] calls Riemann's 'cryptic messages to the future'. The ensuing years saw several fundamentally different schools of thought emerging out of Göttingen, each of which made a plausible case that they were following through on Riemann's conception of function theory and algebraic geometry.[26] By contrast, in the Berlin stream, the most characteristic Riemann techniques were avoided in principle.

The division extended even down to the level of elementary textbooks. As late as 1897 a textbook writer could state:

> Nearly all of the numerous present German textbooks on the theory of functions [of a complex variable] treat the subject from a single point of view – either that of Weierstrass or that of Riemann... In Germany, lectures and scientific works have gradually sought to unify the two theories. But we are in need of a book of moderate length that suffices to introduce beginning students to both methods. I appreciated the need of such a book as I undertook to write this introduction to the theory of functions. Riemann's geometrical methods are given a prominent place throughout the book; but at the same time an attempt is made to obtain, under suitable limitations of the hypotheses, that rigor in the demonstrations that can no longer be dispensed with once the methods of Weierstrass are known. (Burkhardt [1897] p. V)

In the next edition, Burkhardt remarked that his synthesis had met with approval 'outside of the strict disciples of Weierstrass'. (Burkhardt [1906/1913] p. vii)

A similar indication of the schism, and the dominance of Weierstrass' methods, appears in the 1899 description by Stahl of the distinctive approach of Riemann's lectures on elliptic functions (attended by Frege's teacher Abbe):

> ... the peculiarities of Riemann's treatment lie first in the abundant use of geometrical presentations, which bring out in a flexible way the essential properties of the elliptic functions and at the same time immediately throw light on the fundamental values and the true relations of the functions and integrals which are also particularly important for applications. Second, in the synthetic treatment of analytic problems which *builds up the expression for the functions* and integrals solely *on the basis of their characteristic properties* and *nearly without computing* from the given element and *thereby*

---

[26] The details of these subschools would take up too much space here; I go into more detail in Tappenden [----]. For rough orientation, we can distinguish a stream that interpreted Riemann in terms of old-fashioned computational algebraic geometry (Clebsch, Brill, Max Noether...), a stream that took especially seriously the connection to visual geometry and transformations (Klein, Lie...), and a stream that interpreted Riemann in terms of a recognizably contemporary structural algebraic geometry (Dedekind, later Emmy Noether...). There were also some mathematicians (Roch, Prym, Thomae...) who worked in a kind of orthodox Riemannian complex analysis without the kind of reinterpretations defining the other streams.

> *guarantees a multifaceted view into the nature of the problem and the variety of its solutions.* Because of these features, Riemann's course of lectures forms an important complement to the analytical style of treatment that is currently, in connection with Weierstrass' theory, almost exclusively developed. (Stahl [1899] p. III emphasis mine)

Stahl is drawing on clichéd language and rhetoric ('solely on the basis of their characteristic properties' 'nearly without computing',....) favoured by Riemann and students like Dedekind. For example, Riemann sums up his point of view thus, in a methodological overview of article 20 of his thesis [1851]:

> A theory of these functions on the basis provided here would determine the presentation of a function (i.e. its value for every argument) independently of its mode of determination by operations on magnitudes, because one would add to the general concept of a function of a variable complex quantity just the attributes necessary for the determination of the function, and only then would one go over to the different expressions the function is fit for. ([1851] p. 38–39)

In subsequent lines, he makes clear that the 'necessary attributes' were properties like the location of discontinuities, boundary conditions, etc. A few years later, mentioning Article 20 explicitly, he speaks of those methods as bearing fruit in the paper [1857a] by reproducing earlier results 'nearly without computing' (*'fast ohne Rechnung'*). ([1857b] p. 85) In [1857a] itself, Riemann describes himself as having used his methods to obtain 'almost immediately from the definition results obtained earlier, partly by somewhat tiresome computations (*mühsame Rechnungen*)...' ([1857a] p. 67)

Weierstrass, in his lectures, announces his contrasting stance, so carefully echoing this language as to make it clear that he has the Riemann style in mind as the adversary:

> At first the purpose of these lectures was to properly determine the concept of analytic dependence; to this there attached itself the problem of obtaining the analytic forms in which functions with definite properties can be represented... for the representation of a function is most intimately linked with the investigation of its properties, even though it may be interesting and useful to find properties of the function without paying attention to its representation. The *ultimate* aim is always the representation of a function. (Weierstrass [1886/1988] p. 156 emphasis in original)

It is important to appreciate what a simple and fundamental methodological difference is at issue here. On the one hand, Weierstrass holds that there can be no dispute about the kind of thing that counts as a basic operation or concept: the basic operations are the familiar arithmetic ones like plus and times. Nothing could be clearer or more elementary than explanation in those terms. Series representations count as acceptable basic representations because they use only these terms. By contrast, the Riemannian stance is that even what is to count as a characterization in terms of basic properties should be up for grabs. What is to count as fundamental in a given area of investigation has to be *discovered*.[27] In the following remarks, Dedekind sums up the way Riemann's approach to complex function theory understood the quest for the 'right' definition of key functions and objects. As Dedekind sees it, Riemann showed that there is a great *mathematical* advantage to be gained by defining the objects of study in a representation-independent way. Dedekind employed this method in his own profound work in function theory; for example his [1877b] treatment of elliptic modular functions exploits Riemannian methods to powerful effect. As he puts it in an 1876 letter to Lipschitz:[28]

> My efforts in number theory have been directed toward basing the work not on arbitrary representations or expressions but on simple foundational concepts and thereby – although the comparison may sound a bit grandiose – to achieve in number theory something analogous to what Riemann achieved in function theory, in which connection I cannot suppress the passing remark that Riemann's principles are not being adhered to in a significant way by most writers – for example even in the newest works on elliptic functions. Almost always they mar the purity of the theory by unnecessarily bringing in forms of representation which should be results, not tools, of the theory. (Dedekind [1876b] pp. 468–469)

In a later essay, Dedekind puts forward his Riemann-inspired approach – as pushed forward by an emphasis on 'the internal rather than the external':

> [Gauss remarks in the *Disquisitiones Arithmeticae*]: 'But neither [Waring nor Wilson] was able to prove the theorem, and Waring confessed that the demonstration was made more difficult by the fact that no notation can be devised to express a prime number. But in our opinion truths of this kind ought to be drawn out of notions not out of notations.' In these last words lies... the statement of a great scientific thought: the decision for the internal in contrast to the external. This contrast also recurs in mathematics in almost all areas; [For example] (complex) function theory, and Riemann's definition of functions through internal characteristic

---

[27] In Tappenden [----] and [---a] I explore one example of this contrast: the definition and study of elliptic functions. For Weierstrass, the key foothold is a scheme for representing every elliptic function in terms of a distinguished class of series. For Riemann, the keys include the topology of the natural surface on which the functions are defined.

[28] This letter was brought to my attention by Edwards [1987] (p. 14) I've also taken the translation from that article.

properties, from which the external forms of representation flow with necessity. [Dedekind continues, in paraphrase: The contrast also comes up in ideal theory, and so I am trying here to put down a definitive formulation.] (Dedekind [1895] p. 54–55)

Of course, the philosophical question of how to distinguish 'fundamental characteristics' or 'internal characteristic properties' that allow you to 'predict the results of calculation' from 'forms of representation that should be results, not tools, of the theory' is complicated indeed if we see it as an issue in general metaphysics and method. But in the specific cases at issue in complex analysis, the cash value of this contrast was well known, and it would have been transparent to the readers what he was referring to (even if they couldn't give a definition of what he was talking about).[29]

Points of difference as they appeared from the Berlin side (in striking contrast to the above words of Lie, Stahl and Dedekind) emerge in an 1875 letter by Mittag-Leffler, describing his experiences in Weierstrass' seminars on complex analysis:

> ... starting from the simplest and clearest foundational ideas, [Weierstrass] builds a complete theory of elliptic functions and their application to Abelian functions, the calculus of variations, etc. What is above all characteristic for his system is that it is completely analytical. He rarely draws on the help of geometry, and when he does so it is only for illustrative purposes. This appears to me an absolute advantage over the school of Riemann as well as that of Clebsch. It may well be that one can build up a completely rigorous function theory by taking the Riemann surfaces as one's point of departure and that the geometrical system of Riemann suffices in order to account for the till now known properties of the Abelian functions. But [Riemann's approach] ... introduces elements into function theory, which are in principle altogether foreign. As for the system of Clebsch, this cannot even deliver ... [results we won't be discussing here – JT] ... which is quite natural, since analysis is infinitely more general than is geometry.
>
> Another characteristic of Weierstrass is that he avoids all general definitions and all proofs that concern functions in general. For him a function is identical with a power series, and he deduces everything from these power series. At times this appears to me, however, as an extremely difficult path... (Frostman [1966] p. 54–55)[30]

Mittag-Leffler closes by praising the precision, clarity, and 'fear of any kind of metaphysics that might attach to their fundamental mathematical ideas' that he

---

[29] I've discussed this historical debate and its philosophical ramifications elsewhere (Tappenden [2005]) so I'll refer to that paper and move on.

[30] The translation is taken from Rowe [2000].

takes to mark the work of Weierstrass and Kronecker.³¹ The sentiments expressed are characteristic of those held in the Weierstrass circle.³²

Familiar themes found in Frege's writings appear here. In addition to the well-known concern for rigour, Frege also states that geometric interpretations of the complex numbers 'introduce foreign elements' into analysis.³³ The view that 'analysis is infinitely more general than geometry' was a central theme for Frege (as well as Dedekind) and he took the demonstration of this greater generality to be one of his defining objectives.³⁴ In these regards Frege's sentiments fit with Weierstrass'. But there are discordant notes. Weierstrass' position about using only a restricted range of functions is one. His treatment of function quantification pre-supposes the most general notion of function, irrespective of available expressions and definitions. Also, as we saw above, Frege took applications to be of paramount importance in assessing the value of mathematics. Another point of divergence is that Frege *did* work in geometry, and the evidence indicates that he continued to work on geometrical questions past the turn of the century. The mathematics of Clebsch and Riemann – the two mathematicians mentioned by Mittag-Leffler – was the mathematics Frege knew best, and more importantly (so far as his teaching and research into complex analysis is concerned) it is the mathematics he *did*. This raises an issue that Frege scholarship glides over. If we ask 'Just what was Frege trying to lay the logical foundations

---

³¹ The principled separation of metaphysics and mathematics noted by Mittag-Leffler as a characteristic of Weierstrass is another Weierstrass–Riemann contrast that shows itself in connection with Frege. In contrast to Weierstrass' aversion noted here, Riemann read and wrote extensively in philosophy, and in some cases (Herbart's epistemology) he plausibly describes this as shaping his mathematics.

It is well known that Frege said in *Grundgesetze* that he had little hope of gaining readers among those mathematicians who state '*metaphysica sunt, non legentur*'. ([BLA] p. 9) It is not clear to what extent Frege had any specific people or groups in mind, but it is worth noting that possibly the first time Frege uses the turn of phrase '*metaphysica sunt, non legentur*' it is directed, with what feels to be an allusive and knowing tone, at the Weierstrass surrogate Biermann in Frege's draft review of Biermann's account of number in his [1887] (dating uncertain). ([OCN] p. 74) It wouldn't surprise me to learn that this was a recognized catchphrase, so that by speaking obliquely of 'those mathematicians who think...' he was sending a signal whose overtones we now miss.

³² Here is another example, from a potentially long list, of the Riemann–Weierstrass contrast from the Weierstrass point of view. (This is from a retrospective by a Weierstrass student of late 1850/early 1860)

> At the same time, all of us younger mathematicians had at the time the feeling that Riemann's intuitions and methods no longer belonged to the rigorous mathematics represented by Euler, Lagrange, Gauss, Jacobi, Dirichlet and their like. (Königsberger) Königsberger [1919] p. 54

(This quote is from Laugwitz [1999]. I've altered Shenitzer's translation to be consistent with the other translations here.)

³³ [FA] p. 112, [GGII] (p. 155 fn. 1) and elsewhere.

³⁴ I have discussed his concern with the greater generality of analysis in relation to geometry in (Tappenden [1995a]).

of ?' the answer is usually blandly 'arithmetic and analysis' or 'all mathematics besides geometry' or something like that. There is a tacit presumption that just what counts as 'analysis' or 'mathematics' can be treated as unproblematic. But this was under dispute: mathematics in Riemann's sense was a more ambitious discipline. If Frege's foundations were an attempt to ground and diagnose 'analysis', his practice indicates that the target would have been complex analysis *as Riemann did it*. Thus we arrive at the first indication of the point I indicated at the outset: the widely assumed affinity in mathematical attitudes between Frege and Weierstrass is, at bottom, superficial and misleading, while the affinity with Riemann (and with mathematics as conceived and practiced in the Riemannian tradition) is profound despite seeming differences in their standards of rigour.

Specific details: Riemann, Weierstrass, and Epigones on Complex Analysis

> People who know only the happy ending of the story can hardly imagine the state of affairs in complex analysis around 1850. The field of elliptic functions had grown rapidly for a quarter of a century, although their most fundamental property, double periodicity, had not been properly understood; it had been discovered by Abel and Jacobi as an algebraic curiosity rather than a topological necessity. The more the field expanded, the more was algorithmic skill required to compensate for the lack of fundamental understanding. Hyperelliptic integrals gave much trouble, but no one knew why. ... Despite Abel's theorem, integrals of general algebraic functions were still a mystery.... In 1851, the year in which Riemann defended his own thesis, Cauchy had reached the height of his own understanding of complex functions. Cauchy had early hit upon the sound definition of the subject functions, by differentiability in the complex domain rather than by analytic expressions. He had characterized them by means of what are now called the Cauchy-Riemann differential equations. Riemann was the first to accept this view wholeheartedly... [Cauchy even came] to understand the periods of elliptic and hyperelliptic integrals, although not the reason for their existence. There was one thing he lacked: Riemann surfaces. (Freudenthal [1975] p. 449a)

> I have tried to avoid Kummer's elaborate computational machinery so that here too Riemann's principle may be realized and the proofs compelled not by calculations but by thought alone. (Hilbert [1897/1998] p. X)

Weierstrass' 'arithmetization' approach takes as basic the definition of an *analytic function* centered at z_0 as one that can be represented as a power series $f(z) = \Sigma a_i(z-z_0)^i$ where the a_i are complex numbers. The definition is intrinsically local: the series need converge, and hence the function need be defined, only within some given radius. This is not the handicap it might seem to be at first: when the analytic function on an open set extends to a multiple-valued function ('multifunction') on the entire complex plane, this continuation is unique. Note

The Riemannian Background to Frege's Philosophy

though, that a 'multifunction' is not a function, as these are defined in elementary textbooks, since it assigns several values to one argument.

The Riemann approach differs even in its definition of the basic object of study. Functions generally are accepted, with the functions to be studied marked out as those satisfying the *Cauchy–Riemann conditions*:

> With z as the complex variable, and writing the real and imaginary parts of f as u and v (so $f(z) = u + iv$), f is differentiable at (x, y) if these partial derivatives exist and these relations hold:
>
> $$\frac{\partial u}{\partial x} = \frac{\partial v}{\partial y} \quad \frac{\partial u}{\partial y} = -\frac{\partial v}{\partial x}$$

These two definitions of analytic function/differentiable function are now known to be essentially equivalent, but this was not established until after 1900.[35] During most of Frege's productive career, these were seen as distinct and indeed competing definitions, with Riemann's potentially wider.[36] Weierstrass rejected Riemann's definition (though he recognized that the definitions agreed on the most important cases) because he held that the functions differentiable in Riemann's sense couldn't be precisely demarcated.[37]

Fundamental to Riemann's approach is the idea of a 'Riemann surface' on which a multifunction can be redefined so as to be actually be a function, by unfolding it on several separate, though connected, sheets. The device allowed complex functions to be visualized (and in fact it was dismissed by Weierstrass as 'merely a means of visualization'), though its importance went well beyond visualization.

Riemann's geometric approach is further exhibited in his systematic use of *isogonal* (angle-preserving) and especially *conformal* (angle and orientation preserving) mappings (neither Cauchy nor Weierstrass used them systematically). Conformal mappings concern *local* behaviour – the preservation of angles makes sense in arbitrarily small neighbourhoods of a point. By linking conformal mappings and (what came to be understood as) topological properties, Riemann displayed one of his trademark methodological innovations: exploiting interactions between local and global properties.

A third feature of Riemann's approach relevant here is its connection to physics and geometry. (Tight connection to applications was characteristic not just of

---

[35] The qualifier 'essentially' in the text reflects the fact that additional minor assumptions are needed to secure the equivalence. See Gray and Morris [1978].

[36] I should emphasize that what is at issue is not the use of power series expansions. Everybody used them sometimes, even Riemann. What was distinctive of Weierstrass and his tradition was their use as a systematic basis for his theory. (Especially at a time when it had not been rigourously proven that *all* functions satisfying the Cauchy–Riemann equations would be so representable.)

[37] This is recorded in a set of lecture notes by a student who attended Weierstrass' lectures in 1877–78, Pincherle [1880] (p. 317–318). I am grateful to Ferrierós [1999] (p. 36) for drawing my attention to this feature of these notes, and to Bottazzini [1986] p. 287–288 for information about their genesis.

The Riemannian Background to Frege's Philosophy

Riemann but of Göttingen mathematicians as a group.[38]) Complex analysis in the style of Riemann was bound up with its applications in a much more direct and immediate way than Weierstrass' was. The point was similarly expressed by the like-minded Felix Klein (though to be sure in caricature):

> With what should the mathematician concern himself? Some say, certainly 'intuition' is of no value whatsoever; I therefore restrict myself to the pure forms generated within myself, unhampered by reality. That is the password in certain places in Berlin. By contrast, in Göttingen the connection of pure mathematics with spatial intuition and applied problems was always maintained and the true foundations of mathematical research recognized in a suitable union of theory and practice. (Klein [1893b]; quoted in Hawkins [2000] p. 137)

One key bridge to applications is via *potential theory*, which was becoming the core of the theoretical foundations of electricity and hydrodynamics.[39] The formal connection arises directly for Riemann since the equation for potential in two variables is an immediate consequence of the Cauchy–Riemann equations. Consequently, as Ahlfors put it in a retrospective essay, Riemann 'virtually puts equality signs between two-dimensional potential theory and complex function theory.' (Ahlfors [1953] p. 4) On Weierstrass' approach, the connection to these applications is distant, and this distance appears not to have troubled Weierstrass or anyone swimming in his wake.

Both this direct connection to applications and the role of conformal mapping in Riemann's methods flow into a final crucial point – the role of a function – existence principle that has come to be known (following Riemann) as 'Dirichlet's principle'.

A treatment of this principle would require too much space here, so I'll just nod to the longer treatment and some textbooks.[40] All we need to know here is that the principle asserts the existence of certain functions given certain conditions, that it seems most plausible in physical situations, and that it was central to some of Riemann's most important theoretical results.

The situation is complicated by the fact that the Dirichlet principle was not stated sharply enough to delimit its range of validity. Some general forms of

---

[38] On the Göttingen tradition of connections between mathematics and physics see Jungnickel and McCormach [1986] p. 170–185.

[39] The connection between Riemann's complex analysis and physical applications emerges principally through research in electricity and magnetism, and related problems of hydrodynamics. See Riemann [1854/1868a]. The connections between the mathematical theory of potential and the developments in theories of electricity and magnetism have been well documented in secondary literature, so I will just stick to the main point here. On the potential theory itself, see, for example, Kellogg [1929]. A good historical overview of these mathematical events through the lens of potential theory is in Temple ([1981] ch.15)

[40] See Tappenden [----] for a more extended discussion in this specific connection and Courant [1950] for the principle itself.

the principle are false; Weierstrass supplied one famous counterexample. Hence some of Riemann's central arguments contain gaps that in applied cases are filled by physical or geometric intuition, and so the next domino falls: some keys to Riemann's approach remain strictly speaking unproved, leaving genuine questions as to the extent to which intuition grounds fundamental parts of Riemannian analysis.

A diagnostic question then arises: what additional restrictions make the Dirichlet Principle appear evident in physical situations? Can they be formulated sufficiently generally to support Riemann's arguments and make an abstract form of the principle fruitful and interesting, with a broad range of validity? Some of these discussions are especially prone to remind one of Frege's discussions of foundations. He emphasizes repeatedly at the outset of *Grundlagen* that a crucial function of a disciplined proof is that it often reveals the 'limits to the validity' (*Gültigkeitsgrenzen*) of a proposition:

> It not uncommonly happens that we first discover the content of a proposition, and only later give the rigourous proof of it, on other and more difficult lines; and often this same proof also reveals more precisely the conditions restricting the validity *(Bedingungen der Gültigkeit)* of the original proposition. ([FA] p. 3)
>
> Proof is now demanded of many things that formerly passed as self-evident. Again and again the limits to the validity *(die Grenzen der Gültigkeit)* of a proposition have been in this way established for the first time. ([FA] p. 1)
>
> In all directions the same ideals can be seen at work – rigour of proof, precise delimitation of extent of validity *(Gültigkeitsgrenzen)*, and as a means to this, sharp definition of concepts. ([FA] p. 1)

At the same time, the Dirichlet principle was widely discussed in these terms. For one example, in a work on potential theory (Betti [1885] p. VII), the author states that he has avoided the Dirichlet principle not because it is simply invalid, but because the limits to its validity *(die Grenzen seiner Gültigkeit)* have not been established, and this turn of phrase is repeated verbatim in the review of the book in *Jahrbuch über die Fortschritte der Mathematik* for 1886.[41]

Much of the investigation of the Dirichlet principle was carried out in direct connection with physics and applied geometry. We should recall that Frege did not

---

[41] Of course, it wasn't just in connection with the Dirichlet principle that people spoke of 'limits to validity' in these ways. Another example (one that Frege read, cites and discusses) is Riemann's rigourous definition of (what we now call) the Riemann integral. Riemann characterizes his objective repeatedly as the clarifying of 'the extent of validity' *(den Umfang seiner Gültigkeit)* of the concept of definite integral. (Riemann [1954/1868a] (p. 227, twice on p. 239 (a page Frege cites) p. 240 p. 269). In his article on the distribution of prime numbers, Riemann places stress on replacing a function defined only on the upper complex plane with 'an expression of the function which is everywhere valid *(immer gültig)*.' (Riemann [1859/1974] p. 299)

think that there was anything intrinsically improper about appeals to intuition, nor did he think that arguments containing ineliminable appeals to intuition must lack cogency.[42] Indeed, he believed that Euclidean geometry was synthetic a priori and founded on a distinct intuitive 'source of knowledge'. The price of appeals to intuition was a loss of generality; one objective of the *Begriffsschrift* was the diagnostic job of paring off logical arguments from intuitive/logical blends by 'letting nothing intuitive penetrate unnoticed'.

The Dirichlet principle, in a form sufficiently general to support Riemann's proof techniques, was finally proved in Hilbert [1901]. However, in the period 1870–1900 that we are most concerned with, the issue was cloudy. Riemann's methods were used and explored in Germany during 1870–1900 but only by a relatively small band of true believers. Ways to avoid the Dirichlet principle were hammered out, and restricted positive solutions that avoided the Dirichlet principle and sufficed for Riemann's arguments on Abelian functions were carried through.[43] There were also more ambitious attempts to save Riemann's techniques and results by reworking them in novel ways (notably by Dedekind and Clebsch–Brill–M. Noether).

Weierstrass' attitude to Riemann was ambivalent, and his view of those who took up Riemann's mantle (Clebsch, Klein, ...) was harsh. Weierstrass and Riemann were on good terms as young men. But Weierstrass' remarks after Riemann's death tended to be ungracious. Even praise was doled out with sour addenda, and most of his published references to Riemann's methods are belittling. It also indicates what was said in private that Weierstrass' students tended to an unjustly dim view of Riemann's style.

Returning to the Dirichlet principle, note how the history is recounted by Brill and Max Noether, who were perhaps the most rigorous and 'algebraic' of those in the Riemann stream.[44] Writing after much of the smoke had cleared, but a few years before Hilbert's proof, they emphasize the fruitfulness, organic unity and connections to applications of Riemann's approach, while placing the flaw in the original reasoning in what they view as an overly general, uncontrolled concept of function:

> The application of the Dirichlet principle in the generality sought by Riemann is subject to, as we now recognize, considerable misgivings,

---

[42] Note for example: 'It seems to me to be easier still to extend the domain of this formula language to include geometry. We would only have to add a few signs for the intuitive relations that occur there. In this way we would obtain a kind of *analysis situs*.' ([B] p. 7)

[43] By Schwarz [1870], C. Neumann [1877] and later Poincaré.

[44] That is, they are 'algebraic' in the sense of older-fashioned computational algebra as you find in Chrystal [1886]. Dedekind is 'algebraic' in a different, more contemporary sense. Being 'algebraic' in the sense of Clebsch, Brill and Noether is not incompatible with being 'geometric' in a different sense.

> directed against the operation with functions of indeterminate definition in the Riemannian style. The function concept in such generality, incomprehensible and evanescent, no longer leads to reliable conclusions. Recently the exact Riemann path has been departed from in order to precisely bound the domain of validity [*Gültigkeitsbereich*] of the stated theorems. [Schwarz [1870] and Neumann [1877]] have more rigorously, though with circuitous methods described the precise conditions – paying heed to conditions on boundary curves of the surface, discontinuities of the function and so forth – under which the existence proofs that the Dirichlet principle was intended for are possible. It has turned out in fact, that the conclusions Riemann drew for specifically his theory of the Abelian functions remain correct in full generality.
>
> However, we should not set aside Riemann's distinctive style of proof too hastily. It has the virtue of the brevity and relative simplicity of the train of thought; it stands in organic connection to the problems of the mathematical physics from which [Dirichlet's] principle originated; Modelled by nature, Riemann's methods may some day experience a revival in a modified form. (Brill–Noether [1894] p. 265)

These remarks incorporate parts of what might be called 'the conventional wisdom' among mathematicians in the Riemann stream at the time, (apart from the heresy that the problem with Dirichlet's principle was the general concept of function). Similar sentiments were expressed earlier in connection with the Göttingen style:

> ... Riemann makes possible a more general determination of functions, by means of suitable systems of strictly necessary and sufficient conditions. *Independently of the statement of an analytic expression, these permit... the treatment of questions more with pure reasoning than with calculation.* The use of Dirichlet's principle as an analytic instrument as well as [Riemann surfaces] as geometrical support, is characteristic of the theory of functions taught in Göttingen (Casorati [1868] p. 132–3 quoted in Bottazzini [1986] p. 229 my emphasis)

Some aspects of the history of the concept of function have already been well-documented and studied in the philosophical literature. So I should stress at the outset that *in addition* to the already well-known developments concerning the concept of function, *Riemann added something importantly new*. To get our bearings, recall two developments flanking the period 1840–1900. Prior to (say) 1750, a 'function' was essentially a finite analytic expression like a polynomial. Famously, this conception came under pressure when it proved impossible to represent physical problems such as the behaviour of a vibrating string.[45] This initiated an

---

[45] A clear and engaging thumbnail history of some of the questions raised by vibrating strings – that, incidentally, we know Frege read, since he discusses it – is in Riemann [1854/1868a].

evolution toward a conception of function as arbitrary correspondence in various authors. After 1900, with the broader conception of function relatively established, it became the focus of a different skirmish between early advocates of what we now call constructivism (Baire, Borel, Lebesgue) and opponents favouring a conception not tied to definability (Zermelo, Hadamard).[46]

Less well known in philosophical circles is what occurred between these interstices; that is of course the period relevant to Frege. With Riemann the shift to a conception of function that wasn't connected to available expressions becomes systematic and principled: unlike the vibrating-string problems, what was at issue was not a collection of individual anomalies or points of conceptual clarification but a methodical appeal to proof techniques with indirect function – existence principles at their core. Riemann would catalogue the singularities of a function (points where it becomes infinite or discontinuous), note certain properties, then prove that *there must exist* a function with these properties without producing an explicit expression.[47] There was as yet no guarantee that the functions proven to exist could be expressed in any canonical way. Nor need such representations be helpful even if they could be found.[48] It was not just that Riemann had a potentially wider conception of function than Weierstrass; he was committed to methods that only made sense if the wider conception were presupposed.[49] This is part of what gave urgency to Frege's effort to clarify the role of the concept of function in logical reasoning and to clarify the legitimate patterns of function existence argument. If we don't follow Weierstrass and

---

[46] Among the useful discussions are Monna [1972], Moore [1982], Maddy [1997], and Hallett [1984].

[47] So, for example, addressing a topic that had been typically treated computationally, Riemann remarks:

> Everything in the following treatise contains brief hints concerning the application of this theorem which (*as one sees easily with our method that is supported by the determination of a function through its discontinuities and its infinite values*) must form the basis of the theory of the Abelian functions. (Riemann [1865] p. 212 emphasis mine)

[48] In the longer presentation of this material I discuss a high profile example: one of Riemann's innovations elsewhere was a global definition of the ζ-function over the complex plane (minus one point). As an analytic function, this has a power-series representation, but in practice this representation is of no use.

[49] This feature of Riemann's work – the novel systematic use of abstract function existence arguments as a characteristic method – was drawn to my attention by Ahlfors [1953]. It is elaborated throughout Laugwitz [1999]. The only philosopher I know to have recognized the importance of Riemann to the extension of the function concept as it relates to Frege is Bill Demopoulos (drawing on some observations of Bottazzini). ([1994] p. 86) Even Demopoulos' astute observation stops short of a notice of the *systematic* use of function – existence arguments in Riemann brought out by Ahlfors.

restrict our principles for representing functions, what logical principles *do* govern function-existence?⁵⁰

The project of distinguishing geometry and analysis – paring the contribution of intuition from the pure logical content – had intricate motivations and complicated consequences. Many of these – notably the contribution of the foundations of geometry to the emergence of formal semantics – must be left for elsewhere. But it is worth pausing to emphasize how the issues explored so far reveal overlap in the geometry/analysis, logic/intuition and pure/applied divides. As we saw above in the quote from Lie, the value of Riemann's apparent appeals to geometry were the 'fruitful methods' and 'fruitful stimuli for analysis'. The project of disentangling analysis from geometry and the project of providing logical foundations for the 'truly fruitful concepts' in mathematics and natural science are, in complex analysis, the same project.

The history is complicated, and the above is just the beginning of the story in outline. Followers of Riemann disagreed on just how to elaborate their shared positions. In particular, though (apart from Dedekind) the followers of Riemann agreed on the importance of the 'geometrical' point of view, what they took to be the core of 'the geometrical' could be strikingly different. However, the point here has been to emphasize that among topics that were salient to most Riemannians, there was a crucial and deep-rooted *interrelation* among these central methodological themes such as the relation of analysis and geometric intuition, the concept of function and its generality, the fruitfulness of 'geometric' methods and their potential independence from intuition, the opposition to procrustean restrictions to Weierstrassian methods, the role of fruitfulness rather than reduction to antecedently given 'elementary' concepts bequeathed by history, like plus and times, as a guide to the 'internal nature' of concepts and so on. These were not merely discrete characteristic marks of a school but rather aspects of one orientation, in which 'every element is intimately, I might even say organically, connected to the others', to borrow a Fregean phrase.

The concentration on Weierstrass' arithmetization of analysis is a mistake because it considers a single strand of opinion: a forceful conservative thrust that held firmly to a broadly computational view of what truly rigourous mathematics consisted in.⁵¹ To the extent that philosophers have formed a picture of mathematical activity in the late nineteenth century, it seems to draw solely from the research programs and characteristic outlooks of the conservatives, with a consequent impoverishment of our conception of what was mathematically interesting about the nineteenth century and what gave urgency to many of the deepest foundational developments.

---

⁵⁰ This emphasis on introducing functions without reference to specific formulas was taken (especially in Dedekind's hands) in a direction that strikingly anticipates Frege's concern with the mechanics of introducing objects as 'self-subsistent'. I explore this point further in Tappenden [----] and [----a].

⁵¹ Of course, it might be better to say 'overlapping cluster of schools' to avoid a facile identification of Weierstrass with Kronecker or Kummer.

In this environment, Frege's principle that 'fruitfulness is the acid test of concepts' is not merely an idle platitude. It is at a statement of allegiances. So too Frege's use of a general concept of function as a basis of his logic, his project of disentangling intuition from analysis, his quest for general rather than piecemeal definitions, and other features of his philosophical foundations could reverberate significantly with this environment. Well, *was* this Frege's environment? Can we expect that he would have been aware of these developments and anticipate the relevant reactions? The short answer is yes.

III   Beyond the Myth and Countermyth: Riemann and the Riemannian tradition as part of Frege's intellectual context

As we've noted, a variety of Fregean comments – those on fruitful concepts and on delimiting the extent of validity, to mention just two examples – would have seemed to mathematicians around Frege to be loaded remarks alluding to well-recognized disputes. This supports a prima facie assumption that Frege did indeed choose those words and issues deliberately. But leaving general observations aside, what can we say specifically about Frege and his milieu? The following is a quick overview of some of the substance of the intellectual world that can be reconstructed. There isn't space here to lay out all the varieties of fine detail necessary for a lifelike reconstruction of an intellectual environment. My purpose here is just to give an overview of some of what is available, with the full story to appear elsewhere.[52]

a)   Colleagues and Teachers

The mathematicians around Frege – his teachers, colleagues, friends and correspondents – were almost all in Göttingen streams rather than Berlin streams.[53] The most important mathematicians in Frege's environment who concerned themselves with complex analysis – his mentor Abbe, his early geometry teachers Clebsch and Voss, his supervisor Schering, and his colleague Thomae – were all followers of Riemann in one way or another. Abbe, probably the most important

---

[52] There are other details fleshing out the picture, like shared terminology ('Begriffsbestimmung', 'Gebeit', . . .) mutual correspondents, etc. The material in this section should suffice to exemplify how richly the environment can be reconstructed.

[53] To help give a flavour for the intricacy of some of the historical questions – and especially to bring out that 'mathematical practice' is not a monolith – I've concentrated on two competing German centres and roughly contemporaneous styles. However, it is worth noting in addition that there was an even wider gap between mathematics in Germany *tout court* and most of the mathematics in Great Britain. I emphasize this especially because Frege scholarship and Russell scholarship often run on parallel tracks.

intellectual figure in Frege's life, was a devoted student of Riemann at Göttingen. Correspondence between Abbe at Göttingen and a student friend in Berlin in the 1860s reveals Abbe to be a well-informed, enthusiastic partisan of the Riemann sides of fundamental debates with Weierstrass. Frege's only graduate course in complex analysis came from his supervisor Schering, who taught from an annotated copy of Riemann's lectures. This is especially significant since Frege taught classes in complex analysis and advanced topics (Abelian functions) from the very beginning of his time at Jena. Frege's geometry teacher Clebsch was also working out some of the details of Riemann's analysis, within algebraic curve theory.[54]

b) Library Records: Frege's Reading

Surviving library records reveal that complex analysis and especially elliptic and Abelian functions were one of three large clusters of reading activity for Frege between 1873 and 1884. The others were pure geometry and Kantian/anti-Kantian philosophy of science, which reinforces the point that Frege was approaching complex analysis with a geometric basis and a methodologically sensitive eye. Jacobi [1829], a classic treatment that initiated several lines of research into Abelian functions as problems in complex analysis is on a short list of books that Frege borrowed from the library between 1873 and 1879.[55] Two of the others were Enneper [1876] and Schlömilch [1868]. The former is a survey of results about elliptic functions, the latter a book of school exercises in 'algebraic analysis'. A fourth entry is Clebsch–Gordan [1866], which is a text on Abelian functions cowritten by one of Frege's Göttingen teachers. (Frege borrowed Clebsch–Gordan [1866] again in 1883–1884.)[56] Frege also checked out Abel's *Oeuvres Complètes* in 1883. Frege might have been interested in many things there. But in light of his other library borrowings at the same time, it seems reasonable to expect that most or all of Frege's reading from the volume concerned Abel's work on elliptic functions within complex analysis, especially the epochal Abel [1827/1828]. Other borrowings (Gauss's *Disquisitiones Arithmeticae* and Bachmann's book on cyclotomy) had recognized connections to elliptic functions, but in complicated ways that I'll not go into here. If we add the classes Frege had taken on Riemann's theory from Abbe and Schering, the Riemann lecture notes he had access to, and the Thomae monograph he reviewed, what emerges is a picture of Frege immersed in several different approaches to Abelian (and the special case of elliptic) functions/integrals, in the years leading to *Grundlagen*.

---

[54] In addition, there is a lacuna on the opposite side. Frege's environment was largely devoid of representatives of the Berlin perspective. Frege corresponded with Cantor and Husserl, but neither of these could be called orthodox followers of a Weierstrass line. Otherwise, none of his colleagues at Jena was connected to Weierstrass or Berlin, nor were any of his correspondents.

[55] cf. Kreiser [1983] p. 21 for the library records.

[56] cf. Kreiser [1983] p. 25

c) Frege's Lectures and Seminars

I've already noted that Frege regularly taught courses in complex analysis and the advanced subtopics of Abelian integrals and elliptic functions. In fact, Frege offered classes in complex function theory or advanced subtopics 17 *times* between 1874 and 1906.[57] On two occasions ((1903), (1906/07)) a course was titled explicitly 'Complex function theory according to Riemann'. In addition, he offered several courses and seminars on conformal/isogonal mapping. Note too that Frege offered lecture courses on 'Elliptic and Abelian Functions' (Summer 1875) and 'Theory of Functions of a Complex Variable' (Winter 1876/77 and S 1878) and 'Abelian Integrals' (S 1877 and W 1877/78) very early in his teaching career, when he could be expected to have a heavier debt to his teachers, and during which time the generalization of the idea of function in *Begriffsschrift* and the analysis of methods of arithmetic in *Grundlagen* were gestating.

During the late 1880s and 1890s when *Grundgesetze I* was being finished for press and the material for *Grundgesetze II* was presumably under development, there is an especially striking concentration of graduate seminars in Riemannian complex analysis and border areas like conformal mapping and potential theory.[58]

The picture that emerges from the courses for which we do have descriptions supports extrapolations to the many courses for which we have only titles. Altogether it reinforces the observation that Frege spent a large fragment of his teaching career covering the signal topics and techniques of Riemannian complex function theory.

---

[57] References to courses and descriptions are assembled from Kratzsch [1979], Kreiser [1983], Kreiser [1995], and the records, compiled and published by Thomae, of the mathematical seminar at Jena, from Easter to Easter of each year (much of this material has been reprinted in Kreiser 2001 p. 301–320).

[58] In the one case where Frege's graduate seminar on complex analysis has an extended description (1892/1893) Frege hews carefully to the Riemann path: the object of study is defined directly in terms of the Cauchy–Riemann conditions, and multifunctions are unfolded on Riemann surfaces. In 1893 Frege studies conformal mappings in complex analysis. Conformal and isogonal mappings are also the topic in the seminars of 1888/89, 1893/94 and 1897/98. The informative descriptions for the 1888/89 and 1897/98 seminars indicate that one objective was to study conformal mappings to resolve questions about complex integration on Riemann surfaces. The 1888/89 seminar dealt with conformal plane-sphere mappings in connection with evaluating elliptic integrals. In 1901 the seminar may have addressed potential theory.

The 1893/94 seminar in conformal mapping is given the 'red flag' description 'part analytic part geometric'. The 1903/04 seminar on 'mappings' has no description, but it is concurrent with lectures on 'Complex Function Theory according to Riemann'; given that several of the preceding seminars had covered conformal or isogonal mappings, it seems reasonable to expect that this one covered them too, at least in part. The 1896/1897 seminar on mechanics touch on potential theory and the dynamics of compressible fluids. The 1882/1883 and 1900/1901 seminars on mechanics also touches on topics that would have naturally prompted a detour through potential theory, but specific details on proof methods are absent. Of the seminars that didn't touch on complex analysis and near cognate topics, most covered geometry, a further indication of Frege's mathematical inclinations.

d) Frege's research record and *Nachlass*

Frege's only publication in complex analysis is a thumbnail review of Thomae [1876], a work on elliptic functions and their generalizations. (As Frege notes in the review, Thomae's book is avowedly Riemannian in its approach.) However, the *Nachlass* and records of Frege's early lost scientific lectures indicate that his research on the topic was more extensive. In 1875 Frege gave a lecture to the Jena mathematical society entitled 'On some connections between complex function-theory and geometry'.[59] In that context, with that title, the topic would have been an exploration of connections in a Riemannian vein. The *Nachlass* catalogue indicates that Frege continued to carry out research, as he kept notes on power series (p. 103) and analytic functions (p. 96) as well as 17 models of Riemann surfaces (p. 102).

Frege was also active with investigations into potential theory. In 1870 and 1871, he gave several talks to the Jena mathematical society on the derivation of laws of current.[60] This interest was preserved later: the *Nachlass* catalogue lists 9 pages on 'Potential' (p. 102) and 3 notebooks containing 54 pages on 'Hydrodynamics' (p. 102).

e) Frege's Consistent, Decades Long Anti-Weierstrass Stance

Frege harshly criticized Weierstrass' theory of real numbers in *Grundgesetze II*. This is taken to come out of the blue, but in fact Frege is a consistent critic of Weierstrass from *Grundlagen* on. That Frege's critical stance was this long-standing has been overlooked because his earliest shots were aimed at now-forgotten surrogates. Weierstrass was slow to publish, so textbooks that were taken to record his lectures were used as sources. This was true of Kossak [1872] and Biermann [1887], which Frege cites frequently. To read Frege's work as his readers would have, remember that to a nineteenth-century mathematician references to these two sat under a bright sign flashing 'Weierstrass'.

Consider first Biermann [1887]. Frege wrote a cranky draft review [OCN] and fired blunderbuss asides in a draft [DRC] of his review of Cantor [RC] and in his review [RH] of Husserl's *Philosophy of Arithmetic*. Biermann's book was not just any old analysis text pulled randomly off the shelf. It represented itself as, and it was received as, the first published presentation of the foundations of complex analysis worked out in Weierstrass' Berlin seminars.[61] Frege's complaint that

---

[59] Kratzsch [1979] p. 544–55; Schaeffer [1877] p. 24

[60] Schaeffer [1877] p. 18

[61] Contrary to the statement by the editors in [NS] p. 81, Biermann was not a student of Weierstrass, but just someone who had obtained notes of Weierstrass' lectures and used them as the basis of a textbook. However, Frege, like most mathematicians of the time, took Biermann's self-presentation at face value. In [GGII], for example, Frege says that in his criticism of Weierstrass he is drawing on only three sources – Kossak [1872], Biermann [1887] and some handwritten notes from Weierstrass' lectures. ([Gz II] p. 149).

Biermann made mistakes he could have avoided had he understood the definition of number in *Grundlagen*, his grumpy 'Could the author have learned this from Mr. O. Biermann?' in the review of (Weierstrass influenced) Husserl ([RH] p. 205) and Frege's general air of wounded exasperation are in part explained by the fact that these shots at Biermann [1887] were shots at the Weierstrass school as a whole, which had long failed to address what Frege regarded as his well-founded criticism.

Kossak too presented his textbook as based on Weierstrass' lectures, and it was generally so taken.[62] Kossak [1872] is cited four times in *Grundlagen*. One of the citations (p. 74) refers to the by then widespread use of 1-1 correlation as a criterion for numerical identity. Frege's discussion of this is not obviously critical, but under examination turns out to be a jab at Kossak/Weierstrass for inadequate rigor. Three of the citations are clearly critical ([FA] p. 106 p. 112–113). Two are variations on the theme that Kossak/Weierstrass 'proceeds as if mere postulation were equivalent to its own fulfillment' (p. 106, p. 111). One appears in a section on geometric and temporal interpretations of complex numbers rejected because they 'import something foreign to arithmetic'. (§103 p. 112–113) In this context he adds that the Weierstrass account:

> ... appears to avoid introducing anything foreign, but this appearance is only due to the vagueness of the terminology. We are given no answer to the question, what does $i + 1$ really mean? Is it the idea of an apple and a pear, or the idea of toothache and gout?... Kossak's statement once again does not yet give us any definition at all of complex number, it only lays down the general lines to proceed along. But we need more; we must know definitely what 'i' means, and if we do proceed along these lines and try saying it means the idea of a pear, we shall again be introducing something foreign into arithmetic. ([FA] p. 113)

This is an open slap at the Weierstrass program, and to appreciate the passage we need to see what sort of a slap it is. Earlier we encountered a commonplace of the Weierstrass circle: Weierstrass' methods were superior to Riemann's because *even if* the Riemann methods could be worked out rigorously, Weierstrass's method would *still* have the edge because it does not import anything foreign into arithmetic. Frege's rejoinder is that the purported advantage is illusory. Indeed, Weierstrass represents a step *backward*; the geometrical approach to analysis has at least the advantage that it has proven to be a fruitful way to organize the subject. By shunning Riemannian principles to keep out foreign elements and then failing to keep out foreign elements, Weierstrass has given up something of value and attained nothing in return.

---

[62] So, for example, the review of the book in *Jahrbuch über die Fortschritte der Mathematik* emphasizes that Weierstrass was the source. Though this is something 'everyone knew', and therefore by universal instantiation Frege knew it, it is still comforting to be able to tie Frege directly to the information: Schröder [1873] (p. 8) quotes Kossak [1872] (p. 16) and states that the view came from Weierstrass' lectures. This occurs on a page Frege cites ([FA] p. 74 fn.) in a footnote in which Kossak [1872] (p. 16) is also cited.

That Frege had Weierstrass in mind in critical discussions even as early as *Grundlagen* is reinforced by some remarks in 1906, referring back to 1884:

> Why must I always repeat the same arguments? Twenty-two years ago, in my Foundations of Arithmetic §§34–48 I presented at length what must be considered when dealing with this question... At that time, twenty-two years ago and also afterwards, even a Weierstrass could utter a farrago of balderdash when talking about the present subject. ([RT] p. 345)

In his lectures of 1914 (p. 215–223) Frege continues attacking Weierstrass by name. Here he finds room for the heavyhanded, unfunny jokes that became part of his signature in the later years. He takes up a few words of Weierstrass': 'A number is a series of things of the same kind' and characteristically presses them *ad absurdum*:

> A train is a series of objects of the same kind which moves along rails on wheels. It may be thought that the engine is nevertheless something of a different kind. Still that makes no essential difference. And so such a number comes steaming here from Berlin. ([LIM] p. 216)[63]

In the *Grundgesetze II* discussion, Frege restricts himself to Weierstrass' account of natural numbers; Frege doesn't require more because, he asserts, the defects he identifies in Weierstrass' account of natural numbers will carry over to his account of the reals, and because he takes Cantor's account of the reals to have superseded Weierstrass'. So his critique of Cantor is also a supplementary critique of Weierstrass. *Inter alia*, Frege objects that their accounts of the reals do not allow us to make sense of *applications*.

Frege criticized Weierstrass and surrogates because, in his view, they were not as rigorous as they purported to be on the concept of number. But Frege also has independent complaints about real and complex numbers that would survive even if Weierstrass took over Frege's theory of *natural* numbers *in toto*. We should not be blinkered because Frege–like Dedekind and Weierstrass–viewed appeals to geometric intuition as 'introducing something foreign to arithmetic'. Frege differed from these three on the value of geometry as an autonomous discipline and the importance of incorporating the potential for applications into mathematical frameworks. In these respects, Frege was solidly in the orthodox Riemann camp. This makes sense given his view that 'scientific workshops are logic's field of observation.'

---

[63] Lest we fail to twig onto the geographic jab, Frege repeats it a few paragraphs later:

> This afternoon at approximately 5:15 an express train, which is likewise a number, arrives at Sall station from Berlin...the result of multiplying our series of books by the Berlin express would again have to be a series of things of the same kind. ([LIM] p. 216)

f) Frege as Charitable Reader of Riemann

Frege's discussion of Riemann in an unpublished fragment from around 1898–1903 ([LDM] p. 158) is interesting for reasons that are sufficiently involved to require development elsewhere.[64] Here I'll restrict myself to a lighthearted observation. In these lines Frege defends Riemann against the charge of confusing sign and signified. Contrary to Frege's usual practice, he strains to read Riemann's words charitably. This is remarkable, given that we are talking about Frege, especially late Frege. This is one of the only places in Frege's writing where Frege actually goes out of his way, breaking the flow of his own arguments, to say anything complementary about anyone.

g) Infinitesimals, Magnitude, and Negative Numbers

A complicated topic in outline: Several of Frege's manoevres in his account of real numbers – often seen as idiosyncratic – are not unexpected from a Riemannian. Frege's account of the reals as magnitudes has striking affinities to Riemann's account of magnitude as distinguished from number and as conceptually connected with measurement, as articulated in his *Habilitation* lecture (Riemann [1854/1868]).[65] Also, Frege's account of negative numbers in terms of the converse of a relation (also widely regarded as a Fregean quirk) is the account in Abbe's notes on Riemann's lectures on complex analysis. Finally, Frege's puzzling view of infinitesimals – that they are acceptable if introduced by contextual definition – is in opposition to a known Weierstrassian position.

h) Geometry and Spatial Intuition: 'Changes of Space-Element'

Frege's principal non-foundational pre-occupation – analytic geometry in a 'pure projective' vein – is another signal of his preferred mathematical style. Here I'll just note three connected points that are relevant. A) Engagement with geometry is a further signal of Frege's mathematical allegiances. Indeed, in the most sophisticated geometric work Frege was engaged in the late 1890s – the study of 'contact transformations' – we find writers drawing the same Riemann vs. Weierstrass contrasts that were common in complex analysis, in connection with the problem families (Abelian integrals) that Frege is immersed in. B) Here too the contrast of 'conceptual' and 'computational' is widely discussed, in a way that allies the 'geometrical' approach with 'conceptual' thinking. C) Frege's favoured approach to geometry exploited a proof technique called 'changing the space-element', using mappings between different choices of basic geometric building blocks (lines and points, for example, or points and spheres). This development was bound

---

[64] The fragment is written in such a way as to make it fairly clear that it was originally intended as part of *Grundgesetze* II part III before being edited out of the final manuscript. That is, it was meant for the same section of *Grundgesetze II*, in which the uncontrolled rant about Weierstrass appears.

[65] José Ferreirós' contribution to this volume is an eye-opening discussion of further philosophical ramifications of Riemann's work in this area.

up with a kinematical reinterpretation of Kantian intuition, initiated by Helmholtz, and inspired by Riemann's famous *Habilitationsvortrag* that de-emphasized construction and replaced it with transformation invariance.[66] Frege's occasional remarks on geometry and intuition suggest he understood intuition in this transformation-based way.

IV Summing up and Looking Forward: Logic and Mathematical Practice

The work adumbrated above, plus work I present elsewhere on Frege's views on pure geometry, serve us a picture of Frege as professionally engaged with a cluster of questions relative to which the relation between geometry and pure arithmetic and analysis was fundamentally in question, in an environment where these questions were typically addressed in a distinctively Riemannian style. In some of these cases, such as the extent of validity of the Dirichlet principle, the problems remained the object of active investigation throughout the period when Frege was composing *Grundgesetze* I and II. Hence the diagnostic questions of what depends on intuition and what on logic, were especially pressing.

Though a complete treatment requires another paper, it will be worthwhile to reflect quickly on how a richer conception of Frege's context touches on our sense of what Frege was setting out to do, and what the significance of the resulting position might be. We know – it is a cliché – that Frege sought to derive, using logic alone, all those parts of mathematics that are not founded on geometry. To put it another way: Frege sought to derive all of arithmetic and analysis from logic. When such variations on the cliché are uttered, it is standard to assume that the questions 'What is mathematics / what is analysis / what is arithmetic?' are unproblematic. Of course, the philosophical question 'what is mathematics?' may be tricky, but it is assumed there is nothing to worry about in the simple descriptive questions: What is the target? Just what are you setting out to prove? But in a case where there is widespread, principled disagreement over just what 'analysis' *is* we can't be naïve about this question. Does 'analysis' include a definition of a Riemann surface and exploit conformal mapping or not? Is it engaged directly with applications? Does it contain a rigorous version of Dirichlet's principle? Are functions fixed globally by specifying singularities and relying on indirect existence theorems, or defined locally by power series? In short, does 'analysis' mean analysis as understood in Riemann's tradition, or in Weierstrass'? For Frege, 'complex analysis' is *Riemann's* complex analysis. So understood, the Fregean attitude takes a direction similar to that of Dedekind, as successors to the tradition of 'Gaussian' rather than 'Weierstrassian' rigour,

---

[66] This point is developed by Michael Friedman. (Friedman [2000b])

which we might sum up in the slogan: 'Rigour, yes, but also clarity, conceptual simplicity and methodological awareness in lieu of brute calculation'.

There are consequences implicit in the above for our assessment what, for Frege, logic is (to the extent that 'logic' includes an account of what the fundamental logical concepts are). Frege's views on this point are rarely made explicit, and the remarks he does make don't indicate a stable position. He suggests that Dedekind's foundational work does not supported Dedekind's view that arithmetic is logic because (among other reasons) Dedekind's primitives 'system' and 'thing belongs to a thing' 'are not usual in logic and are not reduced to what is recognized as logical' ([BLA] VIII). This is an odd objection to make, and not just because it seems to rest on a brute appeal to a logical tradition that Frege himself is upending. More to the point is a *tu quoque:* the mathematical concept of *function*, which Frege takes as basic and unreduced (with 'concept' defined in terms of 'function'), was then no more usual nor more generally recognized as logical. As we've seen, adopting a general concept of function as basic was seen as methodologically charged in the circles Frege knew. What is Frege's rationale? An explicit answer is not forthcoming. Even when the issue of sharply identifying logical notions is addressed, as in [FGII], Frege says only that the answer won't be easy. In 'Function and Concept', he articulates the logical concept of function, denies that functions are expressions, and describes the pressures from science (such as complex arguments) forcing the extension of the *bedeutung* of 'function', but doesn't state a principled basis for choosing primitives delivering 'function' as the right choice.

But the absence of an explicit rationale notwithstanding, we've already seen Frege's most compelling reason for opting for the concept of function as a basis: its value in 'scientific workshops, logic's true field of observation'. Indeed, he appeals to function/argument decomposition to explain how logical reasoning extends knowledge:

> ... of all the ways to form concepts, [listing characteristics] is one of the least fruitful. If we look through the definitions given in this book, we shall scarcely find one that is of this description. The same is true of the really fruitful definitions in mathematics, such as that of the continuity of a function. What we find in these is not a simple list of characteristics; every element is intimately, I might almost say organically, connected with the others.... [W]ith the more fruitful type of definition ... [t]he conclusions we draw extend our knowledge..., and yet they can be proved by purely logical means ... ([FA] p. 100–101)[67]

This language pops up even in those moments where Frege is taken to be the most 'philosophical', as in his account of the requirements upon an adequate specification of an object (the 'Caesar problem'). Among the (many) reasons Frege cites for crafting definitions as he does is a point about the need for

---

[67] That Frege is gesturing at function–argument decomposition with these metaphors is evident from his use of these metaphors elsewhere, as I explain in (Tappenden (1995b)).

representation-independent definitions, and a range of different presentations of an object, to extend knowledge:

> If one were to say: q is a direction if it is introduced by means of the definition offered above, then the way in which the object q is introduced would be treated as a property of it, which it is not... If this way out were chosen, it would presuppose that an object can only be given in one single way... all equations would come down to this: that whatever is given to us in the same way is the same. But this is so self-evident and so unfruitful that it is not worth stating. The multitude of meaningful uses of equations depends rather on the fact that something can be reidentified even though it is given in a different way. ([FA] §67)[68]

There is more to be said, but this will provide a first foothold, indicating how thoroughly Frege's method was entwined with ongoing mathematical inquiry. His preference for general over piecemeal definitions, his choice of 'function' as a basis for his logic, his treatment of magnitude and infinitesimals, and even his account of objects: these and other points of his foundations are rich with significance for the mathematics around him. These examples reveal the extent to which Frege's 'acid test' of scientific fruitfulness for concepts was embedded in his philosophy as a whole. His philosophical treatment would not have appeared to those around him to be mathematically neutral.

To be sure, any connection to Frege is a bonus: the methodology of Riemannian mathematics, as articulated in diverse ways by subsequent followers, is of considerable philosophical interest independently of any links to the figures studied in 'official' histories of philosophy. But the hook to Frege does induce a re-evaluation of how we conceive the relations between the history of analytic philosophy and the history of science and mathematics. A tacit assumption apparently guiding much recent philosophy of mathematics is that it requires little or no concern for the history and contemporary status of the frontiers of mathematical investigation. Those of us who adopt the opposing stance that our choice of fundamental concepts should be sensitive to the value of those concepts as revealed in practice can not only draw upon the rich Riemannian tradition but can also take heart from finding its echoes at the well-springs of the analytic tradition.

---

[68] The mathematical work that fits most closely these remarks is Dedekind's Riemann-inspired treatment of the theory of ideals, in the wake of Kummer's prior account. I don't discuss this work here, but rather in Tappenden [2005].

'Axiomatics, Empiricism, and Anschauung in Hilbert's Conception of Geometry: Between Arithmetic and General Relativity'

Leo Corry

To what extent the philosophy of mathematics of any individual mathematician is relevant to historically understanding his mathematical work, and to what extent his mathematical work has any bearing in understanding philosophical issues related with mathematics, are questions that have different meanings and have to be approached differently when they refer to different mathematicians. Take, for example, Descartes and Frege. These two thinkers can be considered philosophers in the strict sense of the word, with philosophical interests going well beyond the strict scope of mathematics, each of them in his own way. They devoted much of their time and efforts to develop coherent, well-elaborated philosophical systems, and their writings turned them into philosophers in the eyes of the philosophical community. Their philosophical systems are directly relevant to addressing central questions pertaining to the nature of mathematical knowledge, but they were not intended exclusively as answers to specific problems in the philosophy of mathematics. And besides their intense involvement with philosophical questions, both Descartes and Frege contributed positive mathematical results of various kinds, albeit of different overall impact on mathematics at large, and while working under quite different professional circumstances. A natural question that the historian may be easily led to ask in relation to these two thinkers concerns the mutual relationship between the philosophical systems they developed and the mathematics that each of them produced. One way to answer this question is by investigating, separately, the philosophy and the mathematics of each of them, and then trying to articulate the said relationship.

Descartes and Frege, however, are far from representative of the mainstream mathematician in any given period. Most mathematicians devote little or no effort to philosophical questions in general, and, in particular, they devote little time to formulate coherent philosophical systems. A historian investigating the work of an individual mathematician of the more mainstream kind may attempt to reconstruct her putative philosophy by analysing her mathematical work, and by trying to illuminate the philosophical preconceptions underlying it. The historian may likewise be led to ask for the roots and the background of the philosophical views thus embodied in the mathematician's work. In cases where the mathematician in question has also left some philosophical or quasiphilosophical texts, the historian may try to assess their value and their relationship with her actual mathematical practice. This exercise may be more or less interesting according to the individual mathematician involved, and in many cases it may be of rather limited consequence.

The case of David Hilbert is particularly appealing when seen from the perspective of the spectrum whose two extreme points I outlined above. His contributions to the foundations of geometry led to momentous changes in the most basic conceptions about the nature of this discipline and of mathematics in general. He developed close connections and meaningful intellectual interchanges with influential philosophers in Göttingen, such as Leonard Nelson and Edmund Husserl. He made important contributions to the foundations of logic and of arithmetic, and the finitist program he initiated in this context turns him into a natural focus of philosophical interest. From the point of view of the philosophical discourse about mathematics in the twentieth century Hilbert's name remained intimately linked to the idea of a formalist conception of mathematics as a mainstream interpretation of the nature of this discipline. Hilbert was prone to express ideas of a philosophical or quasiphilosophical tenor and a great many of his pronouncements have remained on the written record. These pronouncements are rich in ideas and they are very illuminating when trying to reconstruct the intellectual horizon within which he produced his mathematics. Still, the picture that arises from this abundance of activities and sources is by no means that of a systematic philosopher. Nor is there any solid reason to expect it to be so. After all, Hilbert was a working mathematician continually involved in many threads of research activity in various fields of mathematics, pure and applied, and he had neither the time nor, apparently, the patience and the kind of specifically focused interest, to devote himself to the kind of tasks pursued by philosophers.

Rather than trying to construe a fully coherent picture of what would be a putative philosophy of mathematics of Hilbert—similar to what one could do for Descartes or Frege, for instance—that would allow analysing his entire mathematical horizon from a single, encompassing perspective, in the present chapter I will suggest that it is more convenient to speak in terms of his 'images of mathematics' and their development throughout the years, and to analyse—in terms of the latter—separate aspects of the enormous body of scientific knowledge that can be attributed to him. In this chapter I will focus the discussion on Hilbert's approach to geometry.

Elsewhere, I have elaborated in greater detail the distinction between 'body' and 'images' of mathematical knowledge,[1] and the possible ways to use these concepts in investigating the history of mathematics. For the purposes of the present discussion it will suffice to point out that this is a flexible, schematic distinction focusing on two interconnected layers of mathematical knowledge. In the body of mathematics I mean to include questions directly related to the subject matter of any given mathematical discipline: theorems, proofs, techniques, open problems. The images of mathematics refer to, and help elucidating, questions arising from the body of knowledge but which in general are not part of, and cannot be settled within, the body of knowledge itself. This includes, for instance, the preference of a mathematician to declare, based on his professional expertise, that a certain open problem is the most important one in the given discipline, and that the way to solve it should follow a certain approach and apply a certain technique, rather than any other one available or yet to be developed. The images of mathematics also include the internal organization of mathematics into subdisciplines accepted at a certain point in time and the perceived interrelation and interaction among these. Likewise, it includes the perceived relationship between mathematics and its neighbouring disciplines, and the methodological, philosophical, quasiphilosophical, and even ideological conceptions that guide, consciously or unconsciously, declared or not, the work of any mathematician or group of mathematicians.

Examining a mathematician's work in terms of the body and the images of mathematical knowledge allows us to focus on the role played by philosophical ideas in his work, without thereby assuming that these ideas must be part of a well-elaborated system that dictates a strict framework of intellectual activity. Rather, one may consider these images as a historically conditioned, flexible background of ideas, in a constant process of change, and in mutual interaction with the contents of the body of mathematics, on the one hand, and with external factors, on the other hand. The images of mathematics of a certain mathematician may contain tensions and even contradictions, they may evolve in time and they may eventually change to a considerable extent, contradicting at times earlier views held by her. The mathematician in question may be either aware or unaware of the essence of these images and the changes affecting them.

The body/images scheme turns out to be useful for analysing Hilbert's conceptions, especially concerning his putative, 'formalist' views on mathematics. As already said, to the extent that Hilbert's name is associated with any particular philosophical approach in mathematics, that approach is formalism. This association, however, is rather misleading on various counts. For one thing, it very often conflates two different meanings of the term 'formalism'. Thus, from about 1920 Hilbert was indeed involved in a program for proving the consistency of arithmetic

---

[1] Corry 2001; 2003.

based on the use of strictly finitist arguments. This program was eventually called the 'formalist' approach to the foundations of mathematics, and it gained much resonance as it became a main contender in a well-known and unusually heated debate known as 'the crisis of foundations' in mathematics. Associating Hilbert with this sense of the word 'formalism' is essentially correct, but it says very little about Hilbert's images of mathematics. The term formalism, at any rate, was not used by Hilbert himself in this context, and it is somewhat misleading. 'Hilbert's Programme' and 'Finitism' have become accepted, and much more appropriate, alternatives.[2] 'Formalism', however, is far from accurately describing Hilbert's *images* of mathematics.

Indeed, a second meaning of the word formalism is associated with the general attitude towards the practice of mathematics and the understanding of the essence of mathematical knowledge that gained widespread acceptance in the twentieth century, especially under the aegis of the Bourbaki group. As these two meanings came to be conflated in an interesting historical process, Hilbert, the formalist in the more reduced sense of the term, came to be associated also with this second sense, not the least because of Bourbaki's efforts to present themselves as the 'true heirs of Hilbert'. Thus Jean Dieudonné explained the essence of Hilbert's mathematical conceptions in a well-known text where he referred to the analogy with a game of chess. In the latter, he said, one does not speak about truths but rather about following correctly a set of stipulated rules. If we translate this into mathematics we thus obtain the conception of Hilbert: 'mathematics becomes a *game*, whose pieces are graphical *signs* that are distinguished from one another by their form.'[3]

On the face of it, one should not be too surprised to realize how widespread the image of Hilbert the formalist became. It is not only the dominance of formalist approaches in twentieth-century mathematics, and more specifically of the Bourbakist approach. It is also that, given this dominance, Hilbert's important early work on the foundations of geometry could easily be misread in retrospect as a foremost representative of this trend in mathematics. And yet, the historical record contains as many important contributions of Hilbert that could hardly be seen as embodying any kind of formalist approach. This is especially—but not exclusively—the case when one looks at his contributions to physics. It is thus remarkable that this side of Hilbert's works was in many cases systematically overlooked as it did not fit his widespread image as the paradigmatic twentieth-century mathematical formalist.

This view of Hilbert as a formalist, in the more encompassing sense of the word, has been consistently criticized for a few years now. Likewise, it is only relatively recently that the real extent and depth of Hilbert's involvement with

---

[2] Detlefsen 1986.

[3] Dieudonné 1962, 551. A similar view is presented in Kleene 1952. Also Weyl (1925–27, 127) refers to the chess metaphor in describing Hilbert's quest to '*formalize* mathematics', but he clearly states that Hilbert followed this approach in order to secure 'not the *truth*, but the *consistency* of the old analysis.'

physics became well known (Corry 2004). Nevertheless, some of his specific contributions to physics were never a secret. This is particularly the case with his solution of the Boltzmann equation, on the one hand, and, on the other, with the formulation of the field equations of general relativity (GTR) followed by a continued involvement with this discipline. Curiously enough, the well-known, enormous impact of GTR on the perceived relationship between geometry and physics has consistently been described in the literature, not infrequently coupled to the claim that GTR turned geometry into a branch of physics. How could then a mathematician like Hilbert be intensely involved in the decisive stages of the development of this discipline around 1915 and at the same time hold a purely formalistic view of geometry from at least as early as 1900? This tension was more easily ignored than explained by anyone who accepted at face value the description of Hilbert as the ultimate formalist.

As will be seen below, and contrary to that view, Hilbert's actual conceptions about the essence of geometry throughout his career fitted very naturally the kind of intimate association postulated by GTR between this discipline and physics, and as a matter of fact this is what led him to be among the first to focus on some of the important issues arising from GTR in regard with this association. It is noteworthy, however, that Einstein himself took for granted the kind of separation between mathematics (particularly geometry) and the physical world that formalist views of geometry tend to favour, and he was prone to attribute such a view to Hilbert. In his talk, 'Geometry and Experience', presented at the Berlin Academy of Sciences on January 27, 1920, Einstein famously asserted that:

> Insofar as the theorems of mathematics are related to reality, they are not certain; and insofar as they are certain, they are not related to reality. (Einstein 1921, 4)

In his view, this relatively recent conception 'first became widespread through that trend in recent mathematics which is known by the name of "Axiomatics". Thus, even though Einstein did not say it explicitly, the context makes it clear the he was referring here to the axiomatic approach developed by Hilbert, as he understood it. For very different reasons both Einstein and Dieudonné coincided in associating Hilbert's conception of geometry with mathematical formalism.

In the present chapter I discuss some of the central images of mathematics, and particularly of geometry, espoused by Hilbert throughout his career. Amid significant changes at several levels, these images never envisaged formalist considerations as a possible way to explain the essence of geometry. In fact, perceptual experience and intuition (*Anschauung*) in the Kantian sense of the term (as Hilbert understood it) are the two main motives of Hilbert's images of elementary geometry. The axiomatic analysis of scientific theories, which provides the methodological backbone and the main unifying thread of Hilbert's overall images of science, was not meant as a substitute for these two main components. Rather, it embodied Hilbert's preferred way to organically combine and articulate them in the framework of a regulative 'network of concepts' (*Fachwerk von Begriffe*) that helps clarify their logical interrelationship.

Perceptual experience and intuition appear in constant interaction in Hilbert's writings and lectures, with their relative importance and the kind of interplay affecting them undergoing subtle changes along the years. One of the interesting consequences of Hilbert's involvement with GTR was that the delicate balance that existed between experience and *Anschauung* in Hilbert's images of geometry was finally disrupted in favour of experience, and decidedly away from intuition.

Before entering into the details of this discussion, and in order to conclude the introduction, it is pertinent to introduce here a quote of Hilbert from around 1919, the time when he began to work out in collaboration with Bernays his finitist program for the foundations of arithmetic. Even if the formalist aspects of this program (in the more restricted sense of the term) may have already begun to emerge at this stage in his work, they were certainly circumscribed to the question of the proof of consistency for arithmetic. As for the essence of mathematical knowledge in general, Hilbert stated a view totally opposed to that attributed to him many years later by Dieudonné. Thus Hilbert said:

> We are not speaking here of arbitrariness in any sense. Mathematics is not like a game whose tasks are determined by arbitrarily stipulated rules. Rather, it is a conceptual system possessing internal necessity that can only be so and by no means otherwise. (Hilbert 1919–20, 14)

Roots and Early Stages

Hilbert's work in geometry, its contents, its methodology and the conceptions associated with the discipline, has as its focal point the introduction and further development of the new axiomatic approach that came to be associated with his name. For all of its innovative aspects, this approach had deep roots in a complex network of ideas developed in the second half of the nineteenth century in research on the foundations of fields as diverse as geometry, analysis, and physics, which deeply influenced him. Moreover, the essentially algebraic outlook that permeated all of Hilbert's work (including his work in fields like analysis and the foundations of physics) played a major role in shaping his axiomatic approach.

Prior to his arrival in Göttingen in 1895, Hilbert had lectured on geometry at Königsberg, and it is interesting to notice how some of the topics that will eventually become central to his mathematical discourse already emerge at this early stage. Thus, for instance, in a lecture course on Euclidean geometry taught in 1891 Hilbert said:

> Geometry is the science dealing with the properties of space. It differs essentially from pure mathematical domains such as the theory of numbers, algebra, or the theory of functions. The results of the latter are obtained through pure thinking ... The situation is completely different

in the case of geometry. I can never penetrate the properties of space by pure reflection, much the same as I can never recognize the basic laws of mechanics, the law of gravitation or any other physical law in this way. Space is not a product of my reflections. Rather, it is given to me through the senses. (Quoted in Hallett and Majer (eds.) 2004, 22)

Separating mathematical fields into two different classes, one having its origin in experience and one in pure thinking, is a motive that had traditionally found a natural place in German mathematical discourse for many generations now. It had been particularly debated, and adopted with varying degrees of commitment, among the Göttingen mathematicians since the time of Gauss (Ferreirós 2006). In the passage just quoted, Hilbert fully endorsed the separation, and thus geometry and all physical domains appear here together on the same side of a divide that leaves in the second side those purely mathematical disciplines for which 'pure thinking', provides the main foundation. Thus, while Hilbert's views images geometry reflect strongly empiricistic conceptions, concerning arithmetic he adopted in the early years of his career an essentially logicist point of view, strongly influenced by Dedekind (Ferreirós 2007).

Within a strongly empiricistic conception of geometry such as expressed here, the axioms play a clearly defined role that is not different from the role they might play for any other *physical* discipline. Thus, it should come as no surprise that, when Hilbert read Hertz's book on the principles of mechanics very soon after its publication in 1893, he found it highly congenial to his own conceptions about the role of axioms in geometry and became strongly influenced by it. The axioms of geometry and of physical disciplines, Hilbert said in a course of 1893, 'express observations of facts of experience, which are so simple that they need no additional confirmation by physicists in the laboratory.'[4]

The empiricist images characteristic of his early courses, especially concerning the status of the axiom of parallels, is also manifest in his consistent references to Gauss's experimental measurement of the sum of angles of a triangle formed by three mountain peaks in Hannover.[5] Hilbert found Gauss's measurements convincing enough to indicate the correctness of Euclidean geometry as a true description of physical space. Nevertheless, he envisaged the possibility that some future measurement would yield a different result. The example of Gauss's measurement would arise very frequently in his lectures on physics in years to come, as an example of how the axiomatic method should be applied in physics, where new empirical facts are often found by experiment. Hilbert stressed that the axiom of parallels is likely to be the one to be modified in geometry if new experimental discoveries would necessitate so. Geometry was especially amenable to a full axiomatic analysis only because of its very advanced stage of development and elaboration, and not because of any other specific, essential trait concerning

---

[4] Quoted in Hallett and Majer (eds.) 2004, 74.
[5] See, for instance Hallett and Majer (eds.) 2004, 119–120.

its nature that would set it apart from other disciplines of physics.[6] Hilbert's empiricist image of geometry is epitomized in the following quotation taken from a 1898–99 lecture course on the foundations of Euclidean geometry:

> We must acknowledge that geometry is a *natural science*, but one whose theory can be *described as perfect*, and that also provides an example to be followed in the *theoretical treatment* of other natural sciences. (Quoted in Hallett and Majer (eds.) 2004, 221. Emphasis in the original)[7]

Grundlagen der Geometrie

A main topic in Hilbert's involvement with geometry between 1893 and 1899 was a detailed enquiry of the mutual relations between the main theorems of projective geometry and, specifically, of the precise role played by continuity considerations in possible definitions of purely projective co-ordinates and a purely projective metric. Foundational questions of this kind had been thoroughly investigated throughout the century by mathematicians such as Klein, Lie, Veronese, and, more recently, Ludwig Wiener (to mention just a few). The role of continuity considerations in the foundations of analysis and arithmetic had been systematically investigated by Dedekind in various works that Hilbert's was well aware of. As Dedekind developed a distinctly axiomatic way to pursue his own analysis of this question, there can be little doubt that his works provided an additional catalyst for Hilbert's own ideas that reached final consolidation by 1899.[8]

That was the year of publication of *Grundlagen der Geometrie*, the text of which elaborated on a course just taught by Hilbert. In the notes to a different course taught the same year, this one on mechanics, we find a balanced and interesting combination of the various topics that inform the basis of Hilbert's views on geometry. In the first place, there is the role of full axiomatization as a means for the proper mathematization of any branch of *empirical* knowledge:

> Geometry also [like mechanics] emerges from the observation of nature, from experience. To this extent, it is an *experimental science*. . . . But its experimental foundations are so irrefutably and so *generally acknowledged*, they have been confirmed to such a degree, that no further proof of them is deemed necessary. Moreover, all that is needed is to derive these foundations from a minimal set of *independent axioms* and thus to construct the whole edifice of geometry by *purely logical means*. In this way [i.e., by means of the axiomatic treatment] geometry is turned into a *pure mathematical science*. In mechanics it is also the case that all physicists recognize its most *basic facts*. But the *arrangement* of the basic concepts is still subject to

---

[6] Hallett and Majer (eds.) 2004, 72.

[7] See also, on p. 302: 'Geometry is the most perfect of (*vollkommenste*) the natural sciences.'

[8] For details, see Corry 2004, 37–40.

changes in perception ... and therefore mechanics cannot yet be described today as a *pure mathematical* discipline, at least to the same extent that geometry is. (Quoted in Corry 2004, 90)

At the same time, however, the choice of axioms is also guided by the pervasive and subtle concept of 'Anschauung', whose actual nature, however, is preferably left out of the discussion:

> Finally we could describe our task as a logical analysis of our faculty of intuition (*Anschauungvermögen*). The question if our space intuition has a-priori or empirical origins remains nevertheless beyond our discussion.

This dilemma, whether the origin of the axioms of mathematics is empirical or is related to some kind of Kantian a-priori intuition, is never fully resolved in Hilbert's early lectures. A strong connection with other natural sciences is a main image of Hilbert's conception of geometry, and it continually strengthens his inclination to emphasize, at the epistemological level, the origins of axioms in perceptual experience. On the other hand, the unity of mathematics, a second main pillar of Hilbert's images of the discipline, underlies the stress on the connection with arithmetic and with a-priori intuition. It thus seems as if having a properly axiomatized version of geometry relieves Hilbert from the need to decide between these two alternatives: axiomatized geometry may equally well serve a thoroughly empiricistic or an aprioristic account of the essence of this discipline.

When Hilbert published his full-fledged axiomatized analysis of the foundations of geometry in *Grundlagen der Geometrie*, in the framework of a *Festschrift* to celebrate the unveiling of the Gauss–Weber monument in Göttingen, it was more than appropriate to open with a festive quotation from Kant. This famous quote states:

> 'All human knowledge thus begins with intuitions, proceeds thence to concepts and ends with ideas'.

One might attempt to analyse in detail the philosophical reasons for Hilbert' choice of this sentence and how the various terms (intuitions, concepts, ideas) relate, or perhaps do not relate, to Hilbert's own conceptions, to his declared views, and to his practice. This would require a thorough discussion of the relevant Kantian texts, and, more importantly, of how Hilbert's contemporaries (and possibly Hilbert himself) understood these texts. I think, however, that this complex exercise would not justify the effort and would not, in itself, shed much light on Hilbert's views. Hilbert clearly wanted, in the first place, to pay a due tribute to the towering figure of his fellow Königsberger at this very festive event. It remains a matter of debate, what this sentence exactly meant for Hilbert and whether or not it faithfully describes his 'true motivations' or the philosophical underpinnings of his work. In fact, there is some irony in the specific sentence that Hilbert chose to use from the Kantian corpus, and in which, of all terms, 'experience' is not mentioned in any way. Intuitions, concepts and ideas – all of these appear in varying degrees of importance in Hilbert's philosophical discourse about mathematics and in his images of geometrical knowledge. But 'perceptual experience' is the one whose paramount epistemological significance was never called into question by Hilbert. It was also the one that his involvement with GTR was to reinforce even more.

In his thoroughgoing exploration of the foundations of Euclidean geometry and of the fundamental theorems of projective geometry and their interdependences, Hilbert saw the culmination of a process whereby geometry turns into a 'purely mathematical' discipline. The above-mentioned, traditional divide with number theory and analysis could thus be overcome and Hilbert's continued quest for unity in mathematics and in the sciences gained additional strength. A fully axiomatized version of geometry thus embodied a network of concepts preserving meaningful connections with intuition and experience, rather than a formal game with empty symbols.

At the technical level, Hilbert undertook in *GdG* several tasks that became cornerstones of all foundational activities in mathematics for decades to come. Thus, Hilbert presented a completely new system of axioms for geometry, composed of five separate groups, and put forth a list of concrete requirements that his system should satisfy: simplicity, completeness, independence, and consistency. We briefly look now at each of these requirements.

Unlike the other requirements, simplicity is one that did not become standard as part of the important mathematical ideas to which *GdG* eventually led. Through this requirement Hilbert wanted to express the desideratum that an axiom should contain 'no more than a single idea.' However, he did not provide any formal criterion to decide when an axiom is simple. Rather, this requirement remained implicitly present in *GdG*, as well as in later works of Hilbert, as a merely aesthetic guideline that could not be transformed into a mathematically controllable feature.

The idea of a complete axiomatic system became pivotal to logic after 1930 following the works of Gödel, and in connection with the finitist program for the foundations of arithmetic launched by Hilbert and his collaborators around 1920. This is not, however, what Hilbert had in mind in 1899, when he included a requirement under this name in the analysis presented in *GdG*. Rather, he was thinking of a kind of 'pragmatic' completeness. In fact, what Hilbert was demanding here is that an adequate axiomatization of a mathematical discipline should allow for a derivation of *all* the theorems already known in that discipline. This was, Hilbert claimed, what the totality of his system of axioms did for Euclidean geometry or, if the axiom of parallels is ignored, for the so-called absolute geometry, namely that which is valid independently of the latter .

Also, the requirement of consistency was to become of paramount importance thereafter. Still, as part of *GdG*, Hilbert devoted much less attention to it. For one, he did not even mention this task explicitly in the introduction to the book. For another, he devoted just two pages to discussing the consistency of his system in the body of the book. In fact, it is clear that Hilbert did not intend to give a direct proof of consistency of geometry here, but even an indirect proof of this fact does not explicitly appear in *GdG*, since a systematic treatment of the question implied a full discussion of the structure of the system of real numbers, which was not included. Rather, Hilbert suggested that it would suffice to show that the specific kind of synthetic geometry derivable from his axioms could be translated into the standard Cartesian geometry, if the axes are taken as representing the entire field of real numbers. Only in the second edition of GdG, published in 1903, Hilbert added an additional axiom, the so-called 'axiom of completeness' (*Vollständigkeitsaxiom*), meant to ensure that, although infinitely many incomplete models satisfy all the other axioms, there is only one complete model that satisfies this last axiom as well, namely, the usual Cartesian geometry.

The requirement on which I want to focus in the context of the present discussion is the requirement of independence, and in particular, the fact that Hilbert analysed the mutual independence of the groups of axioms rather than the mutual independence of individual axioms. The reason for this was that for Hilbert each of these groups expresses one way in which our intuition of space is manifest, and he intended to prove that these are independent of each other. Of course, he paid special attention to the role of continuity considerations and the possibility of proving that continuity is not a necessary feature of geometry. This latter fact was previously known, of course, from the works of Veronese, but Hilbert's study of non-Archimedean geometries appeared here as part of a more systematic and thorough approach.

The focus of Hilbert on the groups of axioms as expression of our spatial capacities or intuitions stresses the non-formalistic essence of the views underlying his entire research. Although the clear, formal building of geometry that emerges from his study is in itself an important mathematical achievement with broad consequences, it by no means indicates an interest in presenting geometry as a purely formal game devoid of inherent meaning. The opposite is true: this successful mathematical exercise was meant to provide conceptual support to the centrality he attributed to empirical and intuitive experience as a basis for geometry. Hilbert did not really elaborate, however, a clear philosophical analysis of space and geometry around these elements. Rather this presentation of geometry successfully embodied a set of images of mathematics where the two elements, empirical experience and some version of Kantian *Anschauung*, could be effectively accommodated.

Roughly simultaneously with his detailed treatment of geometry, Hilbert also advanced a cursory discussion of the foundations of arithmetic, in a talk delivered in 1899 under the title of 'On the Concept of Number'.[9] Hilbert opened his discussion by stating that in arithmetic one is used mostly to the 'genetic' approach for defining the various systems of numbers. He was evidently referring to Dedekind's stepwise construction, starting from the naturals, and successively adding those new numbers that allow extending the operations so that they become universally applicable. The last step in this process is the definition of the real numbers as cuts

---

[9] Hilbert 1900.

of rationals.[10] Hilbert praised the advantages of this approach, but at the same time he intended now to propose that also arithmetic, like geometry, could and should be axiomatically built.

The axiomatic system he proposed for the real numbers is based on ideas he also used in *GdG* and by means of which the real numbers are characterized as an ordered, Archimedean field. To this characterization, however, Hilbert added now, under the heading of 'axioms of continuity', a new condition, namely, the already mentioned axiom of completeness (*Vollständigkeitsaxiom*). Like in geometry, the completeness of his axiom system (not to conflate with the axiom of completeness) was not a property he would know exactly how to handle, and he thus remained silent in relation to it after having mentioned it in the opening passages. Concerning the proof of consistency he simply stated that 'one needs only a suitable modification of familiar methods of inference',[11] but he did not provide further details about the kind of modification that he had in mind.

This talk of Hilbert has been repeatedly mentioned as a harbinger of the views that he would develop later concerning the foundations of arithmetic and of logic.[12] It would be beyond the scope of the present chapter to discuss that point. The lecture is relevant for the present discussion for the contrast it presents between the genetic and the axiomatic points of view. Hilbert found both of them to be legitimate, and as playing important, different roles. At the same time, however, he clearly stated that the logical soundness and the foundational stability of arithmetic is provided, above all *but not exclusively*, by the axiomatic method. At this stage of his career, Hilbert's views on arithmetic were still strongly influenced by Dedekind's logicistic attitudes, and this influence is clearly felt in this talk. However, also this aspect of his conceptions was to change, and in lecture courses he would teach in Göttingen over the next years, he preferred to stress the foundational contribution of intuition, in the sense of *Anschauung*, as part of the stability and soundness that the genetic method provided to arithmetic via the axiomatic method.

Lectures on the Axiomatic Method – 1905

In the period immediately following the publication of *GdG* Hilbert occupied himself briefly with research on the foundations of geometry, and so did some of his students, prominent among whom were Max Dehn and Georg Hamel. At the same time, Ernst Zermelo, who had arrived in Göttingen in 1897 in order to complete his *Habilitation* in mathematical physics, started now to address questions

---

[10] Dedekind 1888. Cf. Ferreirós 1999, 218–224.

[11] Hilbert 1900 (1996), 1095.

[12] See, for instance, Ewald (ed.) 1999, 1090–1092.

pertaining to the foundations of arithmetic and set theory. It was only with the publication of Russell's paradox in 1903 that these latter topics started to receive serious attention in Göttingen. It seems as if Hilbert had initially expected that the difficulty in completing the full picture of his approach to the foundations of geometry would lie in dealing with assumptions such as the *Vollständigkeitsaxiom*, but he now realized that the actual problems lay in arithmetic and even perhaps in logic. It was at this point that he started to seriously consider the possible use of the axiomatic method as a way to establishing the consistency of arithmetic.[13] Still, significant work was done only by Zermelo, who had just started working more specifically on open problems of the theory of sets, such as the well ordering of the real numbers and the continuum hypothesis, and whose famous papers on well ordering would be published in 1904 and 1908.[14]

Hilbert's direct involvement with foundational questions of this kind became increasingly reduced, and after 1905 he devoted very little time to them for many years to come.[15] One of the few, but well-known, instances of what he did in this period of time is a talk presented at the International Congress of Mathematicians held in Heidelberg in 1904, later published under the title of 'On the Foundations of Logic and Arithmetic'. Hilbert outlined here a program for addressing the problem of the consistency of arithmetic as he then conceived it. Hilbert cursorily reviewed several prior approaches to the foundations of arithmetic and declared that the solution to this problem would finally be found in the correct application of the axiomatic method.[16] A somewhat elaborate discussion of the ideas he outlined in Heidelberg appears also in the notes to an introductory course taught in Göttingen in 1905, devoted to 'The Logical Principles of Mathematical Thinking' (Hilbert 1905). These notes are highly interesting since they provide a rather detailed and broad overview of Hilbert's current views on the axiomatic method as applied to arithmetic, to geometry and to physics at large. In particular, and as part of that overview, the notes allow a significant glimpse into the inherent tension among the various elements that inform Hilbert's images of mathematics and his views about the roles of *Anschauung*, empirical experience, and axioms.

An adequate appreciation of how these elements and their interrelations appear in the course notes and, more generally, of Hilbert's conception of the essence and role of the axiomatic method, must pay due attention to the following, illuminating passage:

> The edifice of science is not raised like a dwelling, in which the foundations are first firmly laid and only then one proceeds to construct and to enlarge the rooms. Science prefers to secure as soon as possible comfortable spaces

---

[13] Peckhaus 1990, 56–57.

[14] Zermelo 1908.

[15] Hilbert's gradual return to this field, starting in a limited way in 1914 and then increasingly expanding towards 1918, until it came to dominate his activities after 1922, is described in Sieg 1999 and Zach 1999.

[16] Hilbert 1905, 131.

to wander around and only subsequently, when signs appear here and there that the loose foundations are not able to sustain the expansion of the rooms, it sets about supporting and fortifying them. This is not a weakness, but rather the right and healthy path of development. (Hilbert 1905, 102)

This process of fortifying the 'loose foundations' is attained, of course, by means of an axiomatic analysis of the discipline in question. Thus, the very idea of investigating the foundations of mathematics, and indeed of science in general, is seen by Hilbert as an essentially pragmatic exercise meant to allow the healthy development of any discipline. It is a necessity that arises occasionally, only in case of real necessity. Axiomatic analysis is not a starting point of research in any field of mathematics (certainly not in geometry), and in fact it should not and cannot be done at the early stages of development of any discipline. Rather, it may be of great help only later on, when the theory has reached a considerable degree of maturity.

Of course, the paradigmatic example, but by no means the only one, of correctly and fruitfully applying the axiomatic method is geometry. Hilbert's own *GdG* could be evidently seen as the paramount successful instance of this, whereas the situation with arithmetic was much less clear at this stage. But what is the image of geometry and of arithmetic and of the philosophical underpinnings of these two disciplines that emerged in Hilbert's eyes in view of this situation? Here is what the notes to his 1905 course tell us about that:

> We arrive now to the construction of geometry, in which axiomatics was fully implemented for the first time. In the construction of arithmetic, our real point of departure was in its intuitive (*anschaulischen*) foundation, namely the concept of natural number (*Anzahlbegriff*) which was also the starting point of the genetic method. *After all, the number system was not given to us as a network of concepts (Fachwerk von Begriffen) defined by 18 axioms. It was intuition that led us in establishing the latter.* As we have started from the concept of natural number and its genetic extensions, the task is and naturally remains to attain a system of numbers which is as clear and as easily applicable as possible. This task will evidently be better achieved by means of a clearly formulated system of axioms, than by any other kind of definition. Thus it is the task of every science to establish on the axioms, in the first place, a network of concepts, for which formulation we let *intuition and experience* naturally serve as our guides. The ideal is, then, that in this network all the phenomena of the domain in question will find a natural place and that, at the same time, every proposition derivable from the axioms will find some application. (Hilbert 1905, 35–36)

Thus, for *both* geometry and arithmetic, the axiomatic analysis is meant to allow a systematic and thorough analysis of what in the final account provide the fundamental guide, source and justification, namely, intuition and experience. Hilbert is not very clear about the specific contribution of each of these two main components to every separate discipline. In earlier lectures we have seen him clearly separating between arithmetic and geometry. He then emphasized that while the former is a product of pure thought the latter is based on experience. It is indeed not untypical for Hilbert to change views with time. But as I said above, one should not try to extract a completely coherent and systematic philosophical system from his mathematical practice and from his various pronouncements, but rather attempt to see the elements that conform, sometimes in changing interrelations, his images of mathematical knowledge. And in this regard the above quotation is very revealing and typical, since it brings together very clearly all those important elements: intuition and experience as a starting point, and axiomatic analysis as a clarification tool. A successful axiomatic analysis implies a complete mathematization of the discipline in question, and in this sense logic is granted a main, foundational role for mathematics and for science at large. But this role is not autonomous and fundamental, since logic operates after the basic ideas are already in place creating a 'network of concepts'. In fact, Hilbert indicated an important difference between arithmetic and geometry in relation to the interaction between these various elements take place within them:

> Thus, if we want to erect a system of axioms for geometry, the starting point must be given to us by the intuitive facts of geometry and these must be made to correspond with the network that must be constructed. The concepts obtained in this way, however, must be considered as completely detached from both *experience and intuition.* In the case of arithmetic this demand is relatively evident. To a certain extent, this is already aimed at by the genetic method. In the case of geometry, however, the indispensability of this process [i.e. detachment from both intuition and experience (L.C.)] was acknowledged much later. On the other hand, the axiomatic treatment was attempted here earlier than in arithmetic where the genetic method was always the dominant one. (Hilbert 1905, 36–37)

This detachment from intuition and experience explains the equal mathematical validity and value that has to be attributed to all the possible, axiomatically defined, geometries. And yet, in spite of this, Hilbert explicitly and consistently expressed an inclination to grant a preferred status to Euclidean geometry from among all possible ones. What is the basis of this preference, if from the purely mathematical point of view all geometries are equally legitimate and valid? Hilbert was definitely puzzled about this, and this is no doubt one of the main reasons, as will be seen below, that he welcomed so strongly the rise of GTR with its momentous implications for the relations between geometry and physics. But in 1905, this is what he told his students in Göttingen:

> The question how is it that in nature only the Euclidean geometry, namely the one determined by all the axioms taken together, is used, or why our experience accommodates itself precisely with this system of axioms, does not belong to our logico-mathematical inquiry. (Hilbert 1905, 67)

Thus, five years after the publication of *GdG* and the flurry of activity that followed it both in Göttingen and outside, Hilbert had no doubts concerning the validity of Euclidean geometry as the most adequate description of physical space, but he definitely believed that mathematics itself could not explain the reason for this.

In spite of the successful application of the axiomatic method in geometry, the evidence from the 1905 lecture course clearly indicates that Hilbert did by no means adopt a formalistic view of mathematics in general and of geometry in particular, and did not bar from his lexicon the word *Anschauung* in connection with the foundations of geometry. This does not mean, however, that a thoroughly formalist view could not be derived from the new perspectives opened by Hilbert's innovations. One could find a very different and illuminating example of such a view in the works of a mathematician like Felix Hausdorff, and in the kind of radical views he developed under the explicit influence of *GdG* in a direction initially unintended by Hilbert himself.

Hausdorff indeed postulated the vie of geometry as a fully autonomous discipline, independent of any kind of *Anschauung* or empirical basis.[17] In a manuscript dated around 1904, and properly entitled 'Formalism', Hausdorff praised the full autonomy attained by geometry following Hilbert's work, in the following words:

> In all philosophical debates since Kant, mathematics, or at least geometry, has always been treated as heteronomous, as dependent on some external instance of what we could call, for want of a better term, intuition (*Anschauung*), be it pure or empirical, subjective or scientifically amended, innate or acquired. The most important and fundamental task of modern mathematics has been to set itself free from this dependency, to fight its way through from heteronomy to autonomy.[18]

This autonomy, so fundamental for the new view of mathematics predicated by Hausdorff and widely adopted later on as a central image of twentieth-century mathematics, was to be attained precisely by relying on the new conception of axiomatic systems embodied in *GdG*. As he explicitly wrote in a course on 'Time and Space', taught in 1903–04:

> Mathematics totally disregards the actual significance conveyed to its concepts, the actual validity that one can accord to its theorems. Its indefinable concepts are arbitrarily chosen objects of thought and its axioms are arbitrarily, albeit consistently, chosen relations among these objects. Mathematics is a science of pure thought, exactly like logic.[19]

Pure mathematics, under this view, is a 'free' and 'autonomous' discipline of symbols with no determined meaning. Once a specific meaning is accorded to them, we obtain 'applied' mathematics. Intuition plays a very important heuristic and pedagogical role, but it is inexact, limited, misleading and changing, exactly

---

[17] Purkert 2002, 50, quotes a letter of Hausdorff expressing an opinion in this spirit as early as October 1900.

[18] Quoted in Purkert 2002, 53–54.

[19] Quoted in Purkert 2002, 54.

the opposite of mathematics.[20] Although one can find in some of Hilbert's texts or lecture notes pronouncements that may be seen as fitting this view, the main thrust of his images of geometry is opposed to it, and this opposition became even stronger after 1915.

GTR and Geometry

In the years immediately after 1905 Hilbert directed most of his energies to the theory of integral equations and to physics, including foundational issues of various kinds in the latter discipline. It may thus have come as a nice and unexpected surprise for him to find out that his new focus of interests would eventually bring him back to the foundations of geometry. Indeed, this happened in 1916, as part of his intensive involvement with GTR, and the novel relationship that this theory uncovered between gravitation and geometry. One might think, on the face of it, that Hilbert's involvement with GTR was directly motivated by the strongly geometric content of this theory, but this is far from being the case. Rather, Hilbert came to be interested in GTR in a very roundabout way. As a matter of fact, until 1912, Hilbert's involvement with physics was essentially limited to topics related to mechanics (including fluid mechanics, statistical mechanics, and mechanics of continua). Only after 1912 did the scope of this involvement with physical disciplines significantly broaden so as to include also kinetic theory, radiation theory and, most significantly, current theories of the structure of matter. As part of his involvement with the latter domain, Hilbert came across the electromagnetic theory of matter of Gustav Mie, and, taking it as a starting point, Hilbert attempted to develop his own unified field theory of matter and gravitation. This is what led him around 1914 to increasingly direct his attention to Einstein's recent attempts to complete his generalized theory of relativity, including a relativistic theory of gravitation.[21]

After his initial involvement with GTR, that included a short-lived tension between him and Einstein around the question of priority in the formulation of the explicit, generally covariant field equations of the theory, Hilbert became a main promoter of the theory, which he explicitly and consistently presented as Einstein's brainchild and as one of the most important creations of the human spirit ever. In particular, Hilbert was among the first to teach a systematic course on the theory in 1916–17 and he continued to give public lectures for many years to come, on the implications of the theory for our understanding of space and

---

[20] For the precise quotations, see Purkert 2002, 54.

[21] This is described indetail in Corry 2004, especially in Chapters 5 and 6.

time.²² As part of all this, the subtle balance manifest in Hilbert's early writings between empirical and a-priori intuition as possible sources of geometric knowledge was finally altered, and unmistakably resolved in favour of experience. Moreover, and very importantly, Euclidean geometry had lost its preferred status as the one that naturally accommodates with empirical experience.

Plenty of evidence indicates the strong impact that these developments had on Hilbert. Some of Hilbert's pronouncements to this effect are worth quoting and discussing in some detail. The lecture notes of his 1916–17 course on GTR, for instance, included a section on 'the new physics', in which Hilbert referred to the new relationship between this discipline and geometry. He thus said:

> In the past, physics adopted the conclusions of geometry without further ado. This was justified insofar as not only the rough, but also the finest physical facts confirmed those conclusions. This was also the case when Gauss measured the sum of angles in a triangle and found that it equals two right ones. That is no longer the case for the new physics. *Modern physics must draw geometry into the realm of its investigations.* This is logical and natural: every science grows like a tree, of which not only the branches continually expand, but also the roots penetrate deeper.
>
> Some decades ago one could observe a similar development in mathematics. A theorem was considered according to Weierstrass to have been proved if it could be reduced to relations among natural numbers, whose laws were assumed to be given. Any further dealings with the latter were laid aside and entrusted to the philosophers.... That was the case until the logical foundations of this science (arithmetic) began to stagger. The natural numbers turned then into one of the most fruitful research domains of mathematics, and especially of set theory (Dedekind). The mathematician was thus compelled to become a philosopher, for otherwise he ceased to be a mathematician.
>
> The same happens now: the physicist must become a geometer, for otherwise he runs the risk of ceasing to be a physicist and vice versa. The separation of the sciences into professions and faculties is an anthropological one, and it is thus foreign to reality as such. For a natural phenomenon does not ask about itself whether it is the business of a physicist or of a mathematician. On these grounds we should not be allowed to simply accept the axioms of geometry. The latter might be the expression of certain facts of experience that further experiments would contradict.
> (Hilbert 1916–17, 2–3)

In the course and elsewhere, Hilbert constantly emphasized that both Euclidean geometry and Newtonian physics were theories of 'action-at-a-distance', and that the new physics had indicated the problems underlying such theories. Retrospectively seen, describing Euclidean geometry in these terms may sound somewhat artificial but Hilbert's point was to indicate that the old question of the validity of Euclidean geometry had been rekindled and should be now understood

---

²² See Corry 2004, Chapters 7 and 8.

in two different senses. The first sense is the logical one: is Euclidean geometry consistent? From the mathematical point of view, as Hilbert had already stressed in GdG, Euclidean geometry exists if it is free from contradiction. But from the physical point of view, such an answer is unsatisfactory. What we are interested in is the question of the validity of Euclidean geometry *as a description of nature*. This question, of course, 'cannot be decided through pure thinking'. In the past, even though he could not provide a full, satisfactory philosophical explanation for this, Hilbert had no doubts concerning the primacy of Euclidean geometry. Now conditions had changed and physical theory offered strong reasons to abandon that primacy. The need for such a change posed no problem for Hilbert, especially because of his long-professed, essentially empiricist image of geometry. Moreover, the new insights into the connection between gravitation and geometry fitted easily into ideas originally raised by Riemann, a mathematician whose conceptions of geometry Hilbert widely shared.[23]

But Hilbert's images of mathematics not only provided a natural background that allowed for a smooth adoption of the new conception of geometry implied by GTR. These images and the general mathematical background of Hilbert also led him into a direction within GTR that was quite idiosyncratic by that time. Thus, Hilbert was the first to wonder about the solution of the field equations in the absence of matter. Specifically, he asked about the conditions under which the Minkowski metric becomes a unique solution, hoping that this would happen in the absence of matter and radiation.[24] In contrast, for Einstein the main focus of interest in this context was the question of the Newtonian limit, and therefore the existence of empty-space solutions was not a natural, immediate question to be asked.

The status of Euclidean geometry in connection with the axioms of GTR was a topic that Hilbert addressed in the second of his two communications on the foundations of physics, presented to the Göttingen Scientific Society on December 23, 1916. Hilbert focused on what he called the 'Axiom of Space and Time', a postulate he had previously introduced in his first communication, as an attempt to deal correctly with the question of causality in GTR. Also here we find interesting views about the empirical grounding of geometry, as, for instance, in the following passage:

> According to my presentation here physics is a four-dimensional pseudo-geometry, whose metric $g_{\mu\nu}$ is connected with the electromagnetic magnitudes.... Having realized this, an old question seems to be ripe for solution, namely, the question if, and in what sense, Euclidean geometry – which from mathematics we only know to be a logically consistent structure – is also valid of reality.

---

[23] Hilbert explicitly mentioned this connection in the opening passages of his 1915 communication (Hilbert 1916, 398) and also in the lecture course (Hilbert 1916–17, 168).

[24] See Corry 2004, §8.3.

> The old physics, with its concept of absolute time, borrowed the theorems of Euclidean geometry, and made them the foundation of every particular physical theory. Gauss himself proceeded hardly differently: he hypothetically built a non-Euclidean physics, which, while retaining absolute time, renounced only the axiom of parallels. But the measurement of the angles of a large triangle indicated him the invalidity of this non-Euclidean physics.
>
> The new physics based on Einstein's general relativity takes a completely different approach to geometry. It assumes neither Euclidean nor any other kind of geometry in order to deduce from it the laws of physics....
>
> Euclidean geometry is *a law of action at a distance, foreign to modern physics*. By renouncing Euclidean geometry as a general presupposition of physics, the theory of relativity also teaches us that geometry and physics are similar in kind and, being one and the same science, they rest upon a common foundation. (Hilbert 1917, 63–64. Italics in the original)

Moreover, from the fact that the Euclidean metric, $g_{\mu\nu} = \delta_{\mu\nu}$, cannot be a general solution of the field equations, Hilbert deduced the following important conclusion, that gave additional strength to his empiricist leanings and indicates a possible direction to counter conventionalist or formalist interpretations of geometry:

> This is in my opinion a positive result of the theory, since we can in no way impose Euclidean geometry upon nature by means of a different interpretation of the experiment. Assuming that the fundamental equations of physics that I will develop here are the correct ones, then no other physics is possible, i.e., reality cannot be conceived differently. On the other hand, we will see that under certain, very specialized assumptions—perhaps the absence of matter in space will suffice—the only solution of the differential equations is $g_{\mu\nu} = \delta_{\mu\nu}$. Also this I must take as further support for my theory, since Gauss's angle-measurement experiment in a triangle has shown that *Euclidean geometry is valid in reality as a very good approximation.* (Hilbert 1916–17, 106. Emphasis in the original)

By referring to Gauss's experiment Hilbert was evidently closing a circle that had started way back in his early courses on Geometry. Back then Hilbert had interpreted the outcome of that experiment as the requisite empirical evidence for primacy of Euclidean geometry, but he nevertheless clearly suggested that future experiments could change current views in this regard, and might necessitate correcting our understanding of the role of the parallel axiom. Obviously he had no idea at that time that this assumption would prove correct two decades later, and much less under what circumstances. The new findings of physical science may indeed necessitate a specific choice of the correct geometry of nature, but in order to accommodate these changes in his overall view of mathematics and of science Hilbert could remain true to the empiricist approach that had characterized all of his images of geometry, as well as much of his foundational conceptions of mathematics, from very early on.

The last important phase of Hilbert's career was devoted to the foundations of logic and of arithmetic, and it comprises the years of activity in which the 'formalist' programme for proving the consistency of arithmetic in finitist terms was formulated and initially implemented. As already noted, the presence of Bernays in Göttingen since 1917 was a main factor in rekindling Hilbert's interest in this field. In 1922 Hilbert published his first significant article on the topic: 'New Foundations of Mathematics'.[25] Given his intense involvement during 1916–18 with questions related to GTR, and to the foundations of physics and geometry, one may wonder if, and possibly how, all the significant epistemological ideas developed in this framework played a direct role in the background to the elaboration of ideas related with the finitist program. There seems to be no direct evidence for a positive answer to such a question and to establish a direct, causal connection between Hilbert's activities in the foundation of physics and the transition to the last stage of his career. Nevertheless, it is interesting to examine Hilbert's views on the foundations of geometry after 1920, in order to realize that even at this late point there is no trace of 'formalism' in it, and that, on the contrary, these views become increasingly empiricistic.

One interesting instance of this appears in a series of public lectures given by Hilbert in the winter semester of 1922–23 under the name 'Knowledge and Mathematical Thought'. It is noteworthy that the fourth lecture in this series was entitled 'Geometry and Experience' (*Geometrie und Erfahrung*), exactly like Einstein's 1920 talk in Berlin quoted above. This may have been pure coincidence, but it is nevertheless remarkable that the declared aim of the series of talks was to refute a 'widespread conception of mathematics', and in particular conceptions such as implied by the views alluded to by Einstein in Berlin.[26] In fact, Hilbert thought it necessary to comment on the title of his talk, and he thus said:

> The problem that I want to address here is a very old, difficult and deep-going one. I could also call it: Representation and Reality, Man and Nature, Subjectivity and Objectivity, Theory and Praxis, Thinking and Being. If I have chosen such a title, then I must also stress that I can only treat this problem from a one-sided perspective and within a rather limited scope.... The problem stands in front of us like a high mountain peak that no one has yet fully conquered.... Perhaps we may succeed at least in reaching some important and beautiful observation points. (Hilbert 1922–23, 78)

Hilbert thought that epistemology, in its current state of development, was not yet ready to cope with the new situation created in view of the insights afforded by general relativity,[27] and in particular one cannot see in his own writings meaningful contributions in this direction. Yet he thought it important to stress

---

[25] Hilbert 1922.

[26] Two years earlier Hilbert had given another series of public talks where he pursued a similar aim. See Hilbert 1919–20. He would repeat many of the ideas expressed here in his Königsberg lecture of 1930 (Hilbert 1930).

[27] See, e.g., Hilbert 1922–23, 98.

the essence of these insights and to explain how they affected our conception of the connection between geometry and physics, between geometry and intuition. Typical of the kind of ideas that arise in this context are those expressed in the following passage:

> Some philosophers have been of the opinion—and Kant is the most prominent, classical representative of this point of view—that besides logic and experience we have a certain a-priori knowledge of reality. That mathematical knowledge is grounded, in the last account, on some kind of intuitive insight; even that for the construction of the theory of numbers a certain intuitive standpoint (*anschauliche Einstellung*), an a-priori insight, if you wish, is needed; that the applicability of the mathematical way of reflection over the objects of perception is an essential condition for the possibility of an exact knowledge of nature—all this seems to me to be certain.
>
> Furthermore, the general problem of determining the precise conditions of the possibility of empirical knowledge maintains its fundamental importance. And today more than ever, when so many time-honored principles of the study of nature are being abandoned, this question retains an increased interest.
>
> The general basic principles and the leading questions of the Kantian theory of knowledge preserve in this way their full significance. But the boundaries between what we *a-priori* possess and logically conclude, on the one hand, and that for which experience is necessary, on the other hand, we must trace differently than Kant. For, to take just one example, contrary to what was initially assumed, and to what also Kant claimed, the evidence of the basic propositions (*Grundsätze*) is not decisive for ensuring the success of Euclid's method in the real world. (Hilbert 1922–23, 87–88)

The law of inertia and the laws of Maxwell's theory of electromagnetism were two examples of physical laws that no one had expected to be a priori. But the necessity of turning to experience, as opposed to Kantian-like *Anschauung*, Hilbert added, appeared even in places where one would expect the a priori to be essential for the very possibility of science. This was the case for our conceptions of space and Euclidean geometry, the conception of which had radically changed in the wake of GTR. But in this passage Hilbert also pointed to the similar change that had affected the concept of absolute time that Newton and Kant took for granted but that Einstein's theory of relativity, prompted by the result of Michelson's experiment, had by now completely rejected. Thus, concerning the validity of the assumption of an absolute time, Hilbert said in his typically effusive and all-encompassing style:

> Newton actually formulated this as bluntly as possible: absolute, real time flows steadily from itself and by virtue of its nature, and with no relation to any other object. Newton had really given up any compromise in this respect, and Kant, the critical philosopher, proved here to be rather uncritical, because he accepted Newton without further ado. It was first Einstein who freed us definitively from this prejudice and this will always remain as

one of the most tremendous achievements of the human spirit and thus the all too sweeping a priori theory could not have been driven to absurd more decisively. Of course, a discovery of the magnitude of the relativity of simultaneity caused a drastic upheaval concerning all elementary laws, since now a much closer amalgamation of the spatial and temporal relations holds. We can thus say cum grano salis, that the Pythagorean theorem and Newton's law of attraction are of the same nature, inasmuch as both of them are ruled by the same fundamental physical concept, that of the potential. But one can say more: both laws, so apparently different heretofore and worlds apart from each other—the first one known already in antiquity and taught to everyone in primary school as one of the elementary rules of geometry, the other a law concerning the mutual action of masses on each other—are not simply of the same nature but in fact part of one and the same general law: *Newtonian attraction turned into a property of the world-geometry and the Pythagorean theorem into a special approximated consequence of a physical law.* (Hilbert 1922–3, 90–91. Emphasis in the original)

Anschauung was thus barred from any role in Hilbert's images of geometry and empiricism reigned now alone, as geometry had been definitely turned into a branch of physics. Like Newtonian physics, Euclidean geometry was nothing but a good approximation of truth. Formalism, needless to say, was in no way part of this picture.

Concluding Remarks

We may now return to the questions posed in the opening section, and try to summarize the above discussion by assessing the extent to which the philosophy of mathematics of one individual mathematician, Hilbert, is relevant to historically understanding his mathematical work, and to what extent his mathematical work has any bearing in understanding philosophical issues related with mathematics.

Hilbert's conceptions about geometry, although evidently philosophically well informed, cannot themselves be described as embodying elaborate philosophical views. Rather his changing conceptions are best described as an ongoing dialogue between historically evolving mathematical and scientific theories that Hilbert was involved with, on the one hand, and, on the other hand, a set of fundamental, yet flexible, images of what mathematical knowledge is about, of the relation between mathematics and the empirical sciences, and of the role of empirical perception and *Anschauung* in the various branches of mathematics.

It is indeed certainly necessary to pay due attention to the role played by philosophical conceptions in historically shaping the mathematical work of Hilbert, and its overall intellectual background. But this role needs to be understood in the terms described above: In the case of Hilbert, philosophical ideas provide him, in the first place, with the adequate terms to formulate and understand his own changing conceptions, and also to allow a feeling of continuity amid these changes. An underlying empiricistic drive is highly prominent in Hilbert's images of geometry throughout his career, but this prominence is, to a great extent, an outcome of specific developments in the mathematical disciplines as well as in physics in the relevant period.

Generally speaking—and very conspicuously so in the case of geometry—an overall, consistent philosophical conception does not appear as methodological or epistemological guidelines or as underlying general principles that Hilbert followed in developing mathematical and scientific ideas. Still, it is plausible that in certain, historically localized portions of his scientific career, a more elaborate and consistent philosophical perspective did play a decisive role in shaping his mathematical ideas. On the face of it, a claim in this direction could be made when considering Hilbert's finitistic program for the foundations of arithmetic and the intellectual setting in which it developed. I have not discussed this important part of his career in this chapter, but I suggest that in any such discussion, some of the ideas developed here should also be taken into consideration within the specific circumstances pertaining to the case.

Explorations into the emergence of Modern Mathematics

Methodology and metaphysics in the development of Dedekind's theory of ideals

JEREMY AVIGAD

1 Introduction

Philosophical concerns rarely force their way into the average mathematician's workday. But, in extreme circumstances, fundamental questions can arise as to the legitimacy of a certain manner of proceeding, say, as to whether a particular object should be granted ontological status, or whether a certain conclusion is epistemologically warranted. There are then two distinct views as to the role that philosophy should play in such a situation.

On the first view, the mathematician is called upon to turn to the counsel of philosophers, in much the same way as a nation considering an action of dubious international legality is called upon to turn to the United Nations for guidance. After due consideration of appropriate regulations and guidelines (and, possibly, debate between representatives of different philosophical factions), the philosophers render a decision, by which the dutiful mathematician abides.

Quine was famously critical of such dreams of a 'first philosophy.' At the opposite extreme, our hypothetical mathematician answers only to the subject's internal concerns, blithely or brashly indifferent to philosophical approval. What is at stake to our mathematician friend is whether the questionable practice provides a proper mathematical solution to the problem at hand, or an appropriate mathematical understanding; or, in pragmatic terms, whether it will make it past a journal referee. In short, mathematics is as mathematics does, and the philosopher's task is simply to make sense of the subject as it evolves and certify practices that are already in place. In his textbook on the philosophy of mathematics (Shapiro, 2000), Stewart Shapiro characterizes this attitude as 'philosophy last, if at all.'

The issue boils down to whether fundamental questions as to proper mathematical practice should be adjudicated with respect to general, and potentially extra-mathematical, considerations, or with respect to inherently 'mathematical' standards, values, and goals. Of course, what typically happens lies somewhere in between. Mathematics is not a matter of 'anything goes,' and every mathematician is guided by explicit or unspoken assumptions as to what counts as legitimate – whether we choose to view these assumptions as the product of birth, experience, indoctrination, tradition, or philosophy. At the same time, mathematicians are primarily problem solvers and theory builders, and answer first and foremost to the internal exigencies of their subject.

It seems likely, then, that any compelling philosophical account of mathematics will have to address both general philosophical principles and the more pragmatic goals of the practice. For example, many today hold that ontological questions in the sciences are to be adjudicated holistically, with respect to (possibly competing) standards such as generality, simplicity, economy, fecundity, and naturalness.[1] When it comes to mathematics, it is hard to see how one can take this dictum seriously without first gaining some clarity as to the ways that the objects of modern mathematical discourse influence the subject along these axes.

The work of Richard Dedekind provides a clear example of interplay between general philosophical views and methodological concerns. His work has certainly had a tangible effect on mathematics, inaugurating practices that were to proliferate in the decades that followed. These include the use of infinitary, set-theoretic language; the use of non-constructive reasoning; axiomatic and algebraic characterization of structures, and a focus on properties that can be expressed in terms of mappings between them; the use of particular algebraic structures, like modules, fields, ideals, and lattices; and uses of algebraic mainstays like equivalence relations and quotient structures. These innovations are so far-reaching that it is hard not to attribute them to a fundamentally different conception of what it means to do mathematics. Indeed, we find Dedekind addressing foundational issues in his *Habilitationsrede* (1854), and his constructions of the real numbers (1872) and the natural numbers (1888) are well known. But even his distinctly mathematical writings bear a strongly reflective tone, one that is further evident in his correspondence with colleagues; see (Dedekind, 1968; Dugac, 1976; Ewald, 1996; van Heijenoort, 1967).

In recent years, Dedekind has received a good deal attention from both philosophers and historians of mathematics, e.g. (Corry, 2004; Dugac 1976; Ferreirós, 1999; Haubrich, 1992; Nicholson, 1993; Reck, 2003; Sieg and Schlimm, to appear; Stein 1988). It is interesting to compare the various terms that are used to characterize his work. For example, Gray attributes the success of Dedekind's

---

[1] (Maddy, 2005) nicely surveys the use of terms like these by Quine.

theory of ideals to its handling of ontological issues:

> It occurred in a leading branch of contemporary mathematics, the algebraic theory of numbers. This gave it great weight as an example of how existence questions could be treated. It arose from a genuine question in research mathematics, the outcome of which necessarily involved a point of mathematical ontology. (Gray, 1992, 233)

Here, Gray is using the word 'ontology' in a distinctly methodological sense. In contrast, Stein proclaims that Dedekind is 'quite free of the preoccupation with 'ontology' that so dominated Frege, and has so fascinated later philosophers' (Stein, 1988, 227), and praises Dedekind instead for the way he was able to 'open up the possibilities for developing concepts.' Edwards finds Dedekind's work guided by 'strong philosophical principles,' judging salient characteristics of his work to be 'dictated by [his] set-theoretic prejudices' (1980, 321). According to Corry,

> Dedekind's overall mathematical output reflects a remarkable methodological unity, characterized, above all, by a drive to radically reformulate the conceptual settings of the mathematical problems he addressed, through the introduction and improvement of new, more effective, and simpler concepts. (Corry, 2004, 68)

Corry favours the notion of an 'image' of mathematics to describe the associated world view.

I believe it is fruitless to debate whether Dedekind's metaphysics or his methodology is philosophically prior; that is, whether we should take his general conception of mathematics to explain his methodological innovations, or vice versa. In the end, the two cannot be so cleanly separated. But philosophical discussions of Dedekind tend to focus on the metaphysical side, for example, on the ontological ramifications of his uses of set-theoretic methods, or on the forms of structuralism or logicism that are implicit in his views; see, for example, (Ferreirós, 1999, Chapter VII) and (Reck, 2003). My goal here is to balance these with a discussion of some of the methodological aspects of Dedekind's work. In other words, I will try to clarify some of the more distinctly mathematical concerns Dedekind faced, and explore the mathematical ramifications of the methods he introduced.[2]

---

[2] To be sure, historical treatments of Dedekind often discuss such methodological aspects of his work; see, for example, (Ferreirós, 1999, Chapter III) or (Haubrich, 1992, Chapter 1). Here, I simply explore some of these issues in greater depth.

Jeremy Gray has recently brought to my attention a book by David Reed (2005), which includes a lengthy discussion of the development of algebraic number theory in the hands of Dedekind and Kronecker. Although there is a good deal of overlap, Reed's presentation is largely complementary to mine. In particular, whereas I focus on the relationship to prior mathematical developments, Reed emphasizes differences between Dedekind and Kronecker *vis-à-vis* the subsequent development of class field theory.

I will focus, in particular, on his development of the theory of ideals. Towards the end of the 1850s, both Dedekind and Leopold Kronecker aimed to extend Ernst Kummer's theory of ideal divisors from cyclotomic cases to arbitrary algebraic number fields. Dedekind published such a theory in 1871, but he continued to modify and revise it over the next 23 years, publishing three additional versions during that time. Kronecker's theory was published in 1882, although it seems to have been developed, for the most part, as early as 1859. Despite the common starting point, the two theories stand in stark contrast to one another, representing very different sets of mathematical values. These parallel developments are therefore a wonderful gift of history to philosophers and mathematicians alike. Paying attention to the substance of Dedekind's revisions, and to the differences between his theory and Kronecker's, can illuminate important mathematical issues and help us understand some of the goals that drive the subject.

Below, then, is a preliminary survey of some of the epistemological values that are evident in the development of Dedekind's theory. The analysis is rough, and should be extended in both depth and breadth. Towards greater depth, we should strive for a more careful and precise philosophical characterization of these values, and submit them to closer scrutiny. In particular, it is important to recognize that Dedekind's methodological innovations are not uniformly viewed as positive ones. In Section 3 below, I will discuss some of the aspects of his work that were controversial at the time, and serve to distinguish it from Kronecker's. It will become clear that I am sympathetic to the claim that something important has been lost in the transition to modern mathematics, and so my enumeration of some of the purported benefits of Dedekind's revisions should not be read as wholesale endorsement. A broader philosophical analysis would benefit from contrasting the work of Dedekind and Kronecker, and tracing the influence of and interactions between their differing mathematical styles, through the twentieth century, to the present day.[3]

This research has been supported by a New Directions Fellowship from the Andrew W. Mellon Foundation. I am indebted to Steve Douglas White for numerous discussions of Dedekind's work, which resulted in many of the insights below. A number of topics treated here, especially in Sections 4 and 5, are treated more fully in his MS thesis (White, 2004). I am also grateful to Wilfried Sieg for help with a number of the translations, and Solomon Feferman, José Ferreirós, Jeremy Gray, John Mumma, and Dirk Schlimm for comments. Finally, I am especially grateful to Harold Edwards for many thought-provoking discussions, as well as substantive comments and corrections.

---

[3] Edwards' detailed studies of nineteenth-century number theory, and of the development of ideal theory in particular, provide a strong starting point; see (Edwards 1980, 1983, 1989, 1990, 1992, 1995).

2 The need for a theory of ideal divisors

In this section, I will sketch the historical circumstances that gave rise to Dedekind's theory of ideals.[4] Many of the original works I will mention below are accessible and well worth reading. But there are also a number of good secondary references to the history of nineteenth-century number theory, including (Edwards, 1996) and (Goldman, 1998) and Stillwell's notes to (Dedekind, 1877a) and (Dirichlet, 1863). Edwards' excellent article (1980) focuses specifically on the development of the theory of ideals, and the topic is also covered in some detail in (Corry, 2004). My goal here is not to extend this historical scholarship, but, rather, to supplement it with a more careful discussion of the methodological issues.[5]

The problem of determining whether various systems of equations have solutions in the integers, and of finding some or all of the solutions, dates back to antiquity. In his *Algebra*, Leonhard Euler (1770) took the bold step of using complex numbers towards those ends. For example, in articles 191–193 of Part II, Euler considers the question as to when an expression of the form $x^2 + cy^2$ can be a perfect cube, where c is fixed and x and y range over integers. Euler first notes that the expression can be factored as $(x + y\sqrt{-c})(x - y\sqrt{-c})$. Assuming $-c$ is not a perfect square, Euler asserts that when the two multiplicands have no common factor, in order for the product to be a cube, each factor must itself be the cube of a number of the form $p + q\sqrt{-c}$, where p and q are integers. If $x + y\sqrt{-c}$ is equal to $(p + q\sqrt{-c})^3$ for some p and q, then $x - y\sqrt{-c}$ is equal to $(p - q\sqrt{-c})^3$. Expanding the products and setting real and imaginary parts equal enables Euler to show, for example, that $x^2 + 2$ is a cube only when $x = \pm 5$, and $x^2 + 4$ is a cube only when $x = \pm 2$ or $x = \pm 11$.

There is a gap in Euler's argument.[6] It lies in the assertion that 'integers' of the form $x + y\sqrt{-c}$ behave like ordinary integers, in the sense that if the product of two numbers of this form is a cube, and the two have no non-trivial common factor, then each must itself be a cube. For the ordinary integers, this follows from the unique factorization theorem, since, for the product of two such numbers to

---

[4] Kummer called the ideal divisors described below 'ideal complex numbers,' Kronecker called them 'divisors,' and Dedekind associated them with the set-theoretic notion of an 'ideal.' I will use the term 'ideal divisor' generically, to include all three notions.

[5] Since my goals are not primarily historical, I will generally use contemporary terms to describe the mathematical substance of the developments. For example, where I speak of the ring of integers in a finite extension of the rationals, Dedekind refers to the 'system' of integers in such a field. Readers interested in terminological nuances and historical context should consult the sources cited above. That said, one thing that is striking about Dedekind's presentations is their proximity to modern ones, even down to terminology and notation. For example, we still use his terms 'module' and 'ideal'; and, like Dedekind, we generally use Greek letters $\alpha, \beta, \gamma, \ldots$ to range over algebraic integers, and Fraktur letters **a**, **b**, **c**, ... to range over ideals.

[6] The error is puzzling, especially since in other places Euler seems to know better. See the discussion in (Edwards, 1996, Section 2.3).

be a cube, any prime divisor of either factor must occur in the product a multiple of three times, and hence in that factor as well. In more general rings, a non-zero element c is said to be *irreducible* if it cannot be written as a product ab where neither a nor b is a unit (i.e. a divisor of 1). An integer c that is neither zero nor a unit is said to be *prime* if whenever c divides a product, ab, it divides either a or b. Unique factorization for the ordinary integers follows from the fact that in this setting, these two notions coincide.

In more general rings of integers, however, this need not be the case. For example, among complex numbers of the form $x + y\sqrt{-5}$, 6 can be written as either $(1 + \sqrt{-5})(1 - \sqrt{-5})$ or $2 \cdot 3$, and each of these factors is irreducible. The fact that 3 divides the product $(1 + \sqrt{-5})(1 - \sqrt{-5})$ but does not divide either factor shows that 3 is not prime. Furthermore, the perfect square 9 can be written as the product $(2 + \sqrt{-5})(2 - \sqrt{-5})$, even though neither $2 + \sqrt{-5}$ nor $2 - \sqrt{-5}$ is a square of a number of the form $x + y\sqrt{-5}$, nor do they share a non-trivial factor of that form.

In (1828), Gauss published a proof of unique factorization for what we now call the 'Gaussian integers,' that is, the ring of numbers of the form $\{x + yi \mid x, y \in \mathbb{Z}\}$. This proof appeared in the context of his study of biquadratic residues, i.e. residues of integer primes raised to the fourth power. Thus Euler's and Gauss' work made two things clear:

- Various rings of numbers extending the integers are useful in the study of fundamental questions involving equations and congruences in the ordinary integers.
- Unique factorization is an important feature of some such rings, and has a bearing on the original questions.

Aiming to generalize Gauss' results and obtain higher reciprocity laws, Ernst Kummer studied rings of numbers of the form $a_0 + a_1\zeta + a_2\zeta^2 + \cdots + a_{p-1}\zeta^{p-1}$, where ζ is a primitive p-th root of unity.[7] We now call these *cyclotomic integers*. By 1844, he knew that unique factorization can fail in such rings; the first case occurs when $p = 23$ (for details, see (Edwards, 1980 or 1996)). In 1846, Kummer published a theory that aimed to remedy the situation through 'the introduction of a peculiar sort of imaginary divisors' that he called 'ideal complex numbers.' The paper begins with the observation that in rings of cyclotomic integers with prime exponent, even when a number cannot be decomposed into factors,

> ...nonetheless, it may not have the true nature of a complex prime number, since it lacks the first and most important property of the prime numbers: namely, that the product of two prime numbers is not divisible by any prime different from them. Although they are not decomposible into complex factors, such numbers have, nonetheless, the nature of

---

[7] A number can be represented by more than one expression of this form. Since $\zeta^p - 1 = (\zeta - 1)(\zeta^{p-1} + \zeta^{p-2} + \cdots + 1) = 0$, we have $\zeta^{p-1} + \zeta^{p-2} + \cdots + 1 = 0$; and then ζ^{p-1}, and hence the sum $a_0 + a_1\zeta + a_2\zeta^2 + \cdots + a_{p-1}\zeta^{p-1}$, can be expressed in terms of smaller powers of ζ.

composite numbers; the factors are thus not actual, but *ideal complex numbers*. (Kummer, 1846, 319; my translation)

Given that unique factorization can fail in such a ring, the goal is to find a way to reason about the factorization of an element into ideal primes.

> In order to arrive at a firm definition of the true (generally ideal) prime factors of complex numbers, it was necessary to identify those properties of the complex numbers which persist in all circumstances, and are independent of whether it happens that the actual decomposition exists. ... Many of these persistent properties of the complex numbers are suitable for use in defining the ideal prime factors and, basically, always yield the same result. Of these, I have chosen the simplest and most general. (*ibid.*, 320)

The idea is this. Given a ring of cyclotomic integers that *does* satisfy unique factorization, and given a prime element α of that ring, one can characterize the property of divisibility of an element x by α in terms that do not mention α directly. In fact, Kummer showed that one can find such 'divisibility tests' $P_\alpha(x)$ that make sense even in rings that *do not* satisfy unique factorization, and possess all the central features of the usual notion of 'divisibility by a prime.' Thus one can think of these $P_\alpha(x)$ as representing the property of divisibility by an ideal prime factor α, even in cases where $P_\alpha(x)$ does not coincide with divisibility by any *actual* prime in the ring. Kummer was, in particular, able to show that unique factorization holds, in general, when expressed in terms of these predicates.[8] Thus

> ... it follows that calculation with complex numbers through the introduction of the ideal prime factors becomes exactly the same as calculations with the integers and their actual integer prime factors. (*ibid.*, 323)

So Kummer's 'definition' of the ideal complex numbers amounts to an explicit, algorithmic description of the associated divisibility predicates. Even though Kummer described these ideal numbers as 'peculiar,' he seems perfectly comfortable with this manner of definition, comparing it to the introduction of complex numbers in algebra and analysis, and to the introduction of ideal elements in geometry.[9] He is forceful in emphasizing the importance of these ideal elements to the study of the cyclotomic integers, and hence to number theory more generally:

> ... one sees that the ideal factors unlock the inner nature of the complex numbers, make them, as it were, transparent, and show their inner crystalline structure. (*ibid.*, 323)

Richard Dedekind and Leopold Kronecker later took up the task of extending the theory to the rings of integers in *arbitrary* finite extensions of the rationals.

---

[8] More precisely, Kummer also had to define the notion of divisibility by *powers* of the ideal prime divisors. The *product* of powers of distinct primes could then be understood in terms of their least common multiple.

[9] Kummer refers specifically to the introduction of the 'ideal common chord' of two nonintersecting circles. Such uses of the term 'ideal' in geometry are due to (Poncelet, 1822); see (Rosenfeld, to appear).

(See the beginning of Section 4 for a clarification of the notion of 'integer' in this context.) Despite their common influences and goals, however, the theories they developed are quite different, in ways that I will discuss in Section 3.

Dedekind ultimately published four versions of his theory of ideals (1871, 1877, 1879, 1894). The versions of 1871, 1879, and 1894 appeared, respectively, in his 'supplements,' or appendices, to the second, third, and fourth editions of his transcription of Dirichlet's lectures on number theory (Dirichlet, 1863). The remaining version was written at the request of Lipschitz, translated into French, and published in the *Bulletin des Sciences Mathématiques et Astronomiques* in 1876–1877. It was also published as an independent monograph in 1877, and is, in essence, an expanded presentation of the version he published in 1879.

3 Dedekind's emphasis on conceptual reasoning

Below we will discuss a number of aspects of Dedekind's theory of ideals. There is one, however, that gives it a character that is squarely opposed to that of Kronecker's theory, namely, Dedekind's emphasis on 'conceptual' over algorithmic reasoning.[10]

Whereas Kronecker's theory of ideal divisors is explicitly algorithmic throughout,[11] Dedekind's stated goal was to *avoid* algorithmic reasoning:

> Even if there were such a theory, based on calculation, it still would not be of the highest degree of perfection, in my opinion. It is preferable, as in the modern theory of functions, to seek proofs based immediately on fundamental characteristics, rather than on calculation, and indeed to construct the theory in such a way that it is able to predict the results of calculation... (Dedekind, 1877a, *Werke* vol. 3, 296; trans. Stillwell, 1996, 102; also translated in Stein, 1988, 245)

In this passage, Dedekind is referring to Riemann's approach to the theory of functions of a complex variable, in which functions are characterized by their topological and geometric properties.[12] Dedekind makes a similar claim in a letter to Lipschitz, written around the same time:

> My efforts in number theory have been directed towards basing the work not on arbitrary representations or expressions but on simple foundational

---

[10] (Ferreirós, 1999) explores some of the historical developments that may have contributed to this aspect of Dedekind's thought; see, in particular, Chapter I, Section 4.

[11] The centrality of algorithms is clear not only in Kronecker's mathematical work, but in the only foundational essay he published, (Kronecker, 1887). See also (Edwards, 1989) for a discussion of Kronecker's foundational views.

[12] See (Ferreirós, 1999) and (Laugwitz, 1999) for discussions of Riemann's work that emphasize features that Dedekind may have had in mind.

concepts and thereby – although the comparison may sound a bit grandiose – to achieve in number theory something analogous to what Riemann achieved in function theory, in which connection I cannot suppress the passing remark the Riemann's principles are not being adhered to in a significant way by most writers – for example, even in the newest work on elliptic functions. Almost always they mar the purity of the theory by unnecessarily bringing in forms of representation which should be results, not tools, of the theory. (Dedekind, 1876b, *Werke* vol. 3, 468–469; quoted and translated in Edwards, 1983)

Dedekind returns to this point in (1895), an essay we will discuss below. There, he first quotes an excerpt from Article 76 of Gauss' *Disquisitiones Arithmeticae* (1801), in which Gauss observes that Wilson's theorem was first published by Waring. In that excerpt, Gauss notes that

... neither of them was able to prove the theorem, and Waring confessed that the demonstration was made more difficult because no *notation* can be devised to express a prime number. But in our opinion truths of this kind should be drawn from the ideas involved rather than from notations. (Gauss, 1801, article 76; trans. Clarke, 1966. Dedekind, 1895, quotes the original Latin.)

Dedekind goes on:

When one takes them in the most general sense, a great scientific thought is expressed in these words, a decision in favor of the internal [*Innerliche*], in contrast to the external [*Äußerlichen*]. This constrast is repeated in almost every area of mathematics; one need only think of the theory of [Complex] functions, and *Riemann*'s definition of functions through internal characteristic properties, from which the external forms of representation necessarily arise. (Dedekind, 1895, *Werke* vol. 2, 54–55; my translation)

When it comes to making sense of the passages above, it is easier to say what Dedekind is trying to avoid: definitions of mathematical objects and systems of objects in terms of symbolic expressions and methods of acting upon them (for example, in thinking of functions as given by *expressions* of a certain sort), and proofs that rely on a particular choice of representation when many equivalent representations are available. Saying, in a positive way, how the new methods of reasoning are supposed to accomplish this takes more effort. Dedekind's mathematical work, however, provides us with a sense of what he had in mind. For one thing, his foundational essays (Dedekind 1854, 1872, 1888) all focus on axiomatic characterization of structures of interest; and in (Dedekind, 1888) there is a proof that his axiomatization of the natural numbers is categorical, i.e. it characterizes the structure uniquely, up to isomorphism. We will see below that this focus on axiomatic characterization is central to his work. We will also see Dedekind embrace the ability to introduce new mathematical structures and operations on these structures using set-theoretic definitions, without concern for the way that the elements of these structures are to be represented syntactically, and therefore without providing algorithms that mirror the set-theoretic operations. Indeed,

Dedekind shows no qualms in taking the elements of a structure to be infinitary objects in their own right, or referring, in a definition, to the totality of subsets of an infinite domain. It is clear that Dedekind sees these methods as part of the 'conceptual' approach as well.

It is worth emphasizing that it is not axiomatic methods in and of themselves that distinguish Dedekind's work from Kronecker's. In extending Kummer's work on ideal divisors of the cyclotomic integers, Kronecker (1870) himself gave an early influential axiomatization of the notion of a group. Referring to Gauss' classification of binary quadratic forms, he wrote:

> The extremely simple principles on which Gauss's method rests can not only be applied in the place mentioned above, but also in many others and, in particular, already in the most elementary parts of number theory. This fact suggests, and it is easy to convince oneself of it, that these principles belong to a more general and more abstract realm of ideas. It seems therefore to be appropriate to free the further development of the latter from all inessential restrictions, so that one is then spared from having to repeat the same argument in the different cases of application. The advantage comes to the fore already in the development itself, and the presentation (if it is given in the most general way possible) thereby gains in simplicity and clarity, since it clearly exhibits what alone is essential. (Kronecker, 1870, *Werke* vol. 1, 274–275; trans. Schlimm, 2005; also translated in Wussing, 1984, 64)

Below we will see Dedekind offer similar pronouncements as to the simplicity and generality to be gained from algebraic methods. So the differences between the two do not lie in the use of algebraic notions *per se*, but, rather, in the manner in which these notions are employed.

The progression from Dedekind's first theory of ideals to his last represents a steady transition from Kummer's algorithmic style of reasoning to a style that is markedly more abstract and set theoretic. Thus, it is not surprising that Edwards, who laments mathematics' departure from the explicitly algorithmic styles of Gauss, Kummer, and Kronecker, judges Dedekind's first version of ideal theory to be his best (Edwards 1980, 1992). In contrast, Emmy Noether, who inherited the mantle of structuralism from Dedekind through Hilbert, expressed a clear preference for the last (see McLarty's contribution to this volume). Tracing the development of Dedekind's thinking can therefore help us understand what is at stake.

As noted in the introduction, this essay focuses on Dedekind's point of view. This point of view was controversial at the time, and since then there has been a small but committed minority that agrees with Edwards' contemporary assessment that something important has been lost in turning away from a more explicit, algorithmic standpoint. Although it is hard for mathematicians with a modern training to recapture a nineteenth-century algorithmic sensibility, anyone studying the great works of that century cannot but appreciate the crisp elegance and efficiency of the associated conceptions. (Edwards' expository texts (1984, 1990,

1994, 1995, 1996, 2001, 2005) do a fine job of conveying a feel for this style of thought.) The reader should therefore keep in mind that in the discussion that follows, alternative, Kroneckerian points of view are not adequately represented.

4 The first version of the theory of ideals

For both Dedekind and Kronecker, the first step towards extending Kummer's theory to more general rings of integers involved finding the right definition of an 'integer.' To see that this problem is not a trivial one, consider the primitive cube root of 1 given by $\omega = (1 + \sqrt{-3})/2$. The field $\mathbb{Q}(\omega)$ obtained by adjoining ω to the rational numbers coincides with the field $\mathbb{Q}(\sqrt{-3})$. What should we take the integers of that field to be? From the first representation, following Kummer's lead, we would take the integers to be those numbers of the form $x + y\omega$, where x and y are ordinary integers. From the second representation, we might take them to be the numbers of the form $x + y\sqrt{-3}$. These two choices are distinct; for example, ω itself is an integer on the first choice, but not on the second. The fact that two representations of the same field lead to different choices of integers may, in and of itself, raise concern. But the situation is even more serious: a simple argument shows that with the second choice, *no* good theory of ideal divisors is possible.[13]

The solution is to define the algebraic integers to be those complex numbers θ that are roots of monic polynomials with coefficients in $\mathbb{Z}$, that is, which satisfy an equation of the form

$$\theta^k + a_{k-1}\theta^{k-1} + \cdots + a_1\theta + a_0 = 0$$

for some choice of ordinary integers $a_0, a_1, \ldots, a_{k-1}$. Whereas there are various ways of motivating this choice of definition, Edwards notes (1980, Section 5) that the historical origin is unclear. In their presentations, Dedekind and Kronecker simply put it forth without further comment. In the example above, this definition sanctions the first choice as the correct one.

But even with an appropriate notion of 'integer,' the generalization of Kummer's theory posed a number of difficulties. In (1878), Dedekind reflects on the obstacles that had to be overcome:

> I first developed the new principles, through which I reached a rigorous and exceptionless theory of ideals, seven years ago, in the second edition of Dirichlet's *Lectures on Number Theory*, and more recently in the *Bulletin des sciences mathèmatiques et astronomiques*, presented in a more detailed and in a somewhat different form. Excited by Kummer's great discovery, I had previously worked for a number of years on this subject, though

---

[13] See the footnotes to Sections §8 and §10 of (Dedekind 1877a). The argument is spelled out in detail in (Edwards, 1980, Section 4).

I based the work on a quite different foundation, namely, the theory of higher congruences; but although this research brought me very close to my goal, I could not decide to publish it because the theory obtained in this way principally suffers two imperfections. One is that the investigation of a domain of algebraic integers is initially based on the consideration of a definite number and the corresponding equation, which is treated as a congruence; and that the definition of ideal numbers (or rather, of divisibility by ideal numbers) so obtained does not allow one to recognize the *invariance* these concepts in fact have from the outset. The second imperfection of this kind of foundation is that sometimes peculiar exceptions arise which require special treatment. My newer theory, in contrast, is based exclusively on concepts like that of *field*, *integer*, or *ideal*, that can be defined without any particular representation of numbers. Hereby, the first defect falls away; and just so, the power of these extremely simple concepts shows itself in that in the proofs of the general laws of divisibility no case distinction ever appears. (Dedekind, 1878, *Werke* vol. 1, 202–203; modified from an unpublished translation by Ken Manders and Dirk Schlimm. I have deleted Dedekind's references to page numbers and sections.)

Expanding on the mathematical developments described in this passage results in a 'good news / bad news' story. A set $\{\omega_1, \omega_2, \ldots, \omega_k\}$ of integers in a finite extension of the rationals is said to be a *basis* for the ring of integers in that field when every element of the ring can be expressed uniquely as

$$a_1\omega_1 + a_2\omega_2 + \cdots + a_k\omega_k,$$

where $a_1, \ldots, a_{k-1}$ are ordinary integers. The good news is that when a ring of integers in a finite extension of $\mathbb{Q}$ has a basis of the form $\{1, \theta, \theta^2, \ldots, \theta^{k-1}\}$ for some element θ, Dedekind's theory of higher congruences provides a theory of ideal divisors generalizing Kummer's. The bad news is that not every such ring of integers has a basis of that form; in a sense, Kummer was lucky that this *is* the case with the cyclotomic integers. In general, one can always find algebraic integers θ such that $\{1, \theta, \theta^2, \ldots, \theta^{k-1}\}$ is a basis for the *field*, which is to say, the field is the set of linear combinations of those elements using *rational* coefficients. Furthermore, since every prime ideal divisor has to divide an ordinary integer p that is prime among the ordinary integers, it is enough to characterize the ideal prime divisors of each integer prime p. The good news is that the theory of higher congruences still works as long as one has a θ such that $\{1, \theta, \theta^2, \ldots, \theta^{k-1}\}$ is a basis for the field with the property p does not divide a certain integer value called the *discriminant* of this basis. The bad news is that when the discriminant is not ± 1, there will be finitely many primes p that *do* divide the discriminant, and so are not handled by the theory. These are the 'peculiar exceptions' that 'require special treatment.' The good news is that one can still fashion a workable theory of ideals using a different choice θ for each integer prime p, so it suffices to have, for each p, a choice of θ such that p does not divide the discriminant. The bad news is that there are even cases where for a given p *no* choice of θ will work; Dedekind gives a specific example of a cubic extension of the rationals in which no choice of θ can be used to represent the ideal divisors of 2. The passage above ultimately concludes with a piece of good news, namely, that one can dispense with the theory of higher congruences entirely, and consider more general representations of ideal divisors. In this way, Dedekind happily reports, he has been able to obtain a theory that accounts for the ideal prime divisors, uniformly, and all at once.

Our discussion, to this point, has already highlighted a number of criteria that Dedekind takes to be important for a theory of ideal divisors. These include:

- Generality: the theory should apply to rings of integers beyond the ordinary integers, Gaussian integers, cyclotomic integers, and rings of quadratic integers that were useful to Euler.
- Uniformity: one theory should cover all these cases, and, indeed, one definition of the ideal divisors should account for all the ideal divisors in a given ring of integers. Furthermore, as much as possible, proofs should cover all situations uniformly, without case distinctions.
- Familiarity: the overall goal is to restore the property of unique factorization, which has proved to be important to the ordinary integers, so that one can carry results forward to the new domains.

With respect to the latter, Dedekind's writings show that he is acutely aware of the role that definitions play in structuring a theory. One finds him concerned with such issues of systematization as early as 1854, in his *Habilitationsrede*. There, he characterizes a process of extending operations like addition and multiplication to extended domains, whereby one identifies the laws they satisfy in a restricted domain, and stipulates that these laws are to maintain their general validity (see also the discussion in (Sieg and Schlimm, to appear)).

One reason for this requirement can be found in Dedekind's insistence that definitions and methods of proof used in an extended domain should parallel the definitions and methods of proof that have been effective in more restricted domains. For example, in his presentation of the theory of higher congruences, he highlights the goal of introducing concepts in such strong analogy to those of elementary number theory that only 'a few words need to be changed in the number-theoretic proofs' (1857, *Werke* vol. 1, 40). In his presentations of ideal theory, he is careful to point out where definitions, basic characteristics, theorems, and proofs with respect to algebraic integers and ideals agree with their counterparts for the ordinary integers, and he seems to enjoy citing parallel developments in Dirichlet's *Lectures* wherever he can. The methodological benefits are clear, since it is often easy and efficient to reuse, adapt, and extend familiar modes of reasoning.

Thus far, we have seen little that would distinguish Dedekind's perspective from that of Kronecker, who is equally sensitive to the role of definitions in the structure of a theory, and who, for example, in his *Grundzüge* (1882), explicitly chose his basic concepts in such a way that they would remain unchanged when one passes from the rational numbers to algebraic extensions. From a methodological point of view, perhaps the most striking difference between Dedekind's theory and those of Kummer and Kronecker is Dedekind's use of the set-theoretic notion of an *ideal*. Recall that Kummer reasoned about his ideal divisors in terms of explicitly given predicates that express what it means for an algebraic integer x of the field in question to be divisible by the ideal divisor α. In contrast, Dedekind chose to identify the ideal divisor α with the set, or 'system,' **a** of all the integers x that α divides. It is clear that this set is closed under addition, and under multiplication by any integer in the ring. Thus, Dedekind defined an ideal to be any system of elements of the ring of integers in question with these properties, and, later, showed that every such ideal arises from an ideal divisor in Kummer's sense.

Dedekind went out of his way to explain why this tack is to be preferred. Referring to the general lack of sufficient divisors in a ring of integers, Dedekind writes that Kummer

> ... came upon the fortunate idea of nonetheless feigning [fingieren] such numbers μ' and introducing them as *ideal* numbers. The *divisibility* of a number α' by these ideal numbers μ' depends entirely on whether α' is a root of the congruence $\eta\alpha' \equiv 0 \pmod{\mu}$, and consequently these ideal numbers are only treated as moduli; so there are absolutely no problems with this manner of introducing them. The only misgiving is that the immediate transfer of the usual concepts of the *actual* numbers can, initially, easily evoke mistrust of the certainty of the proof. This has caused us to inquire after a means of clothing the theory in a different garb, so that we always consider *systems* of actual numbers. (Dedekind, 1871, *Werke* vol. 3, 221; my translation)

He returns to this point in 1877, explaining why Kummer's algorithmic treatment is not well suited to the task at hand:

> We can indeed reach the proposed goal with all rigour; however, as we have remarked in the Introduction, the greatest circumspection is necessary to avoid being led to premature conclusions. In particular, the notion of *product* of arbitrary factors, actual or ideal, cannot be exactly defined without going into minute detail. Because of these difficulties, it has seemed desirable to replace the ideal number of Kummer, which is never defined in its own right, but only as a divisor of actual numbers ω in the domain **o**, by a *noun* for something which actually exists. (Dedekind, 1877a, *Werke* vol. 3, 287; trans. Stillwell, 1996, 94)

Thus Dedekind maintains:

- The objects one refers to in a theoretical development (in this case, the ideal divisors) should be identified with *things*, not just used as modes of expression.

Indeed, Dedekind suggests that this is not just desirable, but even necessary when the mathematics is sufficiently complex.

In a sense, replacing the predicates P_α by the systems S_α of integers that satisfy them is mathematically inert. All the effects that mathematical objects can have on mathematical discourse are mediated by the roles they play in assertions; so if all references to ideal divisors are expressed in terms of the property of dividing an element of the ring, it makes little difference as to whether one takes 'x has property P_α' or 'x is an element of the set S_α' to stand duty for 'α divides x.' Both approaches require one to enforce the discipline that all properties of ideal divisors must ultimately be defined in terms of divisibility of integers, and Dedekind expresses concern that insufficient care may lead one astray (see also the discussion of multiplication of ideals in the next section). But, to be clear, this is a concern on either reading of the divisibility relation.

The insistence on treating systems of numbers – viewed as either sets of numbers, or as predicates – as objects in their own right does, however, have important methodological consequences: it encourages one to speak of 'arbitrary' systems, and allows one to define operations on them in terms of their behaviour as sets or predicates, in a manner that is independent of the way in which they are represented. For example, in 1871, Dedekind defines the least common multiple of two modules to be their intersection, without worrying about how a basis for this intersection can be computed from bases for the initial modules. We find a similar use of non-constructivity when he characterizes integral bases as those bases of integers whose discriminants have the least absolute value; he does this without giving an algorithm for finding such a basis or determining this least discriminant. In fact, in both examples just cited, algorithms can be obtained. But Dedekind's presentation sends the strong message that such algorithms are not *necessary*, i.e. that one can have a fully satisfactory theory that fails to provide them. This paves the way to more dramatic uses of non-constructive reasoning, in which one uses facts about infinitary functions, sets, and sequences that are *false* on an algorithmic interpretation. Such reasoning was used, for example, by Hilbert, in proving his *Basissatz* in 1890.

Dedekind's insistence on having an explicit set-theoretic object to stand as the referent for a mathematical term is coupled with the following exhortation:

- The definition of new objects should be independent of the way they are represented, and arguments involving them should not depend on any particular representation.

In this way, the arguments 'explain' why calculations with and properties of the objects do not depend on these choices of representations. This is why Dedekind, in the passage above, complains that his first attempt at a theory of ideal divisors did not 'allow one to recognize the invariance the concepts in fact have from the outset.'

It is precisely these last two criteria that push Dedekind to the use of set-theoretic language and methods. In particular, his treatment commits him to acceptance of the following:

- Infinite systems of numbers can be treated as objects in their own right, and one can reason about the domain of *all* such objects.

This view is commonly accepted today, but it was novel to the mathematics of the 1870s.[14] It is striking that Dedekind adopts the use of such systems without so much as a word of clarification or justification. That is, he simply introduced a style of reasoning that was to have decisive effects on future generations, without fanfare. When, in (1883), Cantor published an introduction to his theory of the infinite, he invoked considerations from history, philosophy and theology to assess the legitimacy of the use of completed infinite totalities, and responded to criticisms, both actual and anticipated. In (1888), Dedekind did spell out, informally, some of the rules he took to govern the use of infinite systems. But there he simply characterized such systems as 'objects of our thought' and took this as sufficient justification of their legitimacy.[15]

5 The second and third versions of the theory of ideals

Dedekind's second version of the theory of ideals was published in 1876–1877. Since the next published version, which appeared in the third edition of the Dedekind–Dirichlet *Lectures*, is essentially a condensed version of that, I will focus on the second version here, and refer to it as the '1877 version.'

There are at least two significant differences between the 1871 and 1877 versions of the theory. The first has to do with the treatment of multiplication of ideal divisors. In the presentation of 1871, after defining the notion of an ideal, Dedekind defines divisibility of an algebraic integer α by an ideal **a** to mean that α is an element of **a**. He then defines the divisibility of one ideal by another, '**b** divides **a**,' to mean that every element of **a** is an element of **b**. This makes sense when one considers that among the ordinary integers, the assertion that b divides a is equivalent to the assertion that every divisor of a is a divisor of b, and, indeed, this was the notion of divisibility used implicitly by Kummer. But it would be much more natural to say that **b** divides **a** whenever there is another ideal **c** such that **bc** = **a**. At this stage in the development, however, Dedekind cannot do this, for the simple reason that he has not yet defined multiplication of ideals. In 1871, unique factorization is expressed by the fundamental theorem

---

[14] Authors like Laugwitz (1999) and Ferreirós (1999) take Riemann's general notions of manifold and function to be conducive to a set-theoretic viewpoint. It is not at all clear, however, how Riemann would have felt about Dedekind's use of definitions involving quantification over completed infinite totalities. There is a tentative discussion of this issue in (Ferreirós, 1999, Section II.4.2).

In 1883, Cantor found a precedent for the use of completed infinites in Bolzano's *Paradoxes of the infinite* of 1851. Surveying the philosophical literature to date, Cantor wrote that 'Bolzano is perhaps the only one for whom the proper-infinite numbers are legitimate.' (Cantor, 1883, §7 paragraph 7; trans. Ewald, 1996).

[15] There is, of course, a good deal that can be said about how to interpret Dedekind's views in this respect; see, for example, (Ferreirós, 1999, Chapter VIII).

that every ideal is the least common multiple of all the powers of prime ideals that divide it, where the least common multiple of any finite set ideals is defined to be their intersection. Multiplication of ideals plays no role in the proof, and is, in fact, defined only afterwards.

In contrast, in 1877, multiplication of ideals is defined much earlier, and plays a central role in the presentation of the theory. Why the change? We have already noted Dedekind's sensitivity to the role that fundamental definitions play in structuring a theory. In modern terms, it is natural to express the goal of the theory of ideal divisors as being that of constructing a semigroup satisfying unique factorization, together with a suitable embedding of the integers of the field in question (up to associates). This is, for example, the characterization used in Weyl's Princeton lectures on algebraic number theory (1940), as well as in more recent presentations, like Borevich and Shafarevich's textbook (1966). On this view, the goal is to define the collection of ideals *with an associated multiplication*, and to show that the resulting structure meets the specification. From that perspective, multiplication is naturally prior.[16]

One might object that one can equally well characterize the goal of a theory of divisors taking the notion of divisibility (and 'prime power') as primitive. Kummer himself stated the requisite properties of the theory in such a way (see Edwards, 1980, Section 3), and Dedekind's 1871 version shows, directly, these these requirements are satisfied by the system of ideals. But this way of proceeding runs against Dedekind's insistence that definitions and methods of proof used in an extended domain should parallel the definitions and methods of proof that have been effective in more restricted domains, since, when it comes to the ordinary integers, divisibility is invariably defined in terms of multiplication.

Nonetheless, in 1877, Dedekind preserved the original 1871 notion of divisibility, making the theory of ideals diverge from the theory of integers almost immediately. The fact that there are two natural notions of divisibility at hand is confusing, but the good news is that, in the end, the two notions coincide. According to Dedekind's introduction to the 1877 version, this is something we come to see 'only after we have vanquished the deep difficulties characteristic of the nature of the subject.' Indeed, establishing the equivalence is tantamount to establishing unique factorization itself. To see this, note that unique factorization for the integers follows from the fact that the notions of 'prime' and 'irreducible' coincide. Passing to the theory of ideals, it is easy to show that every prime ideal is irreducible, and, further, that every prime ideal is prime with respect to

---

[16] Kronecker provided an entirely different, and elegant, formulation of the problem. In his approach, the central task is to find an appropriate definition of the *greatest common divisor* of any finite number of algebraic integers, whether it is actual or ideal. Edwards (1990) observes that there are many advantages to this way of setting up the theory. For example, whereas the factorization of an integer into ideal prime divisors changes as one expands the ring of integers in question, greatest common divisors remain invariant.

the Dedekind–Kummer notion of divisibility. Demonstrating that the two notions of divisibility coincide therefore shows that there is sufficient agreement with the theory of the integers to ensure that unique factorization holds for ideals as well.

To sum up, in 1871, Dedekind did not define multiplication of ideals until the very end of the theoretical development, where it is clear that multiplication has all the expected properties. In contrast, in 1877, he defined multiplication much earlier in the development, at which point showing that multiplication has the expected properties becomes the central task. Dedekind felt that making this change allowed him to characterize the goal of the theory of ideal divisors in a way that highlights parallels with elementary number theory, and made it clear how one can attain these goals by resolving the apparent differences between the two. This highlights a number of benefits of being attentive to axiomatic properties:

- It makes it clear what properties of a particular domain one wants to preserve, when extending it to a more general domain;
- It makes it clear what desired properties are *absent* from a particular domain, which one may wish to add.
- It can help suggest the definitions by which these two goals can be attained.

We will see in a moment that Dedekind compared his construction of the domain of ideal divisors of a ring of integers to his construction of the real numbers in (1872). Indeed, the two run parallel in a number of respects. In his construction of the reals, Dedekind is careful to isolate the order-theoretic properties of the rational numbers that will be preserved, and is concerned that arithmetic identities will carry over smoothly as well (see Dedekind, 1872, end of §6). He is also careful to identify the goal of the construction as satisfaction of the 'principle of continuity,' a property that clearly holds of the geometric line, but does not hold of the rationals. It is these considerations that suggest a suitable definition of the real numbers. Similarly, in his construction of ideal divisors, he is attentive to the properties of the ring of integers in a finite extension of the rationals, qua divisibility, that will be preserved; and the properties, familiar from the integers but generally absent from such rings of integers, which need to be added to obtain unique factorization. Again, these considerations suggest a suitable definition of the ideal divisors.

As far as the importance of defining multiplication from the outset, Dedekind describes the state of affairs as follows:

> Kummer did not define ideal numbers themselves, but only the divisibility of these numbers. If a number α has a certain property A, to the effect that α satisfies one more congruence, he says that α is divisible by an ideal number corresponding to the property A. While this introduction of new numbers is entirely legitimate, it is nevertheless to be feared at first that the language which speaks of ideal numbers being determined by their

> products, presumably in analogy with the theory of rational numbers, may lead to hasty conclusions and incomplete proofs. And in fact this danger is not always completely avoided. On the other hand, a precise definition covering *all* the ideal numbers that may be introduced in a particular numerical domain **o**, and at the same time a general definition of their multiplication, seems all the more necessary since the ideal numbers do not actually exist in the numerical domain **o**. To satisfy these demands it will be necessary and sufficient to establish once and for all the common characteristic of the properties $A, B, C, \ldots$ that serve to introduce the ideal numbers, and to indicate, how one can derive, from properties A, B corresponding to particular ideal numbers, the property C corresponding to their product. (Dedekind, 1877a, *Werke* vol. 3, 268–269; trans. Stillwell, 1996, 57)

Thus, Dedekind maintains that defining multiplication from the start makes it less likely that we will improperly transfer patterns of reasoning that we are accustomed to use with the ordinary integers. The passage above also reiterates the value of uniformity, in the assertion that all the ideal numbers should be introduced by a single definition, all at once, and the relevant operation should be introduced at the same time. In a footnote to the passage above, he compares the situation to the development of the real numbers:

> The legitimacy, or rather the necessity, of such demands, which may always be imposed with the introduction or creating of new arithmetic elements, becomes more evident when compared with the introduction of *real irrational* numbers, which was the subject of a pamphlet of mine... Assuming that the arithmetic of *rational* numbers is soundly based, the question is how one should introduce the irrational numbers and define the operations of addition, subtraction, multiplication, and division on them. My first demand is that arithmetic remain free from intermixture with extraneous elements, and for this reason I reject the definition of real number as the ratio of two quantities of the same kind. On the contrary, the definition or creation of irrational number ought to be based on phenomena one can already define clearly *in the domain R* of rational numbers. Secondly, one should demand that all real irrational numbers be engendered simultaneously by a common definition, and not successively as roots of equations, as logarithms, etc. Thirdly, the definition should be of a kind which also permits a perfectly clear definition of the calculations (addition, etc.) one needs to make on the new numbers. (*ibid.*)

It is clear from the discussion that follows that by 'definition of the calculations' in the last sentence Dedekind means set-theoretic definition, since he refers to his own (set-theoretic) definition of the product of two positive real numbers.

The claims made here represent a central part of Dedekind's world view. From a Kroneckerian perspective, introducing a new domain of 'numbers' is a matter of introducing symbolic expressions, operations governing them, and methods of determining equality, or equivalence, of expressions. For example, if $f(x)$ is an irreducible polynomial with rational coefficients, one can extend the rational numbers, $\mathbb{Q}$, to a field with a root of $f(x)$, by considering the quotient $(\mathbb{Q}[x]/(f(x))$. It is straightforward to extend the field operations to the larger domain and embed the original domain, $\mathbb{Q}$. Non-constructively, it is easy to iterate this process, and obtain a field in which $f(x)$ factors into a product of linear polynomials: if $f(x)$ does not split into linear factors after the first adjunction of a root, pick any non-linear factor of $f(x)$ that is irreducible in the new field, adjoin a root of that, and so on. It is a much greater challenge, however, to show that this can be done algorithmically. To start with, one needs a procedure to determine whether or not $f(x)$ splits into linear factors after the first step, and, if not, to find a non-linear irreducible factor. In fact, Kronecker designed his 'Fundamentalsatz der allgemeinen Arithmetik' (1887) to provide an algorithmic description of a splitting field for any given polynomial. He viewed this theorem as a proper formulation of the fundamental theorem of algebra, since it establishes the 'existence' of the polynomial's roots in a suitable extension of the rationals (see the discussion in (Edwards, 2005), as well as alternative constructions in (Edwards, 2005 and 1984)). Of course, one may later wish to deal with trigonometric functions, or the complex exponential or logarithm; here, again, the challenge is to give the appropriate rules for manipulating the relevant expressions and relate them to the prior ones. There is no further need to say what, exactly, the complex numbers *are*, to define them 'all at once,' or even to suppose that they exist; we simply expand our methods of representation and calculation as the need arises. (In the appendix to his calculus textbook (1994), Edwards illustrates this view with a 'Parable of the mathematician and the carpenter.')

From Dedekind's point of view, this piecemeal approach will not do; this is exactly what the second criterion in the passage above is meant to rule out. Instead, we want definitions that determine *all* the real numbers, *all* the complex numbers, and *all* the functions we may wish to define on them, from the start. Of course, with Kronecker, we can go on to develop means of representing *particular* numbers. For a given choice representations, we can then ask how to calculate the values particular functions that are of interest. But a definition of the entire system should come first, and functions should be defined in a manner that is independent of any choice of representations. Issues having to do with calculation come later.

There is something to be said for such an approach, in that it serves to unify and guide the various extensions. A single uniform definition of the real numbers gives an account of what it is that particular expressions are supposed to represent, and a uniform definition of multiplication tells us, from the start, what it is that particular algorithms, based on particular representations, are supposed to compute. Verification that the algorithms behave as they are supposed to then guarantees that properties that have been shown to hold for multiplication, in general, also hold for the algorithmic instantiation. Of course, the downside is that one may generally define functions without knowing how to compute with representatives of elements in a given domain. Algorithmic issues, like the one suggested above, can be quite difficult, and one might credit the set-theoretic development for encouraging us to look for the appropriate algorithms (or to determine whether or not the functions are computable).[17]

There is evidence that Kronecker would have concurred with these assessments, to an extent. His student, Hensel, later explained that Kronecker was not dismissive of non-constructive definitions and proofs; rather, he felt that a theory is not entirely satisfactory or complete until the algorithmic details are in place (see the introduction to Kronecker, 1901; the relevant excerpt is quoted in Stein, 1988). In other words, Kronecker may have recognized the value of non-constructive methods in providing heuristic arguments, with the caveat that such arguments are not the subject's final goal, which is to have rigorous proofs. In contrast, Dedekind is willing to accept a non-constructive argument as a rigorous proof in its own right.

Summarizing, then, we can ascribe to Dedekind the following views:

- We desire not just an axiomatic characterization of mathematical domains of interest, but constructions that yield *all* the (possible) elements, at once.
- Functions of interest should be defined from the start, uniformly, for all the elements of the domain.
- Non-constructive, set-theoretic definitions are perfectly admissible, and non-constructive proofs can be rigorous and correct.
- Issues regarding algorithms and explicit representations of elements come later in the development. Central properties of the domain in question should be established without reference to particular representations.

We have seen that in 1877 Dedekind provided a set-theoretic definition of the ideals in any ring of algebraic integers, and a set-theoretic definition of the product of two ideals. His goal was then to base the proof of unique factorization on these definitions, and to avoid relying on particular representations of the ideals. He was only partially successful in this respect, and, indeed, it is this concern that accounts for the second major difference between the 1871 and 1877 versions of the theory. In both 1871 and 1877, Dedekind defined a *prime ideal* to be an ideal whose only divisors are itself and the entire ring of algebraic integers. In both presentations, he showed that an ideal **p** is prime if and only if whenever it divides a product $\alpha\beta$ of integers, it divides either α or β. In 1871, however, Dedekind went on to define a *simple ideal* to be a prime ideal that can be represented in a certain way, namely, as the set of all solutions π to a congruence $\nu\pi \equiv 0 \pmod{\mu}$. As was the case with Kummer's theory, this provides an effective test for divisibility by these ideal prime divisors: an algebraic integer π is divisible by the ideal divisor corresponding to μ and ν if and only if it satisfies the associated congruence. This can be extended to provide a test for divisibility by powers of these ideal divisors,

---

[17] For another example, note that the non-constructive development of the reals makes it clear that if $f(x)$ is an odd-degree polynomial with rational coefficients, it has a real root; and if α is the least such root, one can extend not just the arithmetic operations on the rational numbers but also the ordering to the field $\mathbb{Q}(\alpha)$. Seeing that this can be done algorithmically requires a lot more work.

cast in 1871 as a definition of the powers of the simple ideals. Dedekind showed that the notion of divisibility by powers of the simple ideals has the requisite properties; in particular, every element of the ring of integers is determined (up to associates) by the powers of the simple ideals that divide it. This implies that every prime ideal is a simple ideal, and that, in turn, implies that every ideal (other than the trivial ideal, {0}) can be represented by an appropriate μ and ν. Thus, every ideal in the new sense arises as the set of integers divisible by one of Kummer's ideal divisors. This is what Dedekind has in mind when he writes, in the 1877 introduction:

> A fact of highest importance, which I was able to prove rigorously only after numerous vain attempts, and after surmounting the greatest difficulties, is that, conversely, each system enjoying [the new definition of an ideal] is also an ideal [in Kummer's sense]. That is, it is the set **a** of all numbers α of the domain **o** divisible by a particular number; either an actual number or an ideal number indispensable for the completion of the theory. (Dedekind, 1877a, *Werke* vol. 3, 271; trans. Stillwell, 1996, 59–60)

Relying on simple ideals, however, runs counter to Dedekind's goal of avoiding reasoning that is based on particular representations of ideals rather than their 'fundamental characteristics.' By 1877, he has therefore dropped the term. That is not to say that he has avoided the use of such representations in his arguments: the ν and π above become κ and λ in a key argument in the 1877 version (§25), but they are deprived of the honorific status that is accorded by a definition, and the associated calculations are relegated to a pair of 'auxiliary propositions' in the preceding section.

We will see in Section 6 below that Dedekind's inability to dispense with these calculations entirely remained a thorn in his side for years to come. In the meanwhile, his 1877 exposition and the mathematical context suggest a further reason that the calculations have been moved. Contemporary algebraic treatments of the theory of ideals tend to identify the most general classes of structures for which the various results of the theory hold; Dedekind's 1877 treatment is remarkably modern in this respect. Chapter 1 of that version (as well as §161 of the 1871 version) develops general theorems that are true of arbitrary modules.[18] In the 1877 version, he then, very self-consciously, develops the portion of the theory of ideals that only pre-supposes that one is dealing with a ring of integers whose rank as a module coincides with the degree of the extension.

---

[18] In both presentations, Dedekind defines a module to be a system of complex numbers that is closed under sums and differences. But at the end of Chapter 1 of (Dedekind, 1877a), he notes that the 'researches in this first chapter...do not cease to be true when the Greek letters denote not only numbers, but any objects of study, any two of which α, β produce a determinate third element $\gamma = \alpha + \beta$ of the same type, under a commutative and uniformly invertible operation (composition), taking the place of addition. The module **a** becomes a *group* of elements....' In other words, Dedekind observes that the results hold for any (torsion-free) abelian group, viewed (in modern terms) as a free module over $\mathbb{Z}$. Today we recognize that, in fact, they hold more generally for free modules over a principal ideal domain.

Following Dedekind, these structures are still called *orders* today. With a specific counterexample, Dedekind notes that not every order has a theory of ideal divisors (see footnote 13 above, and the preceding text), and then identifies the auxiliary propositions as being precisely the point at which one needs to assume that the ring in question is integrally closed, that is, consists of *all* the integers of the ambient number field. These propositions are clearly necessary for the ring to have a theory of ideal divisors; the subsequent development in 1877 shows that they are also sufficient. This is an instance of another methodological dictum that is in evidence in Dedekind's work:

- One should take great care to identify the axiomatic features of the domain in question that are in play at each stage of the development of a theory.

This ensures maximum generality, and also simplifies the theory by suppressing irrelevant distractions.

The observations regarding the necessity of considering *all* the integers of a field are also present in the 1871 version; the 1877 version simply makes them more prominent. It is easy to sympathize with Edwards, who feels that the resulting reorganization makes the proof of unique factorization seem *ad hoc* and unmotivated. As noted above, Dedekind himself was never fully happy with this version of the proof. But in localizing and minimizing the role of representations and calculations, and making them secondary to structural systematization, Dedekind is exhibiting tendencies that have become hallmarks of modern mathematics.

6 The final version of the theory of ideals

In the supplements to the fourth edition of Dirichlet's *Lectures*, Dedekind published his final version of the theory of ideals (1894), yet again markedly distinct from his prior versions. In (1895), he also described an additional, intermediate version that he obtained in 1887, and that was later obtained, independently, by Hurwitz. Dedekind's goal in (1895) is to explain why he takes the 1894 version to be superior to this intermediate one. The mathematical details of these two versions are nicely summarized in (Edwards, 1980), and, unsurprisingly, Edwards' judgment of their relative merits is the opposite of Dedekind's. Dedekind's analysis is, as usual, rife with methodological claims, and it is well worth recounting some of them here.

Dedekind again finds the central focus of theoretic development to lie in the task of proving equivalence between the two notions of divisibility. He writes:

> In §172 of the third edition of the *Number Theory*, as well as in §23 of my essay *Sur la théorie des nombres entiers algébriques*, I have emphasized that

the greatest difficulty to be overcome for the foundation of the theory of ideals lies in the proof of the following theorem:

1. If the ideal **c** is divisible by the ideal **a**, then there is an ideal **b** which satisfies the condition **ab** = **c**.

This theorem, through which the relationship between the divisibility and multiplication of ideals is ascertained, is, in the presentation of the time, only provable at nearly the conclusion of the theory. This fact is palpable in a most oppressive way, especially since some of the most important theorems could be formulated in appropriate generality only gradually, by successively removing restrictive assumptions. I therefore came back often to this key point over the years, with the intention of obtaining a simple proof of Theorem 1, relating directly to the concept of the integers; or a proof of one of the following three theorems, which, as one easily sees, are of equal significance to the foundation of the theory:

2. Each ideal **m** can, by multiplication with an ideal **n**, be turned into a principal ideal.

3. Every finite, non-zero module **m**, which consists of either integers or fractional algebraic numbers, can via multiplication by a module **n**, whose numbers are formed from those in **m** in a rational way, be turned into a module **mn**, which contains the number 1 and consists only of integers.

4. From m algebraic numbers μ_r that do not all vanish, one can obtain, in a rational way, m numbers ν_s, which satisfy the equation

$$\mu_1 \nu_1 + \mu_2 \nu_2 + \cdots + \mu_m \nu_m = 1,$$

as well as the condition that all the m^2 products $\mu_r \nu_s$ are integers.

Now, these four theorems are completely equivalent, insofar as each of them can be obtained from each of the three remaining ones without difficulty. In such cases it often happens that one of the theorems, due to its simpler form, is more amenable to a direct proof than the others. In the previous example, clearly theorem 4, or also theorem 3, which differs only superficially in the use of the concept of a module, stands out as simpler than theorems 1 and 2, which deal with the more complicated concept of an ideal. (Dedekind, 1895, *Werke* vol. 2, 50–52; my translation. I have omitted two footnotes, and Dedekind's references to statements of the theorems in his previously published works.)

Even in this short excerpt, we can discern a number of general claims, including the following:

- Theorems should be stated at an 'appropriate level of generality,' even though it is often not evident how to do this at the outset.
- An important theorem should have a proof that 'relates directly' to the relevant concepts.
- Sometimes, casting a theorem in a 'simpler' or more general form makes it easier to find a direct proof.

Methodology and Metaphysics in the Development of Dedekind's Theory of Ideals

These can be viewed as calls for a kind of methodological directness, or purity. In the case at hand, even though Theorems 1–4 above are easily shown to be equivalent, Dedekind takes Theorems 3 and 4 to be preferred, because they deal with the more general concept of a module rather than an ideal. (A module, in Dedekind's terminology, is what we would call a $\mathbb{Z}$-module, and has only an additive structure; the concept of an ideal relies on the notion of multiplication as well.)

Once again, a summary of the mathematical developments reported on in (Dedekind, 1895) will help us grasp the methodological import. In 1882, Considering Kronecker's great work on algebraic number theory (1882), Dedekind found what he took to be a gap in a proof of a fundamental theorem. But he was able to prove a special case, a generalization of Gauss' theorem on the product of primitive polynomials, which he later published and referred to as his 'Prague Theorem.' Furthermore, the Prague Theorem implies Theorem 3 above. This was also the tack discovered by Hurwitz, who took the Prague Theorem to be implicit in a work of Kronecker from 1883. But Dedekind felt that the detour through properties of polynomials with coefficients in the ring of integers destroys the theory's uniform character ('einheitlichen Charakter'), and is therefore unacceptable. After the commenting on the quote from Gauss that we discussed in Section 3 above, Dedekind writes:

> As a result, one will understand that I preferred my definition of an ideal, based on a characteristic inner property, to that based on an external form of representation, which Mr. *Hurwitz* uses in his treatise. For the same reasons, I could not be fully satisfied with the proof of Theorem 3 mentioned above, based on [the Prague Theorem], since, by mixing in functions of variables the purity of the theory is, in my opinion, tarnished... (*ibid.*, 55)

So, Dedekind went back to the drawing board. Considering a very special case of the Prague Theorem, he reproduced a proof of a special case of Theorem 4 that he had discovered earlier, which works only for modules generated by two elements. But now, with the Prague Theorem under his belt, he saw how to extend the argument inductively to a full proof of Theorems 3 and 4. That is, he found a direct proof of one of the 'simpler' statements that he was after, one that avoids passage through areas he took to be external to the theory. But even this was not enough to satisfy him:

> Hereby I had finally found what I had long sought, namely, a truly appropriate proof of Theorems 3 and 4, and therefore also the foundation for the new formulation of my theory of ideals. However, I was not entirely satisfied with this inductive proof, since it is dominated by mechanical calculation... (*ibid.*, 57)

With some additional reflection, Dedekind discovered a general identity involving modules,

$$(\mathbf{a} + \mathbf{b} + \mathbf{c})(\mathbf{bc} + \mathbf{ca} + \mathbf{ab}) = (\mathbf{b} + \mathbf{c})(\mathbf{c} + \mathbf{a})(\mathbf{a} + \mathbf{b}),$$

which forms the basis for a proof of Theorem 3 by induction on the number of generators of the module **m**. It is this proof that finally met his approval, and that appears in (1894).

Dedekind's narration is remarkable. We find him, true to form, doggedly determined to eliminate (or hide) any trace of calculations from his proofs, to present his proofs in such a way that they do not rely on any extraneous properties of the structures at hand, and to base his proofs only on the 'intrinsic' properties of these structures rather than particular representations of the elements. It is exactly these features that Emmy Noether praised in her editorial notes in Dedekind's *Werke*, and exactly these features that Edwards finds artificial and disappointing. In the decades that followed, the attitudes that Dedekind expresses here were to have dramatic effects on the course of mathematics, and so it is well worth trying to understand what lies behind them.

7 Coming to terms with methodology

We have considered a number of reasons that Dedekind felt that later versions of his theory of ideal divisors were successive improvements over prior ones, as well as over Kummer's, Hurwitz's, and Kronecker's theories. Probing these judgments and submitting them to careful scrutiny can help us get a better handle on the epistemological goals that drove Dedekind, as well as the benefits later generations saw in the methods he introduced. This illustrates a way in which it is possible to use the history of mathematics to develop a robust epistemology that can account for the broad array of value judgments that are employed in the practice of mathematics. Put simply, the strategy is as follows: find an important mathematical development, and then explain, in clear terms, what has changed. Understanding how these changes are valued, or inquiring as to why these changes are valued, can yield insight as to what is fundamentally important in mathematics.

But, on closer inspection, the appropriate means of proceeding is not so clear. To describe 'what has changed,' one needs a characterization of the local state of affairs 'before' and 'after.' And that raises the question of how the relevant states of affairs are to be described, that is, what features we take to be essential in characterizing a state of mathematical knowledge (or, perhaps, understanding, or practice) at a given point in time. But this, of course, presupposes at least some aspects of a theory of mathematical knowledge. So we are right back where we started: any use of historical case studies to develop the philosophy of mathematics will necessarily be biased by philosophical pre-suppositions and the very terms used to describe the developments (as well as, of course, a bias as to whether these developments constitute 'advances'). Lakatos describes this state of affairs in the philosophy of science more generally:

> In writing a historical case study, one should, I think, adopt the following procedure: (1) one gives a rational reconstruction; (2) one tries to compare

this rational reconstruction with actual history and to criticize both one's rational reconstruction for lack of historicity and the actual history for lack of rationality. Thus any historical study must be preceded by a heuristic study: history of science without philosophy of science is blind. (Lakatos, 1976, 138. The entire passage is parenthetic and italicized in the original.)

This give-and-take is unavoidable: the best we can do is present a theory of mathematical knowledge and see how it fares with respect to mathematical, logical, historical, and philosophical data; and then use the results of this evaluation to improve the philosophical theory.

My goal here is not to develop such a theory. In (Avigad, to appear), I suggest that a syntactic, quasialgorithmic approach should be fruitful. That is, I expect that it will be useful to characterize states of knowledge not only in terms of collections of definitions, theorems, conjectures, problems, and so on, but also in terms of the available methods for dealing with these; e.g. methods of applying definitions, verifying inferences, searching for proofs, attacking problems, and forming conjectures. But rather than extend this speculation now, I would like to consider the ways that our discussion of Dedekind's work might inform, and be informed by, such a theory.

I have noted in the introduction that philosophical discussions of Dedekind's use of axiomatic methods usually have a metaphysical character, whereby the goal is to square the use of such reasoning with an appropriate 'structuralist' view of mathematical objects. But we can also try to understand these axiomatic methods in terms of their impact on mathematical activity, and their benefits and drawbacks. We have come across a number of possible benefits in our analysis:

1. Axiomatization allows one to state results in greater generality.
2. It suggests appropriate generalizations.
3. It suggests appropriate definitions.
4. It allows one to transfer prior results smoothly, or adapt prior proofs to a new setting.
5. It simplifies presentations by removing irrelevant distractions, allowing one to focus on the 'relevant' features of the domain at hand at each stage of a development.

A good philosophical theory of mathematics should help us make sense of these claims.

Similarly, rather than focus on metaphysical justifications for a set-theoretic ontology, we can try to come to terms with some of the mathematical influences of set-theoretic language and methods:

1. They allow us to treat predicates, and properties, as objects in their own right.
2. They allow us to provide uniform definitions of mathematical domains, and to refer to 'arbitrary' elements of those domains and 'arbitrary' functions on those domains.

3. They allow us to obtain results that hold generally, and apply to particular elements and functions that may be introduced in the future, with proofs that are independent of the manner of representing these elements and functions.
4. They can suppress tedious calculational information.

We have seen that these effects are far from cosmetic, and generally force us to relinquish an algorithmic understanding of mathematical assertions and proofs. In other words, the use of set-theoretic methods comes at a serious cost, since these methods ignore or distract our attention from algorithmic issues that are also of great mathematical importance. A good philosophical analysis should put us in a better position to weigh the virtues against the vices.

And what are we to make of Dedekind's general aim of providing a 'conceptual' approach to mathematics? In trying to map out a hierarchy of values and goals in mathematics, one option is to take this to be a goal in its own right: every well-founded hierarchy has to bottom out somewhere. An alternative is to *identify* Dedekind's conceptual approach with the use of axiomatic and set-theoretic methods, which are to be justified by appeal to different mathematical goals, such as simplicity, uniformity, or generality. One might even, paradoxically, try to justify a conceptual approach by its ability to support algorithmic developments, by providing a rigorous framework that can guide the search for effective representations and algorithms. Or one might, instead, try to justify a conceptual approach on pragmatic grounds, in terms of ease of learnability and communicability; on more general esthetic grounds; or on external grounds, like social utility or applicability to the sciences.

In trying to sort this out, it would be a mistake, I think, to expect a simple narrative to provide a satisfactory theory. I believe it is also a mistake to try to fashion such a theory from the top down, before we have begun to make some sense of the basic data. What we need first is a better philosophical analysis of the various value-laden terms that I have just bandied about, and the way they play out in particular mathematical settings. There is enough going on in the development of the theory of ideals, in particular, to keep us busy for some time.

Emmy Noether's 'Set Theoretic' Topology: From Dedekind to the Rise of Functors

Colin McLarty

> 'In these days the angel of topology and the devil of abstract algebra fight for the soul of each individual mathematical domain.' (Weyl 1939, 500)

If Hermann Weyl ever put faces on these spirits they were his good friends, the angel Luitzen Brouwer, and the devil Emmy Noether. Weyl well describes the scope of their ambition. But topology and algebra were not fighting each other. They would come to share the soul of most of mathematics. And the angel and devil had worked together. In 1926 and 1927 Emmy Noether induced young topologists gathered around Brouwer to use her algebra to organize the kind of work on topological maps that Brouwer taught. This created the still current basis for 'algebraic topology.' It was a huge advance for the structuralist conception of mathematics. And it turned Noether from a great algebraist to a decisive figure for twentieth-century mathematics.[1]

The algebra was not her first great work. A recent check of 300 references to Noether in *Mathematical Reviews* found 80% of those with identifiable topic concerned her earlier conservation theorems in mathematical physics. Among many surveys of this work, see (Byers 1996). Yet the algebra is central to her reputation. Alexandroff notes

> Emmy Noether herself was partly responsible for her early work being remembered less frequently than would be natural. For with all the fervor of her nature she was herself ready to forget what had been done in the first years of her mathematical activity, considering these results as standing apart from her true mathematical path—the creation of a general abstract algebra. (Alexandroff 1981, 101).

This algebra would be so general as to apply over all mathematics.

---

[1] (Alexandroff 1932) is a beautiful 50-page introduction to just the topology studied here. Originally commissioned as an appendix to (Hilbert & Cohn-Vossen 1932), this gem highlights Brouwer's and Noether's ideas.

It is widely said that Noether created 'algebraic topology' by bringing groups into *homology theory*. The fullest historical studies up to now are by Jean Dieudonné and Saunders Mac Lane.[2] Mac Lane concludes that Poincaré already knew of homology groups. Dieudonné rejects this, saying Poincaré wrote a great deal about groups and never mentioned homology groups. Mac Lane says no one used these groups until Noether in Göttingen and Vietoris independently in Vienna, about 1926. Dieudonné accepts that. We will see Mac Lane is entirely right about Poincaré. But Noether was not in Göttingen at the crucial moment, nor was Vietoris in Vienna. They were both in tiny Laren, Holland, visiting Brouwer.

The place is important because the event combined Brouwer's ideas in topology (not foundations) with Noether's algebra. The young topologists used some of his technical tools. More to the point, they followed him in giving continuous maps at least as much attention as point sets in topology. His most famous result, his *fixed point theorem*, is explicitly about maps: Every continuous map $f: B^n \to B^n$, from an n-dimensional ball to itself, has a fixed point, a point $x \in B^n$ such that $f(x) = x$. His greatest single paper is titled 'On mappings of manifolds' (Brouwer 1911). His theorems do not all refer to mappings, but his proofs do.[3]

The time is important because Noether was just then developing what she called her set-theoretic foundations for algebra. This was not what we now call set theory. It was not the idea of using sets in basic definitions and reasoning. She took that more or less for granted, as did other Göttingers by the 1920s. Rather, her project was to get abstract algebra away from thinking about operations on elements, such as addition or multiplication of elements in groups or rings. Her algebra would describe structures in terms of selected subsets (such as normal subgroups of groups) and homomorphisms. Alexandroff applied her tools to Brouwer's use of continuous maps—though Vietoris was the first to make it work.

Noether brought something much deeper and more comprehensive to topology than just the use of homology groups. The next section (and a more technical appendix) will show that groups were familiar in homology before her. She brought an entire programme of looking at groups, and other structures in algebra, and other structures outside of algebra like topological spaces, in terms of the homomorphisms between them. She called this 'set-theoretic foundations.' Section 6 describes the programme in general, and Section 6 quotes Noether on her chief tools for making this work: her homomorphism and isomorphism theorems. She credits these to Dedekind and they are probably the most important single case for her famous slogan 'Es steht alles schon bei Dedekind (this is all already in Dedekind).' Section 6 describes what she got from him on this point

---

[2] (Dieudonné 1984), (Mac Lane 1986), and Dieudonné's review of Mac Lane in *Math. Reviews* 87e:01027.

[3] Brouwer's key concept, the *degree* of a map, remains central to the topology of manifolds. It is called the *Brouwer degree* in (Vick 1994, 25).

and how she developed it. It is far from a comprehensive account of Dedekind's influence on Noether and Noether's way of finding things in Dedekind that no one could see before her.[4] Section 6 is a close look at how the young topologists in Laren took these ideas into topology. Then section 6 looks more broadly at the role of Noether's 'set-theoretic foundations' especially in category theory and the current use of 'structure' in mathematics.

1 Homology before Noether

For a rough idea of Poincaré on homology we can consider the torus.[5] Poincaré said a closed curve C on the torus is *homologous to zero*, and wrote $C \sim 0$, if it cut out a piece of the torus.

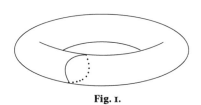

Fig. 1.

A circle around a little piece of the surface cuts out that piece, but a curve drawn around the small circumference of the torus, like the solid and dotted line in the figure, cuts the torus open to become a cylinder without cutting out any piece. It is not homologous to zero.

Let C_α be that curve and C'_α another curve like it, to one side of it, so that C_α and C'_α cut a cylindrical piece out of the torus. For this, Poincaré would write $C_\alpha - C'_\alpha \sim 0$ and say that the difference $C_\alpha - C'_\alpha$ is *homologous to zero*. The minus sign shows the two curves lie on opposite sides of the cylindrical piece. Let C_β be a curve going the whole way around the large circumference of the torus (like the elliptical outline of the torus in the figure). The curves C_α and C_β cut the torus into a cylinder and then a flat sheet, but do not cut out any piece.

The key fact about the homology of the torus is that given any list of three different curves some combination of them will cut out a piece of the torus. Maybe one of them alone will do it, or some two of them, but anyway some combination of them will. At most, two curves can be *independent* on the torus. The two C_α, C_β are independent. So we say the 1-dimensional *Betti number* of the torus is 2.

In a higher-dimensional space, Poincaré said a sum of curves $C_1 + C_2 + \cdots + C_n$ is homologous to zero if they jointly form the boundary of some surface in that

---

[4] On Dedekind, see Jeremy Avigad's contribution to this volume. We do not address Dedekind's work on foundations, including his correspondence with Cantor, which Noether coedited (Noether & Cavaillés 1937), see also (Ferreirós 2005).

[5] See (Herremann 2000), (Pont 1974), (Sarkaria 1999). Poincaré was often vague in his definitions and would change them from one page to the next without comment. We ignore all such issues.

space. He used formal sums with integer coefficients such as

$$3C_1 - 2C_2 - C_3 + 5C_4$$

to say that curve C_3 is used three times, and the opposite to C_2 is used twice and so on. He said two formal sums were homologous:

$$C_1 + 2C_2 \sim C_3 + C_4$$

if and only if their formal difference was homologous to zero:

$$C_1 + 2C_2 - C_3 - C_4 \sim 0.$$

He remarked that 'in homologies the terms compose according to the rules of ordinary addition' (Poincaré 1899–1904, 449–50). He means the associative, commutative, and negation rules:

$$C_1 + (C_2 + C_3) \sim (C_1 + C_2) + C_3$$
$$C_1 + C_2 \sim C_2 + C_1 \qquad C - C \sim 0$$

Poincaré's homology was 'algebraic' as that term was understood at the time, and also in our sense of using group theory. Alain Herreman has nicely laid out the earlier meaning. He quotes Öystein Ore from 1931: 'it might be said that algebra deals with the formal combinations of symbols according to prescribed rules' (Herreman 2000, 26). Ore said that definition was already out of date, but it has appeared as recently as (Mac Lane 1988, 29). It was current in the 1890s and Poincaré's homology was explicitly algebraic in this sense as he manipulated the formal sums of curves and of higher-dimensional cycles.

Further, Poincaré knew his cycles formed groups. The 1-dimensional cycles of a given space, with their formal sums, formed today's 1-dimensional homology group of that space. The 2-dimensional cycles formed today's 2-dimensional homology group. The point is controversial because Poincaré did not write of 'groups' in homology. But homology groups are commutative, and Poincaré systematically avoided referring to commutative groups. For him, 'group' meant a Lie group or at least a transformation group. Even then, if he knew a given Lie group was commutative he preferred not to call it a 'group.' Among his great projects was, in our terms, to find all the commutative subgroups of Lie groups. He worked on them with all the tools of group theory at that time. But he did not call them 'groups.' He called them the *sheaves* (faisceaux) of Lie groups (Poincaré 1916–1956, vol. 5, passim). He felt they were too elementary to be called groups. Detailed evidence of how much Poincaré knew about groups in homology is below in the appendix.

From today's viewpoint, each homology group of a suitable space is *generated* by some finite list of cycles. The 1-dimensional homology of the torus is generated by C_α, C_β as above. Every 1-dimensional cycle on the torus is homologous to a unique sum of multiples of C_α and C_β. This is another way to say the 1-dimensional Betti number of the torus is 2. Poincaré described homology in terms of Betti numbers while he used group theory to calculate with the cycles.

Vietoris rightly wrote to Mac Lane:

> Without doubt H. Poincaré and his contemporaries knew that the Betti numbers and the torsion coefficients were invariants of groups, whose elements were cycles under the operation of addition... Then one worked with the numerical invariants rather than with the invariant groups. That was a matter of 'taste'. (Letter of 16 March 1985 quoted in (Mac Lane 1986, 307).)

Hugo Gieseking in 1912 wrote of 'the Abelian group' of a surface to mean its 1-dimensional homology group (Vanden Eynde 1992, p. 177). Oswald Veblen first published the term 'homology group,' also for the 1-dimensional groups. He lectured on Poincaré's topology in 1916. For the war he became Captain Veblen and eventually published a 'thoroughly revised... more formal' account (Veblen 1922, iii). He gave Poincaré's favourite way to calculate the first Betti number of a polyhedron. The routine naturally produces a group that Veblen said 'may well be called the homology group' (Veblen 1922, 141) noted in (Mac Lane 1978, p. 11). Others took no notice because they already knew that homology cycles formed groups in all dimensions.[6]

Weyl spoke of the group of cycles of a (simplicial) space, and the group of boundaries (Weyl 1923, 393). He spoke group theoretically to say that homology looks at the cycles modulo the boundaries. But he conspicuously avoided forming the quotient groups, the homology groups.[7] That makes all the difference, because it prevented him from saying that continuous (simplicial) maps of spaces induce homomorphisms of homology groups. Of course everyone working in homology had known this fact in some way, and used it, since Poincaré. But no one before Noether saw how powerful, simplifying, and unifying it would be to use the groups in the right way.

2 Noether's set-theoretic foundations

Alexandroff relates Noether's set-theoretic algebra to her influence on topology:

> my theory of continuous partitions of topological spaces arose to a large extent under the influence of conversations with her in December and

---

[6] Thanks to Ralf Krömer for pointing me to Gieseking.

[7] In Poincaré's terminology cycles were identified modulo boundaries in the first place. Today we say Weyl's 'cycles' are curves while Poincaré's are equivalence classes of curves (and analogously for higher dimensional homology). Both senses of 'cycle' are used colloquially today. I thank Erhard Scholz for pointing out that Weyl had a philosophic reason to avoid forming the quotient groups in that he disliked forming sets of infinite sets.

> January of 1925–1926 when we were in Holland together. It was also at this time that Noether's first ideas on the set-theoretic foundations of group theory were developing. She lectured on these in the summer of 1926. Although she returned to them several times later, these ideas were not developed further in their initial form, probably because of the difficulty of axiomatizing the concept of a group by taking coset decomposition as the basic notion. But the idea of set-theoretic analysis of the concept of a group turned out to be fruitful, as shown by the subsequent work of Ore, Kurosh, and others. (Alexandroff 1981, 108) cited (Corry 1996, 247)

The next four sections explain Alexandroff's claims. This one describes the set-theoretic conception and its key tool, the *homomorphism theorems*. The next section quotes Noether's own treatment of these theorems and quotes her one published application to topology. Section 6 compares and contrasts Dedekind on these theorems, before Section 6 returns to Alexandroff and topology in Laren.

Consider the ordinary integers $\mathbb{Z}$ and the integers modulo 3, written $\mathbb{Z}/3$.[8] One natural way to compare them is just to say which integers count as 0 modulo 3. That is the subgroup of integer multiples of 3:

$$3\mathbb{Z} = \{\cdots -3, 0, 3, 6, 9 \ldots\}.$$

Another natural way is to take the homomorphism $q : \mathbb{Z} \to \mathbb{Z}/3$ taking each integer n to its remainder modulo 3:

$$q(3i) = 0, \quad q(3i+1) = 1, \quad q(3i+2) = 2 \quad \text{for all } i \in \mathbb{Z}.$$

The homomorphism determines the subgroup since $3\mathbb{Z}$ is just the set of integers i such that $q(i) = 0$. And the subgroup determines the homomorphism. First the subgroup has a pair of *cosets*:

$$3\mathbb{Z} + 1 = \{\cdots -2, 1, 4, 7, 10 \ldots\}$$

and:

$$3\mathbb{Z} + 2 = \{\cdots -1, 2, 5, 8, 11 \ldots\}.$$

These partition $\mathbb{Z}$ into three disjoint, exhaustive subsets. So define q by the rule

$$q(3\mathbb{Z} + i) = i.$$

Then $\mathbb{Z}/3$ is called a *factor group* of $\mathbb{Z}$ because it comes from dividing $\mathbb{Z}$ up along the lines of the subgroup $3\mathbb{Z}$.

The homomorphism theorem for commutative groups says every onto homomorphism is, up to isomorphism, projection to a factor group. A textbook version from the Noether school was:

Theorem. Whenever a commutative group $\overline{\mathfrak{G}}$ has a homomorphism onto it from a group $\mathfrak{G}$, it is isomorphic to a factor group $\mathfrak{G}/\mathfrak{e}$. Here $\mathfrak{e}$ is the subgroup of $\mathfrak{G}$ whose

---

[8] So $\mathbb{Z}/3 = \{0, 1, 2\}$ with addition given by $1 + 1 = 2$, $1 + 2 = 0$, $2 + 2 = 1$.

elements correspond to the 0 of $\mathfrak{G}$. *(van der Waerden 1930, adapted for commutative groups from vol. 1 p. 35).*

The Noether school contrasted a 'set-theoretic' conception of algebra to what they considered an 'arithmetic' conception in Dedekind's work. Dedekind had noted that divisibility relations between (algebraic) integers can be expressed in terms of *ideals*.[9] To say that 3 divides 6 is the same as saying that the principal ideal (3) contains the principal ideal (6).

$$3|6 \qquad (6) \subseteq (3).$$

Dedekind generalized the divisibility relation to ideals so that whenever one ideal contains another, $\mathfrak{b} \subseteq \mathfrak{a}$, he calls $\mathfrak{a}$ a *divisor* of $\mathfrak{b}$.

Wolfgang Krull said Dedekind's terminology is 'better adapted to the needs of arithmetic, but experience shows it seems strange to every beginner.' The larger ideal is divisor of the smaller one and, worse, the 'greatest common divisor' of two ideals is the smallest ideal that contains them both. Krull said his terminology and exposition will always 'keep in mind (berucksichtigen) the group- and set-theoretic viewpoint.' Given $\mathfrak{a} \subseteq \mathfrak{b}$, Krull called $\mathfrak{a}$ a *subideal* (Unterideal) of $\mathfrak{b}$, or $\mathfrak{b}$ a *superideal* (Oberideal) of $\mathfrak{a}$.[10]

Noether's idea was genuinely deep. She described 'purely set-theoretic' methods as 'independent of any operations' (Noether 1927, 47).[11] These methods do not look at addition or multiplication of the elements of a ring (or a group,etc.). They look at selected subsets and the corresponding homomorphisms. The homomorphism theorem for commutative rings correlates homomorphisms to all ideals. In groups in general, homomorphisms are correlated only to *normal* subgroups. So she looked at groups in terms of their normal subgroups and homomorphisms. She looked at a ring in terms of its ideals and ring homomorphisms, since those correspond in the homomorphism theorem for rings.

Of course, these selected subsets were themselves defined in terms of the addition and multiplication operators, so her methods were not purely 'set-theoretic',

---

[9] An ideal in a ring $\mathfrak{R}$ is a non-empty subset $\mathfrak{a} \subseteq \mathfrak{R}$ closed under subtraction (i.e. any additive subgroup of $\mathfrak{R}$) such that $y \in \mathfrak{a}$ implies $xy \in \mathfrak{a}$, for every $x \in \mathfrak{R}$. Each $x \in \mathfrak{R}$ gives a *principal ideal* (x) containing all the $\mathfrak{R}$-multiples of x. The relation $(x) \subseteq (y)$ says every multiple of x is a multiple of y, thus x is a multiple of y, or y divides x. In the ring $\mathbb{Z}$ of integers every ideal is a principal ideal (n). This is not true in every ring. But it means that in $\mathbb{Z}$ Dedekind's theory of ideal divisors agrees entirely with the classical theory of integer divisors.

[10] All quotes are (Krull 1935, 2). Wolfgang Krull (1899–1971) was strongly influenced by Noether, in 1920, at the start of her long concentration on algebra.

[11] A key element of this was the *ascending chain condition* (Noether 1921). This is called the Noetherian condition on a ring or module in any abstract algebra text today. It describes the ordering of ideals of a ring, or submodules of a module, with no (explicit) reference to operations on the ring or module. Noether established its power in abstract algebra, number theory, and the factorization of polynomials.

but that was her direction. Ore eventually took this up:[12]

> In the discussion of the structure of algebraic domains, one is not primarily interested in the *elements* of these domains but in the relations of certain *distinguished subdomains* ((Ore 1935, 406), quoted in (Corry 1996, 272))

But Ore's results, like the results of Kurosh mentioned by Alexandroff above, have ceased to be of central concern to mathematicians. They focused too much on the order relations among distinguished subsets of a single ring or module, while keeping the idea of homomorphism in the background.[13]

Noether used the homomorphism theorems to prove *isomorphism theorems*, which show that certain relations among the subsets imply that certain morphisms are isomorphisms. These and other ideas served Noether's well-known goal, in Krull's terms:

> Noether's principle: base all of algebra so far as possible on consideration of isomorphisms. (Krull 1935, 4)

Van der Waerden captured the method exactly: To understand ring ideals is to understand their analogy with normal subgroups, for which 'we proceed from the concept of homomorphism!' Over the next two pages he introduced and proved the homomorphism theorem for rings (van der Waerden 1930, 55–7).[14]

3 Noether's 'Abstrakter Aufbau der Idealtheorie'

Noether gave her first clear statement of the 'set-theoretic' conception in (Noether 1927).[15] She was explicit that there are different kinds of algebraic structure, that each has its own homomorphism theorem, and that each homomorphism theorem implies isomorphism theorems for that same structure. We should note that she did not use the term 'homomorphism theorem' although she italicized

---

[12] Alexandroff's claim that she began this in the mid-1920s is confirmed by (Noether 1924), (Noether 1925), (Noether 1926), and the submission date of (Noether 1927). See (Corry 1996, 265 ff.) for the early twentieth-century history of approaches to algebra through substructures and especially Ore's relation to Noether.

[13] Reinhold Baer led Mac Lane to look at transcendence degree of field extensions, and related examples, in these terms in (Mac Lane 1938).

[14] Mac Lane later canonized the homomorphism and isomorphism theorems in the commutative case in his theory of *Abelian categories*. His idea was redone and simplified by Grothendieck into a standard tool in homology theory. The Abelian category axioms are more or less exactly what it takes to state and prove the homomorphism and isomorphism theorems. Indeed, one might say that Grothendieck gave the first purely 'set-theoretic' foundation, in Noether's sense, for any practically important part of algebra.

[15] The paper was written in 1925 and abstracted in (Noether 1924).

the statement. She used the name, and used the theorem even more cleanly, in (Noether 1929, esp. p. 647).[16]

In her paper (Noether 1927) Noether dealt with modules, and spoke of 'remainder class modules' where we have spoken above of 'factor groups.'[17] She said one module M is homomorphic to another $\overline{M}$ when each element of M corresponds to a unique element of $\overline{M}$ in a way that respects the addition law in the modules and the law of ring multiplication. More precisely, where she wrote $M \sim \overline{M}$ we would specify an onto homomorphism $f: M \to \overline{M}$. Noether wrote $\overline{\beta}$ for our $f\beta$, and $\overline{(\beta - \gamma)} \sim (\overline{\beta} - \overline{\gamma})$ for our $f(\beta - \gamma) = f\beta - f\gamma$. Noether also retained the arithmetic terminology, so that a divisor of a submodule is any submodule containing it. Guided by Krull's remarks quoted above, we shall rewrite her theorems in modern notation without significantly altering their meaning.

Noether said that two modules were *isomorphic* if there are one-to-one onto homomorphisms between them in each direction. She began by noting that the isomorphism theorems 'assume only ring- or respectively module-properties and no further axioms.' She then observed that

> If $\mathfrak{A}$ is any $\mathfrak{R}$-module contained in M then one gets a module $\overline{M}$ homomorphic to M — the remainder class module M|$\mathfrak{A}$ — by taking congruence modulo $\mathfrak{A}$ as the equality relation. Each element of M is thereby coordinated with all and only the ones equal to it in $\overline{M}$. To pass from this equality relation in $\overline{M}$ to the identity relation, means to collect all the equal elements of $\overline{M}$ into one class—a *remainder class*—and to conceive these remainder classes as the elements of $\overline{M}$.
>
> *Every homomorphism is generated by such passage to a module of remainder classes*; because if $M \sim \overline{\overline{M}}$ and $\mathfrak{A}$ is the module coordinate to the zero element of $\overline{\overline{M}}$ then as shown above, $\overline{\overline{M}}$ *is isomorphic to the remainder class module* M|$\mathfrak{A}$.

In modernized language, which changes only one word, her first isomorphism theorem then states: *Let $\overline{M}$ be the remainder class module* M|$\mathfrak{A}$ *and $\mathfrak{C}$ a module containing $\mathfrak{A}$. Then there is an isomorphism* $\overline{M}|\overline{\mathfrak{C}} \simeq M|\mathfrak{C}$. Her proof suppresses all mention of the ring or module operations. It deals only with the equality relation:

> This is because congruence modulo $\mathfrak{A}$ implies congruence modulo $\mathfrak{C}$. Elements equal modulo $\mathfrak{A}$ thus remain equal modulo $\mathfrak{C}$. So one can form the remainder class module M|$\mathfrak{C}$—and so equality modulo $\mathfrak{C}$—by first setting the elements equal modulo $\mathfrak{A}$, that is passing to $\overline{M}$, and then collecting the ones equal modulo $\mathfrak{C}$ which expresses equality modulo $\overline{\mathfrak{C}}$ in $\overline{M}$, and thus the formation of $\overline{M}|\overline{\mathfrak{C}}$.

---

[16] Van der Waerden follows the 1929 proofs, especially proving isomorphism theorems from homomorphism theorems in (van der Waerden 1930, vol. 1, p. 136).

[17] A module M over a ring $\mathfrak{R}$ is a commutative group M acted on by $\mathfrak{R}$. Readers not familiar with modules may suppose $\mathfrak{R}$ is the ring of integers $\mathbb{Z}$. Every commutative group M is a module over $\mathbb{Z}$ where each integer acts by multiplication: For example, $3\alpha = \alpha + \alpha + \alpha$, and $-1\alpha = -\alpha$ for every $\alpha \in M$.

Her second isomorphism theorem states that *If $\mathfrak{B}$ and $\mathfrak{A}$ are modules contained in M, and $(\mathfrak{B}, \mathfrak{A})$ is the smallest submodule containing both, then there is an isomorphism*: $(\mathfrak{B}, \mathfrak{A})|\mathfrak{A} \simeq \mathfrak{B}|[\mathfrak{B} \cap \mathfrak{A}]$. Noether proves this by the homomorphism theorem:

> The module $\mathfrak{B}$ becomes homomorphic to $\overline{\mathfrak{B}}$ when $(\mathfrak{B}, \mathfrak{A})$ is set equal to $\overline{\mathfrak{B}}$; and since this makes all elements of $[\mathfrak{B}, \mathfrak{A}]$ and only those correspond to the zero element of $\mathfrak{B}$ the above comment (i.e. the homomorphism theorem) yields the isomorphism.

She quickly repeated the reasoning for commutative rings and adapted it to non-commutative rings (Noether 1927, 39–40).[18]

Noether's footnote to a major theorem explains how it is 'purely set-theoretic.' The theorem is:

> Theorem I. *Assuming the ascending chain condition, every ideal of $\mathfrak{R}$ can be presented as the intersection of finitely many irreducible ideals, that is ideals which cannot be presented as intersections of any proper ideals.*

Her proof again ignores ring operations in favour of the order relation on ideals: if the result is not true, one can construct an infinitely long chain of ascending ideals, contrary to the a.c.c.[19] The footnote says:

> Theorem I obviously holds in just this way for modules, when the ascending chain condition is assumed for systems of modules. That is to say, it has a purely set-theoretic character—at the same time it is the only one we prove using the well ordering of ideals. It deals, independently of any operations, with the following set-theoretic concepts:
>
> Let a set $\mathfrak{M}$ have a distinguished subset Σ of the power set—that is a system of subsets. Assume Σ is well ordered; the elements of Σ may be called Σ-sets. Assume the *ascending chain condition* in Σ: Every chain of Σ-sets $\mathfrak{A}_1, \mathfrak{A}_2, \ldots \mathfrak{A}_\nu, \ldots$, such that each $\mathfrak{A}_\nu$ is a proper subset of $\mathfrak{A}_{\nu+1}$, is finite. A Σ-set $\mathfrak{A}$ is called *reducible* when it is the intersection of two Σ-sets, each properly greater than $\mathfrak{A}$, and in the contrary case irreducible. Then the above translates to: *Every Σ-set can be presented as the intersection of finitely many irreducible Σ-sets.* (Noether 1927, 46–7)

The homomorphism and isomorphism theorems come in again for a theorem, trivial in itself but crucial to polynomial algebra: every irreducible ideal is primary (Noether 1927, 47).[20]

In Göttingen on the 27 January 1925 Noether gave a talk in which she proved the structure theorem for finitely generated commutative groups by abstract methods, and then derived the elementary divisor theorem of arithmetic from

---

[18] Her rings are not required to have a unit. So every ideal is a ring and the second isomorphism theorem gives a ring-isomorphism $(\mathfrak{B}, \mathfrak{A})|\mathfrak{A} \simeq \mathfrak{B}|\mathfrak{B} \cap \mathfrak{A}$.

[19] Here Noether did not assume the well ordering theorem (or axiom of choice) in general but explicitly assumed a well ordering on the elements of a given ring $\mathfrak{R}$. This gives a lexicographic order on finite subsets of $\mathfrak{R}$. Since $\mathfrak{R}$ is assumed to satisfy the a.c.c. every ideal has a finite basis so this gives a well ordering on ideals.

[20] She also proved this in (Noether 1921, 39). By 1925 the proof was simpler and the theory of ideals as a whole was radically simpler and farther reaching.

it—which says that an integer matrix can always simplify into a standard form (a diagonal matrix where each entry on the diagonal divides the next one along) by certain arithmetic steps. She precisely reversed the argument of (Frobenius and Stickelberger 1879) who had more or less founded the organized study of finitely generated commutative groups by deriving the structure theorem from the arithmetic one. Noether claimed her order is clearer and (Lang 1993, p.153) among others today agrees. But she did more than reverse the order.

She made the structure theorem itself less central. Her basic methods apply uniformly to all commutative groups, all rings, etc. So today, even when it is vital that the relevant groups are finitely generated (e.g. in homology theory of manifolds as in (Vick 1994)) the structure theorem may never be mentioned. The finiteness is invoked as little as possible, and replaced as far as possible by simpler general methods.

Noether never published her proof of the structure theorem for finitely generated commutative groups. She published an abstract of it, and its last sentence is her only published reference to homology:

> So the theorem on groups is the simpler one; in applications of the theorem on groups—e.g. Betti numbers and torsion numbers in topology—no recourse to the elementary divisor theorem is needed. (Noether 1926, 104)

She assumed that everyone knows there are groups in homology. Her point is that the groups are simpler than the arithmetic.

4 Dedekind to Noether

Noether's influence on homology was closely tied to what she took from Dedekind on the homomorphism and isomorphism theorems, and to the advances she made from there. In the course of editing his collected works she found that Dedekind's early personal notes on group theory give 'such a sharp work-up of the homomorphism theorem as has only lately again become usual' (Dedekind 1930–1932, vol. 3, p. 446). That is to say, only with herself and her school. But he did it without any name for homomorphisms. The closest he came to defining homomorphisms in these notes is when he said:

> Let M be a group with m objects; and let each object θ in M correspond to an object θ_I in such a way that every product $\theta\phi\psi\ldots\lambda$ of objects $\theta, \phi, \psi, \ldots, \lambda$ contained in M corresponds to the product $\theta_I\phi_I\psi_I\ldots\lambda_I$ of the corresponding objects $\theta_I, \phi_I, \psi_I, \ldots, \lambda_I$. The complex of m_I mutually distinct objects θ_I will be called M_I. (Dedekind 1930–1932, 440)

Then he proved M_I is a group by proving it is closed under products. Dedekind discussed only finite groups. What Noether (quite correctly) recognized as the homomorphism theorem is Dedekind's description of M_I in terms of representative elements from M; and it assumes that M_I is finite.

Noether saw in Dedekind an understanding of everything she intended in her version of the homomorphism theorem for groups. He described the kernel of a homomorphism, its cosets, and the group of those cosets. The restriction to finite groups is incidental to the concepts even if it is thoroughly embedded in the notation and the statements. These ideas had entered mathematical folklore to some extent but (Corry 1996) and the textbooks discussed there support Noether's claim that from Dedekind's time to hers the homomorphism theorem was rarely if ever put this well.

The gap widens around the isomorphism theorems. Noether said that the isomorphism theorems are first found 'in a somewhat more special conception' in Dedekind (Noether 1927, 41n). In fact, Noether there stated the theorems for 'modules' in our current sense—commutative groups acted on by a ring. Her term was 'Modulbereich' and would soon be 'Modul.' Dedekind had a far narrower sense of module, namely as a set of complex numbers closed under subtraction—that is an additive subgroup of the complex numbers. Notice that a module in this sense, if it contains more than 0, cannot be finite and so Dedekind's stated version of the homomorphism theorem cannot apply to it.

This could have been trivial. Dedekind remarked that the isomorphism theorems and other elementary results on modules apply to 'any group,' and his gloss shows that by this he means what we call any commutative group (Dedekind 1996, 82, cf. pp. 65–6). But it relates to a crucial difference: Dedekind did not use quotient groups for his isomorphism theorems because quotient groups are not modules in his sense. A coset of complex numbers is not a complex number. So, while he has the notion of a quotient group, he has a specific reason not to apply it here. For a similar reason (Noether 1921) does not form quotient modules. She does so by 1925.

Dedekind gave much weaker statements of the theorems than Noether. Let us adapt Dedekind's isomorphism theorems to commutative groups. He did not speak of isomorphisms at all but only counted the cosets of various subgroups. Given a commutative group H and subgroup K, let us define with Dedekind (H, K) to be the number of cosets of K in G if that is finite, and 0 otherwise.[21] Let M be another subgroup of G containing K, and I another that need not contain K. Dedekind wrote explicitly numerical equations:

$$(H, K) = (H, M).(M, K) \quad \text{and} \quad (I + K, K) = (I, I \cap K).$$

These simply say $0 = 0$ if any of the subgroups has infinitely many cosets. Noether formed quotient groups H/K of groups by subgroups and wrote group

---

[21] Compare (Dedekind 1930–1932, vol. 3, p. 76), and especially Dedekind's most mature version of the isomorphism theorems (Dedekind 1900, 382 and 384).

isomorphisms

$$(H/K)/(M/K) \simeq (H/M) \quad \text{and} \quad (I+K)/K \simeq I/(I \cap K).$$

These express a great deal whether the quotients are finite or infinite.

Even for finite groups Dedekind's arithmetic captures less information than group isomorphisms, as Dedekind knew. Finite groups may differ widely, while they have the same number of elements. So Dedekind's version is easier and more elegant for some applications, notably in arithmetic, while it is too weak for others. Dedekind never actually lost that information. It has a plain arithmetic meaning, he kept track of it in practice, and indeed all number theorists of the time kept track of such information in one way or another. But he did not express it in his theorems on groups or modules. Noether did.

Noether entirely reconceived the scope of the theorems and the relation between them. Let me be very clear that I do not say these further ideas have no precedent in Dedekind. Noether was famous for saying her ideas *were* somehow already in Dedekind. But she went on in ways that are never stated in Dedekind. They are not brought together as one theme in Dedekind's work, and no one before Noether saw them there.

Dedekind gave the homomorphism theorem as a way of constructing one finite group from another. He stated the isomorphism theorems as a way of counting cosets of his 'modules'—infinite additive subgroups of the complex numbers. Noether stated isomorphism theorems as dealing with isomorphisms and gave a uniform method of proving them from homomorphism theorems for many categories of structures—all groups, commutative groups, groups or commutative groups with a given domain of operators, all rings, commutative rings, rings with operators, and more. The 'set-theoretic' presentation quoted above (from (Noether 1927, note 27, p. 46–7)) shows she foresaw wide application. She later listed six kinds of examples, still not covering all that she used in practice, with an inconclusive tone that makes it clear she could not yet tell the reach of this idea (Noether 1929, 645–7). These examples were all *categories* in today's mathematical sense, with specified structures and specified homomorphisms as objects and arrows, although the term 'category theory' was not yet coined.

When Dedekind hinted at generalizing the homomorphism theorem he suggested just one generalization, which from his point of view was all-inclusive: the case of all groups. But Noether saw that, for example, the homomorphism theorem for rings is *not* included in that for groups. A ring homomorphism is *not* just a group homomorphism that happens to go between rings. It has further properties and the homomorphism theorem for rings addresses those properties.[22]

Noether was more aware of homomorphisms. She not only described algebraic structures 'up to isomorphism' but expressly described structures by way of homomorphisms. She wrote $\mathfrak{M} \sim \overline{\mathfrak{M}}$ where today we write $\mathfrak{M} \to \overline{\mathfrak{M}}$ and her

---

[22] Ring homomorphisms also preserve multiplication. So the homomorphism theorem for rings says onto ring homomorphisms correspond not to all additive subgroups of rings but specifically to ideals.

proofs were built around these homomorphisms. What categorists call 'arrow theoretic thinking' was 'tilde theoretic' in Noether. In fact, by a prescient typographical error one tilde in the first edition of van der Waerden *Moderne Algebra* was printed as an arrow (van der Waerden 1930, vol. 1, p. 33).

5 Blaricum/Laren 1926–27

By the 1920s Professor Brouwer was rarely in Amsterdam. He spent most of his time in Blaricum, then a town of under 3000 people some 15 miles east of Amsterdam. Blaricum adjoins the larger town of Laren, population some 6500 at that time, where Brouwer's friends would stay. The towns were an artistic and Bohemian center surrounded by heath and farmland. Brouwer and Noether had probably first met in Karlsruhe at a 1912 German Mathematical Union (DMV) meeting, and at any rate they were friends by 1919 (information from Dirk van Dalen).

In the academic year 1925–1926 all the promising young topologists of Europe visited Brouwer, including Hopf, Hurewicz, Menger, Ulam, and Urysohn. Of interest to us: Alexandroff arrived in May 1925 and Vietoris that summer. Both stayed through Winter Semester 1926. During this time Brouwer lectured on homology theory (Alexandroff 1969, 117). No trace of the lectures seems to have survived.

One of Brouwer's devices especially attracted Alexandroff, namely the idea of replacing a continuous curve in the Euclidean plane by a finite chain of small steps (Brouwer 1912a). The effect of continuity is achieved by taking infinite sequences of such chains, where the lengths of the steps decrease to 0. Before arriving in Holland Alexandroff generalized this idea into a 'Foundation for n-dimensional set theoretic topology' (Alexandroff 1925b). Like Brouwer, Alexandroff assumes a continuous topological space in which the points will lie.[23] But for Alexandroff this continuous space is an abstract topological space, whereas Brouwer had looked at the real co-ordinate plane.

Intuitively, Brouwer approximated a curve by a finite chain of intervals, and treated an interval as merely a set of two endpoints. Alexandroff approximated a surface by a finite set of triangles (pasted together along their edges) and treated a triangle as merely a set of three vertices. He approximated a 3-dimensional space by tetrahedra pasted together, where a tetrahedron is a set of four vertices, and so on in higher dimensions.[24]

---

[23] This has nothing to do with eliminating continuity in favour of finite sets.

[24] Vietoris used these ideas for *Vietoris homology* that we are about to see (see also (Hocking and Young 1961, 346)). Eduard Čech would use them another way to get *Čech homology* (Hocking and Young 1961, 320ff.).

Alexandroff used this kind of idea to pursue theorems on topological images, that is on continuous onto maps $f: T \to T^*$.[25] These theorems also led to Alexandroff's theory of **continuous decompositions**:

> My theory of continuous decompositions of topological spaces was created in large part under the influence of my conversations with [Noether] in December and January of 1925–1926 when we were in Holland together.[26] (Alexandroff 1981, 108) cited (Corry 1996, 247)

Alexandroff wrote $T^* = f(T)$ to say f is an image, and said:

> Every map $T^* = f(T)$ determines a decomposition of the space T into disjoint subsets $X = f^{-1}(x^*)$ where x^* is any point of the image space T^*. If the map is continuous then all the sets X are closed.
> To approach this we can investigate a priori the decompositions $T = \Sigma X$ of a space into disjoint closed subsets X. (Alexandroff 1926, 556)

So f partitions T into disjoint closed subsets

$$f^{-1}(y) = \{x \in T \mid f(x) = y\}$$

for each $y \in T^*$. An onto group homomorphism $f: G \to H$ partitions G into disjoint cosets

$$K + y = \{x \in G \mid f(x) = y\}$$

for each $y \in H$, where $K \subseteq G$ is the subgroup of all x with $f(x) = 0$. The homomorphism theorem for groups says that the onto homomorphisms from G correspond uniquely (up to isomorphism) to the coset partitions of G. Alexandroff wanted to show that every topological image of T corresponds uniquely (up to isomorphism) to a *continuous decomposition* of T.[27] In fact he can do it only for certain kinds of images. 'Thus one sees that the concepts of continuous map and of continuous decomposition do not correspond to each other in a satisfactory way' (Alexandroff 1926, 559). A satisfactory way would have been a 'homomorphism theorem' for topology. Alexandroff does not use that name but neither did Noether in 1925 (published as (Noether 1927)).

Meanwhile:

> Emmy Noether arrived [in Holland] in mid December. She stayed the whole Christmas holiday in Blaricum and more. At this time she was thoroughly absorbed by her group theoretic lectures in Göttingen' (Alexandroff 1969, 120).

---

[25] To be explicit $f: T \to T^*$ is an image when each $y \in T^*$ has at least one $x \in T$ with $f(x) = y$.

[26] See (Alexandroff 1925a) and (Alexandroff 1926). The first was communicated to the Royal Academy in Amsterdam by Brouwer on 28 November 1925. The second is datelined 'Blaricum bei Amsterdam November 1925' and says 'The first, abstract part of this work is closely related to Frl. E. Noether's recent investigations in general group theory, and partly suggested by them' (Alexandroff 1926, 505). Alexandroff misremembered the timing, but it was Noether's influence.

[27] Precisely defined on (Alexandroff 1926, 557).

Alexandroff was absorbed in her ideas. And Vietoris was talking with Alexandroff if he did not talk with her himself.

Vietoris had a paper communicated to the Royal Academy of the Netherlands by Brouwer on 27 February 1926. In it he cited (Alexandroff 1925a) and (Alexandroff 1926) on images, and an 'oral communication' from Alexandroff on a theorem in topology. In a long footnote Vietoris mentioned a modified version of one theorem, plus 'the proof which was given to me orally in its main points, along with the theorem, by Prof. Brouwer' (Vietoris 1926a, 1008). In this paper, and its more detailed *Mathematische Annalen* version, both written in May 1926, Vietoris generalized Alexandroff's 'set-theoretic foundation' to all metric spaces, and made the first published reference to 'homology groups' beyond the single use by Veblen (Vietoris 1926a, 1010) and (Vietoris 1927, 457).

Like Alexandroff, Vietoris began with the by-then classical homology of polyhedra pasted together from triangles, tetrahedra, and so on. Then he depicted this inside any compact metric space R. In this sense a 'triangle' in R is any three points even if none can be connected to any other by a line in R, a 'tetrahedron' is any four points, and so on.

Since a triangle in R is just any three points it need not have any relation to the topology of R. But the topology of R does determine whether or not it is possible to *subdivide* a triangle $\{a, b, c\}$ in R into six triangles in R each with shorter sides than the original. Is there a 'centre point' with the right distance relations? The picture is obvious although the sides need not exist as lines in R. Vietoris used this to define **fundamental sequences** of polyhedra in R. These are infinite sequences of polyhedra where the side lengths converge to 0. The fundamental sequences in a space R reveal a great deal of its topology, and Vietoris transferred homology to R in this way. Compact metric spaces are much more general than polyhedra. A compact metric space R may have infinitely many holes in it and infinitely many twists. The homology groups need not be finitely generated, and so they reveal much more than the matrix presentations and arithmetic methods that worked for finitely generated groups.

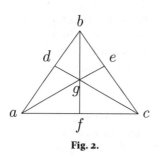

Fig. 2.

Besides that, Vietoris proved theorems on maps $f: M \to M'$. Even if M and M' have finite Betti numbers, the numbers are of little use in studying the map f. Topologists had always studied maps to some extent but in *ad hoc* ways. By treating f as a homomorphism of homology groups, Vietoris made a lot of apparently trivial details overtly trivial in a uniform way. So he cleared up access to the real information.

In correspondence with me (8.2.96) Vietoris emphasized his independence from Noether. She certainly left no trace of using homology groups. Vietoris is certainly the first to publish work with them and I believe he was the first to work with them in any serious way. Just as certainly, whatever the influences on him, Vietoris was original in creating this homology. But he did so in Blaricum/Laren, within months of Noether's visit, in constant contact with Alexandroff. Vietoris's (Vietoris 1926a) grew from his (Vietoris 1926b), which is explicitly concerned with the same problems as (Alexandroff 1925a), which was explicitly inspired by Noether's group theory.

Personal conflicts complicate the picture. Alexandroff made a campaign of slighting (Vietoris 1927) as 'a direct result' of Brouwer's remarks.[28] He never mentioned Vietoris's homology groups. Hopf later said of his own 1928 work 'I do not actually know if the concept 'homology group' had ever before appeared in black and white in the literature' (Hopf 1966, 185). And yet in 1928 he must have known Vietoris's papers. Vietoris spoke on them with Alexandroff as chair at the 1926 Annual meeting of the DMV in Düsseldorf. (Vietoris, letter of 8.2.96, mentions that Noether was in Düsseldorf but he does not know if she heard his talk.)

Alexandroff and Hopf went to Princeton for the academic year 1927–28, on Rockefeller grants arranged by Weyl on Noether's advice. They heard lectures by Alexander, Veblen, and Lefschetz. Hopf said:

> By far the most important to us was Lefschetz—on one hand because he was Alexandroff's ally in the struggle for using algebraic methods in set theoretic topology, and on the other hand since my work, on fixed-point theorems was tied to his groundbreaking work. (Hopf 1966, 185)

Lefschetz was not to use homology groups for several more years. But Hopf became the second to work with them, the following summer. He first published what he called a 'generalized Euler–Poincaré formula' not using homology groups but with an added note:

> As this was being typeset, I found a way to fundamentally simplify the proof of the central theorem, Thm. I, specifically following a suggestion by Fräulein E. Noether. The simplified proof appears in (Hopf 1928) and acquaintance with that note will allow the reader to skip the first three sections of this work. (Hopf 1929, 494)

The simplified version using homology groups actually came out in print sooner, saying:

> I was able to put my original proof of this generalized Euler–Poincaré formula into fundamentally clearer simpler form during a series of lectures I gave in Göttingen in Summer 1928, by introducing group theoretic ideas under the influence of Fräulein E. Noether. (Hopf 1928, 127)

---

[28] (Alexandroff 1927, 552), (Alexandroff 1928, 324), (Alexandroff 1969, 119)

Hopf laid out the algebraic machinery so systematically that the whole idea becomes clear. He showed how to take any of several notions of cycle, and of boundary, and look at cycles modulo boundaries. Of course, the particular results depend on the particular definitions. The method stays the same.

6 The Rise of Functors

In less than a decade Noether's ideas were the textbook basis for homology (Seifert and Threlfall 1934) and (Alexandroff and Hopf 1935). All topologists knew n-dimensional boundaries were a subgroup of n-dimensional cycles, $B_n(T) \subseteq Z_n(T)$. Noether naturally used the homomorphism theorem to replace the subgroup inclusions by the corresponding homomorphisms

$$Z_n(T) \longrightarrow Z_n(T)/B_n(T) = H_n(T)$$

onto homology groups. A continuous map $f: T \to T'$ takes cycles in T to cycles in T', and boundaries to boundaries. Noether used the first isomorphism theorem to replace these relations among subgroups by homomorphisms between homology groups.[29] More fully, for each n, there is a homomorphism of n-dimensional homology groups

$$H_n(f) : H_n(T) \to H_n(T').$$

Brouwer's world of spaces and maps was interwoven with Noether's world of groups and homomorphisms. Two decades after Noether's work, each world would be called a *category* and the interweaving would be called a series of *functors*.

The event suggests two related points bearing on structuralism in the philosophy of mathematics: First, morphisms matter more than structures. Homology does not interweave group structure with topological structure.[30] It interweaves continuous maps and group homomorphisms. And, second, morphisms cannot be defined as structure-preserving functions. Many different classes of functions preserve different aspects, of topological structure—or *reflect* different aspects, which is not the same thing as preserving them. The continuous maps are those that reflect one particular aspect.

---

[29] Here was a glitch because these homomorphisms are generally not onto. Group theorists into the 1950s generally defined homomorphisms as onto, as Noether did in the 1920s. So Noether's students became the first to define homomorphisms as we do today, as in (Alexandroff and Hopf 1935). See (McLarty 1990, 355).

[30] The theory of continuous groups does that.

Brouwer's focus on maps $f\colon T \to T'$ ran against the trend of the time that focused on topological spaces T and subspaces $S \subseteq T$.[31] Yet Brouwer continued an older tradition. The 'Betti numbers' of surfaces were first defined in pursuit of Riemann's complex analysis where analytic functions were studied as continuous maps between Riemann surfaces before continuity on Riemann surfaces even had a clear definition.[32] Topologists long gave names to individual maps, as Alexandroff wrote $T^* = f(T)$ to say f is a continuous onto map. Even the arrow notation $f\colon T \to T^*$ came from topology (Mac Lane 1998, 29).

Algebraists focused very much more on individual groups, rings, etc. and their substructures than on homomorphisms. Dedekind went rather against this trend but we have seen that even he did not use a name for 'group homomorphisms,' let alone name specific homomorphisms $f\colon G \to H$. Noether named many distinct kinds of homomorphisms: homomorphisms of groups, of modules acted on by a group, etc. She still did not name specific homomorphisms or isomorphisms. She would write $\mathfrak{M} \sim \mathfrak{N}$ and say '$\mathfrak{M}$ is homomorphic to $\mathfrak{N}$;' or write $\mathfrak{M} \simeq \mathfrak{N}$ and say '$\mathfrak{M}$ is isomorphic to $\mathfrak{N}$.' This could suggest she thought of homomorphy and isomorphy as relations between structures, and that she merely defined structures 'up to isomorphism' as we say today.[33] But in fact she always referred to specific homomorphisms $f\colon\mathfrak{M} \sim \mathfrak{N}$ and isomorphisms $i\colon\mathfrak{M} \simeq \mathfrak{N}$ between structures. Specific homomorphisms and isomorphisms are tools in her proofs.

Algebraists looked to structures more than homomorphisms because the homomorphisms of algebra *are* defined as structure-preserving functions. A ring homomorphism $f\colon R \to R'$ is defined as a function that carries the operations and constants of R forward to those of R':

$$f(0_R) = 0_{R'} \quad f(1_R) = 1_{R'}$$
$$f(x +_R y) = f(x) +_{R'} f(y) \quad f(x \cdot_R y) = f(x) \cdot_{R'} f(y).$$

The same holds for other kinds of homomorphisms and so it could seem that structure was more basic than morphism.

Continuous maps $g\colon T \to T'$ are more subtle. They do not preserve distance—they can famously 'stretch or contract' parts of a space as in 'rubber sheet geometry.' For a time, notably around Riemann's time, it seemed continuous

---

[31] See much of Poincaré and *inter alia* (Schoenflies 1900), (Dehn and Heegaard 1907), (Schoenflies 1913), (Hausdorff 1914).

[32] See notably (Weyl 1913). Weyl has much to say about the history of the subject and especially Brouwer's role.

[33] It is wide of the mark to say that mathematicians today treat each isomorphism as an identity. Galois theory measures precisely how many ways an isomorphism can fail to be an identity. The practice of defining objects up to isomorphism is more subtle and particularly requires keeping track of morphisms.

maps could be defined as those that preserve limits of infinite series. But already by 1900 more general spaces were in use where that definition did not work. When 1920s topologists converged on the now standard definition, it did not say continuous functions *preserve* anything.

Today a topological space is a set T with certain subsets $U \subseteq T$ called **open**, and the **continuous** functions $g: T \to T'$ reflect open subsets. That is, they carry open subsets of T' back to open subsets of T:

$$\text{If } V \subseteq T' \text{ is open in } T' \text{ then the}$$
$$\text{inverse image } f^{-1}V \subseteq T \text{ is open in T.}$$

Here structure was not prior to morphisms. The open set structure was first identified *because* the desired morphisms reflect it. Preserving it is another matter: maps that preserve the open sets are called **open**, and they are also used in topology but are not the morphisms of most interest; the continuous maps are. Long before these issues were clear, topologists had to be clear about which maps they used.

Noether emphasized homomorphisms, and her influence on homology forced topologists and algebraists to bring their methods together. So algebra shifted to an ever-greater focus on homomorphisms. All this algebra looked too abstract to a tough-minded geometer like Solomon Lefschetz. In the 1920s he made tremendous but obscure progress applying homology in algebraic geometry. It was a major source of his reputation for never stating a false theorem or giving a correct proof. Realizing that his arguments needed serious improvement he went to work on the homology of topological spaces. His first book on it says:

> The connection with the theory of abstract groups is clear.... Indeed everything that follows in this section can be, and frequently is, translated in terms of the theory of groups. It is of course a mere question of a different terminology. (Lefschetz 1930, 29)

But Poincaré had said, in promoting *analysis situs*, that we must not 'fail to recognize the importance of well constructed language' (Poincaré 1908b, 180). Whether or not Lefschetz noticed this passage he shared the thought.

Soon his topology was heavily algebraic (Lefschetz 1942). He kicked off the spate of articles on what became category theory when he asked for an appendix by Eilenberg and Mac Lane (Eilenberg and Mac Lane 1942). This became the standard foundation for algebraic topology and for the huge proliferation of (co-)homology theories in the 1950s and since. The unprecedentedly vast machinery gave unprecedented power in solving concrete problems from topology to abstract algebra to number theory. Lefschetz wrote:

> As first pointed out by Emmy Noether, the proper and only adequate formulation of the relation between chains, cycles, ... requires group theory. (Lefschetz 1949, 11, Lefschetz's ellipsis)

In fact it required Noether's formulation of group theory, and that soon required functors.

7 Appendix: Poincaré on groups in homology

Two basic kinds of evidence show that Poincaré knew of homology groups. First, his homology uses the textbook methods of commutative group theory. Second is his own comparison of homology to **fundamental groups** in homotopy questions. He worked on fundamental groups by a textbook method for producing new groups and he showed the result was homology. He said the rules of homology differ from the homotopy group rules only in that homology is commutative. He knew that his friends, and the textbooks of the time, spoke of 'commutative groups' but he preferred not to call something a group if it was commutative.

Poincaré used algorithms with linear forms and matrices to calculate with homology cycles throughout (Poincaré 1899–1904). This was the standard presentation of finitely generated commutative groups at least since (Frobenius and Stickelberger 1879). Poincaré's algorithm for calculating Betti numbers was Frobenius and Stickelberger's algorithm for calculating group invariants (Poincaré 1899–1904, Complément 2) and (Frobenius and Stickelberger 1879, section 10).[34] All these methods are in the standard textbook (Weber 1899).

Poincaré defined the **fundamental group** of a space U, also called the **first homotopy group** of U today, as the group of paths in U from a fixed base point x_0 back to x_0. Two paths are considered the same if they have a *homotopy* between them: each can be continuously deformed into the other while keeping the base point fixed. The product of one path with another is just the first followed by the second (Poincaré 1895, 239). This is the textbook definition today. He quickly proved it isomorphic to a transformation group permuting the values of integrals or multivalued functions on U. He often described the fundamental group of a given space by generators and relations, which was then the standard presentation of finitely generated (not necessarily commutative) groups. He would find a short list of paths that generate the group (every path is homotopic to some product of ones on the list), and a short list of relations (a few homotopies among products of these paths), which by general group theory imply all the homotopies that hold in the group.

He used fundamental groups to calculate Betti numbers. Given generators and relations for the fundamental group of a space, he would adjoin new relations making the generators commute, then count how many generators remained linearly independent, and that number was the 1-dimensional Betti number (e.g. (Poincaré 1895, 243ff.)).[35] He knew the standard textbook method of his time: If you present a group by generators and relations, and impose further

---

[34] The algorithm was the elementary divisor theorem, which Noether proposed to eliminate from topology by her group theory, as discussed in Section 6.

[35] In later terms, he used the Hurewicz isomorphism of the first homology group with the Abelianization of the fundamental group.

relations you get a new group. But if the new relations imply commutativity Poincaré preferred not to call it a group.

He contrasted homotopy to 1-dimensional homology in precisely these terms: 'in homologies the terms compose according to the rules of ordinary addition; in [the fundamental group] the terms compose according to the same rules as the substitutions of a group.'[36] By the rules of ordinary addition he meant associativity, commutativity, existence of zero and of negatives. He knew these were the 'commutative group' rules. Yet he reserved the term 'group' for groups of substitutions, as in the quote, and he would not call even a substitution group by that name if he knew it was commutative.[37]

---

[36] (Poincaré 1899–1904, 449–50), see also (Poincaré 1895, 241ff) and (Poincaré 1983, 325).

[37] (Poincaré 1901) talked about the '*groupe*' of rational points on an elliptic curve and today we know these points form a finitely generated Abelian group. But Poincaré's operation did not make the curve a group in the sense of algebra. It was not associative and had no neutral element except in trivial cases. Today we alter his operation to get a group in the usual sense. See (Schappacher and Schoof 1996, 62–3).

Tarski on models and logical consequence

Paolo Mancosu

In the last two decades there has been a heated debate on the exact nature of Tarski's theory of logical consequence (Tarski 1936a, 1936b).[1] Since the publication of Etchemendy's papers and book in the 1980s, contributions by, among others, Sher, Ray, Gomez-Torrente, and Bays have provided a much more detailed picture of Tarski's seminal 1936 paper. However, this intense focus on the original publication has led to several disagreements with respect to important interpretative issues related to Tarski's contribution. One of the bones of contention, and the only one I will treat in this chapter, concerns Tarski's notion of model, the key element of Tarski's famous definition of logical consequence. In this chapter I shall offer new arguments (see Sections 2 and 3) to show that Tarski upheld a fixed-domain conception of model in his 1936 paper and that he was still propounding it in 1940.

Let us begin with the definition of logical consequence:

> The sentence X follows logically from the sentences of the class K if and only if every model of the class K is also a model of the sentence X. (Tarski, 1936a, p. 417)

What then is a model? First of all note that X is a sentence and K a class of sentences. In order to define the notion of model for a sentence, Tarski uses the notion of satisfaction of a sentential function. He says:

> One of the concepts which can be defined in terms of the concept of satisfaction is the concept of model. Let us assume that in the language we are considering certain variables correspond to every extra-logical constant, and in such a way that every sentence becomes a sentential function if the constants in it are replaced by the corresponding variables. (Tarski, 1936a, p. 416)

---

[1] For Tarski 1936a (originally in German) I will use the English translation from Tarski 1983. For Tarski 1936b (originally in Polish) I will use the English translation from Tarski 2002. Although the two versions are different at times (see introductory notes to Tarski 2002) I will still refer to 'the' 1936 paper. Page numbers will refer to the translation, when available.

Thus for L a class of sentences we obtain, by replacing non-logical constants by variables, a class L' of sentential functions. The notion of model is introduced next[2]:

> An arbitrary sequence of objects which satisfies every sentential function of the class L' will be called a model or a realization of the class L of sentences (in just this sense one usually speaks of models of an axiom system of a deductive theory). (Tarski, 1936a, p. 417)

The definition of logical consequence provided by Tarski looks prima facie just like the model-theoretic definition of logical consequence we are used to from our courses in logic and model theory. However, whether this is so depends on the notion of model used by Tarski. This is where interpretations differ. Etchemendy in

---

[2] When did the word 'model' become common currency in axiomatics? 'Model', as an alternative terminology for interpretation, seems to make its appearance in the mathematical foundational literature with von Neumann 1925, where he talks of models of set theory. However, the new terminology owes its influence and success to Weyl's 'Philosophy of Mathematics and Natural Science' (1927). In introducing techniques for proving independence Weyl describes the techniques of 'construction of a model [Modell]' (p. 18) and described both Klein's construction of a Euclidean model for non-Euclidean geometry and the construction of arithmetical models for Euclidean geometry (or subsystems thereof) given by Hilbert. Once introduced in the axiomatical literature by Weyl, the word 'model' finds a favourable reception. It occurs in Carnap 1927, 2000/1928, 1930, Kaufmann 1930, and in articles by Gödel (1930b), Zermelo (1929, 1930) and Tarski 1935c and 1936a. The usage is, however, not universal. The word model is not used in Hilbert–Ackermann 1928 (but it is found in Bernays 1930). Fraenkel 1928 speaks about realizations or models (p. 353) as does Tarski in 1936. They do not follow Carnap in making a distinction between realizations (concrete, spatio-temporal interpretations) and models (abstract interpretations). 'Realisation' is also used by Baldus 1924 and Gödel 1929. Among the few variations one can mention 'concrete representation' (Veblen and Young 1910, p. 3 and Young 1911, 43). It should be pointed out here that while the word 'model' was common currency in physics (see 'dynamical models' in Hertz 1956/1894 and Boltzmann 1902), it is not as common in the literature on non-Euclidean geometry, where the terminology of choice remains 'interpretation' (as in Beltrami's 1868 interpretation of non-Euclidean geometry). However, 'Modellen', i.e. desktop physical models, of particular geometrical surfaces adorned the German mathematics departments of the time. Many thanks to Jamie Tappenden for useful information on this issue.

The reader should see Sinaceur 1997, Webb 1995, Guillaume 1994 for the notion of interpretation in mathematics in the nineteenth century and earlier periods.

As for 'structure' it is not used in the 1920s as an equivalent of 'mathematical system'. Rather, mathematical systems have structure. In *Principia Mathematica* (vol. 2 (1910), part iv, *150ff.) and then in Russell 1919 (ch.6) we find the notion of two relations having the same structure. In Weyl 1927, p. 21, two isomorphic systems of objects are said to have the same structure. This process will eventually lead to the idea that a 'structure' is what is captured by an axiom system: 'An axiom system is said to be monomorphic when exactly one structure belongs to it [up to isomorphism]' (Carnap 2000/1928, 127). Here it should be pointed out that the use of the word 'structure' in the algebraic literature was not yet widespread, although the structural approach was. It seems that 'structure' was introduced in the algebraic literature in the early 1930s by Oystein Ore to denote what we nowadays call a lattice (see Vercelloni 1988 and Corry 1996).

For general information about Tarski's metatheoretical conceptions during the period in question see, among others, Bellotti 2002, Czelakowski and Malinowski, 1985, de Rouilhan 1998, Sinaceur, 2000, 2001, Blok and Pigozzi, 1988, Wolenski 1989, Feferman 2004.

his 1988 paper claims that Tarski in 1936 is not working with our notion of model, in that there is no variability of domain of quantification in Tarski's approach:

> 'For the standard account, besides requiring that we canvass all reinterpretations of the nonlogical constants, also requires that we vary the domain of quantification. However, as long as the quantifiers are treated as logical constants, Tarski's analysis always leaves the domain of quantification fixed. Because of this, sentences like (15) will come out logically true on Tarski's account:
>
> (15) $(\exists x)(\exists y)(x \neq y)$
>
> This simply because on the present selection of logical constants, there are no nonlogical constants in the sentence to replace with variables. Thus, such sentences are logically true just in case they happen to be true; true of course in the intended interpretation' (1988, p. 69)

He then later says: 'Not only is Tarski's 1936 account of logical consequence not equivalent to the model-theoretic definition, he clearly avoided such an account with open eyes' (p. 72).

By the fixed domain conception of model one usually refers to an account of logical consequence of the sort described by Etchemendy, i.e. one in which one does not allow domain variation. We will see in Sections 3c and 4 that the fixed conception of model comes in at least two main different versions.

Many scholars are not persuaded by Etchemendy's interpretation. For instance, Sher 1991, Ray 1996 and Gomez-Torrente 1996 have argued against Etchemendy's account of Tarski's 1936 paper. In particular, Gomez-Torrente 1996 is considered by some as having solved all the historical issues. Discussing Etchemendy's historical claims (including the one on non-variability of domains), Scott Soames says

> 'In my view, all of these essentially historical criticisms have been refuted by a variety of scholars, with the most thorough and penetrating refutation I am aware of being given by Mario Gomez-Torrente' (1999, p. 118)

What is Gomez-Torrente's view on the issue of models? According to him the reason why a sentence like $\exists x \exists y \neg (x = y)$ is not a logical consequence of an arbitrary consistent theory rests on the following consideration:

> 'Tarski had in mind mathematical theories in whose canonical formulation the domain of objects of the intended interpretation or interpretations is the extension of a primitive predicate of the language of the theory. Assuming that to specify a domain for an interpretation is nothing but to give an interpretation for such a predicate, Tarski's definition would allow for domain variation in the test for logical consequence. Under this assumption, sentences like '$\exists x \exists y \neg (x = y)$' would not be declared logical truths, because they would be mere unofficial abbreviations for other sentences with relativized quantifiers ('$\exists x \exists y (Nx \& Ny \& \neg (x = y))$' for example). It is natural to picture Tarski as having in mind the idea that

the domain of the intended model or models is denoted by an extra-logical predicate, but without even thinking of formulating or caring to formulate this as an explicit requirement for the application of his definition' (1996, p. 143)

Despite Soames' claim, Gomez-Torrente's solution leaves many questions unanswered. For instance, Bays 2001 marshals several arguments against interpreting Tarski as defending a variable domain conception of model in the 1936 paper.[3]

In this chapter I will provide new evidence and considerations that lead me to the conclusion that Tarski was working with a fixed domain conception of model in 1936. The chapter proceeds as follows. The first section provides background on the notions of interpretation and model up to Tarski 1936. In Section 2, I will criticize Gomez-Torrente's position on the issue of the domain variability of models by showing that it does not account for a wide variety of theories considered by Tarski. Section 3 will provide new evidence from an unpublished lecture by Tarski that in my opinion tilts the balance in favour of the fixed-domain conception of model. Finally, Section 4 will discuss a number of open problems.

1 Axiomatic systems as sets of propositional functions and their interpretations

The study of axiom systems for various disciplines goes back to the third part of the nineteenth century (Pasch, Peano).[4] It is the group of scholars centered around Peano that spent most care in trying to specify what is involved in the axiomatic method. In particular, Padoa and Pieri wrote important articles on the axiomatic method. Pieri (1901) asserted that the primitive notions of any

---

[3] Bays 2001 is an extended attempt to show that a fixed-domain conception of model creates fewer difficulties than Sher, Gomez-Torrente and others claim. In consonance with the spirit of this paper, whose main aim is to determine which notion of model is defended in the 1936 chapter, I will not enter into a detailed discussion of how to account for the Löwenheim–Skolem theorem in this context (but see the comments in the conclusion) or for the logical validity of inferences involving the natural numbers. This is left for another paper. However, Bays' article is the obvious starting point for any such discussion. John Corcoran has since the 1970s claimed as evident that Tarski upheld the fixed-domain conception of model. An overview of his position is now given in the unpublished nine-page typescript 'The Absence of Multiple Universes of Discourse in the 1936 Tarski Consequence-definition Paper'. Unfortunately, given the debates to which the 1936 paper has given rise to, one cannot be left satisfied by claiming either the fixed-domain conception or the relative-domain conception of model as evident. A paper that defends the fixed-domain conception and that comes close, as far as I can tell, to my final concluding remarks is Sagüillo 1997.

[4] See Mancosu, Zach, Badesa (forthcoming) for an introduction to the study of metatheorical properties of axiomatic systems and semantics from 1900 to 1935.

deductive system whatsoever 'must be capable of arbitrary interpretations within certain limits assigned by the primitive propositions', subject only to the restriction that the primitive propositions must be satisfied by the particular interpretation. He applied this notion of interpretation to discuss independence between propositions and referred to Padoa for a more extensive treatment. Alessandro Padoa was another member of the group around Peano. Just like Pieri, Padoa (Padoa, 1901, 1902a) also speaks of systems of postulates as a pure formal system on which one can reason without being anchored to a specific interpretation, 'for what is necessary to the logical development of a deductive theory is not the empirical knowledge of the properties of things, but the formal knowledge of relations between symbols' (1901, 319; transl. p. 121). It is possible, Padoa continues, that there are several, possibly infinitely many, interpretations of the system of undefined symbols that verify the system of basic propositions and thus all the theorems of a theory. He then adds:

> 'The system of undefined symbols can then be regarded as the abstraction obtained from all these interpretations, and the generic theory can then be regarded as the abstraction obtained from the specialized theories that result when in the generic theory the system of undefined symbols is successively replaced by each of the interpretations of this theory. Thus, by means of just one argument that proves a proposition of the generic theory we prove implicitly a proposition in each of the specialized theories.' (1901, pp. 319–320; transl. 121)

The most natural reading of Pieri and Padoa on the issue of interpretations is that once the new specification of meaning for the primitive terms is given, the newly interpreted system gives rise to a new set of propositions, which can be either true or false in the new interpretation. This interpretation foreshadows the conception of axiomatic systems as propositional functions, to be discussed shortly.

Throughout the 1910s the terminology for interpretations of axiomatic systems remains rather stable. Interpretations are given by reinterpreting the meaning of the original constants so that they refer to systems of objects with certain relationships defined on them. Bôcher 1904 suggests the expression 'mathematical system' to 'designate a class of objects associated with a class of relations between these objects'. (p. 128)

During the first 15 years of the twentieth century we encounter a flurry of publications by a group of mathematicians collectively known as postulate theorists (see Scanlan 1991). Inspired by Hilbert and Peano's approaches to axiomatic theories their goal was to investigate systems of objects satisfying certain laws. The laws are expressed in terms of certain undefined primitives and taken as postulates. In 1906–07 Huntington describes the approach as follows:

> 'The only way to avoid this danger [using more than is stated in the axioms, PM] is to think of our fundamental laws, not as axiomatic propositions about numbers, but as blank forms in which the letters a, b, c, etc. may denote any objects we please and the symbols + and x any rules of combination; such a blank form will become a proposition only when a definite

interpretation is given to the letters and symbols – indeed a true proposition for some interpretations and a false proposition for others.... From this point of view our work becomes, in reality, much more general than a study of the system of numbers; it is a study of any system which satisfies the conditions laid down in the general laws of §1.' (1906–07, pp. 2–3)

In 1913 the approach is defined explicitly in terms of the concept of propositional function[5]:

> We agree to consider a certain set of postulates (namely, the postulates stated in chapter II), involving, besides the symbols which are necessary for all logical reasoning, only the following two variables:
>
> 1. The symbol K, which may mean any class of elements A, B, C, ...; and
> 2. the symbol R, which may mean any relation ARB, between two elements
>
> These postulates are not definite propositions – that is they are not in themselves true or false. Their truth or falsity is a function of the logical interpretation given to the variables in such an equation. They might therefore be called 'propositional functions' (to use a term of Russell's) since they become definite propositions (true or false) only when definite 'values' are given to the variables K and R. (Huntington 1913, pp. 525–26)

Notice that this approach differs from the one we found in Padoa, as the basic postulates of the theory are not considered to be propositions.

For the postulates under consideration (defining geometry by means of the relationship of inclusion between spheres) Huntington 1913 gives immediately two interpretations. In the first interpretation K is the class of ordinary spheres including the null sphere; R is interpreted as the relation of inclusion. This interpretation satisfies all the postulates. The second interpretation has K = {2, 3, 5, 7, 10, 14, 15, 21, 210} and R = 'factor of'. This does not satisfy all of the postulates as postulate 4 is false in this interpretation[6].

There are two points that are relevant here. First, the conception of an axiomatic system in terms of propositional functions. Second, the notion of a mathematical system as an interpretation of an axiomatic theory conceived as a set of propositional functions, to be discussed below.[7]

The conception of axiomatic systems in terms of propositional functions is quite widespread in the 1910s and 1920s. It is found, among others, in Whitehead 1907, Huntington 1911 and 1913, Korselt 1913, Keyser 1918a, 1918b, 1922,

---

[5] See Young 1911, p. 172 for an earlier statement by Huntington. Whitehead 1907 is the earliest clear statement of the conception I have been able to find: 'Par conséquent (la classe des points étant indéterminée) les axiomes ne sont pas du tout des propositions: ils sont des fonctions propositionelles. Un axiom, en ce sens, n'étant pas une proposition, ne peut être ni vrai ni faux.'(p. 35)

[6] Postulate 4 says: 'If X is a point of the segment [AB], then [AB] is the 'simple sum' of the two segments [AX] and [BX]'. (Huntington 1913, p. 537).

[7] The quotes are also important for the issue of 'truth in a structure'. I will not deal with the problem in this chapter but see Mancosu, Zach, Badesa (forthcoming).

Langford 1927a, Lewis-Langford 1932, Carnap 1927, 2000 (1928), 1930.[8] Tarski probably encountered it for the first time in Ajdukiewicz 1921, who seems to have proposed it independently of the above sources (Tarski was also familiar with Huntington 1913, which he quotes in Tarski 1929a). Talking about Hilbert's system for geometry (with interpreted logical symbols) Ajdukiewicz says:

> 'Let A(X) denote the logical product of the axioms of geometry whose consistency is to be shown. These axioms are not unambiguous sentences but are susceptible to various 'interpretations', i.e. they are sentential functions defined for a system of variables such as 'point', 'straight' etc. This whole system of variables is represented by the letter X in the symbol A(X). The totality of objects represented by it forms the 'domain' of geometry. The domain of geometry is thus a set of variables whose values are again sets, relations, etc. The axioms are, therefore, neither true nor false but turn into true or false if values are substituted for all variables' (1966/1921, pp. 23–24)

Thus, for instance, when Hilbert replaces 'point', 'straight' and 'point a lies on straight line b' with a set of arbitrary pairs of real numbers (x; y), a ternary relation of real numbers (u; v; z) and 'the equation ux+vy+w = 0', respectively, then he has transformed the axiom system A(X) into A(Ω).

This widespread conception of axiomatic system in terms of propositional functions is at the source of Tarski's notion of model. While in the 1936 article it is in the background (through the process of elimination of the non-logical constants by means of variables) it is explicitly stated in various publications of the period. At the 9th Congress of International Philosophy (1936), and thus at the same time of the publication of Tarski 1936, Tarski presented a paper entitled 'Sur la méthode deductive' (Tarski 1937b). In it he describes the conception as follows:

> 'Let us imagine that in the axioms and theorems of the constructed science, we have replaced everywhere the primitive terms with corresponding variables (in order not to complicate the discussion let us ignore the theorems containing defined terms). The laws of the science have ceased to be propositions and have become what in contemporary logic are called propositional functions. These are expressions having the grammatical form of propositions and which become propositions when one replaces the variables occurring in them by appropriate constant terms. Considering arbitrary objects, one can examine whether they satisfy the axiom system transformed in the way described, that is if the names of these objects, once put in place of the variables, turn these propositions into true propositions; if this turns out to be the case we say that these objects form a model of the axiom system under consideration. For instance, the objects designated by

---

[8] For reasons of space I cannot get into Carnap's conception of model, which is very similar to Tarski's. See Carnap 2000, 1930, 1934a and 1934b and Carnap–Bachmann 1936. For important aspects of Carnap's work during this period see the works by Awodey cited in the references. In general, from the point of view of Tarski's 1936 paper, one should devote a whole paper to the relationship between Carnap and Tarski on the notion of model and logical consequence.

the primitive terms constitute such a model. This model does not play any privileged role in the construction of the science; in deducing this or that theorem from the axioms we do not think at all of the specific properties of this model; on the contrary, from the way in which we reason it follows that not only this special model but every other model of the system of axioms under consideration must satisfy the theorems which we prove.'
(1937b, pp. 331–332)

One finds, almost verbatim, the same characterization in Tarski's 'Einführung in die mathematische Logik' (1937a, pp. 81–82). There is thus no question that this is the conception of model that Tarski in the 1936 paper says comes from the methodology of the deductive sciences. Let me simply remark, to avoid confusion on the part of the reader, that when Tarski speaks of 'arbitrary objects' this is taken to include individuals, classes, relations, etc. Another source of confusion is that Tarski at times speaks of 'concepts' instead of 'objects' (see below). To avoid confusion, I will talk of objects throughout.

2 Tarski on Models and Gomez-Torrente's interpretation

We still need to get a more precise idea about what a model is. In 1935c, Tarski thinks of concrete deductive theories as 'models' ('realizations') of a general deductive theory that contains four primitive predicates (Tarski calls them 'concepts') and specifies that any quadruple of 'concepts' (his terminology) that satisfies the axioms is called a model of the theory (see Tarski 1986, vol. 2, note 1, p. 28). Thus, when dealing with a system of axioms that have only finitely many primitive (non-logical) constants, a model for that system is a finite sequence of objects (of the appropriate kind) that satisfies the propositional functions corresponding to the postulates (see also below, Section 3).

Going back now to logical consequence it is easy to see how the notion of model of an axiomatic system is generalized and put to service in the explication of the relation 'X is a logical consequence of K'. We take every sentence in K ∪ X and we replace the non-logical constants by variables of the appropriate kind. Thus K' ∪ X' is the resulting set of propositional functions. A model of K' is a sequence of objects (of the appropriate kind) satisfying every propositional function in K' (notice that if a sentence L does not have any extra-logical constants then L'=L). Then, X is a logical consequence of K iff every model of K' is a model of X'.

This notion of model is obviously different from the one we are accustomed to. For instance, the domain of the model (the set over which the individual variables range) is not specified explicitly. Moreover, since K and X might contain constants for higher-order objects and quantifications over higher-order objects we must assume that the background logic the models in question have to 'support' is a higher-order logic. The range of these higher-order variables is also not mentioned explicitly. Finally, note that in contemporary model theory we reinterpret the constants over different domains. Here the non-logical constants are not reinterpreted; rather we replace the non-logical constants by variables and consider the satisfaction of certain propositional functions by certain objects (see also note 9 in Bays 2001).

Usually, during the 1930s, Tarski uses as a background logic for his theories a simple theory of types (with or without axiom of infinity).[9] Examples are the theory of real numbers developed in 1931 and the theory of classes developed in Tarski 1933b. This usage is, however, not without exception. Often, he focuses on first-order logic (which can be seen as the first-order fragment of the simple theory of types). This is the case, for instance, when he is studying elementary theories (see the elementary theory of dense orders given in 1936c, vol. II of Tarski 1986, pp. 232–234)[10] Quite often, Tarski is silent about the background system of logic. For instance, the axiomatic theories of Boolean algebra presented in Tarski 1935a and 1938a or the theory of Abelian groups studied in 1938b are given by Tarski without specifying the background logic.

---

[9] In 1939b Tarski also entertains the possibility of having, as an alternative to a type theory, a type-free system resembling Zermelo's set theory as background logical theory.

[10] Concerning the theories Tarski treated in the 1920s, here is a brief attempt to survey the background logic. In the early articles on set theory, the background theory is usually that given by Zermelo's 1908 (see Tarski 1924a, vol. I of 1986, p. 41; Tarski 1924b, vol. I of 1986, p. 67) but no specific background logical system is mentioned. Tarski–Lindenbaum 1926 points out that most of the set-theoretical results of the paper can just as well be obtained in the systems of *Ontology* of Lesniewski or in *Principia Mathematica*. Exceptions are noted on p. 186 of vol. I of Tarski 1986. Of interest in this article is the description of the theory of arithmetic of ordinal numbers (p. 196, vol. I of 1986). This is given in terms of four axioms involving the predicate '<' with *Principia Mathematica* as the background system. In this part of the article Tarski shows his mastery of independence proofs. Remarkable is also the fact that the system of *Principia* and the Ontology of Lesniewski are seen as systems of set theory (see p. 200; the same in 1929b, p. 241, where Tarski remarks that the system of *Principia*, unlike Zermelo–Fraenkel set theory, cannot even prove the existence of a single infinite cardinal number). The next item of interest is 'Les fondements de la géométrie des corps'. As the English translation is quite different from the original 1929 article, it is quite important to refer to the original. The background system here is Lesniewski's ontology. However, in note 1 on p. 229 Tarski specifies that he is using 'class' as it is used in *Principia* and that he is treating spheres as objects of the lowest rank in the system of *Principia*, that is as individuals, and points (classes of spheres) as objects of second rank. He states that the axiom system considered is categorical but the reader should note that the notion of categoricity appealed to is the Veblen notion and not the stronger one that Tarski considers in 1935–36.

In 1930b Tarski begins emphasizing that a theory can become an object of metatheoretical study only if it based on a determinate logic (see 1930b, pp. 347–8 of vol. I of Tarski 1986). Moreover, the metatheoretical enterprise requires a general logical basis, which Tarski says can be taken to be that offered by *Principia* without assuming choice or infinity (he uses infinity but claims that its use is not essential). The 1931 article 'On definable sets of real numbers' (vol. I of Tarski 1986, pp. 519–548) is as far as I know, the first place where Tarski explicitly uses the simple theory of types as the background logical theory for the theory of real numbers (and in the metatheory for studying the system of real numbers). For the importance of simple type theories as a standard logical framework in the 1920s and 1930s see Ferreiros 1999.

It is worthwhile, for later discussion, to single out two important logical systems: STT+I and STT. 'STT' stands for the simple theory of types; 'I' stands for the axiom of infinity. Of course, Principia has a ramified theory of types but Tarski, starting in 1931, always prefers to use the theory of types in its simple form and for this reason I have singled out the above systems. In any case, whenever a deductive theory is formalized within the background of STT+I, or a fragment thereof, the objects constituting the 'universe of discourse' of the theory are taken to be part of the individuals that form the class of lower complexity in STT. Thus, in the 1931 article on the theory of real numbers the real numbers are taken to be individuals.

So far we have established that models are in general finite sequences of 'objects' that satisfy the propositional functions corresponding to a theory in which the non-logical constants have been replaced by variables of the appropriate kind. The conflict of interpretations begins when we ask what is the range of the individual variables. We have seen that in 1988 Etchemendy claimed that Tarski's 1936 account of logical consequence is not equivalent to the usual model-theoretic definition. The reason, according to him, is that there is no mention of domain variability in Tarski's definition of logical consequence and that models share a fixed domain of individuals. In my opinion, the great virtue of Etchemendy's position here is the natural account it yields of Tarski's claim, let us call it (LM), in the 1936 paper (p. 419) according to which if one treats all non-logical constants as logical constants, the notion of logical consequence coincides with that of material consequence. In a variable-domain conception of model it is hard, if not impossible, to make sense of the claim (see also Bays 2001, pp. 1078–79).[11] We have seen that Gomez-Torrente 1996 disagrees with Etchemendy but remarkably he says nothing at all about Tarski's claim (LM). He might be probably tempted to take on this issue the line taken by Ray 1996:

> 'Tarski's 1933 work has, in effect, already shown us how truth-theoretic semantics will fit together with his semantic account of the logical properties for which domain-relativization is crucial. I think it gives us strong (though not historically decisive), additional reason *not* to suppose that Tarski just missed the need for domain relativization when it came to logical consequence, i.e. we have further reason to attribute to Tarski only the lesser of the two errors [the mistake of thinking that logical and material consequence coincide when all terms of a language are treated as logical constants, PM]. This judgment, in turn, tends to undermine Etchemendy's divergence argument, because that argument presupposes that Tarski made the greater error [the mistake of missing the need for domain variation, PM].' (1996, p. 630)

---

[11] I would like to point out that the Polish text makes an 'iff' claim while the German text has an 'if' claim: 'The extreme would be the case in which we treated all terms of the language as logical: the concept of following formally would then coincide with the concept of following materially: the sentence X would follow from the sentences of the class K if and only if either the sentence X were true or at least one sentence of the class K were false' Tarski 2002 (pp. 188–189) For a pointed answer by Etchemendy to Ray see Etchemendy 1999, note 13.

In short, those who claim that there is domain variability in Tarski's account of logical consequence either do not address Tarski's (LM) claim or simply dismiss it as an obvious mistake. Obviously, this is quite unsatisfactory.

Let us now move to the positive part of Gomez-Torrente's 1996 account. While I agree with much of Gomez-Torrente 1996, I find myself disagreeing with his treatment of the issue of domain variation in the definition of model contained in Section 4 of his paper. I will repeat it for the reader's convenience:

> 'Tarski had in mind mathematical theories in whose canonical formulation the domain of objects of the intended interpretation or interpretations is the extension of a primitive predicate of the language of the theory. Assuming that to specify a domain for an interpretation is nothing but to give an interpretation for such a predicate, Tarski's definition would allow for domain variation in the test for logical consequence. Under this assumption, sentences like '$\exists x \exists y \neg (x = y)$' would not be declared logical truths, because they would be mere unofficial abbreviations for other sentences with relativized quantifiers ('$\exists x \exists y (Nx \& Ny \& \neg(x = y))$' for example). It is natural to picture Tarski as having in mind the idea that the domain of the intended model or models is denoted by an extra-logical predicate, but without even thinking of formulating or caring to formulate this as an explicit requirement for the application of his definition' (1996, p. 143)

The claim derives its prima facie plausibility from the fact that many mathematical theories considered by Tarski at the time of writing his 1936 paper contain an extra-logical predicate that is meant to characterize the 'domain of discourse' of the mathematical theory. Gomez-Torrente refers to the theories presented by Tarski in his 1937 introductory book. Consider, for instance, the theory T on the congruency of segments presented by means of a language containing S and $\cong$ as primitives and two non-logical axioms:

1. for any element x in S, $x \cong x$
2. for any x, y, z in S, if $x \cong z$ and $y \cong z$ then $x \cong y$.

A model for T is given by any pair (K, R) satisfying the axioms. Variations of K would correspond to the variation of domains for individual variables in the standard model-theoretic account. Since $\exists x \exists y (\neg(x = y))$ would here be a short hand for $\exists x \exists y (Sx \& Sy \& \neg(x = y))$, this immediately yields that $\exists x \exists y (\neg(x = y))$ is not a logical consequence of T (as it is easy to display a one-element model of the theory).

Despite the prima facie plausibility of such a claim, the examples of mathematical theories provided by Gomez-Torrente as illustration of his point do not settle the issue concerning the fixed domain vs variable domain issue, because those examples (taken from Tarski 1937 and other articles of the period) can be accounted for straightforwardly in a fixed-domain interpretation (see also Bays 2001, pp. 1711–1712). We could think of S as taking different interpretations within the same 'universal' class of individuals provided by the background theory of types.

The problem then is: can these two interpretations be shown to differ substantially so as to adjudicate the issue of where Tarski stands in 1936? I claim they can.

First of all notice that the metatheoretical constraint posed by Gomez-Torrente's interpretation has the following two consequences.

(1) The constraint would force the predicate S to have a non-empty extension. In fact, $\exists x(x = x)$ is a theorem of the background logic and thus the metatheoretical constraint would always force the theory T to have as an axiom $\exists x(S(x)\&x = x)$.
(2) The constraint is such that every theory T formulated with a non-logical constant S characterizing the domain must be inconsistent with the claim $\exists x \neg S(x)$. Since the variable x must be restricted to objects in S, the claim is equivalent to $\exists x(S(x)\&\neg S(x))$. But that's obviously unsatisfiable.

It is not hard to show that Tarski would not accept (1). On p. 145 of his 'Einführung in die mathematische Logik', Tarski is discussing a theory A formulated by means of the predicate Zl (for Zahl) and he remarks that the theory does not prove or disprove the statement $\exists x Zl(x)$. Since, as I pointed out, under Gomez-Torrente's constraint $\exists x(Zl(x)\&x = x)$ (and thus $\exists x Zl(x)$) would have to be an axiom (resp. a theorem) of the theory, it follows that the constraint is incorrect when judged against Tarski's practice.

Concerning 2, in what follows I will show that Tarski would reject that every mathematical theory he is considering must be inconsistent with the statement that there are individuals that fall outside the 'domain of discourse'. While the argument given for (1) already proves the point (since in the case Zl is empty the theory A still proves the theorem $\exists x(x = x)$ forcing an element to be in the complement of Zl), I will use a different set of examples. I will provide evidence that Tarski and other logicians at the time made a distinction between 'range of the quantifiers' (or range of significance of the individual variables) and 'domain of discourse', that is they entertained theories that, while presenting a predicate S for the 'domain of discourse', either prove $\exists x \neg S(x)$ or simply do not decide the issue either way.

In order to simplify the discussion, from now on I will restrict myself to axiomatic systems with two undefined primitive symbols, i.e. two non-logical constants, and treat them in terms of propositional functions f(x) (for a class of objects) and g(x,y) for a relation.[12] The first thing to observe is that, as we have seen, most axiomatic systems studied at the time were intended to characterize 'mathematical systems' (K, R) where K is the value of f(x) and R (generally, but not always, a relation on K) the value of g(x,y), so that all the postulates are

---

[12] Generalizing to systems with n non-logical constants for classes and m non-logical relations is easy. Much of the work on postulate theory at the time involves the study of mathematical systems formed by a class with operations defined on them, say (M, +).

My discussion in the text covers these cases too, as we can think of n-ary operations in terms of n+1-ary relationships. This is already pointed out in Bôcher 1904.

true on (K, R). This fact has led many interpreters (including Gomez-Torrente) to forget that the range of significance of the individual variables occurring in the propositional functions is wider than the 'universe of discourse' K. The distinction was obviously salient to many people involved in axiomatic studies. Langford, one of the foremost postulate theorists, says:

> 'the set [of postulates] (a)-(d) [for dense linear orders, PM] places no restrictions upon things not satisfying the function f, that is, not belonging to the class determined by this function. In connection with any interpretation of the set, we are interested merely in relationships among those things belonging to the class determined by the interpretation put upon f; this class constitutes, as it were, the 'universe of discourse'. If, for instance, we allow f(x) to mean 'x is a planet' and g(x, y) to mean 'x is larger than y' we are concerned solely with size among planets. Nevertheless, the propositions resulting from this interpretation will be significant for other things, and will always be satisfied by them' (Lewis-Langford, 1932, p. 353)

Some interesting metatheoretical applications required keeping the distinction between 'universe of discourse' and range of significance of the individual variables clearly in mind. In an article by Langford (1927a), which Tarski studied carefully and used in his seminars in Warsaw in the 1920s, we find an axiomatization of dense linear orders without end-elements that explicitly requires two axioms determining how objects not contained in the 'universe of discourse' behave. The last two axioms of Langford's axiomatization are very important for our goals.

While the first axiom states that no object x (no restriction to K) is such that xRx, axioms 2 to 8 state properties relativized to objects in K. For instance, axiom 2 says that for all x, y, if x, y are in K and $x \neq y$, then Rxy or Ryx.

Langford then adds:

> 'these properties with some modifications, are the ones usually assigned for this type of order. They are, however, with the exception of (1), confined to assertions relevant to elements in the class K, and it is customary to omit any mention of properties belonging to elements not in K. But it is necessary in the present case to consider such properties – otherwise some important theorems break down.' (1927a, p. 21)[13]

He then adds postulates 9 and 10 to the effect that

(9) 'If x and y do not both belong to K, then Rxy fails.'
(10) 'there are at least n elements not in K.'

This fact is highly relevant for the debate that has surrounded the interpretation of Tarski's notion of model in the article on logical consequence.

---

[13] The reason why Langford thinks he needs to state things that way need not detain us here. On Langford's work and its influence on Tarski see Scanlan 2003.

I will first show that Tarski entertains theories that, like the one given above by Langford, state facts about objects falling outside the 'universe of discourse', i.e. K in the previous example.

Consider a given axiomatic theory with only two primitives. I will denote the axiomatic theory by A(f(x), g(x,y)) to point out that the two non-logical constants have been replaced by variables that stand for classes of individuals and binary relations between such individuals. A model of such a theory, if there is any, is given by a pair (K, R). It is quite possible that this model also satisfies a sentential function such as $\exists x \neg f(x)$. Thus, if $\exists x \neg f(x)$ can be satisfied in a model (K, R) we have to remark that the range of the quantified variables is not something that is given explicitly in the presentation of the model (K, R). We have seen that Langford discussed theories of this sort. Did Tarski?

There are two articles by Tarski written around the period of his work on logical consequence that show that Tarski was in fact careful to keep the above distinction between 'universe of discourse' and range of the quantifiers carefully in mind. The articles are Tarski 1935b and Tarski–Lindenbaum 1936. The context in which the issue emerges is the following. In 1935b Tarski is discussing the problem of the completeness of concepts. One of his main results consists in showing how this notion of completeness of concepts relates to the notion of categoricity. He says: 'As is well known a set of sentences is said to be categorical when two arbitrary 'interpretations' ('realizations') of this set are isomorphic'. In an appended note he refers to Veblen 1904 as the source of the notion. Let us recall that in this article, Tarski is working within the background of a simple theory of types. In particular, there is a 'universal' class of individuals V.

For the theorem he is after, Tarski needs, however, a stronger notion of categoricity than the one given by Veblen. In note 15 he says:

> 'we use the word 'categorical' in a different, somewhat stronger sense than is customary: usually it is required of the relation R [...] only that it maps x', y', z', ... onto x", y", z",... respectively, but not that it maps the class of all individuals onto itself. The sets of sentences which are categorical in the usual (Veblen's) sense can be called intrinsically categorical, those in the new sense absolutely categorical. The axiom systems of various deductive theories are for the most part intrinsically but not absolutely categorical. It is, however, easy to make them absolutely categorical. It suffices, for example, to add a single sentence to the axiom system of geometry which asserts that every individual is a point (or more generally one which determines the number of individuals which are not points).'(Tarski 1983, pp. 310–11, note 1)[14]

---

[14] A possible example to clarify the above distinction is the following. Suppose V, the class of individuals, is countable. Take the axioms for the first-order theory of a dense linear ordering without endpoints. Consider an ordering < on V that satisfies the axioms for the theory. Consider now a countable subset A of V such that A $\neq$ V and define a relationship $<_A$ on it satisfying the axioms of the theory. Since (V, <) and (A, $<_A$) are both countable they are isomorphic, by a classical theorem of set theory. However, no isomorphism between (V, <) and (A, $<_A$) can send V one-one onto V, for

Theorem 4 in Tarski's paper is false, as he warns us in note 19, if categoricity is taken in Veblen's sense.[15]

A similar warning is given in Tarski–Lindenbaum 1936 concerning the applicability of a certain theorem proved in the text. This 'must be restricted to such axiom systems from which it follows that there are no individuals outside of the domain of discourse of the theory discussed' (Tarski 1983, 392).

From the above quotes the following facts are evident:

1. Most theories that are intrinsically categorical (such as geometry and arithmetic) do not prove that every individual is a point (in geometry) or every individual is a number (in arithmetic). One can add such axioms to the original theory and then absolute categoricity and intrinsic categoricity coincide.
2. In general, the 'domain of discourse' of a theory does not coincide with the range of significance of the individual variables (i.e. the range of the individual quantifiers). This only happens when a special axiom is added to the theories in question.

The upshot of the above for the Gomez-Torrente interpretation is that there is no evidence that Tarski thought of most of the theories he treated as pre-supposing a restriction of the range of the quantifiers to the 'universe of discourse' of the theory as he claims that most examples of theories that are intrinsically categorical fail to be absolutely categorical because they do not have as an axiom or as a theorem the statement $\neg \exists x \neg S(x)$. Thus, $\exists x \neg S(x)$ is at least consistent with them or, in some cases, even provable in them. But then one cannot relativize the existential quantifier to that same predicate S.

Now it might seem paradoxical that I appeal to the articles Tarski 1935b and Tarski–Lindenbaum 1936 as evidence against Gomez-Torrente as the results on categoricity stated in those papers are used even by Bays 2001 (p. 1710, note 14) as evidence that Tarski, despite pre-supposing the fixed-domain conception of model in the 1936 paper on logical consequence, also worked comfortably with a variable-domain conception of model:

> 'In [1935b], for instance, Tarski proves two theorems concerning the categoricity of several (second-order) systems of axioms. First, he proves that the axioms for second-order arithmetic are categorical *on the assumption that these axioms include an axiom stating that every individual is a number.* Second, he proves that there is a categorical set of axioms which characterize the real numbers. Clearly, these two theorems cannot be jointly accepted on a fixed-domain conception of model. For, on such a conception, the first

---

by assumption A is strictly included in V. Thus the theory is Veblen categorical but not absolutely categorical.

[15] There are several passages in the article where Tarski keeps reminding the reader that certain results only works if the theory has an axiom that forces the identity of the universe V with say the points of geometry or the natural numbers. See, for instance, Tarski 1983, p. 313, note 2, where Tarski says: 'This is exact only if the axiom system of arithmetic contains a sentence to the effect that every individual is a number'.

result would show that the number of objects in the world is merely countable (since we can find some model of the natural numbers with the whole world as its domain), while the second result would show that the world is uncountable (since it contains enough individuals to construct a model for second-order analysis).' (Bays 2001, p. 1710)

I will come back in the conclusion to these issues raised by Bays.

In my opinion, Tarski 1935a and Tarski–Lindenbaum 1936 show that Tarski in publications around the logical consequence paper is dealing, among other things, with theories expressed within the background of simple type theory and that the range of significance of the individual variables is determined by the class of individuals corresponding to the interpretation of the individual variables assumed for the simple theory of types (that is, the universe V). For this reason, when specifying a model for a theory, there is no need to specify what the range of the individual variables is; as for the 'universe of discourse' of the mathematical theory in question this will be given by a class taken as the value of the variable corresponding to one of the primitives of the theory (K as the value of f(x) in the examples above).[16]

Further evidence can be adduced against Gomez-Torrente's claim. Tarski did in fact come back to explicating what was involved in his notion of categoricity and explicitly denied that his mathematical theories are to be thought of as having a predicate characterizing the domain. This is found in correspondence Tarski had with Corcoran concerning changes for the second edition of *Logic, Semantics and Metamathematics*. In 1979 John Corcoran had raised some objections to Tarski's claim on Veblen vs absolute categoricity to which Tarski replied in 1980. (This material is found in the Tarski papers at the Bancroft Library at U.C. Berkeley.)

Tarski first instructs Corcoran to replace on p. 311 (line 14) 'usual (Veblen)' by 'customary'. Then he adds a clarification in reply to two objections by Corcoran (contained in folder 9.13, sheets dated 8–22–79) where Corcoran objects first, that Veblen categoricity is not the same, as Tarski implies, as intrinsic categoricity and second, that intrinsic categoricity and absolute categoricity are the same. Tarski replies:

> Re your objection to this line and the following ones. I do not claim that my definition of intrinsic categoricity is exactly equivalent to the one of Veblen. To decide this question I would have to know what Veblen understands by a model of a theory. What would be a model for him if the theory discussed were provided, say, with 2 or 100 of unary predicates and (for simplicity) with no n-ary predicate for $n \geq 2$? Or a contrary case: the theory provided with no unary predicates but with some predicate of rank $n \geq 2$? From what you find on the same p. 311 you will see e.g. that I am interested

---

[16] Note, moreover, that not in all cases are we going to have a model given with a 'universe of discourse' over which the primitive relations and/or function symbols are defined. In 1935c, for instance, Tarski gives an axiomatization of dense linear orders without end-elements in which the only primitive is a binary relationship R. A model for such a theory is given by a finite sequence containing only the relation R (the 'universe of discourse' is implicitly defined by the field of R).

> in a theory provided with a ternary predicate as the *only* [underlined by Tarski, PM] primitive predicate – this is not Veblen's case. At any rate I have removed, as you see, the reference to Veblen, so that the objection is not applicable. As regard your second objection – the alleged equivalence of intrinsical and absolute categoricity, your argument is wrong. Consider an intrinsically categorical system with two primitive predicates a unary U and a ternary R; to simplify matters assume that one of the axioms of this system is
>
> $$(x, y, z): R(x, y, z). \supset .U(x).U(y).U(z)$$
>
> Certainly there may be two models of this system ⟨U', R'⟩ and ⟨U", R"⟩ such that there are just two elements of the universal class V which do not belong to U', and just three such elements which [do not, interpolated by PM] belong to U"(*). Then obviously there is no function which maps in a one-one way both V onto V and U' onto U"; thus our system is not absolutely categorical.
>
> (*) I assume that there are no axioms of our system which ascertain anything about elements of V not belonging to U, since this is irrelevant for intrinsical categoricity. I also assume that the axioms [unreadable, PM] the infinity of the set U. (Tarski Papers, Box 9. Folder 9.11, p. 43a.).

Two points are essential for our discussion:

1. Tarski emphasizes that in the case discussed he is 'interested in a theory provided with a ternary predicate as *the only* primitive predicate'. Thus, one cannot claim that Tarski always pre-supposes that there will be a unary predicate characterizing the domain of the theory. By the way, in Lindenbaum–Tarski 1936 (Tarski 1986, vol. 2, pp. 208–209) the theory of geometry appealed to has only a four-placed predicate of congruency between point pairs.

2. The simplification adopted in the discussion is revealing. Tarski says: 'Consider an intrinsically categorical system with two primitive predicates a unary U and a ternary R; to simplify matters assume that one of the axioms of this system is $(x, y, z): R(x, y, z) .\supset .U(x).U(y).U(z)$' This means that in general one does not assume that the objects satisfying the relationship $R(x, y, z)$ have to be in the extension characterized by a unary predicate U. In other words no assumption of cross-binding is given in general between U and R as would be necessary in the Gomez-Torrente interpretation (since the interpretation of U would have to be the domain of discourse of the model over which $R(x, y, z)$ must take its values). Of course, specific mathematical theories might force the cross-binding through the axioms, but that's another issue.

There is one more important issue related to Gomez-Torrente's reconstruction that cannot be left unmentioned. In order for his proposed reconstruction to go through, it is essential that the background logic contains no existential assumptions. He says:

> Tarski naturally intended his definition to be applicable not only to purely logical theories, but also to mathematical theories with special

mathematical primitives and postulates. In the works of this period Tarski considers several mathematical theories formalized using a logical apparatus, or 'logical basis', to use Tarskian terminology, without any cardinality assumption. Generally, this logical basis for formalization is again the calculus of levels, but without the axiom of infinity.(1996, p. 141)

But that is a puzzling comment as Tarski did in fact consider many theories as formalized within a logic with strong existential assumptions. I will come back to this in the next section but the relevance of the point is related to one aspect of the notion of model that Tarski emphasizes (see Tarski 1937a, p. 83). If we have a model of an axiomatic system T then that model also satisfies all the theorems of the axiomatic system. But notice that all the axioms and theorems of the background logic will also be theorems of the axiomatic system T. In particular, if the axiom of infinity is part of the background logic the model would have to satisfy the axiom of infinity. If we restrict the range of the quantifiers to the 'universe of discourse' this would come into conflict immediately with any interpretation in which the 'universe of discourse' is finite. But if we make the distinction between 'universe of discourse' and range of the individual variables then there is no conflict.

Where does this leave us? I think that the evidence adduced so far shows that the original strategy by Gomez-Torrente does not work because it forces in every theory the identity between the intended domain of the theory and the range of the quantifiers. But Tarski thinks that most theories (arithmetic, analysis, geometry) are such that that identity does not hold. This suggests that the range of the quantifiers is always to be taken as the 'universal' domain of individuals. But against this speaks the fact that we have strong categoricity results for arithmetic, analysis and geometry that would seem to force, on a fixed domain conception of model, the (fixed) universe of individuals to be at times countable and at times uncountable. I will discuss in the conclusion whether these absolute categoricity results can be accounted for in the fixed conception of model.

I thus take the above considerations to show the weakness of the Gomez-Torrente strategy but not as decisive as evidence for the fixed-domain interpretation of model, or a variation thereof.

In the next section I will provide a further piece of evidence that speaks in favour of the fixed-domain interpretation.

3   Tarski's 1940 lecture on completeness and categoricity

In this section I will give new archival evidence that speaks in favour of thinking that Tarski had in mind a fixed conception of model in his theory of logical consequence. In 1940 Tarski gave a lecture at the Logic Club at Harvard entitled 'On the completeness and the categoricity of deductive systems' (Tarski 1940).[17] It contains some important (and I claim decisive) evidence concerning the issues I have been discussing.

Tarski's general aim in the lecture is to develop semantical analogues of the notion of syntactic completeness for a deductive theory by means of the notion of semantic completeness and to carry out related investigations on the notion of categoricity.[18]

The motivation for developing these semantical notions of completeness is given by Tarski through some historical reflections on attempts to capture syntactically the notion of logical validity. He begins with a familiar distinction between logical and non-logical sentences as applied to an arbitrary deductive theory:

> In what follows we assume that the concepts (constant terms) of a deductive theory are divided into two classes, the logical and the non-logical, to the first of which belong in any case the constants of the calculus of sentences and the quantifiers. Correspondingly, we divide the sentences of our theory into two classes, the logical and the non-logical, depending on whether they contain exclusively logical constants or not. (p. 3)

The logically valid sentences are a special subset of the logical sentences:

> Among the logical sentences we single out the logically valid sentences. This is usually done in an axiomatic way: the logically valid sentences are defined as those which can be obtained by applying the determined rules of inference to the given logical axioms.
>
> The class of logically valid sentences forms the logical basis of the given deductive theory. (p. 4)

---

[17] A brief summary of this text has been published by Jan Tarski and Jan Wolenski in 'History and Philosophy of Logic' without an exact specification of its location (see Tarski 1995). I was recently able to locate this text in Carton 15 of Tarski's archive in the Bancroft Library at U. C. Berkeley. J. Tarski and J. Wolenski conjecture that this might be the text of the second of a series of lectures given by Tarski at Harvard in 1939. However, this does not seem right. I base my dating of the lecture on the following footnote found in Quine, Goodman 1940: 'The latter notion (synthetically complete), under the name 'completeness relative to logic', is due to Tarski. It is easier to formulate than the older concept of categoricity, and is related to the latter as follows: systems that are categorical (with respect to a given logic) are synthetically complete, and synthetically complete systems possessed of logical models are categorical. These matters were set forth by Tarski at the Harvard Logic Club in January, 1940 and will appear in a paper 'On completeness and categoricity of deductive theories' (Quine, Goodman 1940, footnote 3, p. 109).

[18] Let us recall that a theory is syntactically complete (Tarski says 'absolutely complete or simply complete') 'if every sentence which can be formulated in the language of this theory is decidable, that is, either derivable or refutable in this system' (Tarski 1940, p. 1) Modulo some trivial facts about the theory, the condition is equivalent to stating that a theory is complete 'if for every sentence either it or its negation is derivable' (p. 1) Tarski points out that due to Gödel's incompleteness theorems the propery of 'absolute completeness occurs rather as an exception in the domain of the deductive sciences, and by no means can it be treated as a universal methodological demand' (p. 3).

I hasten to point out that the axiomatic notion of logically valid sentence cannot coincide, exactly for the reasons given below by Tarski, with the class of sentences true in all models.[19] Tarski specifies that he will consider deductive theories for which a concept of derivability has been specified in such a way that 'every logically valid sentence is derivable from any system of logical or non-logical sentences' (p. 4) In short, any theory T will have as consequence all the logical axioms and the theorems derivable from the logical axioms. If the background logic has an axiom of infinity, the theory will (trivially) have that axiom as part of its theorems.

As an example of the sort of deductive theories Tarski has in mind he mentions the logic of *Principia Mathematica*, or a fragment theoreof, with additional non-logical constants for geometry through which a system of axioms for Euclidean geometry is formulated:

> Taking any deductive theory let us consider an arbitrary system of non-logical sentences of this theory. As the theory let us think, e.g. of the system of *Principia Mathematica* or of a fragment of it, but in either case enriched by certain non-logical, viz. geometrical constants and a system of sentences, a system of axioms for Euclidean geometry. (p. 4)

Thus, notice that Tarski is thinking of the background logic for such theories as being a theory of types and allows the possibility that it be a fragment of Principia (in particular, no assumption has been made that the axiom of infinity be among the logically valid sentences). In general, if the logical basis (the class of logically valid sentences) is strong enough, one encounters the phenomenon of syntactic incompleteness:

> If the logical basis of our theory is rich enough, we can formalize within its boundaries the arithmetic of natural numbers and for this reason its logical basis is incomplete: i.e. there are logical sentences which are not logically valid and whose negations are likewise not logically valid.
>
> In other words, there are problems belonging entirely to the logical part of our theory which cannot be solved either affirmatively or negatively with the purely logical devices at our disposal.(p. 4)

Now, it is a well-known fact that one needs the axiom of infinity to develop arithmetic in a theory of types (see *Introduction to Logic*, 1941, p. 81, p. 130; 1937, p. 51, p. 80 (note 1) and 87). This shows that Tarski does not insist on a logic that carries no existential assumptions and thus considers freely mathematical theories that have as a basis a theory of types with the axiom of infinity (contrary to claims by Gomez-Torrente mentioned in the previous section).

---

[19] Since the term 'logically valid sentence' has already been used by Tarski in this lecture to indicate axioms of logic and the theorems derivable from it, we need a new term to indicate the sentences that are true in all models. The 1936 paper suggests, following Carnap, to call a sentence 'analytical' if every sequence of objects is a model of it. And modulo a few assumptions on the languages considered, Tarski claims in 1936 'that those and only those sentences are analytical which follow from every class of sentences (in particular from the empty class)'(Tarski 1936a, p. 418).

It is on account of the incompleteness phenomenon that Tarski introduced the notion of semantical completeness[20]:

> Now let us introduce the concept of semantic completeness. Because incompleteness is such a general phenomenon, the problem arises as to whether this is due to our conception of derivability. The concept of derivability developed in modern logic and reduced there to the concept of constructive rules of inference was intended to be a formal analogue of the intuitive concept of logical consequence. Thus, a doubt arises as to whether this intention has been realized. I discussed this question in some of my papers published some years ago and in this lecture I can only avail myself of the final result of this discussion. It turned out that between the intuitive concept of logical consequence and the formal concept of derivability there was a big gap. If we want to formulate an exact definition of the concept of logical consequence we must apply quite different methods and concepts. (pp. 4–5)

Here I will simply remark that Tarski is not distancing himself from any part of the analysis of logical consequence given in the 1936 paper, and thus this lecture can be considered to be in line with the positions put forward in that paper. Now we finally arrive at Tarski's summary of his approach to logical consequence:

> The most important role is played here by the concept of model or realization. Let us consider a system of non-logical sentences and let, for instance 'C_1', 'C_2'... 'C_n' be all the non-logical constants which occur. If we replace these constants by variables 'X_1', 'X_2'... 'X_n' our sentences are transformed into sentential functions with n free variables and we can say that these functions express certain relations between n objects or certain relations to be fulfilled by n objects. Now we call a system of n objects O_1, O_2 ... O_n a model of the considered system of sentences if these objects really fulfill all conditions expressed in the obtained sentential functions. It is of course possible that the whole system reduces to one sentence; in this case we speak simply of the [sic] model of this sentence. We now say that a given sentence is a logical consequence of the system of sentences if every model of the system is likewise a model of this sentence. (p. 5)

This is very much in line with Tarski's 1936 definition. However, notice that Tarski is here more explicit about the fact that models of axiomatic theories will, in general, be finite sequences, since there are usually finitely many non-logical constants used in the formulation of the theory. This, I also pointed out above by referring to Tarski 1935c.

After remarking on the semantical nature of the concepts involved ('model', 'fulfillment') Tarski goes on to formulate the notion of semantical completeness, which is obtained by replacing the concept of derivability with the semantical

---

[20] Before introducing the notion of semantical completeness, Tarski considers a different concept: 'completeness with respect to the logical basis' or simply 'relative completeness'. This will be treated in Section 3B.

concept of logical consequence in the definition of completeness:

> Thus a system of sentences of a given deductive theory is called semantically complete if every sentence which can be formulated in the given theory is such that either it or its negation is a logical consequence of the considered set of sentences.(p. 5)

Then Tarski claims the following:

> It should be noted that the condition just mentioned is satisfied by any logical sentence: hence we can deduce without difficulty that the concept of semantical completeness is a generalization of the concept of relative completeness: every system that is relatively complete is likewise semantically complete (but it can be shown by an example that the converse is not true).(p. 5)

Tarski makes here an important claim, i.e. 'that the condition just mentioned is satisfied by any logical sentence', and then proceeds to draw a mathematical result from it. The fact that he draws a mathematical result from the first claim shows that the claim cannot be dismissed lightly as an oversight. Thus, I will first analyse what Tarski's claim implies and then give the details of the mathematical result he draws from it.

3A. Tarski's claim

I will refer to Tarski's claim as 'C'. Tarski premisses his discussion of alternative notions of completeness by saying that he will be interested in theories that do have non-logical constants in them. On page 4 he says:

> 'We shall be interested here only in such deductive theories in which non-logical sentences actually do occur, and with respect to their completeness we shall consider exclusively systems consisting of non-logical sentences. To simplify our discussion let us assume moreover that there are no non-logical constants of our theory which do not occur in sentences of the considered system.'

Let L(T) be the language (logical and non-logical) in which the theory is formulated. To clarify what C amounts to, let us simplify the situation and consider as the system of sentences in question the entire theory T, identified here for convenience with all the statements containing non-logical expressions of L(T). T is semantically complete iff for every P expressed in L(T), either P is a logical consequence of T or ¬P is a logical consequence of T. If L is a sentence in L(T) that contains only logical symbols then, according to C, it is automatically the case that either L is a logical consequence of T or ¬L is a logical consequence of T.[21] Thus, either all models of T make L true or all models of T make ¬L true. Hence, each model of T gives us complete information about which logical sentences are true in it and all models of T agree on the truth values of the logical sentences.

---

[21] That every logical sentence is determinate is one of the central claims of Carnap's account of analyticity. I plan to carry out a comparative analysis of Tarski and Carnap in a different paper.

On the face of it this claim commits one to a conception of logical consequence that cannot allow for full domain variation. Let, for instance, T be the axioms for the theory of groups. Consider the logical sentence '∃x∃y¬(x = y)'. According to Tarski's claim either '∃x∃y¬(x = y)' is a logical consequence of T or its negation is. But this cannot be accounted for in a conception of logical consequence that allows for full domain variation. For in that case we could easily find a one-element domain satisfying the axioms and a two-element domain satisfying the axioms thereby showing that neither '∃x∃y¬(x = y)' nor its negation is a logical consequence of T.

3B. Semantical completeness implies relative completeness

Lest the reader thinks that claim C is just an offhand remark by Tarski, or worse a slip of the pen, I hasten to point out that Tarski relates his claim to the notion of relative completeness. Before introducing the notion of semantical completeness, Tarski considers a different concept 'completeness with respect to the logical basis' or simply 'relative completeness'. Consider a deductive theory and an arbitary system [set] of non-logical sentences of the theory. If the logic is strong enough we are faced with incompleteness. Requiring the system of non-logical sentence to be complete would mean that the non-logical sentences can decide the problems that are left undecided by the logic. This is in general too strong a requirement. However, 'we can at least require that the considered system of sentences should not extend the incompleteness of the logical part of our theory' (p. 4)

The condition of relative completeness is then stated as follows:

> 'In order to formulate this requirement in an exact way, let us denote two given sentences as equivalent with respect to the considered system of sentences, when if this system is enriched by the addition to it of the first of these sentences, the second becomes derivable and vice versa. Our requirement can now be stated as follows: For any sentence of our theory, there must be a logical sentence which is equivalent to it with respect to the given system of sentences. If this condition is satisfied, we say that the considered system is *complete with respect to its logical basis*, or simply, that it is *relatively complete*. It is clear that in case the logical basis is itself complete, relative completeness reduces to absolute completeness.'(p. 4)

Let us now show how the condition described in 3A is used by Tarski to show that 'every system that is relatively complete is likewise semantically complete' (p. 5) The argument can be spelled out as follows. Suppose S is relatively complete. As Tarski does let us assume that from S all logically valid sentences can be derived. Let ϕ be an arbitrary sentence in L(S) (L(S) includes also all the logical symbols). We want to show that either ϕ or $\neg\phi$ is a logical consequence of S. By definition of relative completeness for any sentence ϕ there is a logical sentence ϕ^* such that $S \cup \{\phi\} | - \phi^*$ and $S \cup \{\phi^*\} | - \phi$.

If ϕ is a logical sentence then, by claim C, it or its negation is a logical consequence of S, so there is nothing to prove. If ϕ is not logical then let ϕ^* be a logical sentence satisfying the condition given in the definition of relative completeness for S. We consider two cases. Because ϕ^* is a logical sentence either all models of S are models of ϕ^* or all models of S are models of $\neg\phi^*$. First assume all models of S are models of ϕ^*. Then any model M of S is also a models of S $\cup$ $\{\phi^*\}$. Since S $\cup$ $\{\phi^*\}| - \phi$ and the logical system is assumed sound, M is a model of ϕ. So ϕ is a logical consequence of S. Now assume that every model of S is a model of $\neg\phi^*$. Furthermore, by way of contradiction, assume there is a model M' of S such that M' is not a model of $\neg\phi$. Thus M' is a model of ϕ. Because S $\cup$ $\{\phi\}| - \phi^*$ hence by the soundness of the logical system, M' is a model of ϕ^*. This contradicts all models of S being models of $\neg\phi^*$. So $\neg\phi$ is a logical consequence of S. Consequently, S is semantically complete.

Notice that this claim by Tarski gives a general theorem about conceptual relationships between relative completeness and semantical completeness and it is stated in such a way that no qualification on the logic is made. Tarski meant this result to hold, as he explicitly says in a quote given in the main text, at least for the system of *Principia Mathematica* and all fragments of it, including the first-order fragment without the axiom of infinity.

3C. 'Absolute' fixed conception of model vs 'relative' fixed conception of model

The formalization of any theory would have two parts. First, the logical basis expressed in a type-theoretic framework, let us call it STT. Then there would be a set of axioms for the theory formulated by means of non-logical constants.

It is the type-theoretic framework that in the first place decides what the class of individuals (V) is. We can think of this as follows. The type-theoretic framework comes interpreted. In particular, the quantifiers play the role of logical constants and thus, as Etchemendy pointed out, there is no reinterpretation of the quantifiers. For this reason every model of the mathematical theory T in question will either make true '$\exists x \exists y \neg (x = y)$' or its opposite. However, it is not at all obvious that '$\exists x \exists y \neg (x = y)$' will in fact be the sentence that turns out to be a logical truth. That will depend on the class of individuals over which the variables of lowest order are ranging.

Here we have to clarify what has been an ambiguity in the fixed-domain interpretation all along. It seems to me that up to this point the fixed-domain conception has been understood in the literature as follows. The theory of types comes already with the meaning of the quantifiers fixed and that determines the range of the individual variables in all possible theories that can be formulated over the theory of types. The range of the individual variables (thus the class V) is taken to be the 'real' universe of individuals. Obviously, on such a conception '$\exists x \exists y \neg (x = y)$' would have to be a logical consequence of all theories T formulated within the background of the theory of types. I would like to dub this the 'strong' fixed-domain conception of model.

But if we look at the practice of using the theory of types as background for a mathematical theory we notice that something else is going on. Every theory T comes equipped with its background theory of types and with its own interpretation of the theory of types. Thus, for instance if one has decided to study an axiomatic system for Peano arithmetic over a theory of types (as Gödel does in 1931) one can take the meaning of the individual quantifiers to be such that they range over a superclass of the natural numbers or, if one prefers, over just the natural numbers. The same for a theory of real numbers (Tarski 1931) where the class of individuals might be larger than the class of real numbers or be exactly identical with it. Thus, there is a certain flexibility in choosing what the class of individuals is that will be assumed in the background. I would like to call this the 'weak' fixed-domain conception of model. It is this kind of flexibility, I claim, that accounts for the categoricity results that have given so much trouble to the original interpretation of the fixed-domain conception of logical consequence.

I thus think the above argument makes it plausible that in Tarski's definition of logical consequence all models come equipped with a fixed domain in the background and that logical validity is not a matter of truth in all models with variable domains.

4 Conclusion

The contemporary debate on Tarski's notion of logical consequence has ranged widely from historical issues about interpreting Tarski's 1936 text to philosophical debates about what the correct notion of logical consequence should be. Obviously, in this chapter my intention was only to discuss the historical claims that have been made concerning the notion of model in the 1936 article. In particular, I do not touch at all the theoretical issue of whether Tarski's notion of logical consequence is what we want as an account of logical consequence and whether it undergenerates or overgenerates with respect to such a desideratum. Once again, I am aiming here at clarifying what the historical Tarski, rightly or wrongly, thought. However, the evidence provided heightens the problem of finding a coherent interpretation of Tarski's position. This is not an easy task. The problem is that some of the claims by Tarski point in the direction of a fixed-domain conception of model and others point in the direction of a variable conception of model. Is this an unresolvable tension or can one find a point of view that will manage to accommodate the evidence in a unified framework without doing damage to the Tarskian texts? The following reflections are proposed as a tentative suggestion for a possible solution.

Let me briefly summarize the situation. In favour of the fixed-domain conception of model one can adduce a) Tarski's claim that if every constant of a language is taken as logical then logical consequence reduces to material consequence;

b) Tarski's claim contained in the 1940 lecture that for every logical sentence L and for any mathematical theory T formulated in a given background logical theory, either L or ¬L is a logical consequence of T (see claim C in Section 3).

In favour of the variable-domain conception we can mention: c) the claim that intrinsic categoricity and absolute categoricity coincide when the theory in question has as an axiom or consequence a sentence to the effect that the domain of individuals coincide with the intended domain. Second-order arithmetic and analysis are both absolutely categorical in this sense but in the first case the class of individuals, V, turns out to be countable and in the second case uncountable (see quote by Bays in Section 3). Thus there seems to be variability of V after all; d) Tarski's use in his metamathematical work of upwards and downwards form of Löwenheim–Skolem. Consider the upward Löwenheim–Skolem theorem. On the fixed-domain conception of model, if the class of individuals has a specified infinite cardinality, then any theory that has an infinite model will also have models of cardinality higher than the cardinality of the class of individuals. But that's obviously not possible, since no model could have more elements than the class of all individuals.

I will sketch in broad outline here what I think is a plausible solution for accounting for all these claims at once. The most important thing, which is often forgotten, is that the notion of logical consequence for Tarski is always tied to specific interpreted languages. Consider an axiom system for geometry with simple type theory as the logical basis. On the reconstruction I have given, the axiom system can be represented as superposing on the simple theory of types a set of axioms, which for convenience we can abbreviate as AS(**R**), where **R** stands for a vector of predicates and relations symbols needed for the specific axioms. The type theory is already interpreted, e.g. the quantifiers range over a fixed class of individuals. Under general circumstances we let this interpretation be the natural one, e.g. 'all' individuals. However, there are (as pointed out in Tarski and Lindenbaum 1936) special axiomatic investigations in which it is convenient to assume that the class of individuals coincides with the class of points or with the class of numbers. This is exactly what happens, with minor differences in Gödel's 1931 paper and in Tarski's 1931 paper on definability of real numbers. Let us go back to the case of geometry. Let L' stand for the interpreted language in which the range of the quantifiers consist of 'all' individuals. Let L" stand for the interpreted language (same syntax as L') in which the range of the quantifiers is limited to points. Because the two languages are different (despite their having the same syntax) so will be, in general, also the respective extensions of the consequence relationship. But notice the following. Once the interpreted language is given, the logical consequence relationship is reduced to truth in that interpretation of a universal sentence. To judge whether a certain statement F(**R***) (**R*** a subset, possibly empty, of **R**) is a logical consequence of AS(**R**) we ask whether $\forall \mathbf{X}(AS(\mathbf{X}) \to F(\mathbf{X}^*))$ is true in the intended interpretation. How do we account then for Löwenheim–Skolem-type theorems in this context? Simply, we look at them as metatheorems about the various interpretations that can be given to an axiomatic system based on the first-order fragment of the simple theory of types (obviously, we need to limit the type theory otherwise the theorems in question do not hold). Consider the upward Löwenheim–Skolem theorem. Consider a first-order axiomatization of geometry. In the metatheory then we prove that if we have an interpretation of a type theory L' such that there is an appropriate sequence of objects **O'** satisfying the propositional function corresponding to the axioms and that are sets, then we can find an interpreted language L" of type theory and a sequence **O"** (which are sets) such that the cardinality of the domain from which they are taken is higher than that corresponding to the objects **O'**. The use of set theory in the metatheory is absolutely essential. Type theory alone would not be sufficient to generate the required cardinalities.

What I just said can be read in two ways. Either as a claim that for any chosen universe of individuals, which is a set, we can always go on and come up with a new universe of individuals or as claiming that there is a universal class V from which we can manage to carve set-theoretic interpretations of higher and higher cardinality. If we go the second way then it is essential to insist that the domain of all individuals V underlying every single one of these interpretations be a proper class in the set-theoretic sense, i.e. it can have no cardinality. Here the picture would be the following. We assume that we are dealing with a unique interpretation of the type theory that is common to all the possible axiomatic systems that can be superimposed on it. The domain of individuals then is a proper class and every interpreted language L' will have all its variables ranging over the same class V. In order to account, in this picture, for the categoricity results, we would have to assume that we are investigating the consequence of a contrary to fact assumption (this way of reading the situation has been suggested by Marcus Giaquinto). And there is evidence that points this way. For instance, Tarski says explicitly that 'the axiom systems of various deductive theories are for the most part intrinsically but not absolutely categorical'. Examples would include geometry and arithmetic. So, one could argue that the ordinary systems of arithmetic and geometry are only intrinsically categorical because the domain of individuals is wider than the domain of natural numbers (in the case of arithmetic) or the domain of points (in the case of geometry). That is, the identity of all individuals with the numbers or with the points that is forced by adding to the systems of arithmetic or geometry sentences to that effect would result in theories that are consistent but contrary to fact. If this were the case, then the apparent relativization of the universe of individuals V to a countable or uncountable set would just be a contrary to fact assumption.

How about the first alternative? In that case there is no single, given once and for all, class of individuals V such that it is fixed in advance. Every interpretation (i.e. arbitrary choices for what counts as the class of individuals) of the type-theoretic apparatus is fine. Once fixed, that interpretation gives rise to an interpreted language L' with its notions of truth in L' and logical consequence for L'. But notice that despite the variability of V for different interpretations of the type theory, once the interpretation of V is fixed questions of logical consequence for that specific interpreted language are to be answered only by looking at models (i.e. sequences of appropriate objects) coming out from that specific interpretation for V. Thus all models for that interpretation share the same domain V (e.g. the domain of the interpretation). Obviously, in this set up there is no problem in accounting for the categoricity results. In conclusion, according to this second reading, what seems to have been a source of confusion in understanding the notion of a fixed domain for deciding what logical consequence consists in is not having kept clearly in mind that the notion is always relativized to a specific interpreted language. It is with respect to that interpretation that the domain for all the models becomes fixed.

Whichever of the two options outlined above one accepts it will not affect the fact that to determine what logical consequence consists in for a specific language L' we need to focus on that interpretation of L' and all the models in questions will then share the same domain.

Let me conclude by pointing to one further historical problem that would deserve more detailed study.

In his 1988, Etchemendy, addressing the issue of why Tarski moved from his 1936 account of logical consequence to that of 1953 (see Tarski, Mostowski and Robinson [1953]), where he embraces a standard model-theoretic account, says:

> 'How about our second question? Tarski was surely aware of the above flaw in his 1936 definition, and it was no doubt partly responsible for his later giving them up. But this in itself does not explain why he subsequently came to endorse the standard account, in spite of the considerations raised in the earlier article. Unfortunately there is little to go on here, since Tarski never addressed the philosophical issues raised in the 1936 article. Indeed, in his introductory logic text [41m], although he discusses similar issues in some detail, there is no mention at all of analyzing logical truth or logical consequence semantically. Instead he seems to offer a syntactic gloss of the consequence relation (pp. 118–119), perhaps a sign of his dissatisfaction with his 1936 definition. Then later, when he gives the standard model-theoretic definitions in [Undecidable Theories, PM] he says nothing about the divergence of these definitions from the earlier account.' (1988, p. 73)

I agree that it would be interesting to know more in detail how Tarski moved from one account to another. Indeed, the 1940 lecture discussed in Section 3 provides us with the specific information that until 1940 Tarski stood by his 1936 analysis. This shows that Etchemendy's tentative conjecture about what is going on in the 1937 textbook (from which the 1941 English translation is derived) is not warranted; moreover, one should recall that the 1937 book (but not the 1941 English version) is a literal translation from a book written in 1936 in Polish. It is still possible, as Etchemendy speculates, that Tarski's change was influenced by discussions with Carnap and Quine in 1941. And while I have began to make a few forays in this direction (see Mancosu 2005) the results will have to wait for another occasion.

Acknowledgements

I would like to thank John MacFarlane, Aldo Antonelli, Daniel Isaacson, Marcus Giaquinto, Ignacio Jané, José Ferreiros, Johannes Hafner, and Sol Feferman for their invaluable feedback on previous versions of this chapter. I am also grateful to the Bancroft Library at U.C. Berkeley for having granted permission to quote from Tarski's unpublished correspondence to Corcoran and from the 1940 lecture.

A Path to the Epistemology of Mathematics: Homotopy Theory

Jean-Pierre Marquis

> ... the history of topology provides, ..., a typical shortcut of the history of mathematics. Algebraic topology finds its origin, on the one hand, from the examination of problems arising in other parts of mathematics, notably functions of one complex variable and algebraic geometry, and on the other hand, from its own developments (like the *Hauptvermutung*); their solution requires new methods and techniques, and the reflection on these methods leads to new kinds of problems (theory of algorithms or decidability, for instance) or sparks the creation of new notions or theories (such as categories or homological algebra) which will allow instructive synthesis by shedding a new light on questions found in other chapters of mathematics, or that will even provide tools that will lead to significant progress in the study of other types of problems. (Hirsch, 1978, 260–261; our translation)

1 Introduction

Algebraic topology is indisputably one of the greatest achievements of twentieth-century mathematics. If, as Hirsch suggests, algebraic topology is a typical shortcut of the history of—and we would add here *twentieth-century*—mathematics, homotopy theory is a shortcut to the history of algebraic topology itself. The notion of a homotopy between maps has its roots in the late eighteenth century, appeared implicitly in the nineteenth century in the theory of functions of a complex variable, the theory of algebraic functions and the calculus of variations, was used informally by Poincaré in his papers entitled *Analysis Situs* that mark the birth of algebraic topology, and was finally explicitly defined as we know it by Brouwer in 1912. Homotopy *theory* came into existence in the 1930s, after Hopf's introduction of the fibrations that now bear his name and Hurewicz's introduction of the higher homotopy groups together with some of their fundamental properties. From this point on, homotopy theory interacted strongly with the other tools of algebraic topology, e.g. homology theory, cohomology theory, spectral sequences, it moved slowly to the forefront of algebraic topology in general, led to new synthesis in the form of homotopical algebra and is now being applied in a wide variety of fields, e.g. Voevodski's application of homotopical methods in algebraic geometry, for which he obtained the Fields Medal in 2002.

If Hirsch is correct, this is a typical evolution of a successful mathematical field: a notion appears in a given context or given contexts as being part of the solution to a problem or a class of problems, it is then clarified, cleaned up of extraneous elements, developed to a certain extent autonomously and, either at the same time or soon after, it is applied to other, unexpected, problems and fields and, in the best cases, it leads to the development of new notions, new tools, new theories that are then applied to a variety of contexts. This suggests that there is a pattern to the development of mathematics, at least in the twentieth century, or perhaps, going in a slightly different direction, it suggests that there are distinctive elements to mathematics of the late nineteenth and twentieth centuries. The elements mentioned by Hirsch are of course too broad and vague to be of any real value. A more detailed analysis of the various steps, moments, moves and periods is required. Of course, only a start can be made on that here. We do believe that algebraic topology in general, and homotopy theory in particular, do indeed provide a rich and fertile ground for philosophical reflection on the nature of mathematical knowledge and its development.

We will concentrate in this chapter on one specific epistemological element that can be extracted from the history of homotopy theory. Our main objective is to show that a typical component of twentieth century mathematics is the emergence, proliferation and establishment of *systematic mathematical technologies* within mathematics and that this development is not unlike the emergence, proliferation and establishment of scientific technologies in general. We believe that we can see within the history of homotopy theory such examples of these technologies. Furthermore, in the same way that the shift of attention towards the experimental and technological aspects of scientific research in philosophy of science is giving rise to an epistemology of scientific instrumentation and thus a more faithful epistemology of scientific knowledge[1], we claim that similarly in philosophy of mathematics, the recognition and analysis of mathematical technologies and instrumentations should lead us to a modification or a more adequate epistemology of contemporary mathematics. We suggest that *parts* of mathematical knowledge should be thought of as a form of conceptual engineering and that, therefore, mathematical knowledge is as complex and as messy as scientific knowledge in general. If this is correct, the picture of mathematical knowledge and of its development we end up with is radically different from the standard 'axioms-definitions-theorems-proofs of truths picture' of mathematical knowledge we often find in the literature.

---

[1] See, for instance, Galison 1997, Baird 2004.

2   Forms of mathematical knowledge

> The equivalence of all these infinite loop space machines was later proved by May and Thomason... (Kriz, 2001, xxiii)

Mathematical knowledge is a fabulously intricate mixture of know that and know how. In order to prove certain results, to construct specific counterexamples, to compute or solve certain equations, to define a new concept, to transfer various constructions from one field to another, one has to *know how* to do certain things and *know that* properties hold of the objects and procedures one is using and one is working with. Various periods and various people have often insisted more on the know how, the *technè* aspect of mathematics, presenting the latter as an art, others have underlined the know that, the *episteme* aspect of mathematics, presenting it as a science. But as Polanyi has already observed '... mathematics can be equally well affiliated either to natural science or technology.' (Polanyi, 1958, 184) It should be obvious to everyone that the *practice* of mathematics involves a lot of technical expertise[2] and that the *results* of mathematical practice are often considered the epitome of scientific knowledge. Furthermore, it would not be such a great exaggeration to claim that mathematical knowledge is characterized by the continual transformation of know how into forms of know that. This simply means that the methods, techniques and tools developed by mathematicians become *objects* of knowledge themselves.

But between scientific knowledge and technological knowledge, we find intermediate forms of knowledge. Here is how Polanyi puts it:

> We have, correspondingly, two forms of enquiry that lie between science and technology. Technologies founded on an application of science may form a scientific system of their own. Electrotechnics and the theory of aerodynamics are examples of *systematic technology* which *can be cultivated in the same way as pure science*. Yet their technological character is apparent in the fact that they might lose all interest and fall into oblivion, if a radical change of economic relationships were to destroy their practical usefulness. On the other hand, it may happen that some parts of pure science offer such exceptionally ample sources of technically useful information that they are thought worth cultivation for this reason, though they would otherwise lack sufficient interest. The scientific study of coal, metals, wool, cotton, etc. are branches of such *technically justified science*. (Polanyi, 1958, 179. See also Polanyi, 1960–61, 405.)

We submit that it is reasonable to *transpose*, with appropriate adjustments, Polanyi's classification of forms of knowledge to mathematics. We claim that these distinctions can and should be introduced *within pure mathematics* itself. More precisely, we believe that twentieth-century mathematics in general and algebraic

---

[2] It is customary among mathematicians to qualify some mathematical work as being a technical prowess.

topology in particular are marked, on the one hand, by the appearance of *systematic conceptual technologies*, and not just techniques and methods, and, on the other hand, by *technically justified mathematics*. For the purposes of this chapter, we will not distinguish these two forms from now on[3]. We believe, moreover, that historically these developments parallel the developments seen in the natural sciences and technologies and, conceptually, they have much in common.

What is, informally, a systematic conceptual or mathematical technology in pure mathematics? It is a conceptual technology, that is a specific conceptual know how with a specific epistemic goal. Mathematics is filled with these conceptual know hows. But it is systematic in as much as it rests upon a whole mathematical theory or a collection of mathematical theories for its design, definition and applications. The remaining sections of this chapter will hopefully illuminate these claims, as well as illustrate and provide evidence for them.

If there are pieces of mathematics that are viewed as systematic conceptual technologies or technically justified mathematics, we should observe differences in the way a piece of work is *valued* by mathematicians, depending on whether the work is seen as a piece of science or a piece of technology (or, in the case of mathematics, both, something that might be a distinctive feature of mathematics itself). We will use as a springboard a simple list of values proposed by Polanyi for sciences and technologies. According to Polanyi, ... 'a statement is of value to natural science if it (1) corresponds to the facts, (2) is relevant to the system of science and (3) bears on a subject matter which is not without intrinsic interest;' and 'a statement is of value in technology (1) if it reveals an effective and ingenious operational principle which (2) achieves, in existing circumstances, a substantial material advantage.' (Polanyi, 1953, 187.) We should add here that an additional element is that technologies can fall into oblivion simply because they are replaced by other technologies. It is easy to give examples of mathematical knowledge that are valued because (1) they are relevant to the system of mathematics and (2) bear on a subject matter that is not without intrinsic interest. We leave aside the question of relevance to facts, since it is clearly more controversial in the case of mathematical knowledge. We submit that algebraic topology and, in particular, homotopy theory are filled with statements and, more generally, forms of knowledge, that (1) reveal an effective and ingenious operational principle that (2) achieves, in existing circumstances, a substantial *conceptual* advantage. But what is more, these same forms of knowledge are *also* relevant to the system of mathematics, for they often reveal *how* various pieces of mathematical knowledge *are* related to one another, and they certainly bear on a subject matter that is not without intrinsic interest.

---

[3] As Polanyi himself observed, the distinction might in practice be merely rhetorical: Systematic technology and technically justified science are two fields of study lying between pure science and pure technology. But the two fields may overlap completely. (Polanyi, 1958, 179.)

Mathematicians themselves regularly talk about parts of algebraic topology and homotopy theory in terms of technologies, machines, tools and instruments[4]. We will now take a close look at what seems to us to be a representative sample of what can be found in the literature. The list could be extended indefinitely. Needless to say, the fact that mathematicians talk in that way does not constitute a conclusive argument in favour of our claim, but the following quotes provide powerful evidence in support of our thesis. We will comment on the quotes as we go along.

> In this chapter we obtain some results about the homotopy groups of spheres. The method we follow is due to Serre and uses the *technical tool* known as a spectral sequence. This algebraic concept is introduced for the study of the homology and cohomology properties of arbitrary fibrations, but it has other important applications in algebraic topology, and the number of these is constantly increasing. Some indication of the *power* of spectral sequences will be apparent from the results obtained by its use here. (Spanier, 1966, 465) [our emphasis]

Let us immediately underline the elements that stand out. Spanier identifies a method as a technical tool, namely the method of spectral sequences. This is clearly a case of know how. In the next sentence, he claims that it is a concept and states its purpose: although we are talking about a technical tool, it is in the end a form of knowledge. To learn and understand spectral sequences is to *know how* to use spectral sequences. Finally, Spanier argues in favour of the *power* of the technology on the basis of the results obtained with its help. It is not so much the quantity of results that is at stake here, but the conceptual importance of the results and the fact that they cannot be obtained otherwise. Spectral sequences are *valued* because of their power and this, despite the fact that they are extraordinarily complicated and difficult to use. The next quote goes exactly in the same direction and does not require any further comments:

> The book might well end at this point. However, having eschewed the use of the heavy *machinery of modern homotopy*, I owe the reader a sample of things to come. Therefore a final chapter is devoted to the Leray–Serre spectral sequence and its generalization to non-standard homology theories. .... Some applications are given and the book ends by demonstrating the *power* of the machinery with some qualitative results on the homology of fibre spaces and on homotopy groups. (Whitehead, 1978, xv)[our emphasis]

At least one algebraic topologist has used explicitly the analogy between components of algebraic topology and components of the natural sciences.

> Despite the large amount of information and *techniques* currently available, stable homotopy is still very mysterious. Each new computational

---

[4] One mathematician once told me that his introductory (graduate) course in algebraic topology was all about machinery. I suspect that most mathematicians teaching the subject would say something similar.

> breakthrough heightens our appreciation of the difficulty of the problem. The subject has a highly *experimental* character. One computes as many homotopy groups as possible with *existing machinery*, and the resulting *data* form the basis for new conjectures and new theorems, which may lead to better methods of computation. In contrast with physics, in this case the experimentalists who gather data and the theoreticians who interpret them are the same individuals. (Ravenel, 1986, xvi)[our emphasis]

We have here a specific admission that methods of computation are replaced by new, more powerful, methods. We will see in the next sections what Ravenel has in mind when he talks about *existing machinery*. But it follows immediately that we should take seriously the idea that mathematicians build conceptual machinery, evaluate the quality of this machinery and they replace existing machinery by new ones. The next quote is even more explicit.

> The study of the homotopy groups of spheres can be compared with astronomy. The groups themselves are like distant stars waiting to be discovered by the determined observer, who is *constantly building better telescopes* to see further into the distant sky. The telescopes are spectral sequences and other algebraic constructions of various sorts. *Each time a better instrument is built* new discoveries are made and our perspective changes. The more we find the more we see how complicated the problem really is. We can distinguish three levels in the subject. The first (comparable to observational astronomy) is the collection of *data* about homotopy groups by *various computational devices* (...). While this aspect of the subject is not fashionable and is seldom discussed in public, it is vital to the subject. Without *experimental data* there can be no valid theories. (...) The second level of ideas in homotopy theory is the identification of certain patterns known as periodic families. This may be compared to the discoveries of Kepler and Halley. (...) The third level (comparable to cosmology) is the formulation of general theories about the mechanisms which produce the observed phenomena. (...) As in theoretical physics one can make various models of the universe based on certain oversimplification or idealizations. While these constructs have obvious limitations, their study is instructive as it leads to some insight into the nature of the real world. We will discuss several of these models now. (Ravenel, 1987, 175–176)[our emphasis]

Gathering data about homotopy groups of spheres is a highly technical endeavour. It is very hard. Conceptual machines, instruments, probes and tools have to be built and used properly. This is the kind of highly systematic know how that we want to focus on.

A few words about algebraic topology might help illuminate these quotes further. As its name indicates, algebraic topology is the study of topological spaces and continuous transformations by algebraic means. The Graal of algebraic topology is the classification of spaces under continuous deformations. The general strategy is to associate to a space various algebraic structures, e.g. groups, modules, rings, algebras, etc., in such a way that a continuous map of spaces is transformed into a homomorphism of the appropriate kind, e.g. a homomorphism of groups, or modules or rings, etc. and homeomorphisms of spaces are transformed into isomorphisms of the associated algebraic structures. In other words, the algebraic structures associated to a space are invariant. Thus, one tries to encode topological properties by algebraic means in such a way that whenever there is a difference between the corresponding algebraic structures associated to two spaces, then one can conclude that the spaces are different. We submit that finding systematic ways of encoding topological properties in algebraic structures is a form of conceptual engineering and what is elaborated to do the encoding constitute examples of systematic conceptual technologies. Thus, homology theories, cohomology theories, homotopy groups, spectral sequences, fibrations, etc. are all instances of systematic conceptual technologies.

These technologies are valued when they reveal an effective and ingenious operational principle and they achieve a substantial conceptual advantage. Mathematicians rarely praise purely *ad hoc* solutions, no matter how clever these solutions are.

In the remaining parts of this chapter, we will concentrate on one specific concept in the history of homotopy theory, namely the concept of fibration that has an interesting history of its own and, furthermore, illustrates some of the key features of these systematic technologies.

3   Forms of mathematical knowledge: fibrations

> The concept of fibration has been one of the most important mathematical tools in the twentieth century; born in geometry and topology, it has gradually invaded many other parts of mathematics. (Dieudonné, 1989, 383)

A brief history of homotopy theory with fibrations in mind

Homotopy theory starts from an extremely simple geometric idea: the continuous deformation of a curve into a curve, or a path into a path. It is extraordinarily easy to give a vivid illustration of a specific homotopy between two curves. But it is a different matter to know *how* to *define* the notion precisely and it is even less clear *why* such a definition ought to be given. For one thing, the notion is so intuitively clear that it does not seem necessary to provide a precise formal definition. Furthermore, once a rigorous definition has been provided, it is not clear what has been gained thereby, apart from rigour for its own sake. Even when someone understands the precise definition clearly, it does not mean that one understands the point of the notion, *why* it *is* an important notion. The latter makes sense only when the *role* played by that notion in a broader context is understood. We believe that this is true of many other similarly simple mathematical notions: to understand a mathematical notion in a given context, one has to understand its *function* in that context. This means that for many mathematical notions, to understand that notion, it is irrelevant to specify what it is 'made of', or its underlying 'ontology' but rather what it is *used for*: when, why and how.

It is certainly not our goal to sketch the whole history of homotopy theory. This would be a daunting task that would require a book, perhaps many books. With the concept of fibration in mind, we can roughly distinguish the following periods in the history of homotopy theory:

1. The prehistory: from Lagrange until and including Poincaré;
2. The introduction of the concept and its first uses: Brouwer's explicit definition of a homotopy of paths in 1912 and its applications by Brouwer himself;
3. The birth of homotopy *theory* as such: from Hopf's study of maps between spheres between 1926 and 1935, the definition of higher homotopy groups by Hurewicz in 1935 to Serre's computations of classes of homotopy groups of spheres using fibrations and spectral sequences in 1951;
4. The development of simplicial homotopy theory by Kan and others in the mid-1950s until Quillen's introduction of homotopical algebra with its underlying notion of model categories published in 1967.

Again, this is extremely schematic and does not do justice to the extraordinarily complex development of the field. A history of computations of homotopy groups of spheres, for instance, would be divided differently[5]. But our goal here is to provide the general background in which the notion of fibration appeared and played a key role. Let us now turn to some of the details of this history and its key developments, especially the first three phases.

The concept of a homotopy of paths appears implicitly in the works of Lagrange, Cauchy, Riemann, Puiseux, Jordan, Klein and Poincaré[6]. Poincaré's work has to be set apart, for although he does not define formally a homotopy of paths and for this reason has to be put in the first period of the history, he is the first one to see how the concept can be intrinsically useful to reveal important properties of a manifold. Before Poincaré, the notion appeared in specific (non-topological) contexts, e.g. the calculus of variations, integration of a complex function of a complex variable, algebraic functions, etc. In these contexts, the focus of attention of the mathematicians was, for instance, certain specific functions and their integration and a homotopy of paths was simply an obvious requirement that had to be met by those functions. There was no reason to define the notion of homotopy precisely, since it did not play any mathematical role in these contexts. It may very well also be that invariance under a change—as a *general* and significant method—was being assimilated slowly in the nineteenth century. Even Klein, who introduced the idea of invariance of geometric properties via transformation

---

[5] Toda's paper, precisely on the topic of the history of computations of homotopy groups of sphere, cuts the history into slices of ten years. This is as if the field had no internal conceptual dynamics. Whitehead's historical paper does not have much more conceptual perspective. See Toda 1982 and Whitehead 1983.

[6] We rely essentially on Vanden Eynde 1999 for this period.

groups in elementary geometry, does not make the notion of homotopy precise and, like many of his contemporaries, confused homotopy with homology[7]. Furthermore, it is far from clear that mathematicians of that period would have had the means, that is the concepts and appropriate language, to define the concept of homotopy explicitly.

Some of these remarks apply to Poincaré as well. Although Poincaré gave birth to algebraic topology in his paper *Analysis Situs* and its five complements, published between 1895 and 1904, and although Poincaré is certainly the first mathematician to see that continuous deformations can actually reveal properties of a manifold in which they are defined, it is clear that Poincaré's focus of attention is on the concept of homology. Indeed, in the 1895 paper, Poincaré defines what he calls the fundamental group of a manifold, what will later become the first homotopy group, but as he says clearly in §13 of that paper, the information obtained from the fundamental group is used to determine what he calls the 'fundamental homologies' and it is not considered intrinsically. Furthermore, it is not before 1904, in the fifth complement, that Poincaré shows that the fundamental group can be different from the first homology group, thereby showing that the two concepts differ. Adding to these ingredients the fact that general topology was still not available as a language to define this concept in all its generality and Poincaré's own informal style, we can see why the concept was not made explicit by him, despite the fact that it was used explicitly for topological reasons[8].

Brouwer was the first mathematician to give a precise formal definition of a homotopy of paths. His definition is almost identical to the one we find in contemporary algebraic topology textbooks. However, Brouwer thought it was sufficient to give the definition *in a footnote* of his paper on continuous transformations of spheres in themselves. Here is Brouwer's definition:

> By a continuous modification of a univalent continuous transformation we understand in the following always the construction of a continuous series of univalent continuous transformations, i.e. a series of transformations depending in such a manner on a parameter, that the position of an arbitrary point is a continuous function of its initial position and the parameter. [Brouwer, 1912a, 1976, 527, ft 4]

The process of continuous deformation is now made explicit. The definition tells us how an arbitrary point 'moves' from its initial position. Nowadays, this is stated thus: Let X and Y be topological spaces, I be the standard unit interval $[0, 1]$ and f and g be continuous maps $X \to Y$. The map f is said to be *homotopic* to g, denoted by $f \simeq g$, if there exists a *homotopy* of f to g, that is, a continuous map $H : X \times I \to Y$ such that $H(x, 0) = f(x)$ and $H(x, 1) = g(x)$. In the body

---

[7] Vanden Eynde attributes this confusion to the fact that in the context of Riemann's work, the distinction does not have to be made. See Vanden Eynde, 1999, pp. 75–76.

[8] The *name* 'homotopy' was introduced by Dehn and Heegaard in their article on Analysis Situs published in the German encyclopedia of mathematics in 1907. However, the name does not designate the concept we now know.

of the same paper, Brouwer defines what we now call the *homotopy class* of a map: 'we shall say that two transformations *belong to the same class* if they can be transformed continuously into each other.' (Brouwer, 1912a, 1976, 528.) In other words, the existence of a homotopy defines an equivalence relation between maps. The notion of homotopy class of a map was in itself very important, since it provided a novel classification of maps, different from the classification that homology was about to deliver.

Why did Brouwer define the concepts of homotopy of paths and of homotopy class of a map? First, it is interesting to note, as is emphasized by Freudenthal in his commentaries to Brouwer's topological papers (see Freudenthal 1976, 436), what is absent in Brouwer's work: there is no mention and no use whatsoever of the tools of homology[9]. Thus, Brouwer's focus of attention is radically different from Poincaré's. Second, the concept of homotopy played a key role in his proofs of some of his infamous theorems, e.g. the fixed-point theorem, the invariance of dimension, etc. Brouwer's main tool, together with simplicial approximation, is the notion of degree of a map[10]. Informally, the degree of a map $f : S^1 \to S^1$ of the circle into itself, denoted by deg(f), is the number of times $f(z)$ turns around S^1 when z turns once around S^1. The concept can be defined for any map $f : S^n \to S^n$ of the n-sphere into itself[11]. The crucial property of the notion of degree of a map is that it is homotopy invariant, that is, if $f \simeq g : S^n \to S^n$, then deg(f) = deg(g). Furthermore, homotopies of maps play an essential role in the method of simplicial approximation. We now see that the focus of attention is on the homotopy class of maps and not, as in the case of Poincaré, the group of such homotopy classes. Furthermore, in contrast with his predecessors, the deformation *is* the key property and not simply an obvious condition in the background of the problem. Nonetheless, Brouwer did not deem it necessary to include the definition in the main part of his paper.

The notion of homotopy remained a footnote until Hopf made essential use of it in his work on continuous mappings of spheres from 1925 until 1935. We have seen earlier that the notion of the degree of a map is homotopy invariant. In 1912, Brouwer conjectured that the converse of this statement was also true, that is, for any $f, g : S^n \to S^n$ such that deg(f) = deg(g), $f \simeq g$, but sketched a proof only for $n = 2$. Hopf proved the conjecture in 1925. Furthermore, Hopf's proof yields an

---

[9] 'In retrospect, it therefore seems legitimate to consider Brouwer as the cofounder, with Poincaré, of simplicial topology. More precisely, it may be said that Poincaré defined the *objects* of that discipline, but it is Brouwer who imagined *methods* by which theorems about these objects could be *proved*, something Poincaré had been unable to do. (...) It is all the more surprising then that Brouwer did not attempt to use his techniques in order to put Poincaré's 'theorems' in simplicial homology on less shaky foundations. (...) At any rate, Brouwer never showed any interest for homological concepts in his '*n*-dimensional manifolds.' (Dieudonné, 1989, 168)

[10] Hopf gave the actual definition of degree of a map with the help of homology groups in 1930. See, for instance, Whitehead 1978, 13 or Spanier 1966, 196.

[11] Brouwer defines the notion of degree for any continuous map $f : M \to N$, where M and N are compact, connected, oriented n-dimensional 'manifolds' (in a restricted sense).

isomorphism between the n-th homotopy group of the n-sphere and the integers, i.e. in Hurewicz's notation $\pi_n(S^n) \approx \mathbb{Z}$. Brouwer had worked with maps between manifolds of the same dimension. For $n < m$, it was known that the homotopy groups $\pi_n(S^m)$ are trivial, in other words, for any continuous map $f : S^n \to S^m, f$ is homotopic to a constant map. Before 1930, almost nothing was known about continuous maps $f : S^m \to S^n$ for $m > n$. What was known is that homology was *useless* in that context. To use homological information, one would look at the induced homomorphism $f_* : H_\bullet(S^m) \to H_\bullet(S^n)$ between homology groups. But for $p > 0$, either $H_p(S^m) = 0$ or $H_p(S^n) = 0$ (for $H_p(S^n) \neq 0$ if and only if $p = n$), and thus, in both cases, f_* is a trivial group homomorphism. Something else has to be used.

This is where the notion homotopy class of a map turned out to be informative and, thus, played a crucial role in our understanding of the situation. In 1930, Hopf proved that there are infinitely many homotopy classes of maps from S^3 to S^2. This was to be interpreted shortly after as saying that $\pi_3(S^2) \approx \mathbb{Z}$. Hopf's proof came as a total surprise. In particular, Hopf defined a continuous map $f : S^3 \to S^2$ that is not homotopy equivalent to a constant map, now known as the *Hopf fibration* or the *principal Hopf bundle*. As its name already indicates, it will play a role in our story since it is an early example of a fibration. (For a detailed description of the map, see Aguilar et al. 2002, 129–130 or Hatcher 2002, 377–378.) Then, in 1935, Hopf generalized his results to maps $f : S^{2n-1} \to S^n$. Hopf showed that for $n = 4$ and $n = 8$, the maps $f : S^7 \to S^4$ and $f : S^{15} \to S^8$ are what we now call fibrations. These yield the isomorphisms $\pi_7(S^4) \approx \mathbb{Z} \oplus \mathbb{Z}/4\mathbb{Z} \oplus \mathbb{Z}/3\mathbb{Z}$ and $\pi_{15}(S^8) \approx \mathbb{Z} \oplus \mathbb{Z}/8\mathbb{Z} \oplus \mathbb{Z}/3\mathbb{Z} \oplus \mathbb{Z}/5\mathbb{Z}$ (these are taken from Toda 1962). In fact, Hopf did much more than give these specific maps, for he obtained results for n even, introducing along the way a construction that was going to be extremely influential and important afterwards. It is certainly fair to say that Hopf showed that homotopic methods could provide important information about spaces that seemed to be inaccessible otherwise and, in this sense, launched homotopy theory. Hopf's work convinced mathematicians that homotopy classes of maps could be used effectively to obtain information about various spaces. A technology was on its way. It had to be developed systematically. The Polish mathematician Witold Hurewicz took care of that.

Before we move to Hurewicz, let us briefly go back to the fundamental group introduced by Poincaré, for we now can define it. Let (X, x_0) be a *pointed space*, that is a space with a privileged point $x_0 \in X$ and $(S^1, *)$ be the circle (as a subspace of $\mathbb{R}^2$) with privileged point $* = (1, 0)$. A loop in X at x_0 is a continuous mapping $\alpha : S^1 \to X$ such that $\alpha(*) = x_0$. Poincaré showed how loops α_1 and α_2 can be composed: first go around α_1 and then around α_2[12]. This gives a law of composition for loops, denoted by $\alpha_1 \vee \alpha_2$. This law is *not* commutative: for given a $z \in S^1$, $(\alpha_1 \vee \alpha_2)(z)$ will not, in general, be equal to $(\alpha_2 \vee \alpha_1)(z)$. Clearly, there is a *constant loop* $\alpha_0 : S^1 \to X$, defined by $\alpha_0(z) = x_0$ and given a loop α_1,

---

[12] We will leave it to the reader to provide the formal details. This is another case of an extraordinarily simple geometric idea that has to be turned into a genuine mathematical concept.

we can define the inverse loop α_1^{-1} of α_1 as the loop going exactly along the same path but in the direction opposite of α_1. These data do *not* yield a group, however. It is precisely at this point that the notion of homotopy of loops enters the scene and plays a key role. A homotopy between loops $\alpha_1, \alpha_2 : S^1 \to X$ is a continuous map $F : (S^1, *) \times I \to (X, x_0)$ such that $F(z, 0) = \alpha_1$, $F(z, 1) = \alpha_2$ and $F(*, t) = x_0$ for all $t \in [0, 1]$. We can consider the set of equivalence classes of loops, denoted by $[S^1, *; X, x_0]$. A tedious but straightforward verification shows that the equivalence class $[\alpha_1 \vee \alpha_2]$ depends solely on the classes $[\alpha_1]$ and $[\alpha_2]$. Therefore, we can define a product between equivalence classes of loops by putting $[\alpha_1] \cdot [\alpha_2] = [\alpha_1 \vee \alpha_2]$. It can be verified that the product thus defined does indeed yield a group, named by Poincaré the *fundamental group* of the space and it is denoted by $\pi_1(X, x_0)$[13]. Thus, in this context, the notion of homotopy allows one to define a group structure on a space. It is again by moving to the homotopy classes of maps that we succeed in obtaining relevant and useful information about a space.

It seems entirely natural and a promising idea to generalize Poincaré's construction by considering the set $[S^n, *; X, x_0]$ of equivalence classes of maps $(S^n, *) \to (X, x_0)$ for $n > 1$. This is precisely what Čech did and presented in a very short note in 1932 at the International Congress of Mathematicians. Čech showed that the higher-dimensional homotopy groups, as they are called, are abelian. Because of that, no one thought they would be of any use and Čech himself abandoned this line of research. It was expected that only non-abelian groups would provide genuinely new information, that is information going beyond what homology groups revealed. This expectation was based on what Poincaré had already shown: the abelianization of the first homotopy group (of a variety in the case of Poincaré) is isomorphic to the first homology group of that space.

When Hurewicz turned his attention to the topology of deformations, he had already done important work in dimension theory and descriptive set theory. In particular, he had assimilated various concepts and methods of point set topology, the most important being that of a function space and its topology[14]. Indeed, the very first sentence of his first paper on homotopy groups sets the stage:

> When investigating continuous mappings of a space X into a space Y, it proves very useful to interpret the collection of those mappings as a topological space in its own right. In the most important cases, the components of this function space coincide with the Brouwer classes of mappings that are continuously deformable into each other (homotopic). (Hurewicz 1935, in Kuperberg, 1995, 350)

---

[13] In fact, more is true. One can thus define, as it is now done, the *fundamental groupoid* of a space. The latter notion could have appeared much earlier and naturally in the history of mathematics, but it did not.

[14] It is worth noting that Hurewicz used Fréchet's work on functional spaces. As he indicated himself in a footnote, the functional spaces Y^X are usually *metric* spaces, the resulting metric depending on the choice of a metric in Y. But he indicated immediately that when X is compact, the resulting topology of Y^X does not depend on the choice of the metric in Y.

Hurewicz made essential use of function spaces in his definition of the higher-dimensional homotopy groups. Given two spaces X and Y, it is possible to define a topology on the set X^Y of functions $Y \to X$. In 1935, Hurewicz had to assume that X was a metric space and Y was compact. This is not as such a considerable restriction, but one of the first questions left open by Hurewicz's work was whether the homotopy groups could be defined for any topological space X and Y. The answer was given in the early 1940s when the compact-open topology was introduced[15].

Instead of starting with the set $[S^1, *; X, x_0]$ of homotopy classes of loops, Hurewicz started with the *loop space* $\Omega(X, x_0)$, i.e. the function space $(X, x_0)^{(S^1,*)}$ with the appropriate topology. A *path* in the loop space $\Omega(X, x_0)$ is a continuous map $I \to (X, x_0)^{(S^1,*)}$. With the appropriate topology, each path is equivalent to a homotopy $(S^1, *) \times I \to (X, x_0)$ between loops at x_0. Hurewicz observed that the components of the space $(X, x_0)^{(S^1,*)}$ are therefore the same as the homotopy classes of maps from $(S^1, *) \to (X, x_0)$. The loop space is itself a pointed space: it is the space $(\Omega(X, x_0), \alpha_0)$ where $\alpha_0 : S^1 \to \{x_0\}$ is the constant loop. We can therefore consider its fundamental group $\pi_1((\Omega(X, x_0), \alpha_0)$. The beauty of this construction is that it can be repeated inductively: $\Omega^n(X, x_0) = \Omega(\Omega^{n-1}(X, x_0), \alpha_{n-1})$, where $\alpha_{n-1} : S^1 \to \{\alpha_{n-2}\}$ is the obvious constant map. The n-th homotopy group $\pi_n(X, x_0)$ is then the *fundamental group* of the $n-1$ loop space of (X, x_0), i.e. $\pi_n(X, x_0) = \pi_1(\Omega^{n-1}(X, x_0), \alpha_{n-1})$.

With this definition in hand, Hurewicz stated without proofs various important properties of the homotopy groups and applying these results to the homotopy groups of topological groups, he obtained a new proof of Hopf's result on $\pi_3(S^2)$ as well as many others.

Hurewicz published three more papers on homotopy groups, all of which established important properties (with proofs in these cases) of homotopy groups, for instance their connections with homology groups in the second note, and also using homotopy groups to prove properties of spaces and even to define classes of spaces, e.g. the aspherical spaces in the fourth note. But the next important concept was to appear *at the very end* of the third note. It is the notion of *homotopy type*[16]. Two spaces X and Y are said to have the same *homotopy type*, or to be *homotopy equivalent*, if there are maps $f : X \to Y$ and $g : Y \to X$ such that $fg \simeq 1_Y$ and $gf \simeq 1_X$. Before the introduction of this definition, homotopic information was used to classify *maps*, now, it is used to classify *spaces*. Hurewicz used the notion in the fourth paper on homotopy theory to classify aspherical spaces.

With the publications of these four papers, homotopy theory was now on firm grounds and could start a life of its own. The basic definitions and their relevance to algebraic topology were explicit and clear; some of the information it delivered

---

[15] See, for instance, Aguilar *et al.* Chap. 1 for the definition of the compact-open topology.
[16] Hurewicz defined it for compact spaces only.

was only available through its channels. There was only one fundamental glitch: homotopy groups were extraordinarily hard to compute. Freudenthal took an important step in 1937 with the introduction of the notion of the suspension of a space. The next step would await the clarification and the use of fibrations together with spectral sequences. A detailed and illuminating discussion of spectral sequences would require too much space, and we will therefore limit ourselves to fibrations. But this is no great loss, since fibrations occupy a central role in contemporary homotopy theory, as we will see.

Fibrations: an historical sketch

The history of fibrations intertwines with the history of fibre bundles (vector bundles, sphere bundles) and fibre spaces. We will leave the history of fiber bundles aside since they are related more to differential geometry than to algebraic topology, and concentrate on the homotopical aspects of the story. (The interested reader should consult Dieudonné 1989 and Zisman 1999 for the history of fiber bundles.)

A special case of fibration appeared, according to Zisman 1999, as early as 1879 in a note published by Emile Picard. Other specific cases showed up again in the work of Seifert in 1931, 1935, Hurewicz in 1935, Hopf in 1935 and Borsuk in 1937. Then, in 1940 and 1941, five mathematicians, namely Hurewicz and Steenrod working together, Ehresmann and Feldbau also working together and Eckmann, identified a property, namely the *homotopy lifting property* (HLP)[17], that allowed them to obtain new and interesting results about homotopy groups. With these results in hand, it seemed reasonable to define a new structure by a property general enough to 1) include the spaces they were interested in as well as others that seemed important *and* 2) that would yield a simple proof of the HLP for a large class of spaces. Mathematicians had found a property that played a key role in the proofs of important results but that did not seem to characterize an entity as such. The search for a general property that would fit the bill was launched. Hurewicz and Steenrod in 1940, published in 1941, were the first to introduce *fibre spaces* with these goals in mind.

Before we look at fibre spaces, let us state the homotopy lifting property. Let $p : E \to B$ be a continuous map and C a class of topological spaces. Then p is said to satisfy the *homotopy lifting property (HLP) with respect to* C, if for every $X \in C$, every map $f : X \to E$ and every homotopy $H : X \times I \to B$ such that $H(x, 0) = (p \circ f)(x)$, there is a homotopy $\tilde{H} : X \times I \to E$ such that $p \circ \tilde{H} = H$ and $\tilde{H}(x, 0) = f(x)$. A simple diagram, when read properly, allows us to grasp the whole definition at a glance: given all the data, p satisfied the HLP with respect to

---

[17] Hurewicz and Steenrod (1941) called it the *covering homotopy property (CHP)* and it is sometimes called this in various books and articles.

$\mathcal{C}$ if the following diagram commutes

$$\begin{array}{ccc} X & \xrightarrow{f} & E \\ j\downarrow & \tilde{H}\nearrow & \downarrow p \\ X\times I & \xrightarrow{H} & B \end{array},$$

where $j: X \to X \times I$ is the inclusion $j(x) = (x, 0)$.

Notice that the HLP is a property of a *map*. As such, it is a simple property and it is hard to see *why* it is important. To state this, we need to introduce a bit of terminology: let $b_0 \in B$, then $F = p^{-1}(b_0)$ is said to be the *fiber* above b_0; E is called the *total space* and B is called the *base space*[18]. Fibres in this context customarily have additional structure, e.g. F is itself a topological space. Clearly, we have an inclusion $i: F \to E$. The crucial fact is that when a map p satisfies the HLP with respect to a class $\mathcal{C}$, it is possible to construct isomorphisms $\pi_n(F, x_0) \xrightarrow{i_*} \pi_n(E, x_0) \xrightarrow{p_*} \pi_n(B, b_0)$ for any $x_0 \in F$[19]. Furthermore, the HLP allows defining a homomorphism $\partial : \pi_n(B, b_0) \to \pi_{n-1}(F, x_0)$, which in turns yields the so-called homotopy exact sequence

$$\cdots \pi_n(F, x_0) \xrightarrow{i_*} \pi_n(E, x_0) \xrightarrow{p_*} \pi_n(B, b_0) \xrightarrow{\partial} \pi_{n-1}(F, x_0) \to \cdots.$$

In other words, the homotopy groups of the fibres are systematically connected to the homotopy groups of the total space and the base space all the way down to the path-components π_0. When E, B or F satisfy further specific conditions, one uses these connections to establish other useful isomorphisms, e.g. $\pi_n(E) \approx \pi_n(B) \oplus \pi_n(F)$ for $n \geq 2$. Thus, the HLP plays a crucial role in the construction of various maps that in turn allows one to obtain significant results about homotopy groups.

Hurewicz and Steenrod stipulated that E is a *fibre space over B relative to p*, where $p: E \to B$ is a continuous map, E is a topological space and (B, δ) is a *metric space*, if there exists an $\varepsilon_0 > 0$ such that for all open subsets $U_{\varepsilon_0} = \{(e, b) \in E \times B : \delta(e, b) < \varepsilon_0\}$ there is a continuous function $\phi: U_{\varepsilon_0} \to E$ such that for all $(e, b) \in U_{\varepsilon_0}$

$$p \circ \phi(e, b) = b \text{ and } \phi(e, p(e)) = e.$$

A map ϕ is called a *slicing function* (nowadays, we would say that it is a local section). Hurewicz and Steenrod then gave a list of examples of fibre spaces: these include all the fibre bundles as defined then by Whitney (sphere bundles), product spaces, covering spaces, the Hopf maps and projection maps of a Lie group onto a quotient by a closed subgroup. Theorem 1 of their paper is the HLP

---

[18] This terminology comes from the theory of fibre bundles and was introduced by Whitney in 1937.

[19] In their paper, Hurewicz and Steenrod constructed an isomorphism between the *relative* homotopy group $\pi_n(E, F, x_0)$ and $\pi_n(B, x_0)$, for any $x_0 \in F$. Since we haven't said a word about relative homotopy, we refrain from formulating these results in those terms.

with respect to *all* topological spaces, which is achieved by assuming that the homotopy $H : X \times I \to B$ is *uniformly* continuous. They indicate after the proof that the uniformity requirement is unnecessary if X is assumed to be a compact metric space. They then proceed to use the HLP to prove important properties of homotopy groups, now defined for arbitrary topological spaces[20]. They also show, for the first time, that if the base space is arcwise connected, then the fibres all have the same homotopy type.

At the same time, unknowingly of Hurewicz and Steenrod's work because of the Second World War, Ehresmann and Fledbau were defining fibre bundles more or less as we know them now. They then proved the HLP for bundles with respect to finite complexes and then used the HLP to deduce various isomorphisms. Thus, their goal was not to define a structure whose main purpose is to capture the HLP, but they are well aware of its importance and the fact that it ought to be proved for the class of structures they are interested in.

Meanwhile in Zurich, Eckmann defined what he called *retrahierbare Zerlegungen*, retractable partition. We will not give Eckmann's definition here, for it is given under more restrictive assumptions than Hurewicz and Steenrod's and can be shown to be a special case of theirs. But the general strategy is the same: after presenting his definition, Eckmann proceeds to prove the HLP with respect to compact spaces, followed by a proof (of what will become) the homotopy exact sequence and obtains various results about homotopy groups.

The situation did not change much during the 1940s. In 1943, Ralph Fox generalized Hurewicz and Steenrod's definition by removing the restriction on the base space B. Then, Jean-Pierre Serre shocked the world of homotopy theory with the publication of his thesis in 1951.

Serre's attitude towards fibre spaces is remarkable and left an ineffaceable imprint: he says explicitly that since the only thing he needs to establish his results is the HLP, he *defines* a fibre space as a map $p : E \to B$ that satisfies the HLP with respect to *finite polyhedra*. Serre immediately points out that fibre spaces in this sense include (locally trivial) fibre bundles, Hurewicz and Steenrod's fibre spaces, principal G-bundles and the class of spaces he is going to use later in his paper, namely *path spaces*. The latter, which have become important in their own right, are defined as follows (Serre, 1951, 479): let X be an arcwise connected space and $A, B \subset X$; then the function space $E_{A,B} = \{f : I \to X : f(0) \in A \wedge f(1) \in B\}$ with the compact-open topology is called a *path space*. Define $p_{A,B} : E_{A,B} \to A \times B$ by $p_{A,B}(f) = (f(0), f(1))$. Serre showed that this map satisfies the HLP with respect to *all* spaces. Then, Serre considered the special case when $A = \{x_0\}$ and $B = X$. In this case, the map $p_{A,B} : E_{A,B} \to A \times B$ becomes $p_{x_0,X} : E_{x_0,X} \to X$ with fibres $F = \Omega(X, x_0)$, the loop space at x_0. The fibration $p_{x_0,X} : E_{x_0,X} \to X$ occupies a key role in the whole work: by applying a spectral sequence to it, it is possible to find the homology groups of the loop space $\Omega(X, x_0)$ from the homology groups of the

---

[20] As underlined by Zisman 1999, the restrictions imposed on the spaces in the original definition of homotopy groups presented by Hurewicz five years earlier are lifted without a single comment.

base space X. Furthermore, since $p_{x_0,X} : E_{x_0,X} \to X$ is a fibration, it is possible to construct the homotopy exact sequence, i.e. we get

$$\cdots \to \pi_n(\Omega(X)) \xrightarrow{i_*} \pi_n(E_{x_0,X}) \xrightarrow{p_*} \pi_n(X) \xrightarrow{\partial} \pi_{n-1}(\Omega(X)) \to \cdots.$$

This automatically yields the following connection between the homotopy groups and the loop space of a space:

$$\pi_n(X, x_0) \approx \pi_{n-1}(\Omega(X), \alpha_1) \approx \cdots \approx \pi_{n-p}(\Omega^p(X), \alpha_p) \approx \cdots \approx \pi_0(\Omega^n(X)).$$

Using spectral sequences and knowledge of homology groups, Serre then proceeds to prove general results about homotopy groups of spheres, for instance:

1. for all $i > n$, if n is odd, the groups $\pi_i(S^n)$ are finite;
2. if n is even and $i = 2n - 1$, then $\pi_i(S^n)$ is the direct sum of $\mathbb{Z}$ and a finite group.

As we have just seen, for Serre, a fibre space is a map $p : E \to B$ satisfying the HLP with respect to finite polyhedra (it was soon shown afterwards to be equivalent to satisfying the HLP with respect to all CW-complexes, a large and useful category of topological spaces, especially in homotopy theory). Curtis and Hurewicz soon after independently gave new but equivalent definitions of fibre spaces for which they proved that a map $p : E \to B$ was a fibre space in this new sense if and only if it satisfies the HLP with respect to *all* topological spaces. (See Curtis 1956 and Hurewicz 1955[21].)

But Serre's point of view prevails to this day. A *Serre-fibration* is defined to be a map $p : E \to B$ satisfying the HLP with respect to hypercubes I^n, whereas a *Hurewicz-fibration* is a map $p : E \to B$ satisfying the HLP with respect to all topological spaces. Other types of fibrations have been defined, e.g. weak fibrations and quasifibrations. The latter are quite interesting in themselves: a *quasifibration* is a continuous surjective map $p : E \to B$ such that for each point $b \in B$:

1) the map $\pi_i(E, F, y) \to \pi_i(B, b)$ is an isomorphism, for any $i \geq 1$ and any $y \in F = p^{-1}(b)$;
2) $\pi_0(F) \to \pi_0(E) \to \pi_0(B) \to 0$ has to be exact as a sequence of pointed sets.

Thus, one retains only what one was able to prove from the HLP and use *that* as the defining property. The surprising fact is that there are non-trivial quasifibrations.

But this is not the whole story, far from it. In 1967, Quillen changed the scenery completely, a development that led to what deserves to be called *abstract homotopy theory*. The nature and the consequences of this radical shift will be briefly explored in the next and last section.

---

[21] A terminological remark: Hurewicz's 1955 paper defines *fiber* spaces, whereas he previously used *fibre* spaces. Both terms are found in the literature right from the beginning.

Fibrations: their form and functions

We will confine ourselves in this section to the essential elements required for our analysis, leaving the historical details and most of the mathematical background behind.

With hindsight and from a purely conceptual point of view, homotopy theory becomes relevant in a mathematical situation whenever either one is not interested so much in a specific map but rather in the class of homotopically equivalent maps, or one is not interested so much in a specific space or any homeomorphically equivalent space but only in the homotopy type of the space. Thus an abstract homotopy theory should allow us, first, to define appropriate equivalence relations between maps of objects and appropriate equivalence relations between objects themselves and, second, apply the tools of homotopy theory, e.g. homotopy groups, the homotopy exact sequence, etc. to these objects and maps. What is 'appropriate' here is in a sense dictated by the topological case: whatever abstract sense one gives to homotopically equivalent maps and homotopy type, one should be able to recover the standard topological meaning of these expressions. This is precisely what Quillen succeeded in doing in 1967. Using Quillen's framework, it is possible to define a homotopy theory in various contexts and to compare homotopy theories, e.g. state precisely when two homotopy theories are in fact the same.

Quillen made essential use of category theory in his work, in particular the idea of model category (see Quillen 1967, 1969) which there is not space to define properly here. The axioms of a model category have two functions[22]. First, and this is clearly a standard feature of the axiomatic method, if a property of a model category can be proved from the axioms, then it holds of any model category and therefore they can be applied directly to any specific case. Second, and this is perhaps more peculiar, the axioms ought to be thought of as a *conceptual design* for a homotopy theory. They stipulate the conditions under which a homotopy theory can be built. It is the whole point of the definition: to be able to define the notion of homotopy in that context and then use all the machinery of homotopy theory to obtain significant results. Thus, although one might want to say that a property is true of model categories if it follows from the axioms, I seriously doubt that anyone would want to claim that the axioms *themselves* are true. What matters most or what is valued most, I believe, is rather the fact that the axioms, in the words of Dwyer and Spalanski,

> give a reasonably general context in which it is possible to set up the basic machinery of homotopy theory. The machinery can then be used

---

[22] This is certainly true of other axiomatic definitions, e.g. Eilenberg and Steenrod's axioms for homology and cohomology. I do believe that the attitude towards the axiomatic method as a method of capturing essential ingredients present in various contexts instead as a way of presenting intuitive truths about a fixed domain of objects emerged during the last quarter of the nineteenth and the beginning of the twentieth century. It is thus a characteristic feature of twentieth-century mathematics.

> immediately in a large number of different settings, as long as the axioms are checked in each case.... Certainly each setting has its own technical and computational peculiarities, but the advantage of an abstract approach is that they can all be studied with the same tools and described in the same language. (Dwyer and Spalinski, 1995, 75)

Given a model category C, it is possible to construct the *homotopy category* Ho(C) associated to C. Presenting the construction would require introducing many other concepts and thus, considerably more space. Suffice it to say that, from the conceptual point of view, the construction of the homotopy category guarantees that the weak equivalences of C are turned into genuine isomorphisms in Ho(C).

Furthermore, Quillen proposed a criterion to determine when two homotopy categories, say Ho(C) and Ho(D), are the same, that is equivalent as categories. Such an equivalence is now called a *Quillen equivalence.* Quillen equivalence determines a *homotopy theory*, i.e. a homotopy theory is more or less all the homotopical information preserved by a Quillen equivalence. Using this terminology, Quillen has shown, for instance, that the homotopy category of the category of simplicial sets and the homotopy category of topological spaces with the appropriate model structure are equivalent, thus they are models of the same homotopy theory. Another interpretation of this result if given by Dwyer and Spalanski: 'this shows that the category of simplicial sets is a good category of algebraic or combinatorial 'models' for the study of ordinary homotopy theory.' (Dwyer and Spalanski, 1995, 122.)

Let us come back to fibrations. In the context of model categories, we have left behind the topological setting. It is one setting among many others. Fibrations are defined together with cofibrations in the axioms. Their properties are stipulated by the axioms. The HLP still plays a key role: faithful to Serre's approach, it is used to define model categories. It is not given with respect to a class of objects, but with respect to the class of cofibrations (and with respect to fibrations). This should be no surprise by now. But the exact role of fibrations in homotopy theory is still being clarified:

> A closer look at the notion of a model category reveals that the weak equivalences already determine its 'homotopy theory', while the cofibrations and the fibrations provide additional structure which enables one to 'do' homotopy theory, in the sense that, while many homotopy notions involved in doing homotopy theory can be defined in terms of the weak equivalences, the verification of many of their properties (e.g. their existence) requires the cofibrations and/or the fibrations. (Dwyer *et al.* 2004)

Here lies the divide: a systematic technology, in contrast to a technique, depends upon scientific knowledge for its design. Homotopy theory rests upon a portion that should be qualified as being 'scientific', certain fundamental mathematical laws captured by what Dwyer *et al.* call 'homotopical categories', which are defined by a class of weak equivalences satisfying some simple properties. But to actually carry on homotopy theory, to construct the various structures required, one needs some machinery and this is precisely where fibrations (cofibrations) come in.

4 Concluding remarks

Fibrations are sometimes introduced as the appropriate homotopic generalization of the concept of fibre bundle. Here is a typical example:

> Of course, from a homotopy viewpoint, having homeomorphic fibres and the rest of the rigid structure of a fibre bundle is overkill. In this section we will study a generalization (and its dual) of the concept of fibre bundle in which the fibres over points in a common path component are not homeomorphic but merely homotopy equivalent,... (Selick, 1997, 53.)

The generalization in question is the concept of fibration. This certainly suggests that this is how one should view or understand what fibrations are about: they constitute the generalization of the concept of fibre bundle with the right homotopy theoretic property. Although fibrations do indeed have that property and it is indeed homotopically important, anyone who thinks that this is the point of fibrations would miss the crucial element. It is worth reading the whole quote from Selick's book for he himself is entirely aware of this point:

> One of the features of a fibre bundle $p : X \to B$ (...) is that the 'fibres', $F_b = p^{-1}(b)$, are homeomorphic for all points b in a common path component of B. From the homotopy point of view, the key property of a fibre bundle is that for any pointed space W, there is an exact sequence $[W, F_b] \to [W, X] \to [W, B]$, where b is the base point of B. Of course, from a homotopy viewpoint, having homeomorphic fibres and the rest of the rigid structure of a fibre bundle is overkill. In this section we will study a generalization (and its dual) of the concept of fibre bundle in which the fibres over points in a common path component are not homeomorphic but merely homotopy equivalent, and although the overall structure is much less rigid than that of a fibre bundle, it is still sufficient to give the exact sequence $[W, F_b] \to [W, X] \to [W, B]$. (Selick, 1997, 53)

Selick is in fact clear: from a homotopical point of view, the important property of fibre bundles is that they are fibrations, i.e. one can construct the desired exact sequence. The fact that fibers over points in a common path component are homotopically equivalent is a crucial *theoretical* indication that the concept is just right, that it captures the right kind of information, but it would be wrong to conclude that from an epistemic point of view, fibrations are *merely* generalizations of fibre bundles.

The concept of fibration is essentially a *relational* concept. What characterizes a fibration is the *class* of spaces with respect to which it satisfies the HLP. Although it is possible, as we have seen, to define a fibre space as an object with an intrinsic property and prove that such spaces satisfy the HLP, mathematicians prefer to define fibrations directly by specifying the class of spaces with respect to which their maps satisfy the HLP. Thus, the concept is tailored to their specific *needs*. If you want more maps as fibrations, use Serre-fibrations (e.g. if you are interested in locally trivial bundles); if you need fewer maps as fibrations, use Hurewicz-fibrations; if you can't use the HLP but the basic isomorphism of the homotopy exact sequence is available, use quasifibrations.

Within the world of mathematical concepts, fibrations have a different epistemological status than, say, fibre bundles or principal bundles, but also different from the homotopy groups. Fibre bundles are fundamentally geometric and, as such, model various properties of what one might think of as space. Homotopy groups should be thought of as *measuring instruments* since they provide information about certain crucial aspects of spaces. Although the latter are groups in the standard axiomatic sense, they are epistemologically radically different from the groups of the nineteenth century. Groups in the nineteenth century were always acting on something, either a set or a space; they were *transformation* groups of a space or *permutation* groups of a set of roots of a polynomial equation. Homotopy groups (and here we might as well mention homology and cohomology groups) do not act on anything. They are not *defined* in the same way nor are they *used* in the same way. The purpose of these *geometric* devices is to classify spaces in different *homotopy types*. Many concepts and methods of point-set topology, e.g. compactness, are simply irrelevant to homotopy types (compactness is not an homotopy invariant notion). Homotopy theory contributed to a large extent to the sharp separation between algebraic topology and point-set topology in the 1950s. Points of spaces, as defined in the usual set-theoretical way, do not play an essential role in homotopy types. This is in sharp contrast to the role they have in homeomorphism types. Fibrations also play a role in the separation between algebraic topology and point-set topology, since, as we have seen, they can be defined in various contexts, e.g. categories, and used to develop homotopy theory and homotopy types in these contexts. But in contrast to homotopy groups, fibrations *cannot* be thought of as *measuring instruments*. Fibrations are devices that make it possible to apply the measuring instruments and other devices; they have to be seen as an ingenious and extraordinarily useful tool for the construction of informative structures.

Knowledge of fibrations is clearly knowledge of their *usage* and that knowledge resembles more technological knowledge than scientific knowledge. It should be clear at this stage that fibrations reveal, to paraphrase Polanyi, an effective and ingenious operational principle that achieves, in existing circumstances, a substantial conceptual advantage.

Furthermore, fibrations are not merely a technique, but a set of rules one follows to solve a problem or compute a certain quantity. We can talk about the 'concept' of fibration and fibrations are thought of as a certain structure. They certainly deserve the label 'systematic technology' since, in the case of a fibration of a space, fundamental concepts of topology have to be applied properly and, in the more general case of model categories, fundamental concepts of category theory have to be applied properly.

If fibrations are to be thought of as tools, then there is no point in thinking about them in terms of truth, but rather in terms of efficiency. As such, like any technology and, more to the point, like any technological knowledge, it might very well become useless or obsolete, although it might be hard to imagine how this could be now.

Alternative Views and Programs in the Philosophy of Mathematics

Felix Hausdorff's Considered Empiricism

Moritz Epple

1 Introduction

This chapter presents some of the epistemological views of a rather unconventional mathematician, philosopher, and writer of the period around 1900, Felix Hausdorff. Hausdorff was born in 1868, and together with his wife, Lotte Hausdorff, née Goldschmidt, and her sister, he took his own life on 26th of January 1942 in order to avoid being sent to a concentration camp.[1]

Today, Hausdorff is mostly known for his ground-breaking monograph on set theory, the *Grundzüge der Mengenlehre*, published in 1914, shortly before the outbreak of World War I.[2] Among other things, this monograph contained the first extended discussion of the notion of a topological space. In the interwar period that followed, Hausdorff's *Grundzüge* and its successor, simply entitled *Mengenlehre* and published in 1927, became probably the most often cited book in this area of mathematics.[3] Mathematicians and historians of mathematics also know Hausdorff from his further work in analysis and descriptive set theory, such as his discovery of a famous paradox in the foundations of measure and integration theory. Here, Hausdorff showed that there exist decompositions of the surface of an ordinary sphere in Euclidean 3-dimensional space into three mutually congruent subsets (i.e. subsets that can be transformed into each other by means of suitable rotations around the centre of the sphere) and a denumerable set of points such that after further rotations around the

---

[1] For biographical information on Hausdorff, see (Eichhorn, 1994) and the biographical article in the forthcoming vol. I of Felix Hausdorff's *Gesammelte Werke*. These will, in the following, be cited as (Hausdorff, Werke). Page numbers, however, refer to the original pagination, which is reproduced in this edition. Further biographical information given below derives from a lecture course given by Egbert Brieskorn at Bonn University in 2000 and 2001 (unpublished).

[2] See the re-edition in (Hausdorff, Werke II, 2002) and the editors' introduction to this volume.

[3] See, e.g., Walter Purkert's analysis of citations in the Polish journal *Fundamenta Mathematicae*, in (Hausdorff, Werke II, 2002, 58).

centre of the sphere, just *two* of these three subsets already form the complete surface of the sphere (up to a denumerable set of points).[4] A refined variant of this paradox later became known as the Banach–Tarski paradox. Another ground-breaking contribution of Hausdorff was the introduction, in 1919, of a new notion of dimension for certain metric spaces, according to which the (Hausdorff) dimension of a space could be a non-integral real number (Hausdorff, 1919/2001).

Despite these (and more) influential contributions in crucial branches of modern mathematics, in the early 1930s Hausdorff the set theorist may have been less well known to the general public than another author who had written under the name of Paul Mongré. The widespread *Handbuch des jüdischen Wissens*, a dictionary on Jewish culture published in 1936 in the Philo-Verlag, Berlin, for instance, did not list Hausdorff among the 46 Jewish mathematicians mentioned in a corresponding article. However, among the over hundred Jewish philosophers we find a Paul Mongré, born in 1868. He is listed among the group of 'Nietzscheans', and we find the same Paul Mongré in a long article on 'Schrifttum', 'writings', where he is listed as an author of philosophy, lyrics, and drama.[5] This article even reveals that the civil name of Paul Mongré the writer was Felix Hausdorff – nevertheless, mathematics was *not* mentioned among Hausdorff's occupations or writings.

Hausdorff's publications as Paul Mongré.[6]

1897 Sant' Ilario – Gedanken aus der Landschaft Zarathustras. Leipzig: C.G. Naumann.
      Selbstanzeige of the above.
1898 Das Chaos in kosmischer Auslese – Ein erkenntniskritischer Versuch. Leipzig: C.G. Naumann.
      'Massenglück und Einzelglück'. NDR 9, pp. 64–75.
      'Das unreinliche Jahrhundert'. NDR 9, pp. 443–452.
      'Stirner.' Die Zeit 213, 29.10.1898, pp. 69–72.
1899 'Tod und Wiederkunft.' NDR 10, pp. 1277–1289.
      Selbstanzeige of *Das Chaos in kosmischer Auslese*.
1900 Ekstasen. Leipzig: H. Seemann Nachf.
      'Nietzsches Wiederkunft des Gleichen.'
      Die Zeit 292, 5.5.1900, pp. 72–73.
      'Nietzsches Lehre von der Wiederkunft des Gleichen.'
      Die Zeit 297, 9.6.1900, pp. 150–152.
1902 'Der Schleier der Maja.' NDR 13, pp. 985–996.
      'Der Wille zur Macht.' NDR 13, pp. 1334–1338.

---

[4] See (Hausdorff, 1914a/2002, 469–472); (Hausdorff, 1914b/2001).

[5] Reprint: Jüdischer Verlag im Suhrkamp Verlag: Frankfurt am Main, 1992.

[6] In the table, the abbreviation NDR means *Neue Deutsche Rundschau*, a leading literary journal of the time.

	'Max Klingers Beethoven.'
	Zeitschrift für bildende Kunst 13, pp. 183–189.
	'Offener Brief gegen G. Landauers Artikel 'Die Welt als Zeit'.'
	Die Zukunft 10, pp. 441–445.
1903	'Sprachkritik.' *NDR* 14, pp. 1233–1258.
1904	'Gottes Schatten'. *NDR* 14, pp. 122–124.
	'Der Arzt seiner Ehre. Groteske.' *NDR* 15, pp. 989–1013.
	(Reprints Leipzig 1910, Berlin 1912)
1909	'Strindbergs Blaubuch.' *NDR* 20, pp. 891–896.
1910	'Der Komet.' *NDR* 21, pp. 708–712.
	'Andacht zum Leben.' *NDR* 21, pp. 1737–1741.
1912	'Biologisches.' *Licht und Schatten* 3, Heft 35 (unpaginated).

The above table lists Hausdorff's publications written under the name of Paul Mongré. The first publication on the list presents Mongré as a follower, maybe even as a competitor, of Friedrich Nietzsche. The later publications – many appeared in renowned journals – reveal their author as an epistemological critic (a proponent of 'Erkenntniskritik'), as a poet, as the author of a play ('Der Arzt seiner Ehre', a grotesquerie attacking the outdated fashion of dueling) and not least as a contributor of several essays on Nietzschean themes, moral issues, and cultural matters of the early 1900s.

Is the case of Mongré the writer and Hausdorff the mathematician a case of a double identity, a multitalented author whose intellectual horizon was just too broad to fit under one roof, or are there intellectual connections between the two?[7]

In the following, I want to discuss some of the threads that join the two authors, Mongré and Hausdorff. This will be done from two perspectives. On the one hand, I will describe how a genuine epistemological interest brought Mongré, the Nietzschean, to study Georg Cantor's theory of sets in 1897/1898. In a way, this is a simple historical observation.[8] The second line of argument, however, goes beyond that. On this level, the objective of the present chapter is to sketch the epistemological position that *both* Mongré, the philosophical critic, *and* Hausdorff, the mathematician of the early 1900s, drew from their engagement

---

[7] For some topics of the following discussion, the reader should compare (Mehrtens, 1990, Sect. 2.3), entitled 'Metaphysikkritik und Mengenlehre: Felix Hausdorff'. In my view, Mehrtens tends to construct a somewhat too homogeneous Hausdorff; the steps in his intellectual development are not followed in much detail. In particular, Mongré cannot be said to be just a 'working mathematician' during the period discussed here. Nevertheless, Mehrtens's main claim that Hausdorff/Mongré is a paradigmatic figure in the emergence of mathematical modernism is certainly valid. A rich and helpful, if unpublished, essay on Hausdorff's intellectual development is (Brieskorn, 1997). – Below, I will follow Hausdorff's own naming convention and speak of him either as Mongré or as Hausdorff, as appropriate.

[8] The crucial point of this connection has already been pointed out by Walter Purkert in (Hausdorff, Werke II, 2002, 2–7).

with epistemology and from their acquaintance with set theory and modern mathematical axiomatics (which Hausdorff learned from David Hilbert's *Grundlagen der Geometrie* of 1899).

Hausdorff called his position 'considered empiricism', in German: 'besonnener Empirismus', sometimes also 'geläuterter Empirismus'. Briefly put, this position claimed that beyond mathematics, no scientific knowledge can claim to be more than a more-or-less plausible, more-or-less economic, and more-or-less complex system of beliefs compatible with the empirical information we may have. On the other hand, Hausdorff argued, no empirical information completely determines such a system of scientific beliefs, thus empiricism has to be 'considered', or 'reflected': It must be aware of the irreducibly arbitrary elements of any one given set of scientific beliefs.

In a certain respect, this position is more precise than other, present-day variants of the idea that 'theory is underdetermined by experiment'. In Hausdorff's view, the irreducible gap between empirical information and scientific conceptions resides in the special role of *mathematics* (and in fact, of mathematics conceived as higher set theory) in the sciences. While in the last analysis, mathematics itself is an autonomous creation of disciplined thinking, and not an empirical science, wherever empirical phenomena are described in mathematical terms, scientists will have different possibilities of producing suitable mathematical descriptions of these phenomena. Even where the mathematization of a physical phenomenon may seem straightforward or granted by a long-standing tradition, a critical analysis of the axiomatic basis of this mathematization will reveal, according to Hausdorff, the possibility of alternative mathematizations. Consequently, he held that any such mathematization – and hence any form of mathematized scientific theory – relies on conceptual decisions, not on absolute truths.

A *philosophical* implication of this view was the rejection of (metaphysical) realism. In any case, no argument in its favour could be drawn from scientific conceptions of the world. Rather than traditional metaphysics or even transcendental philosophy, it was mathematical analysis that provided the main tool for a critical epistemology of the sciences. 'True empiricism', Hausdorff wrote in one of his private notes in the early 1900s, 'is formalism'.[9] For philosophers of science at the beginning of the twenty-first century, this position may not sound too surprising or disturbing. Around 1900, it certainly was. The philosophical movement that we know as 'logical empiricism', and that was to emerge from a group of mathematicians, philosophers and other scientists in Vienna, was only to begin some 10–15 years later.

The chapter proceeds in four steps. First, the Nietzschean views of the young Hausdorff/Mongré are introduced, and some topics that led him to the epistemology of time, space, etc. are pointed out. In the second step, I sketch

---

[9] 'Der Formalismus ist der wahre Empirismus', in Nachlass Felix Hausdorff, Universitäts- und Landesbibliothek Bonn, Kapsel 49, Faszikel 1079, page 1. – Further references to this collection will be given in abbreviated form as in: (Hausdorff papers, 49/1079, 1).

Mongré's attack on metaphysical realism, based on an argumentative strategy using set-theoretical ideas. The third step sketches Hausdorff's 'considered empiricism'. In the last section of the chapter the relations of Hausdorff's position with some views of Henri Poincaré and Moritz Schlick are briefly discussed.[10]

2 Applied mathematics and Nietzschean thought

In 1897, the year in which Mongré's *Sant' Ilario* appeared, Hausdorff belonged to the growing group of *Privatdozenten*, experienced researchers without a permanent position in academia.[11] His professional field was applied mathematics. Both his dissertation (in 1891) and his *Habilitation* (in 1895) had been written under the Leipzig astronomer Heinrich Bruns.[12] These studies dealt with the mathematics of optical properties of the atmosphere as a prerequisite for deriving astronomical data from observations on the surface of the earth. (As an aside, note that this work made Hausdorff well aware of the basic fact that there is no such thing as a straightforward path from raw observations to empirical data as used in a science such as astronomy: mathematical modelling and approximations intervene already at this level.[13])

Professionally, this work did not bring Hausdorff much success and was rather narrowly related to the state of astronomy of the day.[14] This may be one of the reasons why Hausdorff began his activities as Paul Mongré. In fact, he had been involved in more than mathematics from the time when he took up his university studies in Leipzig in 1887.[15] He heard lectures in physics, criminology, history (including a course on the history of socialism given by the Jewish historian Adolf Warschauer), and philosophy, where the Kantian Friedrich Paulsen made a strong impression on him. As his later work shows, Otto Liebmann, who had

---

[10] The present chapter grew out of talks given at Bochum University in 2001 and at a conference in Tübingen organized by Michael Heidelberger and Gregor Schiemann in February 2005. For the purposes of these talks and this chapter I have restricted myself to discussing aspects of the published writings of Hausdorff/Mongré. For a finer discussion including detailed references to Hausdorff's unpublished papers, the reader is refered to my forthcoming introduction and commentary of (Hausdorff, Werke VI). Much of what follows is my individual take on issues that were discussed at length among the editors, including in particular Egbert Brieskorn, Walter Purkert, Erhard Scholz and Werner Stegmaier. I owe sincere thanks to all of them.

[11] For the particular role of this status among Jewish scientists, see (Volkov, 2000, 157–161).

[12] The publications resulting from this work are (Hausdorff, 1891), (Hausdorff, 1893), (Hausdorff, 1895).

[13] An interesting study of this interplay in Maxwell's work on experiments with viscous fluids has been presented in (Sichau, 2002).

[14] See (Hausdorff, Werke V, 2006).

[15] The following biographical information is mostly drawn from a lecture course given by E. Brieskorn, see note 1.

first published his influential collection *Die Analysis der Wirklichkeit* in 1876, was also an important inspiration for some of his ideas. Both Paulsen and Liebmann favoured a Kantianism close to the empirical sciences and critical of the neo-metaphysical tendencies that were not uncommon in German philosophy at the time.

More important than these interdisciplinary studies, however, may have been Hausdorff's relation with an intellectual group that became a forum for modernist trends in science, literature, music and arts, the Akademisch-Philosophischer Verein. In this group, founded in 1866/1867 by some of the Leipzig empiricists (including Richard Avenarius, Gustav Theodor Fechner, Wilhelm Wundt, and others), topics such as Darwinism, psychology, Schopenhauer and Nietzsche, Wagner's music, Marxism, the emancipation of women, contemporary theatre and belletristics were fiercely discussed in regular evening meetings. In Hausdorff's time, several leading members of the Verein were advocating a Bohemian way of life and it may be that Hausdorff was drawn to this as well.

In this circle, it seems, Hausdorff encountered Nietzschean thought. His first publication under the pseudonym of Paul To-My-Liking, as we may translate Paul Mongré, was a collection of aphorisms modelled on Nietzsche's paradigm and printed by the publishing house that also published Nietzsche's works. Its full title is *Sant' Ilario: Gedanken aus der Landschaft Zarathustras*; they were written during a stay on the Mediterranean coast in Italy that was intended to improve Hausdorff's health. Under the headings 'Spätlings-Weisheit, Einer und Nullen, Vom Normalmenschen, Pour Colombine, Müssiggang und Wetterglück, Denken-Reden-Bilden, Splitter und Stacheln, Von den Märchenerzählern, Zur Kritik des Erkennens', Mongré collected 411 aphorisms, some of them short, some several pages long. At the end a number of poems rounded off the collection.

Despite all these Nietzschean motifs, Mongré kept a sceptical distance from his model. In a few places this would turn out to be important. While this is not the place to survey the contents of the volume in detail it should be stressed that the aphorisms present their author as a sceptic about any kind of traditional morality, towards nation and religion, and as a passionate proponent of cultivated, aestheticized individualism. Writers such as Goethe and and Gottfried Keller and musicians such as Richard Wagner found his admiration. The acquisition of Bildung amounts for him more-or-less to the ability to be productive in literature, music, the arts or other fields of intellectual activity, not to the possession of traditional knowledge or to the fulfilment of traditional values.[16]

The last section of the collection, entitled *Zur Kritik des Erkennens*, contains the germ of Hausdorff's epistemological views, as well as a first sketch of his views on science. Here Mongré claimed that while empirical science was a significant human achievement alongside aesthetic production and emotional life, any kind of naive or metaphysical realism would be misguided in science – just as moral

---

[16] See, e.g., aphorisms no. 22, 35, 279, 289, and the whole section 'Denken-Reden-Bilden'; cf. also the discussion in (Mehrtens, 1990, 167–170).

dogmata would be with regard to true life. In science as well, individualism was his motto: Radicalizing Kant and taking up similar motifs in both Liebmann and Nietzsche, Mongré claimed that every individual consciousness produces a well-ordered 'Cosmos' out of transcendental 'Chaos'. 'Cosmos' meant the world we experience, ordered by natural laws and full of meaning and aesthetic value; 'Chaos' denoted a world devoid of laws and meaning, but full of even the fanciest possibilities. Epistemological idealism, he wrote, teaches 'that the teleological, causal, richly organised course of empirical reality says nothing at all about the transcendent [course of reality]; that a completely arbitrary *Geschehen* 'in itself' is sufficient to guarantee for us, as a subjective effect, a determined *Geschehen*; that even the most perfect cosmos arises just by passing the wildest chaos through a sieve [of consciousness].'[17] A background to these bold claims were discussions in the physiology and psychology of sense perception that also had impressed, among others, Otto Liebmann and Friedrich Nietzsche. A recurrent motif was the idea that, for physiological and evolutionary reasons, different species of animals might have different time perceptions, different 'basic subjective measures of time'. (With reference to the zoologist Karl Ernst von Baer, Otto Liebmann had called this a 'fiction well within the physically conceivable'.[18]) Hence 'time' was not a property of 'the world' in itself, but a form of perception characteristic for a given biological species. In a way, Mongré's view was a radical, perspectivist extrapolation of this idea to all aspects and levels of perception and cognition. Empirical science was just this: one out of many possible perspectives on the world, certainly successful, but nonetheless a perspective, nothing more. A perspective peculiar to human beings, but not even for them the only perspective available.

In these remarks on science, mathematics began to play a special role. On the one hand, Mongré made it quite clear that in science he valued quantitative measurement higher than qualitative explanation: 'Every qualitative explanation of facts comes to the point where it has to stand a quantitative test against other, equally plausible explanations, i.e. the test of its congruence with experience down to the hundreth part of a second and the millionth part of a millimeter.'[19] Moreover, he warned against 'mythological' interpretations of relations in experience that could be expressed mathematically. A case in point were interpretations of the notion of gravity: The 'simple fact' of experience that the motion of material points can be expressed in a simple fashion by using first- and second-order differential quotients of the variables of motion (as in analytical mechanics) does not entitle the physicist or philosopher to talk of 'forces' between material points

---

[17] (Hausdorff, 1897/2004, aphorism no. 408). Hausdorff's term 'Geschehen' denotes events, happenings, but also the *course of* events, 'that which happens'. A precise English equivalent seems not to exist. As Mongré uses the term rather often I leave it untranslated, here and in the following.

[18] (Hausdorff, Werke VII, 2004, 823). To trace the varieties and uses of this idea in late nineteenth century scientific literature would provide the topic of an interesting study.

[19] (Hausdorff, 1897/2004, 339). The passage is included in an attack on Schopenhauer's criticisms of mathematics (aphorism no. 399).

as elements of reality (Hausdorff, 1897/2004, aphorism no. 400). The 'last purpose of mathematics', Mongré wrote in consequence of such reflections, might be a 'self-critique of science'.[20]

Of course such a phrase requires explanation. What Mongré had in mind may probably best be exemplifed by his discussion of time. Many of the aphorisms of the section 'Zur Kritik des Erkennens' revolve about the perception of time by an individual consiousness. Evidently, one of his primary goals was to imagine different courses of time or, perhaps more precisely, to imagine different ways of perceiving time. With a reference to a then probably well-known idea of microscopic universes (e.g. consisting of dust particles dancing in the sun) Mongré argued: 'Every intelligence not completely heavy like lead and thick like pitch will succeed in this abstraction of space, [...] in the recognition of the relativity of all measures of space, but one must go a step further in composing and leave open similar possibilities for time.'[21] In every second of human time, myriads of generations of microscopic beings might be born and die again, and beyond possibly a little shimmering nothing would be perceived by us even with the sharpest microscopes. In the same way, we might ourselves only be existent (and perceiving) in certain intervals or moments of a richer course of time, most of which would then be totally unknown to us. This possibility, Mongré argued, might indeed be the reason for our perceiving a chaotic world as an ordered cosmos: The 'world' might be like an orchestra, whose instruments 'rage undirected in a disharmonious mess', while we might be like auditors who listen to this chaos only during certain scattered moments that appear to us like a continuous sequence, with the result that we hear 'a rhythmically clear, pure and harmonious melody' (Hausdorff, 1897/2004, aphorism no. 408).

The intention of such ideas, it seems, was to free the reader of the idea that the structure of time as she or he perceived it had anything to do with an absolute temporal order independent of any perception, and that for this very reason, features of the world as we experience it might be just artefacts of our selective perception. Mathematics came into play in the discussion of another contemporary imagination of the course of time, Nietzsche's famous doctrine of recurrence. Mongré devoted a substantial section to its discussion. He distinguished two interpretations of the thesis. The first interpretation did not involve any empirical consciousness of recurrence: In 'absolute time', an arbitrary interval of 'full time' (i.e. a temporal 'continuum of world states' as experienced by a consciousness, see below, Section 3) could be repeated over and over again without any noticeable effect. This idea would be spelled out and extended in Mongré's second book.

---

[20] 'Uns fehlt eine Selbstkritik der Wissenschaft; Urtheile der Kunst, der Religion, des Gefühls über die Wissenschaft sind so zahlreich wie unnütz. Vielleicht ist dies die letzte Bestimmung der Mathematik.' (Hausdorff, 1897/2004, aphorism no. 401.)

[21] 'Diese Raumabstraction, [...] die Erkenntnis der Relativität alles Raummasses, wird jeder nicht ganz bleischweren und pechzähen Intelligenz gelingen; man muss aber einen Schritt weiter dichten und auch für die Zeit solche Möglichkeiten offen lassen.' (Hausdorff, 1897/2004, aphorism no. 404.)

For now, Mongré concentrated on the second interpretation of Nietzsche's idea, which he claimed was Nietzsche's own: the empirical course of time as a whole might be 'representable as a closed line'. This, Mongré argued, was a possible hypothesis, 'conceivable under the plausible assumption that in historical consciousness, discontinuities have happened, caused, e.g, by geological revolutions or catastrophies of cosmic character.' A less discontinuous historical memory or shorter recurrence times (a larger 'curvature of time') should have empirical effects on human consciousness, but even this could be imagined by means of a 'glance of phantasy into the possible'. Consequently, and given the possible differences of the time-scales of perception in different species, it might just be a consequence of human organization that we do not actually experience such effects (Hausdorff, 1897/2004, aphorism no. 405).

While Mongré thus argued that recurrence was a 'brilliant speculation' and a (contingent) empirical possibility, he rejected the attempts at a scientific proof of recurrence that were found in Nietzsche's *Nachlass* (Hausdorff, 1897/2004, aphorism no. 406). Simplifying somewhat, one may say that Nietzsche had attempted an argument from ergodicity. His premises were (a) there are only finitely many possible physical states in which the world may happen to be, while (b) time is infinite. Consequently, so Nietzsche seems to have hoped, the recurrence of similar time intervals (i.e. of similarly ordered time sequences of physical states) is a necessity.

This is a delicate argument even if Nietzsche's premises are granted. In particular, in order to claim the status of a proof, the agument requires mathematically precise notions of time and of the orbit of the world through the space of its possible physical states. In the *Sant' Ilario*, however, Mongré simply attacked the first premise of Nietzsche's argument, pointing out that according to all scientific conceptions of their day, even the set of possible physical states of just 3 material atoms (billiard balls) in ordinary (Euclidean) space had ∞^3 elements (meaning that the set of *relative* configurations of three material points, disregarding speeds, is a 3-dimensional manifold). Mongré further argued that ∞^2 'times' such as ours (a one-dimensional continuum) would be necessary to exhaust all these possible configurations, a far cry from recurrence. Nietzsche's idea, Mongré concluded, might be an empirical possibility, but in no way it was a proved necessity. Mathematical analysis showed that it could well be otherwise.

Of course Mongré's argument was not much less naive than Nietzsche's own and even defective in a certain sense. When writing the *Sant' Ilario*, Hausdorff was obviously unaware of Cantor's basic result that there exist bijections between point continua of different dimensions (e.g., continuous maps from a one-dimensional point continuum onto a 3-dimensional one). While this did not justify Nietzsche's claim on recurrence, it did undercut his own criticism. The point, however, is that Hausdorff realized his error within a very short time. Just one year later, when he published his epistemological monograph *Das Chaos in kosmischer Auslese*, we find him making use of Cantor's theory of point sets in order to refine his criticism of Nietzsche's argument.[22] Hausdorff returned to Nietzsche's argument on recurrence more than once, advancing its set-theoretic analysis. Doing so, he also recognized that Nietzsche's argument was problematic even given its premises. A rather sharp formulation of the problem may be found in a manuscript dated 22. 11. 1908. There, Hausdorff translated the problem of recurrence into the following mathematical problem: Do there exist mappings from the real numbers $\mathbb{R}$ (interpreted as time) to a set of two elements (the minimal number of states of the world to account for change):

$$f : \mathbb{R} \to \{1, 2\},$$

which are completely recurrence-free in the sense that for no two intervals $I, J \subset \mathbb{R}$ for which f takes both its values infinitely often, f behaves 'similarly' in I and J (Hausdorff meant, in more modern notation: for no such intervals I, J does there exist an order-preserving mapping $\phi : I \to J$ such that $f \circ \phi = f$ holds on I).[23] If such mappings exist, this would imply a rather strong rebuttal of Nietzsche's argument: Even with the minimal number of states of the world to account for change at all, a course of time would be conceivable that does not repeat any arbitrarily small segment of time in a similar fashion. Hausdorff hoped to give a positive answer to his mathematical problem but he got stuck with a related problem on subsets of the reals that he could not solve; presently, I am unaware of a solution of this last problem.[24]

In this critical analysis of Nietzsche's idea of recurrence, one may grasp how Mongré viewed mathematics as a tool for a self-critique of science: It could help to recognize metaphysical elements both in the belief in a linear conception of time *and* in the belief in a 'recurrent' time. It could do so by sketching possible alternative structures of time, making use of mathematical constructions, interpreted by imagining corresponding subjective experiences.

3 Transcendent nihilism

The encounter with set theory changed Hausdorff's outlook on the philosophical problems he had dealt with in his aphorisms. What he found in this theory

---

[22] Compare (Hausdorff, 1898/2004, 193–194). The different treatments of Nietzsche's theory of recurrence have first been used by Walter Purkert as a means to date the encounter of Hausdorff with set theory, see note 8.

[23] See (Hausdorff papers, 49/1078, 12–15), for the precise condition, see p. 14. Pages 12 and 13 do not carry dates, they may have been written earlier.

[24] 'Die Frage, ob schon mit 2 Zeichen eine derartige Besetzung der Zeit möglich ist, hängt von der Frage ab, ob es *gestufte dichte* Theilmengen des Continuums gibt.' (Hausdorff papers, 49/1078, 15.) In his first set-theoretic publication, Hausdorff had called a subset of $\mathbb{R}$ 'gestuft' if and only if it does not contain any two order-similar segments, cf. Purkert's introduction to (Hausdorff, Werke II, 2002, 9).

was a general tool to support the perspectivist claims of his epistemology. In his monograph *Das Chaos in kosmischer Auslese*, we find a much more rigid line of argument aimed at showing that metaphysical claims about the structure of the world 'in itself' were misguided.

Mongré concentrated his attacks on the two basic notions of Kantian epistemology, time and space; a further continuing thread of his discussion concerned what he called the 'materialist' conception of the world, i.e. the view that 'all *Geschehen* is nothing but the motion of material points' (Hausdorff, 1898/2004, 13). From the outset, he noted that his position was in accord with one of the basic tenets of Kantianism, i.e., the thesis of 'transcendental idealism' claiming that space and time (and hence motion) are not forms of the world in itself, nor aspects of a reality independent of human experience. However, he pointed out that in his own time, several philosophers attempted to re-establish new kinds of realistic metaphysics and that, therefore, new arguments in favour of transcendental idealism might be of some value (Hausdorff, 1898/2004, 7). Moreover, he claimed that his argumentative strategy was new. It consisted in an indirect argument:

> 'We want to combat realism in its own territory, to reduce it to absurdity by means of its own claims. To this end we isolate any property or relation attached to the world of our consciousness, transfer it to absolute reality and then try to modify it as strongly as possible without changing the given empirical effects.' (Hausdorff, 1898/2004, 8.)

Three steps of this strategy can be distinguished: the *isolation* of a property or relation of empirical consciousness, the *transfer* of this property to 'absolute reality', and then a *variation* of the property in the domain of the 'absolute'. If this variation could be imagined without troubling the empirical consciousness, then the particular property or relation isolated was recognized *not to be* a knowable property of absolute reality.

In order to show Mongré's stategy at work, let me summarize his critical arguments against the idea that anything might be knowable about an absolute temporal order of the world. In order to pursue the argument, Mongré first laid out a framework for speaking of time consciousness and 'absolute' time. 'In our representation of time, two different single ideas, which I want to call the *content of time* and the *course of time*, are joined together in an easily separable manner'.[25] The first idea consisted in 'a continuous series of states of the world'; for future reference, let us call this series X. Mongré explained a 'state of the world' to be a 'filled time segment [*erfüllte Zeitstrecke*] of length zero', comparable to a point in a line. A 'filled time segment', in turn, was what a receptive and productive consciousness could experience or imagine in time:

> 'The year x, with all its large and small content, from the changes in the galaxies to the swarm of infusoria in a water drop, from the slow

---

[25] 'In unserer Zeitvorstellung sind zwei ungleichartige Einzelvorstellungen, die ich kurz *Zeitinhalt* und *Zeitablauf* nennen will, auf leicht trennbare Weise miteinander verknüpft.' (Hausdorff, 1898/2004, 11.)

weathering of a rock face to the innumerable collisions of the molecules of a gas – that would be such a filled time segment.' (Hausdorff, 1898/2004, 12.)

In mathematical terms, X, filled time or the content of time, could thus be conceived as a one-dimensional continuum of world states, a 'time line' – the linear ordering of this continuum being perceived (or imagined) by a consciousness. In fact, Mongré's discussion suggested: by an individual consciousness. At this point, he avoided addressing the question of whether different individuals might have the same empirical ideas about time and temporal order; his later discussion would show that he still claimed that these could very well be different. Mongré had thus isolated the perceived temporal ordering of world states as the property to be analysed.

The 'course of time', on the other hand, was a 'mysterious formal process, by means of which every state of the world experiences the sequence of changes from future to present and past.'[26] To mathematize the mystery of this formal process somewhat, Hausdorff now introduced the 'auxiliary idea' of 'absolute time', another one-dimensional continuum, say T, whose ordering was the ordering of time 'in itself'. Using this auxiliary idea, the course of time could be represented as a 'motion of the point of presence' ('Bewegung des Gegenwartpunkts') in X, i.e. as a mapping

$$A : T \to X, \tau \mapsto x_\tau.$$

In supposing T to be a linearly ordered continuum (of the order type of the real numbers $\mathbf{R}$), just like X, Mongré had performed the (hypothetical) transfer of a given empirical property to absolute reality. While the functional notation above is ours, the mathematical representations of both the empirical and the absolute aspects of temporal order are Mongré's. The problem now was the following. During the 'motion of the point of presence', a motion 'on the time line' X but happening 'in absolute time' T, a consciousness C would obtain certain empirical ideas about the world, including a conception of empirical time. Can C know anything about absolute time, T (or the course of time, A)?

Next came the crucial step of the argument, the variation of the relation between absolute and empirical time. Hausdorff repeated this step twice, first in a somewhat sloppy form, then in a more rigorous fashion. A naive realist, he began, 'would not be able to imagine anything other than a motion of the point [of presence] in one and the same direction with constant velocity'. He then referred to the idea that this velocity might change.[27] He then continued: 'For the mathematician, who immediately concludes from arbitrary velocity to arbitrary place, from an arbitrary differential quotient to an arbitrary function, it would be

---

[26] '... eines räthselhaften formalen Processes, durch den jeder Weltzustand die Verwandlungsfolge Zukunft, Gegenwart, Vergangenheit erfährt.' (Hausdorff, 1898/2004, 11.)

[27] This was probably another reference to the ideas about a subjective measure of time in the wake of Karl Ernst von Baer, see above.

just a single step from here to the theorem we want to establish.' Mongré called the theorem thus alluded to his 'fundamental theorem':

> 'The point of presence moves on the time-line in a completely arbitrary, continuous or discontinuous fashion. The transcendent succession of world states is arbitrary and does not fall into our consciousness.'[28]

Of course, Mongré was aware that the argument about varying speeds of time was not quite sufficient to make room for 'completely arbitrary', and even discontinuous motions of the point of presence. Therefore he presented a 'proof' of his fundamental theorem in the 'form of a mere syllogism':

> 'Let us suppose a succession A of world states, chosen in a determinate fashion, generates our empirical world phenomenon. If now another succession B is chosen, and if this change would fall into our consciousness, then this would have to be noticed sometime, e.g. during the empirical time segment t [i.e. an interval t of 'filled time', M. E.]. But as we ourselves, including the contents of our consciousness, are part of the totality of world states, i.e. the time line, this would imply that the transposition A–B would have brought something into the time segment t which was not there before; hence this transposition would have changed not only the ordering, but also the inner structure of the states of the world, contradicting our assumption.'[29]

It followed that under the given assumptions C could know nothing at all about the (transcendent, or absolute) course of time.

On the following pages, Mongré illustrated this idea by spelling out various special cases of a variation of the course of time, e.g. (transcendent) repetitions of time intervals (here he extended his remarks of *Sant' Ilario*), interchanges of whole time segments (destroying the absolute meaning of the terms 'earlier' and 'later', 'past' and 'future'), etc. Of course such a dissolution of the conception of time affected ethical views as well. Both ideas of salvation (in some unknown future) and of utilitarianism (it cannot be excluded that a single moment of individual suffering is repeated over and over again in absolute time) were strongly criticized in a separate chapter 'Against Metaphysics'.

Toward the end of his book, Mongré carried his criticisms of time even further. It cannot even be excluded, he argued, that the totality X of states of the world

---

[28] 'Der Gegenwartpunkt bewegt sich auf der Zeitlinie in ganz beliebiger, stetiger oder unstetiger Weise. Die transcendente Succession der Weltzustände ist willkürlich und fällt nicht in unser Bewusstsein.' (Hausdorff, 1898/2004, 16.)

[29] 'Nehmen wir an, eine bestimmt gewählte Succession A der Weltzustände erzeuge unser empirisches Weltphänomen. Wird nun eine andere Succession B gewählt, und fiele diese Veränderung in unser Bewusstsein, so müsste das irgendwann einmal, etwa während der empirischen Zeitstrecke t, zu spüren sein. Da wir selbst aber, mit all unserem Bewusstseinsinhalt, dem Inbegriff aller Weltzustände, der Zeitlinie, eingegliedert sind, so hiesse das, dass durch die Vertauschung $A-B$ etwas in die Zeitstrecke t hineingekommen wäre, was vorher nicht darin war; diese Vertauschung hätte demgemäss nicht nur die Reihenfolge, sondern, gegen die Voraussetzung, auch die innere Structur der Weltzustände verändert.' (Hausdorff, 1898/2004, 16–17.)

is a much more complex set than 'we' (or any other individual consciousness) perceive(s). Similarly, absolute time [T] could be thought of as any sufficiently large set. The 'time line' then might be a 'time plane' or some even more complex set in which different 'courses of time' might be 'realized' for different beings. In fact, we should 'associate with every individual its own point of presence'.[30] Other consciousnesses might perceive the same subsets of X as 'we' do, but in different time orderings, or their points of presence might move in completely different subsets of X, and in different subsets of absolute time T that did not need to be sets of the type of the linear continuum. An argument of the same kind as the one given above would then imply that there exists no possibility for any one such consciousness to know anything about at least those others who move in different subsets of either T or X. In this way, Mongré recovered the radical perspectivism of his earlier aphoristic book, and rephrased it using set-theoretic language, in parts in a rather poetic fashion.

Similar arguments were applied to the conceptions of space. However, Mongré remained more cautious when it came to 'proving' his ideas in this case. Rather than giving a 'syllogistic' argument as for time, he proceeded step by step, beginning with a discussion of simple variations of the relation between (metaphysically) 'absolute' and 'empirical' space.[31] As in the case of time, the starting point was to transfer the traditionally assumed structure of empirical space as a 'continuous, Euclidean, three-dimensional space' to 'absolute' reality. Mongré explicitly noted that even this step involved, at the time of writing, an ambiguous mathematization as 'the question of the number and content of the necessary and sufficient axioms for our conception of space is not considered as solved among mathematicians'.[32] Assuming that many of his readers were not acquainted with the general notion of a mathematical mapping, Mongré then illustrated this notion for mappings from absolute space (say, S) to empirical space (say, Σ) by means of a metaphorical analogy: Imagine each point of absolute space as a drawer in an 'absolute' wardrobe, while each point of Σ represents a piece of the

---

[30] See (Hausdorff, 1898/2004), Chapters 4, 'Die Mehrheit der Gegenwartpunkte' (quotation on p. 69) and 7, 'Die Zeitebene'.

[31] The reason for this reluctance was Mongré's caution with respect to mental effects of physical changes in distant parts of space. While he held that for time, all mental effects were local, i.e. determined by the state of the world in the moment in which the effect occurs, he allowed for the possibility that mental effects could arise if, for instance, some distant masses were annihilated, i.e. the mind might react instantaneously to changes in the gravity potential. While Mongré stated that he did not believe in this possibility, he admitted to have no rigorous argument to exclude it. See (Hausdorff, 1898/2004, 75–79). – In fact, a related problem also concerns Mongré's views on time. In several places, he seems undecided about how to interprete the notion of a 'moment' in time. It could be an 'instant', similar to a point on a line, but it could also be a infinitely small time element, a 'time differential'. Whether or not mental effects of the course of time can be considered completely local depends on this decision. For a discussion of this problem, see the commentary to *Das Chaos in kosmischer Auslese*, page 13, lines 10–13 (Hausdorff, Werke VII, 2004, 821 f.).

[32] (Hausdorff, 1898/2004, 79.) With this remark, he referred to the work of Riemann, Helmholtz, Lie 'and others'.

former's contents. Redistributing these pieces in the set of drawers meant varying the relation between absolute and empirical space; an arbitrary distribution of empirical space points in absolute space meant an arbitrary 'transformation or mapping', say, σ, from S to Σ (Hausdorff, 1898/2004, 82). An alternative view, used by Mongré, was to consider variations σ' of the initial mapping σ between S and Σ as composed of σ and a transformation $\varphi : S \to S$:

$$\sigma' = \sigma \circ \varphi.$$

(Here again, the symbolism is mine.) As in the case of time, Mongré claimed that *arbitrary* variations of the supposed mapping between S and Σ might occur without a consciousness being able to notice it (Hausdorff, 1898/2004, 82–83; this claim he sometimes referred to as his 'transformation principle').

The simplest admissible, i.e. empirically consequenceless variations φ of space structure, well known in the physical and philosophical literature of the time, were translations and rotations.[33] Next Mongré considered whether homogeneous dilations or contractions of absolute space would have empirical effects – another *topos* of several metageometrical debates at the time. He pointed out that this idea only made sense if appropriate physical changes (e.g. of the gravitational constant) were assumed to accompany a dilation or contraction of space. However, as these physical changes only concerned 'absolute' reality they could indeed be assumed without consequences for empirical phenomena. As a fourth kind of spatial variation φ, Mongré mentioned reflections.

Only for these kinds of variations of the spatial structure of absolute reality did Mongré believe he had rigorous arguments showing that a consciousness would necessarily remain unaware of them. Nevertheless, he extended his discussion to include other, more drastic variations φ, such as inhomogeneous dilations, changes of the geometrical structure from a Euclidean space to a space of constant positive or negative curvature, or even changes of local Riemannian curvature. In all cases, Mongré's emphasis was to point out that one might imagine a behaviour of physical bodies (with respect to 'absolute' space) that would account for these variations while leaving our experiences intact.[34]

Obviously, in such arguments Mongré took up and extended similar arguments that could already be found in Helmholtz's papers on non-Euclidean geometry.[35] There, Helmholtz had argued that, given a certain physical behaviour (certain

---

[33] (Hausdorff, 1898/2004, 83.) Mongré added some remarks emphasizing that in physics (as an empirical science) it might well make sense to speak of empirical effects of absolute rotation along the lines of Newton's bucket argument, even if 'transcendent' rotations would have no empirical effects at all. See comments to *Das Chaos in kosmischer Auslese*, page 83, lines 13–15 (Hausdorff, Werke VII, 2004, 852 f.).

[34] See (Hausdorff, 1898/2004, 93 ff., 105, 115 ff., 119 f.).

[35] He explicitly refered to Helmholtz's lectures 'Ueber den Ursprung und die Bedeutung der geometrischen Axiome' (Helmholtz, 1870) and 'Die Thatsachen in der Wahrnehmung' (Helmholtz, 1878); see (Hausdorff, 1898/2004, 107).

place-dependent deformations) of measurement devices, geometrical measurements in Euclidean space might have given the same results as measurements with rigid measuring devices in non-Euclidean spaces and vice versa. However, Hausdorff found the scope of these earlier arguments rather limited: 'Concerning non-Euclidean geometry, we may say that it is most laudable as a liberation from the limitations of Euclidean thinking, but it is much too narrow for delineating possible structures of absolute space.' (Hausdorff, 1898/2004, 121.)

One of the areas in which Mongré found it necessary to go beyond all earlier arguments was the idea of the 'continuity' of space (and time), i.e. as we can say in retrospect, of its topological type (particularly, on the small scale). Taking up notions from Cantor's theory of point sets, he asked whether 'in place of the usually assumed point continuum an everywhere dense point set or a semi-continuum [...] would be of the same service' in describing space. In 1879, Cantor had called a subset P of an open interval (a, b) of real numbers 'everywhere dense' if and only if for every subinterval (c, d) of (a, b), the intersection of P and (c, d) was non-empty (Cantor, 1932, 140). Probably, Mongré had a generalization of this notion to subsets of the 3-dimensional continuum $\mathbb{R}^3$ in mind. To explain Cantor's notion of a semi-continuum, one has first to recall Cantor's definition of a continuum: A subset C of $\mathbb{R}^n$ was called a continuum if and only if C was connected (in Cantor's terminology, this meant that for any two points p, q in C and every $\varepsilon > 0$, there existed a finite sequence of points $p, p_1, p_2, \ldots, q$ such that the (Euclidean) distance between any two successive points was less than ε) and perfect, i.e. C coincided with the set of its limit points. A semi-continuum was then defined to be a connected, but not perfect subset of $\mathbb{R}^n$ with cardinality of the continuum and with the additional property that any two points p, q in C were connected by a continuum (Cantor, 1932, 208). Of course, to characterize all such sets or similarly defined point sets provided a formidable task at the time. Consequently, a critical analysis of the possibilities for describing the continuity of point sets or space was still unavailable: 'But continuity, in the physical as well as in the mathematical domain, is a difficult problem about which discussion has not even seriously started, much less finished, and can be asked for a report of its results.'[36] This critical assessment may have been one of the reasons that would later bring Hausdorff to reconceptualize the issue by means of the introduction of topological spaces.[37]

After his discussion of space, Mongré returned to the problem of time and extended his earlier arguments in several respects, e.g., by speculating about motions of 'points of presence' in a more complex set of world states, and about motions in which absolute time had a different structure from a linearly ordered continuum *and* in which space was different from ordinary conceptions. (In mathematical

---

[36] 'Aber die Continuität, auf physischem wie auf mathematischem Gebiete ist ein schwieriges Problem, über das die Discussion noch nicht einmal recht angefangen hat – geschweige denn beendet und um Mittheilung ihrer Ergebnisse zu befragen wäre.' (Hausdorff, 1898/2004, 122.)

[37] See (Purkert *et al.*, 2002), especially pp. 690–699.

terms, both the earlier assumptions about X and T were weakened, as were the assumptions about absolute space). By means of rather poetic illustrations of such situations, Mongré argued that no consciousness would ever be in a position to distinguish among these possibilities. Hence the only option for a critical mind with respect to the question of the structure of (absolute, as opposed to empirical) space and time was 'transcendent nihilism'.

In the course of this argument, he also extended his 'transformation principle' for the relation between (mathematized versions of) empirical and absolute space, now including time. Having introduced symbolic coordinates x, y, z for absolute space, t for absolute time, ξ, η, ζ for empirical space and τ for empirical time, he stated the most radical form of his principle as follows: 'The succession of time changes arbitrarily from space point to space point, and the structure of space from instant to instant. Then all four quantities $xyzt$ are arbitrary functions of the four others, $\xi\eta\zeta\tau$, or vice versa; let the summarizing expression be allowed: in every point of transcendent spacetime, every point of empirical spacetime may be realized.' (Hausdorff, 1898/2004, 142.)

The meaning of course still was: Whatever the 'realization' function might be, a consciousness would be unable to recognize absolute spatio-temporal structures above and beyond its empirical findings. Once again, the consequence was: No knowledge about absolute space and time is possible. This was written in 1898, several years before relativity theory required a rethinking of the relations between time and space.[38]

4 Considered empiricism

With (metaphysical) realism about time, space, motion, causality, etc. ruled out, what remained to be said about the scientific conceptions of these matters?

In the years after the publication of *Das Chaos in kosmischer Auslese*, this question came more and more to the fore in Hausdorff's epistemological thinking. At the same time, he reoriented his mathematical work toward set theory. In 1901, he gave his first lecture course on set theory (one of the first such courses ever taught), and he published his first paper on order types of sets. In 1908, a major article in the *Mathematische Annalen* summed up his work in this area.[39] A profound impression was made on Hausdorff by David Hilbert's *Grundlagen der Geometrie*.

---

[38] A full discussion would have to address Mongré's reduction of the argument to the possibility of the realization of different world experiences (empirical times) in different subsets of absolute time, framed in terms of his so-called principle of indirect selection, discussed in Chapter 6 of *Das Chaos in kosmischer Auslese*.

[39] 'Grundzüge einer Theorie der geordneten Mengen', *Mathematische Annalen* 65 (1908), 435–505; see (Scholz, 1996) on this work and Purkert in (Hausdorff, Werke II, 2002, 7–16) for further information about Hausdorff's turn to set-theoretical research.

A letter to Hilbert of October 1900 and many sheets in Hausdorff's *Nachlass* document the attention that Hausdorff gave to this new style of mathematical thinking. He reworked the foundations of geometry for himself, including topics that Hilbert had left out such as an axiomatic basis for projective geometry.

Interestingly enough, Hausdorff did not, in this period, attempt to outline an axiomatic foundation of set theory. Even in 1914, in his *Grundzüge der Mengenlehre*, he remained cautious in this regard. While noting that Ernst Zermelo's work had put such an approach on the agenda of mathematics, the axiomatization of set theory was still in progress and thus not (yet) suited as a basis for an introduction to higher set theory.[40] For him, Hilbert's axiomatic approach was not a method for providing a foundation of mathematics as a whole but rather a method to systematize and to analyse theories of higher order, such as geometry, the theory of probability, or, within set theory proper, the theory of order types and the theory of topological spaces for which he would become famous. The crucial role of axiomatics was to define the basic notions of such a theory – such as the notion of a topological space – in the proper fashion, i.e. as sets (whatever sets were in themselves) with certain properties (and additional structure).

A closely related view of the axiomatic method was brought to bear on the issues of epistemology. The main issues for Hausdorff were still the notions of time and space. Three texts, all produced in Leipzig in 1903 and 1904, document his corresponding reflections. The first is an article based on his inaugural lecture as an extraordinary professor, entitled 'Das Raumproblem' (Hausdorff, 1903). This article also was the first article in which Felix Hausdorff, rather than Paul Mongré, presented his philosophical views. The second text is the manuscript of a lecture course on *Zeit und Raum*, given at Leipzig university in the winter semester 1903/1904 (Hausdorff papers, 24/71, 68 pp.). The third is a 12-page. fragment entitled 'Der Formalismus' of the same time, probably a draft of a chapter for a monograph on time and space that Hausdorff never completed (Hausdorff papers, 49/1067, pp. 1–12). A long series of related, shorter manuscripts and notes complements these sources.[41]

The article 'Das Raumproblem' may be taken to exemplify Hausdorff's new approach. In it, Hausdorff set out by distinguishing three levels of the notion of space: the *mathematical* notion of space, 'a certain free creation of our thinking, subject to no other restriction than logic'; the *empirical* notion of space, 'a system of real experiences, the phenomenon of a space-filling external world, passing by in our consciousness', and *objective* or *absolute* space, 'a certain behaviour of things independently of our consciousness', pre-supposed for explaining our (scientific) views and intuitions of space (Hausdorff, 1903, 1–2).

---

[40] See (Hausdorff, 1914/2002, 2). In retrospect, this reluctance was not without wisdom – Hausdorff certainly belonged among those who saw the flaws in Zermelo's axiomatics.

[41] The most important of these texts will be edited in (Hausdorff, *Werke* VI, forthcoming).

Hausdorff then explained the 'space problem' – a problem whose 'solution' required the contributions of 'no less than five sciences: mathematics and physics, physiology, psychology and epistemology' – to consist in the following question: 'Can the mathematical, the empirical, the objective space be defined in a unique fashion, excluding all variations, or do we have a choice among different, equally legitimate hypotheses?' His first answer concerned empirical space. On this level, we 'clearly' do not have a freedom of choice as, he wrote, we have to accept 'the experiences and phenomena of consciousness which constitute it [i.e. empirical space] as *fait accompli*, [...] in pure receptivity, whether we naively observe or whether we make arbitrary experiments'. The situation concerning the other two levels of the notion of space he regarded as quite different: 'Here the [...] freedom of choice exists, and with the most dramatic *Spielraum*: namely a freedom of choice between infinitely many hypotheses, none of which is more legitimate than any other.' Here and in the following, I leave the German term *Spielraum* untranslated as this metaphor plays a crucial role in Hausdorff's thinking. Its meaning comes close to that of a 'range of possibilities', a range that leaves room for play, and for active choices.[42]

For *objective* or *absolute* space, the existence of this *Spielraum* of equally justified hypotheses was fatal. The notion of absolute space 'has to be discarded' as a well-definable notion (Hausdorff, 1903, 17). Of course, this view was nothing but a repetition of Mongré's 'transcendent nihilism'. In his lecture Hausdorff illustrated it (and the transformation principle behind it) by means of the relation between a drawn geographical map and the 'real' territory mapped. In the case of space, the transformation principle amounted to the claim that the 'method of projection' of absolute space onto our mental map of space is in principle unknowable, and hence we have no means of knowing what (and whether) this map is actually mapping.[43]

This left only the mathematical notion of space and the question of its relation to empirical space. On this level, the freedom of choice – a well-known mathematical fact since the advent of non-Euclidean geometry, Hausdorff said (Hausdorff, 1903, 2–3) – was *not* fatal. On the contrary, mathematics furnished complexes of freely chosen axioms, hypothetical-deductive systems of geometry, some of which *did* lead to suitable descriptions of empirical phenomena that are useful in the (empirical) sciences. The main task of mathematics in this connection then was to explore the *Spielraum* for conceiving such systems. This exploration had a positive aspect, i.e. the proliferation of suitable systems of mathematical notions and an analysis of their mutual relations, and a critical aspect to which I will shortly return.

Just as the notion of space had three levels for Hausdorff, the range of possible mathematical conceptions of space was determined in three ways: 'by the

---

[42] All quotations in this paragraph are from (Hausdorff, 1903, 1–2).

[43] While the illustration is new, the argument was explicitly related to Mongré's earlier discussion, see (Hausdorff, 1903, note 28).

Spielraum of thought, the *Spielraum* of intuition, [and] the *Spielraum* of Erfahrung' (Hausdorff, 1903, 3). As the history of geometry during the nineteenth century from the first non-Euclidean geometries to the modern axiomatic conception of geometry in the style of Hilbert's *Grundlagen der Geometrie* had shown, Hausdorff argued, the *Spielraum* of thought was very large indeed. Given *Mongré's* interest in the possibilities of mathematizing space as a means for imagining different structures of *absolute* space (if only for destroying the idea), *Hausdorff's* willingness to endorse an axiomatic analysis of the architecture of the fundamental notions of geometry in order to delineate the possible mathematical descriptions of *empirical* space is not difficult to understand. For him, Hilbert's style of laying the foundations of geometry provided exactly what he needed – an instrument for outlining a spectrum of different mathematizations of space in a systematic fashion. While for Mongré the availability of different mathematizations had implied the rejection of the idea of a unique transcendental structure of space, for Hausdorff the same fact now meant that an empirical science of space could choose from many different mathematical frameworks.

The *Spielraum* of experience, however, was more restricted. Indeed, given Hausdorff's claim that empirical space is (almost by definition) unique, he required an argument that this uniqueness did not single out just one of the possible mathematical notions of space. This argument was the quantitative, and sometimes even qualitative imprecision of observation and experiment. While traditional Euclidean geometry could certainly be viewed as 'empirically valid', the same could be said about 'those non-Euclidean geometries whose divergence from Euclidean geometry remains below the threshold of observability' (Hausdorff, 1903, 4). Hausdorff emphasized that this imprecision of experience did not only concern the very small, but also the very large, about which we do not have too many experiences.

A substantial section of his paper consisted of a discussion of the main properties of the ordinary conception of (Euclidean) space, and the possible variations of this conception *within the bounds of experience*. Usually, space was considered as *flat*. However, the classical non-Euclidean spaces (i.e. hyperbolic, elliptic and spherical spaces) showed that it might be of a small constant curvature. At this point, Hausdorff might have referred to Karl Schwarzschild's recent evaluation of astronomical data in order to give bounds on the curvature of such spaces compatible with observation, but Schwarzschild's paper seems to have escaped his notice (Schwarzschild, 1900). All these spaces still had the property of *allowing free mobility of rigid bodies*, another assumption that could be given up by considering spaces of variable Riemannian curvature or even spaces without a Riemannian metric. Clifford was his witness for the first possibility, Minkowski's *Geometrie der Zahlen* and a paper by Georg Hamel (1903) for the second (Hausdorff, 1903, note 21). A further property of Euclidean space that could be given up without contradicting empirical evidence was *simple connectivity*: The spaces of constant curvature that Wilhelm Killing had termed Clifford–Klein space forms provided possible structures of space whose topological behaviour in the large was different from both Euclidean and the classical non-Euclidean spaces.⁴⁴ Finally, the *continuity* properties of space and the notion of *dimension* called for analysis. As for dimension, the situation required a kind of compromise: While philosophers (hopefully) accepted that no 'necessity of thought' implied that space has three dimensions, 'the mathematician will, while rejecting all spiritualist magic tricks, admit that there is no *Spielraum* of experience for a variation of the Three'. However, the number of dimensions depends on the notion of continuity: Space-filling curves showed that space *might* be conceived as one-dimensional if one was prepared to give up the usual notion of continuity. In another direction, examples by Dedekind and Hilbert had shown that most of Euclidean geometry held in certain denumerable sets that filled (ordinary Euclidean) space 'like infinitely subtle dust' – real space might be described by such a set, or by removing such a set from the ordinary continuum (Hausdorff, 1903, 13).

One feels tempted to ask whether Hausdorff, at this point or later, felt inclined to modify the notion of dimension in such a way that he could get rid of its discreteness. When he did so in (Hausdorff, 1919) this changed the status of the above compromise. With a more flexible notion of dimension, there is no empirical forcing of the number of dimensions of space – provided one is willing to allow suitable topological structures of space. But precisely this connection already seems to have occupied Hausdorff in the present passage. The argument invites one further comment. It actually shows that the empirical indeterminacy of mathematical descriptions should be conceived in a reflexive manner. *Which* range of mathematical possibilities for some notion is left open by empirical information *depends* on the mathematical framework used. Using the traditional notion of dimension, empirical evidence force the choice of 3 spatial dimensions; with Hausdorff's or any other notion, it does not. In this way, an empircal forcing of some mathematical possibility may just be *apparent*, a sign of a defective understanding of the possible mathematical structures that might be brought to bear on the situation.

In sum, none of the traditional properties (not even dimension) was forced by empirical evidence. There remained one possible site of uniqueness: intuition. After having admitted that neither a necessity of thought nor the facts of observation excluded the possibility of non-Euclidean geometries, the 'philosophers' sought to defend their traditionalist views, Hausdorff wrote, by hard-necked arguments about (Kantian) intuition. With some heavy irony, Hausdorff pointed out that these arguments were 'less than an argument about words, namely an argument about persons and personal gifts [...]' (Hausdorff, 1903, 5). With a view to the many unusual mathematical ideas about space Hausdorff wrote that 'the mathematician with a strong imagination will know how to inspire the constructions of his thinking with the liveliness of intuition, while spirits of lesser

---

⁴⁴ Hausdorff's reference was (Killing, 1893). On the history of Clifford–Klein space forms and their cosmological interest see (Epple, 2002). Once again, Hausdorff did not mention the only astronomer who had actually adressed this issue in an empirical context (Schwarzschild, 1900).

flying force or of more abstract inclination will not be able to follow him into his dominion of concrete creation and life'.[45] Consequently, such philosophical fights were weak retreats. Much less than experience did intuition limit the choice of mathematical descriptions of space. It only limited the individual *Spielraum* open to the imagination of an mathematician, a philosopher, or any other person.

In fact, both with a view to (existing) empirical and (supposed) intuitive limitations, mathematical thought could show how to explore and extend the range of empirically valid conceptions of space. This was the *critical* task of mathematical analysis. It helped not only to overcome metaphysical apriorism, but also to dissolve misunderstandings about the scope of individual intuition and imagination, and to reject the idea that empirical evidence uniquely determined the scientific conception of space.

It is quite clear how the general strategy of such an argument could be carried over to other basic notions of scientific theory, e.g., time, or motion. As the lecture course on 'Zeit und Raum' of 1903/1904 shows, an axiomatic analysis of the ordinary notion of time could be given that paralleled more or less that of the notion of space. Even more than in the case of space, the idea of continuity and set-theoretic possibilities for framing 'non-standard' notions of continuity and of temporal order moved into the focus, questioning (or rather, relativizing) the usual role the real numbers were given as a mathematical model of empirical time. Just as Euclidean space, this model of time was neither empirically forced nor intuitively unique or endowed with an absolute meaning.

Given these analyses, and taking into account that space and time provide two of the most basic notions of all mathematized science, one may now understand why Hausdorff viewed formalism (i.e. an axiomatic analysis of formal notions as in Hilbert's approach to geometry and not so much formalism as represented in the later debates of metamathematics) as an ingredient of true empiricism: It provided the basis for a 'considered' judgement about the possibilities of scientific thought given the empirical information available to us. Much more than philosophy, mathematics thus took the role of a critical epistemology of space, of time, and of other scientific notions – substantiating Mongré's earlier remark that the 'last purpose of mathematics' might be a 'self-critique of science'.

5 Discussion

From the perspective of today's philosophy of science, several claims made by Mongré/Hausdorff sound quite familiar (maybe even – depending on one's

---

[45] 'Der phantasiestarke Mathematiker wird den Gebilden seines Denkens auch die Lebendigkeit der Anschauung einzuhauchen wissen, während Geister von schwächerer Flugkraft oder mehr abstrakter Richtung ihm in sein Reich konkreter Schöpfung und Belebung nicht zu folgen vermögen.' (Hausdorff, 1903, 6.)

own particular position – a little outdated). But they do not sound familiar as *Hausdorff's* claims. We would rather expect them to be expressed by some follower or member of the early Vienna circle around Moritz Schlick, or (for certain claims) by a reader of Henri Poincaré's writings on the problem of space and the foundations of geometry.[46]

When, in 1917, Moritz Schlick published his short monograph *Raum und Zeit in der gegenwärtigen Physik*, he included two sections on 'The geometrical relativity of space' and 'The mathematical formulation of spatial relativity' (Schlick, 1917, 6 ff. and 13 ff.). Schlick started out by asking a very similar question to Hausdorff's of 1903: 'The most fundamental question which one may pose about time and space is, in quite popular and preliminary formulation: Are Space and Time something *real?*' (Schlick, 1917, 6; emphasis in the original.) Schlick then rendered the question of the *reality* of space and time as a question about the *measurability* of spatial and temporal properties or relations. Schlick was making here a similar move to that of Mongré, the transcendent nihilist: in science, it only makes sense to speak of reality in a strictly empirical sense, not in a metaphysically absolute sense. Schlick then went on to draw on certain well-known considerations about geometrical deformations of space: Would a simultaneous dilation of all metric distances be measurable or perceptible? Just as his main witness for this discussion, Henri Poincaré, had done, Schlick answered that this would be impossible, provided suitable adjustments of physical parameters (such as the constant of gravitation) and physical hypotheses were made. In other words: our scientific conceptions of space, time and physics leave room for 'certain dramatic geometrico-physical transformations' without any implications for the empirical state of affairs. Extending this argument beyond simple dilations (similarity transformations), Schlick felt justified in stating 'in a mathematical way of expression': 'Two worlds, which are transformed into each other by means of an arbitrary (but continuous and unique) point transformation, are identical with respect to their physical objectivity (*Gegenständlichkeit*).' (Schlick, 1917, 13.)

In consequence, the idea that the points of space and (following the unification of space and time in the wake of Einstein's theory of relativity) of time have some 'absolute' characteristics, independent of the physical behaviour of objects and processes we experience, had to be abandoned: 'Space and Time are never by themselves objects of measurement; together they form a four-dimensional scheme, into which we range physical objects and processes on the basis of our observations and measurements. We choose the scheme in such a way (and we can do so, as we deal with a product of abstraction) that the resulting system of physics receives the simplest possible architecture.' (Schlick 1917, 34.) This last condition was for Schlick indeed the only criterion that prevented the construction of physical theories from becoming arbitrary. The above transformation

---

[46] As an indication of this lack of awareness we may refer to the otherwise well-documented overview (Torretti, 1978). Torretti refers to Hausdorff's considerations on space just once in passing – in a footnote (Torretti, 1978, 383, note 17), and in a mathematical rather than epistemological context.

principle implied that if the criterion of simplicity was suspended, 'arbitrarily many equally complicated systems of physics would exist which all would do justice to experience in the same degree' (Schlick 1917, 16).

When writing down these ideas, which in many ways provided starting points for the methodological considerations of the Vienna circle philosophers, Schlick was unaware of the fact that, 20 years earlier, Hausdorff had formulated an even more radical transformation principle (for space *and* time, and allowing even discontinuous and non-unique transformations) with the same anti-metaphysical implications. At first sight, one might think that Schlick's insistence on the criterion of simplicity as a means for achieving uniqueness among the many mathematically possible and empirically adequate physical theories about spatio-temporal events (an argument certainly drawn from Mach's neo-positivist considerations on 'economy of thought') was a distinctive feature of his own views. But Hausdorff, being a devoted reader of contemporary reflections on science, was perfectly aware of this additional aspect of his own views. For him it was an accidental property of nature *as we experience it* that in certain domains of physics, and on the basis of certain (i.e. the traditional) assumptions about the geometry of space and the structure of time, 'the observed motions can be derived with great simplicity as mathematical consequences of few theoretical assumptions, and that the further consequences of this theory do not contradict experience' (Hausdorff, 1898, 104). For Hausdorff, the best example of this situation was still the Newtonian theory of the solar system.

That the empirical world allows for such simple physical descriptions had, however, to be considered nothing but a happy accident. While some areas of physics (such as mechanics) might lend themselves to simple descriptions, others might not, and rather than using the possibility of simple theories as a basis for renewing metaphysical claims, one should take them as increasing our respect for an empiricist understanding of science: 'In the end, transcendent idealism', Mongré wrote in another section of his epistemological monograph, 'wants nothing but the establishment or re-establishment of a pure, irreproachable *empirical monism*, basic for all our wanting and acting; that world view which today is only held in abstract dilution and without decided consciousness by representatives of natural science, if at all, while otherwise the best brains endorse a coarser or finer kind of metaphysics.'[47]

Hausdorff, who followed the development of the theory of relativity with great interest, read Schlick's booklet after its publication and could not fail to be reminded of his own earlier arguments. He wrote to Schlick and made him aware of Mongré's *Das Chaos in kosmischer Auslese*. Schlick acknowledged Hausdorff's

---

[47] 'Der transcendente Idealismus will schliesslich nichts anderes, als die Herstellung oder Wiederherstellung eines reinen, unantastbaren, für unser ganzes Wollen und Wirken grundlegenden *empirischen Monismus*, derjenigen Weltanschauung, die heute in abstracter Verdünnung und ohne entschiedenes Bewusstsein allenfalls bei den Vertretern der Naturwissenschaft herrscht, während sonst überall die besten Köpfe einer gröberen oder feineren Metaphysik anhängen.' (Hausdorff, 1898/2004, 184 f.)

contribution in the third edition of *Raum und Zeit in der gegenwärtigen Physik* (Schlick, 1920). In a telling footnote added to his reflections on 'the geometrical relativity of space', Schlick wrote:

> 'Unfortunately, only after the publication of the second edition of this writing have I learned about the most astute and fascinating book [*Das Chaos in kosmischer Auslese*]. The fifth chapter of this monograph gives a very perfect presentation of the considerations that follow in the text above. Not only Poincaré's reflections, but also the extensions added above have been anticipated there.' (Schlick, 1920, 24, note 1.)

As indicated in the footnote, Schlick's own main reference for his theses on the geometrical relativity of space was Henri Poincaré. In a series of papers on the philosophy of geometry and space published from the late 1880s and later – mostly incorporated into the volumes *La science et l'hypothèse* (Poincaré, 1902), *La valeur de la science* (Poincaré, 1906), and *Science et méthode* (Poincaré, 1908a) – Poincaré developed a view on geometry and space that partially overlapped with Hausdorff/Mongré's views. Like Mongré, Poincaré interpreted Helmholtz's popular essays on the foundations of geometry as implying that no purely empirical consideration could enforce the choice of a Euclidean or a non-Euclidean geometry of space. Only the combination of a physical theory and a system of geometry could be compared with experience. Different choices in one component could always be compensated by adjustments in the other. Poincaré went on to argue that it did not make sense to speak of a geometrical structure of space in absolute terms. Like Mongré, he based his argument on a version of a transformation principle:

> 'Two worlds [. . .] would be indistinguishable if one could pass from one to the other by means of an arbitrary point transformation. Let me explain. I suppose that to every point of the first [world] there corresponds one and only one point of the second, and vice versa. On the other hand, [I suppose] that the coordinates of the first point are continuous functions – otherwise completely arbitrary – of the coordinates of the second. I further suppose that to every object of the first world there corresponds an object of the same nature in the second world, placed exactly in the corresponding point. Finally I suppose that this correspondence is preserved forever if it is realized in the beginning. We would then have no means to distinguish the two worlds from each other.' (Poincaré, 1906, 64.)

In a later paper on 'The relativity of space' (also published in Poincaré, 1908a), Poincaré reinforced this view, emphasizing again the necessity of a joint consideration of physical events and properties of space. He pointed out that while contemporary physics actually seemed to talk about certain deformations of bodies with respect to absolute space and time (namely, when using the Lorentz–Fitzgerald contraction hypothesis), nobody would be able to tell, for instance, with what velocity the earth actually moves through this space. In fact, one could even 'imagine a completely arbitrary deformation' of space if one would include in this deformation the physical behaviour of 'all bodies without exception', including our own body, light rays, etc. (Poincaré, 1908a, 101.) This time, Poincaré went on to consider the question which geometry should be (and actually is) chosen by human beings. As is well known, he developed a partly psychological, partly evolutionary explanation for the fact that we do in fact choose a Euclidean geometry of space, taking up similar motives from his earlier essays collected in *Science et hypothèse*. However, at the end of the argument he emphasized that this choice was by no means a *forced* choice. No imprecise experience, he claimed, could ever force the adoption of infinitely precise geometrical postulates. Within certain 'bounds of elasticity', it was mainly convenience and the tendency to stick to old habits that suggested Euclidean geometry to us (Poincaré, 1908a, 120 f.).

Once again we find, in a somewhat different shading, reflections close to Hausdorff's idea of a considered empiricism. From Hausdorff's unpublished papers it is clear that he learned about Poincaré's considerations some time after publishing *Das Chaos in kosmischer Auslese*. A short handwritten note, entitled 'Transformationsprincip', documents that 'after having found the principle in others (Poincaré)' he felt the need to rethink it once again.[48]

However, one should not overlook the different perspectives that Hausdorff and Poincaré developed, faced with the indeterminacy of geometry. While, in the end, Poincaré sought for a kind of psychological or evolutionary explanation of our (or maybe his) habit of describing space in Euclidean terms, Hausdorff's intention remained to criticize the tendency to misunderstand this habit as being either metaphysically grounded or empirically enforced.

Poincaré adapted his arguments to the theories of Lorentz and Fitzgerald; Schlick stated his case on the basis of the various stages of Einstein's theory of relativity. Both held that, in the long run, there was one reasonable choice of a physics combined with geometry. Hausdorff, on the other hand, did not continue to publish on the epistemology of space and time after 1905. The question arises, therefore, whether Hausdorff's 'considered empiricism' can be considered more than just a temporarily meaningful epistemological reaction to a very open situation in the history of our knowledge about space, time, and physics – after the decline of the Euclidean paradigm, but before the advent of a new and better physical theory about space and time. Against the backdrop of a naturalized epistemology, has Hausdorff's position become obsolete?

I do not think so. Hausdorff's epistemological strategies can be reiterated wherever mathematical notions are in play in scientific theories – be it in those of his day or in present ones. Scientists *have to* make choices when developing a

---

[48] (Hausdorff papers, 49/1079, 3). What struck Hausdorff most was Poincaré's insistence on restricting the principle to continuous bijections, while he had stated it for transformations that might be discontinuous and even non-unique. Hausdorff wondered whether the need for transforming *the physics* alongside space (and time) would indeed justify this restriction, asking himself whether a 'somewhat drastic materialism' had led Poincaré to this idea.

mathematical description of the world – which number system should be adopted? which assumptions about continuity are in play when describing time or space? what axioms should be chosen for any mathematical notion involved in scientific theory-building? The naive and simple answer – science adopts those choices that are best tried and tested and that lead to the simplest theories – may be pragmatically justified, but epistemologically it invites exactly the kind of criticism that Hausdorff directed at the notions of his day. To think that such choices are metaphysically grounded or empirically forced truths makes science blind to possible alternative mathematizations. To explore the range of these alternatives requires hard mathematical work. One reading of Hausdorff's ground-breaking monograph *Grundzüge der Mengenlehre* of 1914 would be that it provided exactly that: a detailed study of order structures that might come into play in an analysis of temporal order, and a study of neighbourhood structures that might come into play in an analysis of spatial continuity. As the brief discussion of dimension above shows, elaborations of the framework of mathematical analysis may even reopen seemingly empirically determined choices of certain elements of scientific theories.

Certainly the thrust of Hausdorff's considered empiricism, and at the same time an orientation that sets him apart from many later philosophers, was this critical open-mindedness in theorizing, and a distrust of all attempts to limit, by traditionalist, metaphysical, or intuitionist claims, the activity of exploring the wealth of possible scientific conceptions of our empirical world.

Practice-related symbolic realism in H. Weyl's mature view of mathematical knowledge

Erhard Scholz

Introduction

Hermann Weyl's views of mathematical knowledge went through various transformations. He described most of them himself in his retrospective (Weyl 1954). In 1905, at the beginning of his university studies, he was thrown (by Hilbert's views on the foundations of geometry) from a youthful and naive Kantianism to a 'positivism' in the sense of H. Poincaré and E. Mach. Five years later, he came under the influence of Husserl's phenomenology and turned away from positivism. At Zürich he came into close contact with F. Medicus, an expert in the philosophy of post-Kantian German idealism and an editor of J. G. Fichte's works. After Weyl came back from service in the German army in 1916, his philosophical outlook turned radically towards realism in the sense of German idealist philosophy, formed under the impression of his way of reading Fichte and, a little later, under the personal influence of L.E.J. Brouwer. In 1926 he had the chance to rework his philosophical outlook when he wrote his contribution *Philosophie der Mathematik und Naturwissenschaften* for the handbook of philosophy edited by M. Schröter and A. Bäumler (Weyl 1927a).[1] During this work Weyl became more closely acquainted with Leibniz' philosophy, among others. He broadened and refined his philosophical views and started to reconsider his earlier exaggerated rejection of Hilbert's formalist views in the foundations of mathematics.

---

[1] Bäumler was the main editor, but M. Schröter was in charge of the section which Weyl contributed to. Later Bäumler became an ardent protagonist of Nazi ideology, while M. Schröter distantiated himself from Nazi philosophy and worked for a publishing house during the Nazi era.

The book was written at the time of the pathbreaking invention of the new quantum mechanics by W. Heisenberg, M. Born, P. Jordan, and E. Schrödinger. It would have been too early to draw philosophical conclusions from these physical insights. In the years to come, Weyl not only contributed to the conceptual framework of quantum mechanics but also drew his own consequences for the understanding of nature and the way in which mathematics could and can contribute to it. After the Second World War Weyl continued to participate in the philosophical debate about mathematics and natural science. Among other activities, he refined and extended his book for an English edition (Weyl 1949b) and discussed the questions from his peculiar perspective, that he now broadened and presented in moderate and sometimes even modest language, even where he still preferred drastic alternatives to generally accepted views (Weyl 1948, Weyl 1953). When I use the attribute *mature* views in this context, it has to be understood in the sense of post 1926/27. In this chapter I shall refer mainly to Weyl's late contributions, those written after 1948.

The meaning of *symbolic realism* will, hopefully, become clearer during this chapter. To give a short description in advance, recall that Weyl liked to consider science as a 'symbolic construction' (with allusions to both Leibniz and Poincaré) and mathematics as its symbol producing core. Criticizing Hilbert, he insisted that there were more than purely formal aspects involved in reflecting on mathematical knowledge. During his 'maturation phase' in the 1920s he gave up his earlier strongly idealistic understanding of the 'realism' of mathematical concepts and symbols (defended between 1917 and 1921) and developed a more refined view of the quasi-realistic character of mathematical knowledge through its link with broader scientific practices, in particular physics and mathematized technologies, and the existence of a semantical 'input' derived from it.

It may be worthwhile noticing that Weyl's later 'symbolical realism' contained an approach to Hilbert's 'formalism' which differed from logical positivism and the later analytical philosophy of science. Because of the striking difference between the *view* of mathematics given in Hilbert's foundational contributions and the latter's actual practice as a mathematician,[2] Weyl's mature form of symbolic realism was probably even closer to Hilbert's own understanding than the picture of Hilbert's formalism present in large parts of the analytical philosophy literature. Moreover, it was no longer built on strongly idealist ontologies of mathematical and physical reality, as had been the case for Weyl's earlier *Sturm und Drang* realism. It was rather cautiously based on cooperation with other, empirically bound sciences, tending towards coherence with cultural practices, reflected by him as existential experience of the individual in a world of irritating insecurities.

---

[2] See (Corry 2004).

Thus this chapter will present an argument that may be organized around three theses:

(1) H. Weyl lived through – and suffered from – a detachment from classical metaphysics in parts of European intellectual culture during the early twentieth century. In addition he battled with the aporetic problems of self-interpretation of the mathematical sciences arising from the foundational debate in mathematics, from special and general relativity, and from quantum mechanics (Section 1).
(2) Our protagonist searched for a solution to the problems of self-interpretation of mathematics that arose from the detachment from classical metaphysics in a kind of *symbolic realism*. In his later years, Weyl even reflected the problems of the mathematical sciences in the medium of existential philosophy that, in his view, gave an adequate expression to the crisis of metaphysics in high modernity (Sections 2 and 3).
(3) From our own perspective, it is worthwhile adopting the core of Weyl's symbolic realism and integrating it into a broader cultural philosophy of practice (Section 4).

1 Detachment from classical metaphysics

Looking back in the late 1940s at what had happened in and with the mathematical sciences in the first third of the twentieth century, Weyl could be much clearer in several respects than at the time when he wrote his book on the philosophy of mathematics and the natural sciences, which happened exactly during the transition to the new quantum mechanics. Now he could present the turn from the early and the classical modern mechanistic conception of nature to a 'purely symbolical' one of high modernity with even greater perspicuity than when he had been in the midst of it. He characterized the traditional 'mechanistic construction of the world (*mechanistische Weltkonstruktion*)' (Weyl 1948, 295) by two complementary ingredients,

— a *spatio-temporal science of geometry and motion*, which was understood by important early protagonists of this view (among them Kepler, Descartes, and Newton) as reflecting God's spirit,
— and an *atomism* in the explanation of matter, passing from Demokritos through Gassendi and Galilei to Huygens and Newton.

Thus the traditional mechanistic world construction integrated, in a kind of 'consensus' as Weyl expressed it (Weyl 1948, 295), ontological *idealism* with respect to space and time and *materialism* with respect to matter structures. The traditional world view was built upon a balance between these two components and allowed different specifications.

Weyl contrasted this classical view of the relationship between mathematics and nature with the modern one:

> ... in place of a real spatio-temporal-material being we are left only with a *construction in pure symbols*. (Weyl 1948, 295, emphasis here, as in the sequel, in the original)[3]

He immediately made sure, however, that one should not understand this symbolism in a formalist sense. Even the 'pure symbols' had origin and meaning.

> I want to turn towards mathematics to enquire about the meaning and origin of the symbols, and there we will detect *man*, inasmuch as he is a creative mind, as the masterbuilder of the world of symbols. (ibid.)[4]

Moreover, the creative power of the producing mind is not arbitrary. Although Weyl agreed that the traditional ontological bound of mathematics had been cut, he still supposed a binding law to exist. Without attempting here to analyse where it came from, he insisted on a necessary restriction on creativity:

> Only by committing the liberty of mind to lawfulness, is the mind able to comprehend, in reproducing them, the constraints of the world and of its own being in the world. (ibid.)[5]

Already here, we find a clear expression that detachment from the bounds of classical metaphysics did not at all lead Weyl to admit the arbitrariness of symbolic construction. It remains to find out where he saw the 'lawfulness' arising from. Let us first see, however, how Weyl described the modern condition for the mathematical sciences in the foundations of mathematics, relativity and quantum mechanics.

It is worthwhile noticing that, more than two decades after having written (Weyl 1927a) and after a cautious and limited rapprochement to Hilbert's foundational positions, he continued to be nearly as sharp in his philosophical rejection of transfinite set theory as he had been in the 1920s. He drew a direct line between classical metaphysics that had 'written a cipher referring to transcendent reality by posing God as absolute being ...' and repeated his old verdict against the transfinite as an actual infinity in the sense of mainstream modern mathematics:

---

[3] '... anstatt eines realen räumlich-zeitlich-materiellen Seins behalten wir nur eine *Konstruktion in reinen Symbolen* übrig.'

[4] 'Ich werde mich dann zur Mathematik wenden, um Auskunft über Sinn und Ursprung der Symbole zu erhalten, und wir werden da den *Menschen*. sofern er schöpferischer Geist ist, als den Baumeister der Symbolwelt entdecken.'

[5] 'Nur indem die Freiheit des Geistes sich selber bindet an das Gesetz, begreift der Geist nachkonstruierend die Gebundenheit der Welt und seines eigenen Daseins in der Welt.'

> Mathematics too has executed a jump to the absolute in, I would say, naive objectivity, without being aware of the dangers . . . . (Weyl 1948, 327f.).[6]

We see that even the late Weyl warned against undisclosed metaphysical remnants in modern mathematics nearly as strongly as he had done thirty years before as a young philosophical and mathematical radical. He argued that the acceptance of classical first-order predicate logic, with its unrestricted logical use of the existential quantifier and the principle of the *tertium non datur*, already signifies a kind of metaphyical appellation to an 'infinite all-ness (unendliche Allheit)' that 'is not from this world'.

> [He who does so] is already standing on the other shore: the number system, an open domain of possibilities which can only be conceived in the process of becoming, has turned for him into an embodiment of absolute existence. (Weyl 1948, 328)[7]

Although an understanding of transfinite set theory as a symbolical logical possibility *without* any ontological commitment – and thus in strong contrast to G. Cantor's philosophical interpretation of it – had already been expressed by Felix Hausdorff about 40 years earlier in a move to rid mathematics of classical metaphyics,[8] our protagonist did not perceive of such a possibility of thought. He continued his polemics by warning that on such a 'logical transcendentalism resides the power of classical mathematics' (ibid., 329). He himself had been obliged to acknowledge this power in his path-breaking work on the representation of Lie groups and in other parts of his mathematical work.[9] Notwithstanding these experiences, he repeated old phrases, although now openly attributed to his former ally L.E.J. Brouwer: If mathematics attempts to master the infinite by finite tools, it achieves so only by a 'fraud' in its logical-transcendent form,

> . . . — by a gigantic, although highly successful fraud, comparable to paper money in the economic realm. (Weyl 1948, 330)[10]

A similar paper-money metaphor had been used by Weyl in his famous radical article on the *crisis* in the foundations of mathematics (Weyl 1921, 156f.).[11]

---

[6] '[Die Metaphysik] schreibt eine auf das Transzendente verweisende Chiffre, wenn sie Gott als absolutes Sein setzt, ... Auch die Mathematik hat, ohne sich der Gefahr bewußt zu sein, in naiver Sachlichkeit, möchte ich sagen, den Sprung zum Absoluten vollzogen.'

[7] 'Wer die an die unendliche Allheit appellierende Alternative . . . als sinnvoll hinnimmt, steht bereits am jenseitigen Ufer: das Zahlsystem ist ihm aus einem offenen, nur im Werden zu erfassenden Bereich von Möglichkeiten zu einem Inbegriff absoluter Existenz geworden, das 'nicht von dieser Welt ist'.'

[8] (Stegmaier 2002, Scholz 1996)

[9] (Hawkins 2000, Coleman/Korté 2001)

[10] 'Ist es das Ziel der Mathematik, das Unendliche durch endliche Mittel zu meistern, so erreichte sie das in ihrer logisch-transzendenten Form, wenn wir Brouwer glauben wollen, nur durch einen Betrug – durch einen gigantischen, freilich höchst erfolgreichen Betrug; vergleichbar dem Papiergeld auf ökonomischen Gebiet.'

[11] (Hesseling 2003)

2 In search of a 'post-classical' metaphysics

H. Weyl's position on the foundations of mathematics has often been described and discussed,[12] therefore I continue directly with his characterization of the role of modern mathematics in physical knowledge. Here the metaphysical implications are more directly visible.

Weyl did not accept a purely formal hypothetical role for mathematical theory in the natural (or other) sciences. He even started the discussion of this question with a remark that emphasized a metaphysical aspect, although one different from the classical one characterized above (in the sense of reference to the 'mind of God' or comparable transcendent referents).

> One cannot deny that a theoretical desire is living in us that is simply incomprehensible from a purely phenomenological point of view. Its creative urge is directed upon the symbolic representation of the transcendent and is driven by the metaphysical belief in the reality of the external world (...). (Weyl 1948, 333)[13]

For Weyl this was the crucial point. In his view, mathematics did more than offer mere tools for the formation of mathematical models of processes or structures, in a purely pragmatic sense. A good mathematical theory of nature was the result of such a 'productive urge' and expressed, if well done, an aspect of transcendent reality in 'symbolical form'. In his view, the modern criticism of traditional metaphysics would never be able to achieve a complete purge of *all* metaphysical elements in the construction of knowledge by the mathematical sciences. Any such knowledge at least requires a *productive force (Schaffensdrang)* driven by a *metaphysical belief* in some *transcendent* world core, without which no meaningful communicative scientific practice would be possible.

In brackets following the remark just quoted, Weyl added that this urge towards a symbolical representation of the transcendent is driven not only by a belief in the external world, but simultaneously (gleichartig) by beliefs in the 'reality of one's own self, of the foreign thou, and of God'. Weyl did not endow his 'God' with any peculiar feature that preformed the a priori forms of scientific knowledge. His 'God' seemed not far away from Spinoza's; we would not change the argument much, if we substituted other historical names for it, which are legend, such as the 'tao' (Laotse), 'transcendent chaos' (F. Hausdorff), or even

---

[12] See (Mancosu 1998, 65–85) and the following English translation of Weyl's paper; more details in (Coleman/Korté 2001, Feferman 2000, van Dalen 1984, van Dalen 2000, Hesseling 2003, Majer 1988).

[13] 'Es ist nicht zu leugnen, daß in uns ein vom bloß phänomenalen Standpunkt schlechterdings unverständliches theoretisches Bedürfnis lebendig ist, dessen auf symbolische Gestaltung des Transzendenten gerichteter Schaffensdrang Befriedigung verlangt und das getrieben wird von dem metaphysischen Glauben an die Realität der Außenwelt (neben den sich gleichartig der Glaube an die Realität des eigenen Ich, des fremden Du und Gottes stellt).'

less transcendent sounding ones like 'nature/matter' in the original sense of K. Marx.[14] All of these refer to 'some transcendent core of the world' beyond the individual self (eigenes Ich) and a communicative other (fremdes Du) without giving rise to a claim of being able to preform scientific knowledge of the world.

This very general remark about the constitutive role of mathematics in the drive towards a representation of transcendent reality was made more explicit by a discussion of its different appearances in relativity theory and in quantum mechanics. This discussion made it also clear that Weyl used the expression 'transcendent reality' where in other discourses sometimes the term 'objective reality' is used. Weyl's more classical term indicates (among others) that we can never grasp the *something* lying beyond the symbols directly, although we are able to acquire some indirect (symbolical) knowledge about it.

Weyl had extensively discussed the role of mathematical symbolism in relativity theory already in his book (Weyl 1927a). Now in 1948, he briefly summarised how the *principle of relativity* served as the main tool for assuring the best possible symbolic representation of some 'transcendent reality' in the mathematical theory of nature. Natural laws have to be invariant not only with respect to all transformations between possible inertial observer systems, as in special relativity, but even with respect to all admissible, i.e. continuous, transformations between nets of subdivisions of localizations of events in the space–time continuum. This was Weyl's way of expressing in even more fundamental terms what Einstein had also aimed at in his search for the general covariance of natural laws, in the sense of G. Ricci Curbastro and T. Levi-Civita. Phrased in more recent language, Weyl underpinned the *covariance* principle by postulating *invariance* of the mathematical description of the physical world with respect to morphisms of the groupoid of constructively defined topological transformations of the space–time continuum.

Only by this move does the 'projection of the given', as Weyl called it, on the background of the a priori possible constructed by the mathematical mind acquire a well-determined form. He thus put the logico-symbolically possible in the place of a cognitive a priori, which latter lost its formerly assumed (or even 'demonstrated') necessary form. After the modern revolution the a priori has, according to Weyl, to be considered as a *product of the creativity of the free mind*. Of course, this generated new problems, and Weyl thought to compensate for this loss by the principle of invariance.

---

[14] We definitely have to distinguish between K. Marx's natural philosophy and twentieth-century versions of 'dialectical materialism' with its strong input of classical or, even worse, pre-classical scholastic metaphysics in which the determative power of 'God' reappeared in the guise of 'matter' attributes. This has already been shown by A. Schmidt (1962), long before the downfall of the institutional strongholds of 'dialectical materialism' as an official philosophy/ideology in parts of the world of our past.

> [I]t is quite evident from the liberty of mind, that in its constructions some *arbitrariness* is necessarily inbuilt; this can, however, afterwards be compensated for by the principle of invariance. (ibid, 336)[15]

We see how, in Weyl's analysis, the principle of invariance took over the position of the former metaphysical binding law for the construction of empirical knowledge, which had been anchored by philosophical minds of early modernity in the 'spirit of God' and was substituted in Kant's criticism by his a priori categories and forms of intuition. In his way, Weyl alluded to a kind of *relative* a priori. This seemingly contradictory term relates, on the one hand, to the a priori function of the symbolical knowledge of mathematics with respect to the empirical one. On the other hand, it reflects the shift away from necessary structures derivable once and for all and indicates the dependence on a historically achieved and changing stage of knowledge to which it is *relative*. Such a relative a priori constituted a bridge, so to speak, between the productivity of the 'free mind' and the 'urge' to give an account of some 'transcendent' core of reality which appears to the knowledge-acquiring subject through empirical phenomena. So far, Weyl essentially repeated in condensed form what he had written already in his book during the 1920s.

The epistemological consequences of quantum physics for the role of mathematics in the enterprise of understanding nature were stronger than those of relativity. Weyl's philosophical views were strongly influenced by the rise of the new quantum mechanics; he revisited his expectation of mathematical theory in natural sciences and his characterization of the concept of matter.[16] He subscribed to an enlightened Copenhagen–Göttingen interpretation of quantum mechanics, as far as its interior scientific semantics was concerned. In this sense he supported Born's probabilistic interpretation of the Schrödinger 'wave' function from the outset against Schrödinger's, Einstein's and others' attempts to keep to a classical field theoretical interpretation of it. He was not satisfied, however, with the philosophical self-interpretation of the new quantum mechanics or its interpretations by contemporary philosophers.

Weyl accepted the Heisenberg–Bohr insight of the essential role of complementarity in the sense of pairing of conjugate observables in the mathematical description of quantum systems. Very early (indeed, in autumn 1925) he convinced himself that Heisenberg's commutation relation

$$QP - PQ = [Q, P] = i\hbar$$

for any pair q and p of conjugate variables represented by symmetric operators Q and P, such as a linear space co-ordinate and its corresponding momentum, was *the* constitutive insight of the new theory and probably a clue for an elaboration of

---

[15] '[E]s ist einigermaßen aus der Freiheit des Geistes verständlich, daß in seine Konstruktion unvermeidlich *Willkür* eingeht, daß aber diese nachträglich durch ein Prinzip der Invarianz unschädlich gemacht werden kann.'

[16] (Scholz 2004*a*)

a mathematical theory of quantum reality. Already by late 1925 Weyl had started to think about a derivation of Heisenberg's relation more fundamentally as the property of a normal form of projective representations of abelian groups. His (still preliminary) thoughts were finally published in (Weyl 1927b). They formed the starting point for the Stone/von Neumann representation theorem in quantum mechanics. Moreover, they contained an idea for a quantization procedure relying on properties of Fourier transformations on abelian groups, which in the 1970s was taken up in a generalizde form as 'Weyl quantization'.[17]

Weyl was not happy, however, to say the least, with the more general 'philosophy of complementarity' that had been proposed by Niels Bohr in the 1920/30s and continued to be propagated by him after the war. No doubt, there was a certain plausibility in describing oppositional features of life and culture in terms of 'complementarity'. Among them Weyl counted as well-chosen examples the pairs freedom of will – natural causality, living an experience – cognition or reflection of it, moreover in ethics, justness versus love. Nevertheless he warned:

> May it not be that the idea of complementarity, which in quantum physics corresponds to a state of affairs with an exact mathematical expression, could be misused in a similar way as is the case for the idea of relativity? (...) I want no more than to pose a question. (Weyl 1948, 338f.)[18]

Slightly streamlining, we may describe the Copenhagen–Göttingen interpretation of quantum mechanics as a package made up of a positivistic-formalist interpretation of the mathematics of quantum mechanics, with an open-minded philosophy of complementarity in the interpretation of the empirical phenomena of quantum physics.[19] Both together formed something like a 'hard core' of the interpretation. To this, some of its protagonists, in particular N. Bohr, added a complementarity outlook on broader, cultural philosophy in the sense of *Lebensphilosophie*. Others, like W. Heisenberg, stipulated, in their more personal reflections, a kind of platonist ontology underlying quantum reality, while he demanded adherence to a positivist outlook for the crowd of working physicists. Weyl modestly but clearly rejected the *Lebensphilosophie* part of the parcel, although he had great admiration for Bohr as a philosopher of science, inasmuch he was rethinking basic concepts of physics in the light of the new empirical evidence of quantum physics. He also did not subscribe to the formalist ('positivist') interpretation of the role of mathematics in the construction of the theory, which was defended by M. Born, W. Heisenberg and J. von Neumann, in line with the mainstream Göttingen spirit of the time, and was accepted by N. Bohr

---

[17] (Mackey 1988, Mackey 1993)

[18] 'Aber besteht nicht doch die Gefahr, daß hier mit der Idee der Komplementarität, die innerhalb der Quantenphysik einem mathematisch genau zu präzisierenden Tatbestand entspricht, ein ähnlicher Mißbrauch getrieben wird wie mit der Idee der Relativität? ... Ich werfe nicht mehr als eine Frage auf.'

[19] (Chevalley 1995, Chevalley 1993)

as part of the compromise with Heisenberg and Pauli in the discussions on the interpretation of quantum mechanics.[20]

The positivistic-formalist agreement on the role of mathematical theory served well to cover up underlying differences in the metaphysical outlooks of the Copenhagen–Göttingen contributors to quantum mechanics. In fact, it was quite useful as a substitute for a philosophy that allowed one to avoid spelling out subtle differences of a metaphysical kind. From a historian's point of view, such differences appear to be indissolubly linked to social, personal and intellectual experiences of the individual contributing scientists. Differences of this kind are a natural ingredient and flavour of cultural life. Weyl, for his side, was not content with any of these views. He rather looked for other allies, closer to his own philosophical experiences. In *this respect* he now sided with Oskar Becker, who in the late 1920s had been a kind of spokesman of Husserl's phenomenology among mathematical scientists, although at that time he had had differences with him over a proper understanding of geometry after the advent of relativity theory.[21]

In an article also written at the end of the 1920s, called *Das Symbolische in der Mathematik*, Becker had characterized the first step taken by Hilbert, von Neumann and Northeim toward an axiomatic characterization of quantum mechanics (von Neumann e.a. 1928) as a kind of return to the ancient magical origins of the mathematical sciences. Weyl quoted Becker literally:

> In a way one jumps into the 'interpretation' of nature with the complete, ontologically incomprehensible 'mathematical apparatus'; the apparatus works a like a magic key which opens up the field of physical problems – but it does so in the sense of a symbolical representation only, not in the sense of an interpretation really 'discovering' the phenomena (Becker 1927/28)[22]

Becker continued, and Weyl quoted him, apparently with a recondite smile:

> The basic direction of such a symbolical approach comes from time immaterial, archaic, even 'pre-historic': the most modern 'exact' science returns again to magic from which it originally descended. (ibid)[23]

Apparently Weyl sided with Becker's move to break the hermetic nature of the formalist approach to quantum mechanics and the clear characterization the formal system as a symbolical representation of external ('transcendent') reality. On the other hand, he indicated that one might not necessarily be glad about such a return to the 'archaic origins' of science. To express this uneasiness, he

---

[20] (Beller 1999, Hendry 1984)

[21] (Mancosu/Ryckman 2002)

[22] 'Man springt also gewissermaßen mit dem vollständigen, ontologisch unverständlichen 'mathematischen Apparat' in die 'Deutung' der Natur hinein; der Apparat ist wie ein magischer Schlüssel, der das physikalische Problem erschließt, — aber nur erschließt im Sinne einer symbolischen Repräsentation, nicht im Sinne einer die Phänomene wirklich 'entdeckenden' Interpretation.' Quoted in (Weyl 1953, 535).

[23] 'Die Grundrichtung dieses symbolischen Weges ist uralt, arachaisch, ja geradezu 'prähistorisch': die modernste 'exakte' Wissenschaft wird wieder zur Magie, aus der sie ursprünglich abstammt.' (ibid.)

quoted E.T. Bell, who had criticized the reappearance of numerology in the works of A. Eddington and others. Weyl did not disclose, in this text, where he positioned himself in this respect. It is clear, however, that Becker's joy about the resurrection of archaic metaphysics could not be shared without reservation by those who wanted to uphold an enlightening role for science.

If we take other texts into account, it becomes clearer that Weyl liked Becker's allusion to an *ahnendes Erkennen* (apprehensive cognition) of the transcendent, here in the sense of the quantum reality of a 'material agency' (Weyl 1953, Weyl 1924). On the other hand he had strongly experienced, during his long phase of involvement with mathematical physics, that a primarily intuitive apprehension as indicated in Becker's text was highly precarious. In a commentary at the turn to the 1930s on the status of unified field theories and quantum phyiscs he had made it clear that the mathematical methods of quantum mechanics only acquired strength and true meaning through their specific connections with the empirical practices of quantum mechanics, in particular those of spectroscopy (Weyl 1931). Thus the connections to empirical practices had to be taken into consideration, not only by those who favoured an empiricist philosophy of science, but also in any metaphysical reflection on quantum physics that wanted to draw consequences of the crisis of classical metaphysics.

3   Weyl's symbolic realism

In a text resulting from two lectures given at the *Eranos* meeting 1948, an interdisciplinary meeting at Ascona, Weyl gave a beautiful review of the modern mathematical sciences as an expression of the symbol-producing activity in modern culture (Weyl 1948). In these lectures Weyl compared, among others, the way relativity theory had transformed classical mechanics with the consequences of the rise of quantum mechanics. Relativity had managed to transform classical mechanical knowledge by a kind of *Aufhebung* in the Hegelian sense, i.e. it revoked and lifted the former notions all at once (Weyl 1948, 339). The classical concepts of space, time, position, momentum, energy, etc., had been deeply transformed, but could be related back to the classical ones, once the new stage of knowledge had been established. The classical perspective of a physical reality lying behind the empirical phenomena could now be expressed by fields on Minkowski space or, in general relativity, on a Lorentzian manifold. Classical mechanics was then preserved in a well-defined transition of the theoretical structure in the sense of limiting processes.[24]

---

[24] Velocity $v \to 0$ in special relativity; in general relativity more involved, but conceptually completely clarified (in years of work by Einstein) by a double limiting process, first a weak-field approximation of Newtonian gravitation, then $v \to 0$. Cf. (Renn/Sauer 1999).

Such an analysis had been used by F. Gonseth in his dialectical epistemology, to which Weyl referred in his lecture (ibid.). Weyl accepted that such a 'dialectization' of knowledge in the sense of Gonseth made sense for the transition from classical to relativistic mechanics.

> In the new relativistic picture the original concepts are 'lifted' in Hegel's double meaning of the word. That may be correct in a historical sense; but it would be nothing but a 'historical' dialectic. This is because it is possible to give a completely clear and intuitive description of relativistic space and time, which is able to specify the meaning of the concepts of velocity etc. in the new frame without any reference to the absolute standpoint. (Weyl 1948, 3339f.)[25]

In quantum physics things turned out to be more complicated and open ended, much closer to what Weyl demanded from a dialectical relationship that was more than a 'historical' one. In the characterization of the epistemic constellation of quantum mechanics Weyl followed Bohr's analysis of the necessary distinction of a quantum physical process itself and the description of the measuring process by classical experimental language.

> The case may be different in quantum theory. Here one has to distinguish sharply between the hidden physical process which can only be represented by the symbolism of quantum physics, although it may be referred to by such words as electron, proton, quantum of action, etc., and the actual observation and measurement. According to Bohr, we have to talk about the latter in the intuitively comprehensible language of classical physics; or ought we better say: in the language of everyday life? (Weyl 1948, 340)[26]

Weyl had thought about the question whether it was only due to the early stage of development of quantum physics that one had to rely on these two levels of language. Although he did not share the Copenhagen–Göttingen conviction that quantum mechanics was a 'complete theory', an epistemological construct introduced by Heisenberg in the battle against critics like Einstein and Schrödinger, Weyl came to the conclusion that the two conceptual levels might well constitute an unresolvable opposition in quantum physical knowledge, different from the transformation that relativity had achieved for the pre-relativistic concepts.

---

[25] 'In dem neuen relativistischen Bilde sind die ursprünglichen Begriffe, in Hegels Doppelsinn des Wortes, 'aufgehoben'. Das mag historisch zutreffend sein; aber es wäre eben doch nur eine 'historische' Dialektik. Denn von der Relativitätstheorie von Raum und Zeit läßt sich eine vollkommen klare und anschauliche Darstellung geben, die, ohne Anleihen bei dem absoluten Standpunkt zu machen, genau bezeichnen kann, welche Bedeutung die Begriffe Geschwindigkeit usw. in dem neuen Rahmen haben.'

[26] 'Die Sache liegt vielleicht anders in der Quantentheorie. Hier muß man scharf scheiden zwischen dem verborgenen physikalischen Vorgang, der nur durch den Symbolismus der Quantenphysik erfaßbar ist, auf den aber auch mit solchen Worten wie Elektron, Proton, Wirkungsquantum usw. hingewiesen wird, und der tatsächlichen Beobachtung und Messung. Über die letztere müssen wir nach Bohr sprechen in der anschaulich verständlichen Sprache der klassischen Physik; oder sollte man besser sagen: in der Sprache des täglichen Lebens?'

> It may very well be that we can never dispense with our natural understanding of the world and the language in which it is expressed, perhaps a little purified and enlightened by classical physics, and that the symbolism of quantum physics will never be able to offer a substitute for it. In this case we would have here a true dialectic which cannot be resolved/lifted by any historical development.... (Weyl 1948, 341)[27]

He observed an analogous constellation in the foundations of mathematics, brought about by Hilbert's foundational program. There we need, as Weyl stated in agreement with Hilbert, 'signs, real signs, written on paper by the pen or on the blackboard by chalk' (Weyl 1948, 341), not ideas or forms of pure consciousness, but concrete signs, in some material realization. Any attempt to dissolve these material signs by means of a physical analysis of the chalk as constituted by 'charged and uncharged elementary particles' would lead to a resolution into quantum physical symbolism. This would obviously result in a 'ridiculous circle', as Weyl pinpointed:

> ... [T]hese symbols are, in the end, again concrete signs written in chalk on a blackboard. You realize the ridiculous circle....

Weyl drew the natural conclusion, that this circle can only be avoided

> ..., if we accept the way in which we understand things and people dealing with them in everyday life as an irreducible foundation. (Weyl 1948, 342)[28]

We can see in this argument a slighty ironical allusion to the 'vicious circle' that Weyl had struggled with, three decades earlier, in the foundations of mathematics. By this formulation he had denoted, following Poincaré and Russell, the impredicativity problem in the symbol construction of classical analysis. It had driven him, at that time, towards looking for ground in post-Kantian idealist philosophy. Now it was the reality problem of quantum physics that led to a *ridiculous circle* in ontology, if one wanted to restrict the consideration to the realm of signs only. While the 'vicious circle' in the foundations of analysis could be abandoned by a restriction to constructive practices expressed in a semi-formalized arithmetical language, i.e., inside mathematics proper, the 'ridiculous circle' could only be broken if one accepted everyday practices and natural language, not only as a practical basis but even as an 'irreducible foundation' of science, as Weyl stated with an inkling of Göttingen foundationalism in his choice of words.

---

[27] 'Aber es mag bei alledem doch dabei zu bleiben haben, daß wir das natürliche Weltverständnis und die Sprache, in der dieses Verständnis sich ausspricht, vielleicht eine wenig gereinigt und geklärt durch die klassische Physik, nimmer entbehren können und der Symbolismus der Quantenphysik keinen Ersatz dafür zu bieten vermag. Dann handelte es sich um eine echte, durch keine historische Entwicklung aufzuhebende Dialektik ....'

[28] '... die Symbole aber sind letzten Endes wieder konkrete, mit Kreide auf die Tafel geschriebene Zeichen. Sie bemerken den lächerlichen Zirkel. Wir entrinnen ihm nur, wenn wir die Weise, in der wir im täglichen Leben die Dinge und Menschen, mit ihnen umgehend, verstehen, als ein unreduzierbares Fundament gelten lassen.'

When Weyl referred to everyday practices, he did not think of Peircean pragmatism or anything like. Even after fifteen years of life in the United States he continued to think in terms of European philosophy. In the years of the Second Great War, and with another deep world crisis as a result of it, Weyl had turned towards the existential philosophy of M. Heidegger and K. Jaspers. In fact, in his Ascona talk Weyl explained how he saw the epistemic situation of quantum physics related to the existential constellation of the modern individual, described by Heidegger in *Sein und Zeit* (Heidegger 1928), 'being thrown' into the world, no longer bound by some transcendent power.

Heidegger's philosophizing started from the 'Dasein (being there)' of man, awoken to the consciousness of the self, who deals with things in unrefined terms of everyday knowledge. Weyl now described the turn towards scientific knowledge in Heideggerian terms: The latter kind of 'objective' or 'scientific' knowledge pre-supposes a radical detachment from everyday knowledge. He remarked that Heidegger had turned from the question of how to 'prove' the existence of an objective, external world to the philosophical question of

> ... why the 'Dasein' as a being-in-the world has a tendency to bury the external world in nothingness from the viewpoint of the theory of knowlege, and then to prove it afterwards by indirect argumentation (Weyl 1948, 343f.).[29]

Weyl rephrased the Heideggerian experience of the individual of being thrown into an insecure and apparently meaningless existence without resort to a transcendent reality in the following terms:

> After the original phenomenon of being-in-the-world has been suppressed, ones tries to glue the remaining isolated subject to the torn patches of the world; but it remains patchwork ... (Weyl 1948, 344).[30]

In this way Weyl argued that the disruption of the modern existential condition, Heidegger's 'Geworfenheit (being thrown)' is mirrored in the epistemic structure of the modern mathematical sciences.

He was not satisfied with this state of affairs and indicated that the stipulation of a 'natural understanding of the world' as it worked in everyday life was in itself highly problematic and worthy of 'further inquiry' (ibid.). In fact, he continued to think about these questions. In his article on symbolism in the mathematical sciences, quoted already above (Weyl 1953), he extended the perspective to a broader view on cultural philosophies, without negating his allegiance to the existential philosophy of Jaspers and Heidegger.

---

[29] 'Zum Problem der Außenwelt bemerkt darum Heidegger, daß man nicht zu beweisen hat, daß und wie eine Außenwelt existiert, sondern aufzuzeigen, warum Dasein als Sein-in-der-Welt eine Tendenz hat, die Außenwelt erkenntniskritisch ins Nichts zu begraben und dann sie nachträglich indirekt zu beweisen.'

[30] 'Nachdem man das ursprüngliche Phänomen des Seins-in-der-Welt unterdrückt hat, sucht man das zurückgebliebene isolierte Subjekt mit den abgerissenen Weltfetzen wieder zusammenzuleimen; aber es bleibt ein Flickwerk.'

Now he turned to authors who had dealt with the role of language, signs, and their symbolic function, the representation by signs: W. von Humboldt, H. von Helmholtz, H. Noack, K. Voßler and others. He quoted the first volume of Cassirer's theory of symbolic forms and Wittgenstein's *Tractatus*.[31] Wittgenstein's youth work appeared a bit strange to him, because of its solipsistic perception of language. In contrast to the argumentation deployed in the *Tractatus* Weyl insisted that 'the existential origin' and the task of language has to be looked for, at first instance, in *communication* (Weyl 1953, 527).

Weyl characterized the *symbol* as a sign, different in function from names and images by carrying meaning in the scientific or wider communication and as an object that has to be prepared and worked with in intellectual practices. The latter aspect made it comparable to tools in the material practices of artisans. He was glad that mathematics uses written signs reproducible, in principle, without limits.

> Visible configurations of a certain stability are used as signs (rather than sounds or clouds of smoke; at least persistent as is necessary for the execution of certain operations on them). (Weyl 1953, 528f.)[32]

In contrast to an idealist view, which he exemplified on this occasion by a reference to his former ally Brouwer, Weyl sided here with Hilbert and emphasized the tool character of the signs. Here the 'concrete activities of people' comes into the play and allows us to adopt even an 'anthropic' perspective with respect to mathematical knowledge.

> Here the mathematician, with his formulae made up of signs, does not work so differently from the carpenter in his workshop with wood and plane, saw and glue. (Weyl 1953, 529)[33]

Thus in Weyl's final reflections we find that mathematical symbols are understood within the context of a communicative practice that has strong parallels to material practices, in the way that the symbols are handled, and with multiple links to the other scientific and technical activities. Nevertheless, the main goal of the practices is to establish meaning and semantical connections between them and the world beyond the signs. The signs offer the material for a symbolic [re-]construction of some 'objective world', as in relativity and/or quantum mechanics, where the problematic can most strikingly be studied and exemplified. Here two points are important to realize, according to Weyl:

---

[31] (Humboldt 1836, Helmholtz 1887, Noack 1936, Vossler 1925, Cassirer 1922, Wittgenstein 1922)

[32] 'Als Zeichen dienen sichtbare Gebilde von einer gewissen Beständigkeit (nicht etwa Schälle und Rauchwolken; zum mindesten so lange standhaltend, als zur Ausführung der an ihnen vorzunehmenden Operationen benötigt wird).' Note the slightly alienated allusion to Faust's declaration '*Namen* sind Schall und Rauch (*names* are sound and smoke)' (J.W. von Goethe, my emphasis).

[33] 'Da geht der Mathematiker nicht viel anders mit seinen aus Zeichen gebauten Formeln um wie der Tischler in seiner Werkstatt mit Holz und Hobel, Säge und Leim.'

(i) The symbol is neither taken from 'the given (dem Gegebenen)' nor is it a part of the reality represented by it.

(ii) The symbolic construct is *neither* a reality lying at the base of the appearances, *nor* has the bound to the observable been cut.

To make clear that (i) and the first part of (ii) ('neither ...') stand in stark constrast to classical notions, Weyl presented the example of a light ray. C. Huygens 'could with good consciousness still say that a monochromatic light ray *in reality* consists of an oscillation of the light ether ...'. The modern physicist, on the other hand, represents the ray by a 'formula, in which a certain symbol F, called electromagnetic field strength, is expressed by an arithmetically constructed function of four other symbols x, y, z, t' (Weyl 1953, 529). A plane-wave solution of the Maxwell equation is obviously a symbolic construct indicating something in the world, but is neither part nor 'lying behind' or 'at the base' of the optical or, more broadly, electromagnetic observations.

The symbolic construct is neither arbitrary nor self-relying in its meaning. The second part of (ii) is established in an interdisplinary exchange between the sciences. Weyl argued:

> Of course, the bond between the symbol and the given in the observation need not be cut; the physicist understands how the symbolism is 'meant', when he confronts the laws expressed in it with his experience. (ibid.)[34]

We see that Weyl tried, as much as he could, to distance himself from classical metaphysics, in particular its reference to the kind of transcendent reality that was stipulated there. He definitely refused, however, to cut the bonds to all kinds of metaphysics. He rather substituted strong references to symbolical and material practices in place of the old realism.

Moroever, at the core of his mind and heart he remained a believer in some kind of Eckehardt–Fichtean God and, as we added, a Spinozean one, accessible through the experiences of the self and the symbolic cognition of some objective or 'transcendent' reality established in scientific practices. This point is mainly important for a proper and respectful historical understanding of our protagonist. Weyl's turn towards scientific material practices as the most important base for the realism inherent in symbolic knowledge, has to be considered the essential feature of his mature and late work. It seems justified to use the denotation *symbolic realism* for such an approach. We may have reasons to relate to it in our own reflection and work.

---

[34] 'Natürlich braucht darum das Band zwischen Symbol und wahrnehmungsmäßig Gegebenem nicht durchschnitten zu werden; der Physiker versteht, wie der Symbolismus 'gemeint' ist, wenn er die in ihm niedergelegten physikalischen Gesetze an der Erfahrung prüft.'

4 'Symbolic realism' as an ingredient of a trans-modern philosophy of practice

In his symbolic realism, Weyl insisted upon *meaning* acquired in the complex context of scientific, technical and social practices, as an important ingredient of mathematical knowledge. It is not inscribed uniquely, *ex ante*, and forever in the symbolical knowledge of the mathematical sciences. On the contrary, this meaning is constituted in a permanent process of elaboration, communication, and usage of mathematical knowledge. It is multifaceted, in enduring change, and often arrived at *ex post*, i.e. long after corresponding mathematical structures have been established and studied as such.

Such a view does not lead to arbitrariness and allows us to avoid a formalist or neo-positivist reduction of mathematics. The endowment of meaning to symbolic knowledge and its diverse possible usages is deeply bound to goal-oriented practices of science, techniques, education, sometimes arts, and culture more broadly. In this way, it is embedded in a social and cultural discourse on values that are should be part of a *philosophy of practice*. That goes well with radical agnosticism in questions of *fundamental* ontology, inherited from the criticism of the enlightenment and its modern continuators.[35]

Such a philosophy of practice can stand on its metaphysical components, without being in danger of falling back into an unbroken, or even naive continuation of traditional metaphysics that was driven into crisis and even into dissolution, with good reasons, by modernity. In this sense I propose to consider it as a *transmodern* philosophy that is in the making. This term must not be confused with the label 'postmodern/postmodernity' that has completely different meaning and connotations. It would go far beyond the goals of this contribution to fully discuss the connotations intended in such a qualification as 'transmodern'. It may suffice here to indicate that I intend this expression to include a conscious reference to the criticism of technoscientific practices of high modernity, as spelt out by I. Illich in his quest for *conviviality*.[36]

Ivan Illich (1926–2002) attempted to shatter the naive belief in the conception of a uni-directional 'progress' in modern society based on elaborate technoscientific practices and industrial systems as such. He recalled that non-modern practices may be comparable, sometimes and in many places even preferable for a convenient satisfaction of social and cultural needs. He did not call for abandonment of technoscientific-industrial practices en gros, but demanded their critical evaluation and reorientation according to their potentials to contribute to achieve human goals in nature ('conviviality').

---

[35] The discussion of domain specific 'ontologies' is another question.

[36] (Illich 1973)

Surely, Illich's criticism of the conditions of life in the modern sector of our emerging world society went deeper than Weyl's and was written from the background of having lived through large parts of his life in another segment of the world than Weyl, and a generation later, but there is is no reason to dismiss such considerations as foreign to the Weylian perspective. Weyl experienced the cultural development and the social, military, and technical history of the first half of the twentieth century as a deep and multiple crisis. He perceived the outcome of the second world war as a dangerous constellation for mankind as a whole. Like other scientists of the post-second-war period, he considered the development of nuclear weapons to be the watershed of a development of modern technoscience into a stage in which it started to have at least as much destructive powers, as it could serve as a potential for conviviality (an 'improvement of the conditions of life' in more classical terms).

In the manuscript for a talk on *The development of mathematics since* 1900, given about 1949, Weyl warned that the seemingly abstract and detached knowledge of the mathematical sciences may have contributed to give such strong powers to human society that in a kind of revenge of the 'Gods' it might lead to self-destruction, rather than to an improvement of the conditions of life. He referred here to a passage of Aristotle's metaphysics in which the quest for a kind of (metaphysical) 'pure' knowledge, detached from human goals, is discussed, which is of use only for the gods themselves. Such a 'stepping beyond' might be considered as a self-adulating conceit, comparable to what the ancient Greek called 'hybris' and expected to be sanctioned by a 'revenge of the gods'.[37] Weyl transferred Aristotle's warning from metaphysics to mathematics and the mathematical sciences and argued strongly in favour of a symbolic knowledge that is aware of its cultural connectors and practical meanings.

> ... I am not so sure whether we mathematicians during the last decades have not 'stepped beyond' the human realm by our abstractions. (...) For us today the idea that the Gods from which we wrestled the secret of knowledge by symbolic construction will revenge our $υβρις$ [hybris] has taken on a quite concrete form. For who can close his eyes against the menace of our self-destruction by science; the alarming fact is that the rapid progress of scientific knowledge is unparalleled by a congruous growth of man's moral strength and responsibility, which has hardly chance in historical time. (Weyl Ms 1949a, 7, English in original)

Another half a century later, deeper inside the transition to a world-wide society stricken by unjust social divisions of power, labour and resources, the nuclear menace persists, mitigated only slightly and probably only temporarily. In addition to this, much broader and multifarious corridors of destructive practices in nature and society endanger a decent and enduring development of humanity in our terrestrial mesocosmos. In this context, *our context*, it may be more than useful to

---

[37] Aristotle: Metaphysics 982b. Cf. W.D. Ross' translation in (Aristotle 1984). Weyl's own translation of Aristotle's hybris warning was sharper than the one by Ross, compare the quote in (Scholz 2005).

take up Hermann Weyl's thoughts on a symbolical realism for the mathematical sciences and to fuse them with Ivan Illich's challenge to reorient *all* our practices in accordance with conviviality.

5 Acknowledgments

Without the initiative of the organizers of the Sevilla Colloquium, José Ferreiros and Jeremy Gray, and their friendly support I would not have dared to put Weyl's later reflections on mathematics an physics at centerstage of an investigation of mine. Moreover, José added to the final form of the paper by his detailed reading and comments, as did my colleague Gregor Schiemann from the Wuppertal department of philosophy.

From Kant to Hilbert: French philosophy of concepts in the beginning of the twentieth century

Hourya Benis Sinaceur

The legacy of the Kantian theory of knowledge was very much alive among French philosophers of science at the beginning of the twentieth century. To adopt, not the letter of the Kantian system, but its spirit of critique, seemed a good perspective from which to undertake a rational study of science in general, and of mathematics in particular. The critical attitude does indeed invite us to turn away from things in themselves, inaccessible to the human mind, and to keep at arm's length metaphysical questions, examining instead the defining conditions of knowledge.

There is no better way to conduct this examination than in the exercise of a scientific activity. Several brilliant philosophers were to engage in this exercise as they completed their philosophical education through the study of mathematics. They were greatly encouraged by the teaching of Léon Brunschvicg (1869–1944) at the École Normale Supérieure and at the Sorbonne; they were tempted by the example of Gaston Bachelard (1884–1962). A whole generation set about bringing to bear the power of critique on the new disciplines that were transforming traditional concepts of the nature and object of mathematics: non-Euclidean geometries, axiomatic methods, algebraic and topological structures, etc.

On the one hand, Brunschvicg praised Kant for having brought scientific objectivity to bear on human reason, so destroying the idea of an absolute rationality prior to rational activity. On the other hand, he deplored the fact that Kant had established immutable a priori forms, contradicted as these were by the *evidence* of the *indefinite progress* of mathematics. The decisive discoveries did not obey the schema of categories. It was therefore necessary to rethink the Kantian epistemology taking into account the historical character of mathematical results. If the search for truth is an essential task of philosophy, so an examination of the history of mathematics is indispensable, mathematics being, according to Brunschvicg, the discipline that had brought 'the most scruple and subtlety' to this search.

Brunschvicg thus proposed to separate out the critical attitude from the Kantian canvas of a priori forms of intuition and categories of reason.[1] Furthermore, he presented this attitude in a positivist light, insisting, like Auguste Comte (1798–1857), on the necessity of starting from the actual fact of science, and drawing philosophical lessons from its history. In place of the determination of the conditions of the possibility of knowledge in general, Brunschvicg substituted a kind of half-historical, half-philosophical, enquiry into the development of particular sciences. Reason being intimately linked to scientific activity, the inquiry had as its aim to discover in this activity that which Gaston Bachelard emphasizing the intimate link between rationality and historicity, called 'the events of reason'.[2]

Those who followed Brunschvicg held onto several of his points:

- To consider the advent of the critical idea as a 'decisive date in the history of humanity'[3], because it sets out to focus on the power of intellectual and scientific creativity;
- therefore to take the Kantian epistemology as a starting point, but to amend it, modify it, or go beyond it, since it is necessary to
- take account of the indefinite progress of mathematics,
- and underline the unpredictable and complex nature of its results,
- and so give priority to the *development* of mathematical knowledge over the consideration of fixed and timeless frameworks of knowledge,
- to arrange the history of mathematical theories from the point of view of critical examination, that is to submit the chronological succession of results to a 'reflective analysis', which would bring out the internal rationality of their connections,
- to show that these dynamic connections cannot be reduced to static logical relationships.

Brunschvicg made the dynamism of mathematics correspond to the inexhaustible dynamism of the mind and attributed to consciousness the generative power of creation and progress. It is this fundamental choice that brings Brunschvicg's philosophy into line with the western tradition of the philosophy of subject, and it was a choice firmly disavowed by his successors.

A pupil of Brunschvicg, Jean Cavaillès (1903–1944) started by accepting the parallels between mathematical progress and the progress of consciousness, and, more generally, between the enriching of experience and the expansion of consciousness. This is particularly apparent in the closing pages of *Méthode axiomatique et formalisme*[4], the book that formed Cavaillès' thesis under the

---

[1] Brunschvicg, [1924].

[2] Bachelard, [1945], no 1–2, reprinted in *Léon Brunschvicg. L'œuvre. L'homme*, Paris, A. Colin, 1945, p. 77.

[3] Brunschvicg, [1924], p. 229.

[4] Cavaillès, [1938a], in *Œuvres complètes de philosophie des sciences*, Paris, Hermann, 1994, pp. 177, 179.

supervision of Brunschvicg. But in his last work, *Sur la logique et la théorie de la science*, written in prison and published posthumously[5] three years after he was put to death by the Nazis, Cavaillès proposed to substitute for the philosophy of consciousness a 'philosophy of concept'. This programme, which the author did not have the chance to expound fully, nonetheless left a deep and enduring mark on the landscape of the philosophy of mathematics in France. It to some extent overshadowed Brunschvicg's legacy and reverberated throughout French philosophy of science. Figures with starting points as diverse as Georges Canguilhelm's (1904–1995), Gilles Gaston Granger's (1920–) and Jean Toussaint Desanti's (1914–2002) forged the essential part of their arguments from the perspective that Cavaillès had begun to open. Thinkers who ranged widely across philosophical ideas without focusing exclusively on this or that science, such as Michel Foucault (1926–84) or Gilles Deleuze (1925–95), also hailed the virtues of the concept in the construction of a *structural* theory of knowledge. But if so many people pondered the programme floated by Cavaillès like a bottle in the sea, none of them was to accept all the consequences of its author's conceptual idealism.

Here I propose a kind of 'tableau' of the French philosophy of concept, centred on the dominant figure of Jean Cavaillès. The work of Cavaillès[6] remained, for at least thirty years (1940–1970), a source of inspiration all the more diffuse for being dense and difficult. I want to show how the results of axiomatics and mathematical logic, as developed by David Hilbert's (1862–1943) proof theory (Beweistheorie), refined by the objections raised by L.E.J. Brouwer (1881–1966), and extended by A. Tarski's (1901–83) formal semantics, served as a reformation of Kantian epistemology, and led to the placing of concept above consciousness. Naturally, I will attempt to make clear what should be understood by the term 'philosophy of concept'. But we can note right away that while Anglo-Saxon analytical philosophy turned away from consciousness to invest in the objective

---

[5] Cavaillès, [1947b], reprinted in *Œuvres* ..., pp. 473–560.

[6] In addition to the works cited in the preceding notes, it is necessary to add:

— 'Réflexions sur le fondement des mathématiques', Travaux du IXe Congrès international de philosophie, t. VI, Paris, Hermann, 1937. Reprint in *Œuvres* ..., pp. 577–580.

— *Remarques sur la formation de la théorie abstraite des ensembles, étude historique et critique*, Paris, Hermann, 1938. Reprint in *Œuvres* ..., pp. 221–374.

— 'La pensée mathématique' (en collaboration avec Albert Lautman), *Bulletin de la Société française de philosophie*, **40**, n° 1, 1946, pp. 1–39. Reprint in *Œuvres* ..., pp. 593–630.

— 'Transfini et continu', Paris, Hermann, 1947. Reprint in *Œuvres* ..., pp. 451–472.

See also Hourya Benis Sinaceur:

— 'Structure et concept dans l'épistémologie de Jean Cavaillès', *Revue d'Histoire des Sciences*, XL-1, 1987, pp. 5–30, 117–129.

— *Jean Cavaillès. Philosophie mathématique*, Paris, Presses Universitaires de France, 1994. This work contains a more complete bibliography of the writings of Cavaillès and a list of works about him.

factuality of language, it was concept that French philosophy made the source of scientific objectivity.

My tableau will be far from complete. It will nonetheless bear the principle features that characterize the radical change of perspective achieved by Cavaillès. These features are: 1) the collapsing of the transcendental onto the factual, 2) the substitution of that which unfolds for that which is, and of a developmental 'logic' for the usual static one, 3) the taking into account of the symbolic material of mathematics through the new notion of 'formal content', and 4) a specific theory of mathematical signs 5) the elimination of the subject in favour of the object, and the objectification of the concept, the internal dialectic of concepts. It is in so far as they contribute towards underlining one or more of these features that I will also bring in the works of G.G. Granger and J.T. Desanti.[7] These two authors expressly lay claim to Cavaillès and to the philosophy of concept in their works on the epistemology of mathematics.

1 Kant appropriated: the rejection of the transcendental and of the a priori

Until the 1960s, and occasionally to this day, the Kantian doctrine remained a backdrop to epistemological discussions; it was at once a support and a foil. Taking Brunschvicg's example, Cavaillès and Granger engaged in the determined project of reforming of Kantianism, Foucault in a collapsing of the transcendental onto the empirical. Reflection was focused on the notion of the 'synthetic a priori' and, to a lesser extent, on notions of experience, of object and subject, of concept and reason. To be sure, these notions no longer necessarily held the specific meaning Kant had given to them; they were reinterpreted as they were filtered through the bundle of notions brought forward to sketch out the framework of the epistemological project: notions of the act, of activity, of work, of effectiveness, of event or 'moment', of development, of dynamism, of dialectic.

1.1 The philosophy of mathematical practice

Mathematics has traditionally played a paradigmatic role in the development of theories of objective knowledge. Cavaillès embraced tradition in affirming

---

[7] See J. T. Desanti's writings, I have essentially used *Les idéalités mathématiques*, Paris, Le Seuil, 1968. Among G. G. Granger's works, I have restricted myself to:
— *Pensée formelle et sciences de l'homme*, Paris, Aubier, 1960.
— 'Pour une épistémologie du travail scientifique', *La philosophie des sciences aujourd'hui*, sous la direction de Jean Hamburger, Paris, Gauthier-Villars, 1986.
— *Pour la connaissance philosophique*, Paris, Odile Jacob, 1988.
— *Formes, opérations, objets*, Paris, Vrin, 1994.
See also Hourya Benis Sinaceur, 'Formes et concepts', in *La connaissance philosophique. Essais sur l'œuvre de G. G. Granger*, Paris, Presses Universitaires de France, 1995, pp. 95–119.

that 'mathematical knowledge is central to knowing what knowledge is'.[8] Nonetheless, he specifies that 'critical reflection on the very essence of mathematical work' ... leads to 'digging beyond what can strictly be called mathematics, in the ground common to all rational activities'.[9] And there, he sets himself apart from the classical understanding through his reference to an activity and to work. Cavaillès has in mind a theory of knowledge that acknowledges the *practice of mathematics*, the work of the mathematician. Similarly, Granger underlines that 'the epistemological attitude looks to the *practice* of science, in its process of creation and bringing about'.[10] Again, Desanti is interested in '*productive practices*' in the field of mathematics.[11] So it is not a case of determining a priori categories of thought, but of explaining, from the inside, thought in action, of grasping the actual process of production, by tracing the stages and the tangled and many-pathed routes that lead to a novel result. One does not occupy oneself with the empty forms that reason, according to Kant, is bound to impose on empirical facts in order to transform them into objects of knowledge, but rather with the substance itself, with the contents of knowledge, which are at once the object and the product of mathematical practice. Mathematical knowledge is something original, which constitutes a positive reality, existing of itself, irreducible to anything other than itself. It is necessary to put oneself into this specific reality without bringing in any pre-conceived philosophical idea. The question one is looking to answer is no longer 'How can mathematical knowledge be possible a priori?', but simply this: 'How does mathematical knowledge come about?'. It is to bring back to the level of *phenomenon* the question that Kant asked at the 'transcendental' level (that is to say at the level of first principles, universally applicable to *all* knowledge in general).

1.2 Proof

To be sure, a certain dose of pragmatism enters this interest in the actual and material aspect of mathematical activity. But for Cavaillès, as for Granger and Desanti, it is not a case of clarifying the psychological, sociological, cultural or anthropological context of the activity. The mathematical activity is considered in itself, in abstracto; that is to say, its effects and results are thought of independently of real circumstances, contingent and possibly resistant to a totally rational explanation. Pragmatism serves to forbid an external and general discourse *about* mathematics, to try to show mathematics in the process of happening, and to expose in detail situations and problems.

---

[8] 'La pensée mathématique', p. 625. In a similar way, Granger thought that 'mathematics remains a paradigm on which can be modelled every objective thought, however distant.', *Pour la connaissance philosophique*, p. 113.

[9] *Méthode axiomatique et formalisme*, p. 29.

[10] *Pensée formelle et sciences de l'homme*, p. 9.

[11] *Les idéalités mathématiques*, Avant-propos, p. VII.

Across the firm, known facts of mathematical situations, Cavaillès has his eye on what it is that constitutes the *validity* of results. In abstracto, a result is unconditionally affirmed if it is demonstrated. And, in this case, it is 'absolutely intelligible'. Cavaillès' insistence on the role of demonstration is correlative with putting into question intuitive evidence as a reliable way of accessing the truth. Cavaillès takes on board the attitude of the mathematical movement which, since the nineteenth century, had been emphasizing the objective dependencies between definitions and theorems. The first chapter of *Méthode aximatique et formalisme* surveys the contributions of numerous mathematicians, among others Gauss, Cauchy, Grassmann, Bolzano, Frege, Dedekind, Riemann, Pasch, and Hilbert. Cavaillès notes, for example, that the definition of magnitudes by Grassmann and the discovery of non-Euclidean geometries mark a break from intuition and tangible experience. He recalls the exhortation of Moritz Pasch in a famous passage of *Lectures on the new geometry*[12]: 'The process of proof must be entirely independent of the *meaning* of the concepts, just as it must be of the figures: only the *relationships* established between the principles or definitions should be taken into consideration'. He lingers longer over Hilbert's *Foundations of geometry*, in which points, lines and planes no longer have any intuitive meaning. Indeed, application of the axiomatic method consists in revealing the different possible geometrical architectures as a function of the statements chosen as axioms. A theorem is not an absolutely true proposition, it is one proposition demonstrated from others, taken as axioms. Cavaillès emphasizes that Hilbert's merit is to have made geometry the equal of the axiomatic arithmetic of Grassmann, Frege, Dedekind and Peano. Which is to say, in brief, that geometric space is not *the representation* of real space. It is a mathematical concept, tied to mathematical experience and not to real experience. A representational philosophy, such as classical philosophy, is therefore no longer relevant.

1.3 Truth from the perspective of structural mathematics

Thus, demonstration is at the same time the norm of exploration and the norm of the production of truth. The only place truth is to be found is in a system defined by a set of relationships that the mathematician proposes as, or accepts as, fundamental. Varying the axioms opens perspectives that act retrospectively on our understanding of known truths, or even of a whole branch of mathematics. Hilbert had illustrated this in his *Foundations of geometry*, in specifying, for example, the conditions and the limits of validity of Desargues' theorem, or of Pappus–Pascal's. The same proposition can result from different proofs, or might not be demonstrable in certain systems. Truth, therefore, is not monolithic, it has multiple aspects; and above all, it is not bound by the dogmatism of the obvious.

Modern mathematics in fact plays down the role of obviousness in the multiplicity of possible geometries: non-Euclidean, non-Archimedean, projective,

---

[12] *Méthode axiomatique et formalisme*, p. 64, p. 70.

Hermitian, metric, etc. It dissolves the ideal of a universal truth in the construction of non-standard models of arithmetic, of real analysis, of set theory. It puts structures in place of objects, and builds concepts that bring together varied structures: arithmetical, algebraic, topological, etc. Cavaillès, Granger and Desanti are strongly influenced by this architectonic work that shows mathematics as a stratified network of concepts, 'an indivisible whole, an organism whose vital force depends on the connections of its parts'.[13] In 1936, long before Bourbaki popularized the word in an article now become emblematic[14], Cavaillès described to his friend Albert Lautman the effect on him of the texts that he was studying. He said he was literally submerged by 'architectural images of mathematical development'.[15]

Architecture eclipsed the object; or, more exactly, it itself became the object. In French historical epistemology, the object does not have the characteristics of permanence, separate identity and indivisible unity of a *substance*. Conceived in line with the lessons of axiomatics and the *'begriffliche Mathematik'* developed at Göttingen, an object is some element of a domain associated with a system, in which the relationships define a structure, a concept. The mathematician constructs concepts, of a group, for example, or a ring, a field, vector space, metric space, etc. The concept brings together in a complex *functional unit* a collection, which can vary, of schemas of operations prescribed by axioms. It is an anchor-point for reasoning, but, as Desanti wrote, nothing in it is fundamental or foundational.[16] It is not a given fact, but a construct. And when, for example, one talks about *representing* a group, one means that the concept of a group brings about a *figurative effect* in which the operative possibilities indicated by the concept can be realized. From the *anti-psychological* perspective of the French philosophy of mathematics, representation is not essential, but only secondary.

2   Mathematical progression

2.1   Progressive reason

Mathematics shows an 'original dynamism' that escapes 'all prior order'.[17] Cavaillès broadly adopted this idea of Bruschvicg's. He brought it closer to some ideas of Brouwer on the 'auto-deployment' of mathematics, notably in the article

---

[13] Hilbert, [1902], p. 113.
[14] N. Bourbaki, 'L'architecture des mathématiques', *Les grands courants de la pensée mathématique*, Paris, Cahiers du Sud, 1948, pp. 35–47.
[15] Letter of 13 June 1936, in Benis Sinaceur (1987), p. 120.
[16] Desanti, [1968], p. 230.
[17] Brunschvicg, [1912], pp. 562–577.

'Mathematik, Wissenschaft und Sprache', published in 1929.[18] He himself said several times that mathematics constituted a 'real progression', both uninterrupted and unpredictable. On the one hand, there is the continuity between the past and the future: each stage of the progression results from earlier stages. On the other hand, each novelty is complete, in so far as 'one cannot, by simple analysis of notions already used, find within them the new notions'.[19] Mathematical truths are not analytical truths. A real progression stretches into the future.

'Progression belongs to the essence' wrote Cavaillès[20], meaning that nothing comes at a single stroke but by successive steps. It is therefore useless to look for conditions other than the particular conditions of a given mathematical situation. These conditions are not formal. They prevail because they are elements or aspects of an earlier mathematical situation. They are conditions *internal* to the practice and always susceptible of modification. And above all, the conditions and the system they pertain to are *not separable but reciprocally correlative*.

2.2 The mathematical experiment

The biological metaphor of a burgeoning and organic growth in mathematics favoured an 'experimental' conception of mathematical knowledge. Intuitionist ideas also played a role. 'There are no non-experienced truths and logic is not an absolutely reliable instrument to discover truths', wrote Brouwer.[21] Cavaillès thought, on this score, that the activity of mathematicians was an experimental activity. He intended to write a book on *L'expérience mathématique*, in which he would probably have shown at the same time the absence of any break from real-world experiment and the specific difference between the two. Unlike empirical experiment, mathematical experiment is knowledge. That is why it is an experiment of truth. But like any experiment, it is to venture, to test, to try, to risk, and also a thing of custom, acquisition, practice, and expertise. Knowledge proceeds by tentative advances and by reorganizations 'based on experience'.[22] It finds *strategies* (not *principles*) for its action in that action itself. The theory of mathematical knowledge is a theory of action, more precisely a *theory of the rules of the action*.

Here, it is necessary to point out a harmony noted by Cavaillès between Brouwer's intuitionism and certain themes from phenomenology. Cavaillès, Granger and Desanti had read the work of Husserl, notably the *Logische Untersuchungen* and *Formale und transzendentale Logik*. For all that they vigorously

---

[18] Cavaillès alluded to this article in *Sur la logique et la théorie de la science*, p. 497.
[19] 'La Pensée mathématique', *Œuvres...*, p. 601.
[20] *Sur la logique et la théorie de la science*, p. 552.
[21] Brouwer, [1948], p. 488.
[22] The expression is due to Gonseth, [1939], p. 38.

From Kant to Hilbert: French philosophy of concepts in the beginning of the twentieth century

argued over certain aspects, the phenomenological direction left its mark in their works. Here are the principal traces:

- First of all there is the primacy of the *Sachverhalt* [situation], which determines the decision, as I pointed out at the start, to turn one's attention towards the practice of the mathematician and the contents of mathematical knowledge.
- There is also the importance given to meaning. On the one hand, epistemology consists (according to Granger and Desanti):
  - of digging in the archaeological subsoil of those structures now in use to find their ancient roots (which is where history comes in),
  - of making explicit the latent meanings (which is where the hermeneutic method comes in), which the mathematician (as a mathematician) cannot explore without losing the thread of what he is doing,
  - and to 'shed light on the relationships a posteriori necessary to the organization of concepts'.
- Above all there is the analysis of mathematical thought as essentially constituted from procedures of idealization and thematization. I will come back to these procedures later, which Husserl himself described in terms of the contributions of the axiomatic method.
- There is an attention to language, paid particularly by Granger, for whom 'it is without a doubt due to Husserl that epistemology has been reoriented down the difficult path of research on two levels: that of language and that of the object'.[23] Granger brought phenomenology closer to the Anglo-saxon[24] analytical current and was clear that the most significant contribution to epistemology consists of the linguistic analysis of knowledge.[25] He thus took an original path in the field of French epistemology.
- Finally there is a whole vocabulary that would be barely intelligible without reference to the intentionality that Husserl forges from premises that he found in Brentano. In Cavaillès, the words 'act', 'move' and 'directedness' betray the impact of what seemed to him an authentic discovery. Granger made providential use of the notion of the 'categorial outline of the object' to understand the multiple polarity of concepts. Desanti, more directly than he would admit, made broad use of intentional structures and the phenomenon of the horizon.

Despite all these points, faithfulness to Husserl is fundamentally contradicted by the rejection, more radical for Cavaillès than for Granger and Desanti, of the subjective perspective that attributes to consciousness the initiative in the formation

---

[23] *Pensée formelle et sciences de l'homme*, p. 14.

[24] Granger is one of the first French authors to have explained and commented upon the works of Wittgenstein. The notion of 'a language game' naturally drew his attention. He compared it with the Husserlian 'eidetic variation' and made precise the difference: Husserl sought to determine the essentials in creating acts of thought; Wittgensttein sought to determine the function of language in creating the universe of linguistic behaviour.

[25] *Pensée formelle et sciences de l'homme*, p. 12–13.

of mathematical processes. The rejection of consciousness dismisses *de facto* a representational philosophy, which we have already seen does not correspond to axiomatic practice. But it leads to a major difficulty. How are we to sustain a philosophy of practice, of experiment and of action while excluding the subject that experiments and acts?

For Cavaillès, this problem is eliminated by his definition, entirely desubjectivized, and surprising, of what an experiment is. In fact Cavaillès says:

> 'By experiment, I understand a system of moves, governed by a rule and under conditions independent of the moves ... I mean that each mathematical process is defined in relation to an earlier mathematical situation on which it partially depends, and in relation to which it maintains an independence such that the result of the action must be observed in its accomplishment ... That is to say ... the act having been accomplished, by the very fact of it appearing, takes its place in a mathematical system extending the earlier system'.[26]

Cavaillès emphasizes that a mathematical result is always situated in an historical context, which it extends and modifies. He suggests that, as a consequence, a mathematician's act is to be explained in terms of mathematics and not by the mathematician's psychology. Finally, for him, an experiment obeys a system of rules dictated by a *state* of the problem to be resolved. To say that mathematicians' activity is an experimental activity, is simply to say that it is *subject to objective conditions*. These conditions are *internal* to mathematics and are concretely incarnated each time in a package of results, methods and problems. For example, for set theory or set topology in the last twenty years of the nineteenth century, as Granger notes, 'the attentive study of the state of analysis after Cauchy is manifestly indispensable; but one can reasonably doubt that the examination of the situation of the means of production in France and in Germany, and of the development of the battle of classes and of ideologies are of any great help.'[27]

2.3 The historical method

Mathematics is a progression and the mathematical experiment is the continuation of a specific history, which is not to be confused with empirical history. The philosopher who seeks access to this experiment cannot get there except by way of this history, which he shall explore. The historicity of the experiment is not the cause, but the consequence of the moving, successive, winding, at times abrupt, character of the progression, the course of which never ceases to modify and deepen its own traces. There is a history because reason is (of its essence) progressive; and not the other way round.

---

[26] 'La pensée mathématique', pp. 601–602.

[27] 'Are there internal dialectics in the development of science?', *Formes, opérations, objets*, chapitre 18, p. 348.

From Kant to Hilbert: French philosophy of concepts in the beginning of the twentieth century

It has been said and said often that history is to philosophy what a laboratory is to the scientist: a place where practical observations can be made and where the tools of analysis can be crafted. In a polemical spirit against the logicism of Bertrand Russell, promoted in France by the work of Louis Couturat (1868–1913), Brunschvicg insisted on the importance of the historical study of mathematics. He presented history as a *method* for revealing the *multiple*, albeit structured, character of mathematics, for explaining its progress, and, on this basis, for elaborating a supple and open theory of rationality. Bachelard, Cavaillès, Gonseth, Granger, and Desanti systematically applied this method and spent a long time in the laboratory of reason that is the history of mathematics. But the history that these philosophers wrote is far from the landscapes familiar to mathematicians. None claimed to be rivalling the 'Historical notes' of Bourbaki's *Éléments de mathématiques*. Instead it was a case of approaching mathematics as an experiment, that is to say not just by its successes but also by its speculative attempts, its difficulties, its setbacks, its errors. To try to capture its branching, generative evolution. And above all to discern in the succession of events connections that give them significance and make them intelligible. In this way the philosophers defended a structuralist concept of history.

Cavaillès, for example, held that 'one can, by studying the contingent historical development of mathematics as it presents itself to us, perceive necessary facts beneath the string of notions and processes'.[28] It was from this perspective that he undertook the history of abstract set theory. On the one hand, as scrupulous historian, he reviewed an impressive number of mathematical works and memoirs. On the other hand, as philosopher, he summed up in a few words the originality of the Cantorian creation: it is not, according to him, in the *objects*, but in the *methods*, not in the consideration of sets of points, already entailed in the analysis of the representability of function in trigonometric series, but in the two processes of the diagonal and transfinite iteration. All these recounted historical details have, therefore, to converge to show what it was that made the invention of these processes necessary. By the same token, in the history of axiomatics, it is necessary to emphasize the constraints that lead to crossing the boundaries of more restrained theories, to establish the new more general procedures of reasoning. Taking inspiration from the reflections of Dedekind[29] and Hilbert[30], Cavaillès exposed the working of these procedures, which he called paradigm and thematization. Given the repercussions of these notions for French philosophers, I will come back to them in section 4.2.

---

[28] 'La pensée mathématique', p. 600.

[29] Notably in the Habilitation lecture given in 1854 before Gauss and first published only in 1932 by Emmy Nœther in volume III of his *Gesammelte Mathematische Werke*, O. Ore, R. Fricke and E. Nœther eds, Braunschweig pp. 428–438.

[30] Notably 'Axiomatisches Denken', *Mathematische Annalen*, L. 18, 1918 and 'Über das Unendliche', *Mathematische Annalen*, t. 95, 1925.

History is thus an essential instrument of philosophy. It allows us to stand at a distance from the present, the better to home in on originality. It offers reason the chance to test its own critical power and to undo the old prejudices of traditional philosophy. It shows flux rather than things, processes rather than entities, multiplicity rather than monolithic unity, singularities rather than the universal. History is the refracting prism of philosophical notions: being, subject, object, concept, intuition. And above all, history sets philosophers (of mathematics) a major problem: how to conceive of the link between the contingence of events and the internal necessity of the development of notions? The solution to this problem, if it exists at all, is not easy.

3 Form and content

3.1 The autonomy of content

Despite its external links with the physical, social, economic or political real world, mathematical activity produces a reality whose relative autonomy has been highly promoted by French philosophers. This had been the view of Brunschvicg. Cavaillès and Granger had probably noticed its relationship with the Vienna Circle's's neo-positive thesis of the autonomy of science. This is straightforwardly suggested by the fact that Cavaillès wrote in 1935 a brief account of the activities of the Circle, and that Granger had good knowledge of the work of Wittgenstein.

Mathematical reality is made up of 'objective contents' endowed with an autonomous dynamism. For Cavaillès, these contents 'are, in their progression, themselves the essential'.[31] They have a 'creative autonomy'.[32] This progress is driven less by the relationship between object and subject than in a *relationship between the object and it itself*.[33] The result is an increasing complex unification. As Hilbert underlined in his lecture to the International Congress of Mathematicians in Paris (1900), complexity and unification go hand-in-hand. Cavaillès takes advantage of this lesson from modern mathematics to criticise Kant. 'The synthesis which Kant finds in thought requires nothing more or different to be provided but just itself, made multiple by its moments and its progress: that which is unified is not first of all present as varied', he writes.[34] The rhapsody of variety is internal to the development of mathematics. It is not given by the external, it is constructed according to a specific rhythm with unexpected bifurcations. The unpredictable character of mathematics may be caught by the Kantian notion

---

[31] *Sur la logique et la théorie de la science*, p. 486.

[32] *Sur la logique et la théorie de la science*, p. 501.

[33] Desanti thought, moreover, that the object occurs in a mobile and self-regulated relationship with itself, *Les idéalités Mathematiques*, p. 100.

[34] *Sur la logique et la théorie de la science*, p. 510.

of synthetic judgement. But Cavaillès focuses on the *content of judgement*, and disregards the *faculty of judgement*. He is looking for a theory of content, not a theory of judgement. Here he distinguishes himself notably from his teacher Brunschvicg and adopts a perspective that owes much to Gottlob Frege.

3.2 Formal content

In evoking in Kantian terms the genesis of the notion of group, Brunschvicg remarked that 'this notion does not have, as the notions of negative numbers or imaginary numbers did retain, the external appearance of a concept to which an object might correspond; ... it presents itself as an intellectual relationship, establishing ... the most significant trait of modern intellectualism, ... the primacy of judgement over concept'.[35] Cavaillès, Granger and Desanti understood that structural mathematics leads to nullifying the exteriority between concept and object (in Kant's sense), that is to say between form and matter. But they drew an opposite conclusion to Brunschvicg. For them, content is primary. But in fact, mathematical contents are always already formal, and the forms are called upon to become contents for the construction of more abstract forms. The very wide and differentiated application of group theory showed, for example, the incarnation of structure in many varied aspects of mathematics: in algebra (the group of permutations of the roots of an equation), geometry (the group of transformations), matrices, topology, etc. The notion of group is *generic*: it talks not about a determined structure, but a type of structure. This is what explains its usefulness in the search for solutions in transversal problems. Put another way, to make axiomatic is not simply to give form; it gives rise to content.[36]

The border between form and content was crossed another way by Hilbert's metamathematics (Beweistheorie), which took the formal tool *par excellence* for forming mathematical content and made it a mathematical object (mathematical content): 'a formal demonstration is just as much a visualisable and concrete object as is a number' wrote Hilbert.[37] Further, Tarski's formal semantics had perceived various layered levels in the formal. It had accentuated and made a systematic method of the back-and-forth movement between forms: the structure under consideration, and contents: the various models, known or possible, of this structure.

The entire second part of Cavaillès' *Sur la logique et la theorie de la science* is a – difficult – commentary on the philosophical consequences of Hilbert's theory of demonstration, and Carnap's and Tarski's introduction of the semantic point of view. The conceptual framework and the language are borrowed from Husserl.[38]

---

[35] *Les Étapes de la philosophie mathématique*, §349, p. 550.

[36] See notably Granger, *Pensée formelle et sciences de l'homme*, p. 170.

[37] 'Über das Unendliche', p. 179.

[38] This can seem surprising, but it is necessary to recall that Tarski's formal semantics developed an affinity with the philosophical semantics of traditional philosophy fashionable with Bolzano and

Contents and forms are described in terms of objective meanings, as introduced in the *Logische Untersuchungen*. Like Husserl, Cavaillès establishes a correlation between meaning and the act by which it exists. This constitutes a serious difficulty for a philosophy of content, which wants to expel the idea of consciousness. Desanti and Granger faced this difficulty. The first deliberately associated with structural analysis an archaeological analysis, developed along the characteristic lines of phenomenology: the relationship between the explicit and the implicit, the link back from operations to acts, the position of objects, thematization, etc. He noticed that these themes imply the pre-supposition of a 'mathematical activity', correlative to the domain where actions, properties and objects would link up. But he suggested that one can understand this mathematical activity without necessarily referring it to a transcendental subject. I will shed more light on this point of view in Sections 5 and 6 of this chapter, as well as in the conclusion.

For his part, Granger grasped the *internal* link between form and content through the original notion of 'formal content'.[39] He intended to substitute this notion for that of the synthetic a priori, the key notion around which revolved all attempts to realign Kantism. His argument asks us to consider that the opposition between form and content had been understood on an ontological level (by Aristotle) or on an epistemological level (by Kant), whereas it consists of an *opposition of meaning*, that is a functional distinction between two elements correlative in a symbolic universe. The revelation of a form always coincides with the use of symbols, and the opposition of form and content fundamentally accompanies every act of meaning.

4 The theory of the sign and Kantian schematism

4.1 The being of the sign

Formal content in fact arises in a universe of signs governed by explicit rules. The signs relate back to content, to varied meanings, and they are themselves the object, the material of formal transformations. Transformations have laws (which define structures and structures of structures) and limits (impossibility theorems). Laws and limits determine the generic specifications of possible content.

The principal element of Kantian epistemology comes from schematization which is the function of presenting of a concept with an intuition. For Kant, the concept needs to be made schematic: each concept has its schema, that is to say a procedure, a rule for the application of formal conditions of intuition to the empirical material of intuition, so as to give the concept a relationship with objects, that is to say, to give meaning to the concept.

Husserl. Husserl thought that 'everything in the domain of logic is contained in the categories that, correlatively, present the signification and the object'. *Logische Untersuchungen*, I, §29.

[39] The notion of formal content, in *Formes, Opérations, Objets*, Chapter 2.

Hilbert had noted that it was necessary to have symbolic mediation and pointed out the primacy of the sign in mathematics. Mathematical objects are the signs themselves: digits, geometric figures, indeterminate equations, formulae, matrix tables, homomorphical diagrams of morphisms, diagrams of proofs, etc. 'In the beginning was the sign'.[40] Called upon by the critiques of Poincaré, Kronecker and Brouwer to make room for intuition in the formal procedures of his structural and metamathematical methods, Hilbert made the sign the intuitive basis, the content of mathematical thought, and made a plea for an intuitive metamathematics. Signs permitted a concrete handling of the finite sequence of formulae that constituted the diagram of a proof. Signs are external objects; they constitute the irreducible given facts, prior to all thought, of operating intuition. Hilbert thus collapsed the sign, the carrier of meaning, onto the material prior to thought. To rescue intuition, Hilbert sacrificed the meaning of signs to perception, and came back to a type of philosophy where matter is considered as external and prior to thought.

Cavaillès grasped the strategic importance of what he called Hilbert's 'theory of the sign'. Unlike the majority of historians, who followed the example of Bernays' explanation[41], Cavaillès did not argue for Hilbert's declared adherence, from the 1920s, to a position close to Kronecker's. It was to the theory of signs in and of itself to which he turned his attention, and not to the thesis that the integer number is a datum. He emphasized that this theory was not just a psychological description, but picks out an essential characteristic of mathematical terms: the expression of a mathematical situation is itself a mathematical situation. Moreover, it is this characteristic that gives rise to the reflective disposition of modern mathematics. It is the signs, expressions and formulae that make up mathematical reality. The work of the mathematician is to experiment on its formulae. The sign is the substance of thought, but is not prior to thought. Unlike Hilbert, Cavaillès has no polemical reason to place so much emphasis on just the material, tangible, visualizable aspect of the sign. He therefore emphasized the function of the sign as a meeting place for form and matter. The sign has two faces – perceivable, because it really is the mathematician's material in his work, and formal, because it is defined by the rules of its use and its possible transformations. In the realm of perception, the sign is a sign just because it points to one (or more) *meanings*. The sign is the *symbolic matter* and not empirical matter; sign and sense are indivisible. In the realm of the intelligible, the sign is operative content, it acts. The intelligible content is not inert or isolated or fixed. It is a sign for the operations it points to and that act back on its meaning. The sign is what is permanent in that which is changeable in it and by it: the meaning.

So, Cavaillès adjusted Hilbert's theory. He reinstated the sign in the sphere where it belongs, that of meaning, and gave it a philosophical status more

---

[40] 'Neue Begründung der Mathematik. Erste Mitteilung', *Gesam. Abh.* **III**, p. 163.
[41] Hilbert's Untersuchungen über die Grundlagen der Arithmetik, *Ges. Abh*, III, p. 203.

adequate to symbolic mathematics. But, at the same time, he had found what was needed to amend Kant's schematization theory. The sign is a mixed, tangible-intelligible, concrete-abstract, intuitive-formal thing. To it, therefore, belongs the schematic function; it itself is the rule for applying form to content. There is therefore no need for a scheme in the Kantian sense. And intuition has no need to be split into the a priori and the tangible content. The sign condenses the matter: the sign's received meaning; and reveals the form : the new meaning.[42]

4.2 Constitutive properties of mathematical thought

The autonomy of the sphere of meaning was brought to light by the semantic tradition that was partially conveyed by Husserl's phenomenology. In taking the semantic point of view of mathematical content, French epistemology never intended to deal with the question of the *ontological* status of meaning. From mathematical thought it did not describe the thing in itself, but its attributes. Moreover, the hypothesis of a mathematics of itself, that is to say of a region of ideal objects to which mathematics could refer, seemed superfluous. Cavaillès is very clear in his refusal of Platonic realism, that is to say, of the philosophical option of the prime supporters of semantics, Bolzano and Frege. 'I believe', he wrote, 'that a concept of systems of mathematical objects is in no way necessary to guarantee mathematical reasoning'.[43] When we consider this system of objects, all that we think of them as, are rules for reasoning demanded by the problems that arise; and it is these unresolved problems that push us to propose new objects or to change the meaning of the previous objects. This point of view is also that of Granger and Desanti (despite the metaphor of the expressions 'ideal objects' and 'idealities' constantly used by Desanti).

The progress of meaning acts to multiply it; the lines of generalization and of abstraction of modern mathematics appear like lines of meaning. Cavaillès sees two principal axes of the ordered proliferation of meaning. Horizontally, idealization consists in the adjunction of ideals: imaginary numbers, Kummer's ideal numbers, points at infinity in the plane.[44] Idealization frees meaning from particular constraints: operations are dissociated from the elements of the field on which they operate. Vertically, thematization[45] superposes different levels: autonomous operations are transformed into objects of a higher operational field. These two constitutive modalities of thought intersect again and again in the non-uniform,

---

[42] The relationship, called dialectic, of matter and form, of the concrete and the abstract, is presented alike in the commentaries by P. Bernays [1976] on the metamathematical writings of Hilbert, in the work of Gonseth [1939], and in that of Granger.

[43] 'La pensée mathématique', p. 603.

[44] Cavaillès was inspired here by Hilbert's article 'Über das Unendliche'.

[45] This term is taken from Husserl, who created it in contact with mathematics. In his *Formale und transzendentale Logik* (Halle, Niemeyer, 1929), he remarked that the '*thematische Einstellung*' is a solid mathematical tradition.

non-linear timeline of the mathematical experiment. Tangled layers of meaning give the sign its semantic substance, which grows more substantial all the time.

We can illustrate this by taking Granger's examples of conics. The idealization, which Cavaillès also calls 'the paradigm moment', came into play when instead of making conics from different cones, with acute, obtuse and right angles at their point, Apollonius got them from the *same* cone, by varying the plane of the section. The production of content, the conics, comes from an *internal principle of variation*, and not from an *external principle of unification*. In thematization, we consider the form itself as variable: for example, Desargues defining, by the operation of projection, *the* conic itself as an invariant. The specification of curves then depends on the choice of type of transformation; for example, the class of parabola is correlative with the invariance of a line chosen to represent the points at infinity.

Granger accentuates the *semiotic* reform, sketched by Cavaillès, of the transcendental aesthetic. He makes explicit the status of formal contents as correlates of acts of meaning, clearly taking on the inheritance of phenomenology. But at the same time, he is less concerned with the acts themselves than with their structure. He thinks that mathematical axiomatization is a way of determining the objective categories of thought. Therefore he translates the action–meaning correlation into the operation–object duality, underlying the ambiguity of the sign and the back-and-forth movement of meaning between idealization and thematization. This 'duality principle' seems to him to be the essential feature of the act of cognition in general. It consists, according to him, of a 'primitive and radical functional category of knowledge', which constitutes an ultimate condition. This category therefore *takes the place of the transcendental subject*. Granger thus completes Cavaillès' programme of abandoning the philosophy of the subject in favour of the philosophy of concept.

5 The subject displaced, the concept objectivized

5.1 The rational chain-sequence of contents

Formal content forms the *objective* network of thought. But the chain *sequence* that links the contents together has the same objective status. Content and chain sequence are homogeneous terms: an item of content is the result of earlier links, which can be updated from a new perspective. 'Doing mathematics' is to extend it, and to understand mathematics is to redo it. There is a global continuity to mathematics, despite local discontinuities. There is a coherence that surmounts or covers the aleatory ingredients of history. The author can change, the moment can differ: in the long run, the sequence is always taken up again and continued in an objective interdependence with what came before. There is an *internal necessity* to the chain sequence, whatever the historical moment at which it is being extended, and in whatever way someone is extending it. As Desanti expressed it through a paradigmatic image, on one side there is Archimedes, Leibniz and Riemann as so many 'contingent apparitions', and on the other side we have the quadratures, the definite integrals and the Riemann sums as so many 'necessary chain-sequences'.[46]

As I pointed out earlier, the link between the two series is problematic.

One solution, which is the one adopted by Cavaillès[47] and Granger, is to consider that necessity is not a priori, as was the case in classical rationalism, but a posteriori. That does not mean to say that the necessity of the chain-sequence is chronologically posterior to its actualization, or that it is injected retrospectively thanks to some 'rational reconstruction'. It rather means that necessity is there, from the start, but that it only *appears* after the event. It was hidden and only comes to light bit by bit, by the measure of successes. But it is in this that the distinctive mark of mathematics can be seen. Cavaillès and Granger think that what characterizes mathematical progression is not the contingent aspect, which it shares with all other products of culture, but the rational structure of its chain-sequences. Besides, the rational interdependence of mathematical moments goes beyond purely deductive logic. Mathematical demonstration is more than a simple logical deduction. Mathematics cannot be reduced to logic, as Frege and Russell had held. The French philosophers and mathematicians sided with Kant on this point and adopted the anti-logicism of Brunschvicg, Poincaré or Brouwer, who emphasized so strongly the originality of the mathematical experiment. And they took the semantic turn all the more quickly as they saw an opportunity to support, against Kant, the idea of an unbreakable intricacy linking content and form.

5.2 The motor of progress: constraints of mathematical problems

The interiority of the chain-sequence to the contents that it produces destroys the idea of a creating subject.

Hilbert, all the while professing Kantianism, was firmly opposed to the subjective idealism of those who would contradict him: Kronecker, Poincaré and Brouwer. The philosophical idea underpinning his proof theory was, he said, to draw up an inventory of the rules that our thoughts follow in order to function effectively, with a view to freeing us from the arbitrary, from sentiment and from custom, by protecting us from subjectivity.[48] Hilbert therefore searched for rules where others looked to custom, to chance inspiration, arbitrary convention, or, like Brouwer, 'the exodus of consciousness from its deepest home'[49], oscillating between rest and feeling, towards a limitless introspective deployment of innate intuition.

---

[46] *Les idéalités mathématiques*, p. 32.

[47] For the details, see Benis Sinaceur, *Jean Cavaillès. Philosophie mathématique*, pp. 30–33.

[48] 'Die Grundlagen der Mathematik', *Abhandlungen aus dem mathematischen Seminar der Hamburgischen Universität*, **VI.** Band, Leipzig, Teubner, pp. 79–80.

[49] Brouwer, [1948], p. 483.

Like Hilbert, Cavaillès, Granger and Desanti picked out as ways of thinking not empty forms, but effective, repeatable, combinable procedures that create content. Earlier I pinpointed the principal axes (idealization and thematization). For Cavaillès, the chain sequence did not depend on the initiative of a consciousness, be it empirical or transcendental. It is useless to postulate the Kantian 'I think' as an agent of rational *unification*. Because a content is already the bringing together of a polymorphism, partly actual and partly potential, of properties or methods; and the various items of content hold the one to the other via internal links: related through problems, structural identities, methodological or functional analogies, inversion or duality of perspective. As for Brunschvicg's *generative* consciousness, it is threatened by psychologism. According to Cavaillès, progress comes about from the *endogenous* development of content. Items of content, he writes, 'literally are the essential in their development, and the primordial pseudo-experiment of consciousness disappears before the autonomous dynamism which they reveal and which leaves no place for anything other than them'.[50]

In this passage, Cavaillès accomplishes the preliminary step to his demand on the philosophy of concept: he clearly eliminates the subject as an actor in the emergence of content and turns his attention towards the functional and symbolic unity offered by content. The linking of the sequence does not maintain its authority because of a legislative or creative consciousness; in its historicity, it possesses an authority of its own. Its accomplishment brings it up to date and legitimizes it at the same time. Because it is problems that need to be solved and difficulties to be overcome that engage solutions. It is the power of a method that surpasses the original field of its application and moves towards new territory and new problems. There is a reciprocal and objective conditioning between the methods and the extensions of the domains provoked by their application. The rational grows and branches of itself, according to the local constraints.[51] It has a self-organizing, expansive force.

As J.-T. Desanti explained metaphorically, 'the consciousness of the object lives on the life of the object itself'.[52] The subject, 'reduced to the status of an anonymous spectator, [is] nothing other than the manner, different each time, by which the object is manifested'.[53] Here the roles have been reversed: it is the subject that's the instrument of the object, the subject that acts as the medium for the object to express itself.

---

[50] *Sur la logique et la théorie de la science*, p. 486.

[51] A good example of a local constraint is recalled by Granger, who remarks that in algebra nothing in the primitive properties that define a field let one predict that a finite field must be commutative. A certain amount of work is need to establish this property of finite fields.

[52] *Les idéalités mathématiques*, p. 91–92.

[53] *Les idéalités mathématiques*, p. 290.

5.3 Concept without subject

The paradox of the Kantian epistemology had been effectively revealed by Cavaillès and Desanti. It consisted in abstracting from all content of knowledge to free up a formal framework, which, while being all the while inaccessible to consciousness, nonetheless posits consciousness as subjective structure of objectivity.

In Kant's philosophy, pure concepts of understanding determine the rules (the schemas) of subjective unification of representations related in their diversity. Nonetheless, without intuition, they are empty; just as intuition, without concept, is blind. Concepts serve, through the mediation of their respective schemas, to work out the substance of intuition as objective experience. The synthetic unity of the varied, in intuition, and the analytic unity of concept come together in the 'I think', a transcendental act of unification by consciousness. The 'I think' is neither the intuition nor the concept of an object, but the form of consciousness that comes with these two types of representation in the guise of a subjective condition of knowledge. Therefore concepts are functions of the spontaneity of the understanding of the subject.

In French philosophy of science following Cavaillès, the link between concept and subject is broken. Concept moves to the side of object and content, reversing the normal paradigm of knowledge. In determining the objective structures of objectivity, the examination was directed at the objects themselves: the mathematical concepts. A truly Ptolemaic revolution.

The objectivity of the concept had been underlined by the work of Bolzano and Frege, who relieved it of all reference to activity of the mind. For his part, in his *Logical Investigations*, Husserl insisted upon the autonomy of the sphere of meaning; here belonged concepts as 'units of ideal meaning' representing 'constituent moments' of the construction of systematic theories similar to mathematical theories.[54] The mathematical theories that Husserl speaks of, and that Cavaillès had studied in his two theses, deal with the structures of abstract axiomatics. In German, mathematicians generally called them concepts (*Begriffe*) and spoke of an architecture of concepts that organized the flourishing field of new inventions : the concepts of group, field, ring, ideal, algebra, etc.

I have shown the reverberations of this *begriffliche Mathematik* on the philosophical programme proposed by Cavaillès.[55] The roles were redistributed: instead of setting consciousness and concept on the one hand in opposition to intuition and object on the other, Cavaillès ignored the constituent function of consciousness and brought over to the side of the object (to the side of content) concept and intuition. The mathematical experiment is knowledge, and, in that respect, it is not an experiment of consciousness, but an experiment of concept. Thus concept,

---

[54] *Logische Untersuchungen*, I.

[55] Hourya Benis Sinaceur, 'Structure et concept dans l'épistémologie de Jean Cavaillès', *Revue d'Histoire des Sciences*, XL-1, 1987, p. 5–30.

considered as an evolutionary crystallization of meaning, becomes a driving force. From this perspective, what becomes of intuition?

5.4 The parallelism of intuition and concept

Cavaillès was of the opinion that it is difficult to go further than Kant in the analysis of the role of intuition, which is 'not the contemplation of a completed event, but the apprehension in the performance of the act of the very conditions which make it possible'. This is how we can escape, he says, 'from the irrationality of that which is pointed up by the internal necessity of construction'.[56] But he deplores the influence of poorly understood Kantian epistemology: mathematicians (Kronecker, Poincaré, Brouwer and Hilbert) believe themselves faithful to Kant when they look for a zone of irreducible intuition, a kind of minimal reserve of guaranteed first entities. We have already seen that there are no such absolute objects at the start; the point, the continuum or the number are nothing but elements in the series 'from where they take their meaning and which goes beyond them'.[57] An object only has meaning by its function in a system of relationships and by the processes that these relationships put into motion. The idea of an irreducible intuition is contrary to thought, which is by nature systematic *and* progressive. According to Cavaillès, the intuition of Poincare's pure number is nothing but the substantiation of reasoning by complete induction, which could be analogously applied to other 'numbers', for example Cantor's transfinite ordinals. Also, the mathematical continuum differed from the intuitive continuum. Richard Dedekind was the first mathematician to point that out in his essay on the theory of numbers.[58] Later on, this his *Foundations of Geometry*, Hilbert constructed an algebraic model of the geometric continuum, such that the numbers in this model were sufficient for all Euclidean constructions, without needing recourse to what Hermann Weyl called the 'spatial sauce spread between them'.[59]

For French epistemologists, mathematical intuition, which is distinct from perceptual intuition, is not something that can be isolated as itself, and so does not offer a permanent ground on which to build objects; because it itself shifts and lines itself up in parallel with the construction of the objects, which are not *given facts* for thought, but are rather *produced* by earlier conceptual chain-sequences. Bachelard notes that an intuition reveals itself progressively in a *discursive* manner, by variation of the examples where the associated notions act.[60] Desanti

---

[56] *Méthode axiomatique et formalisme*, in Œuvres..., p. 35.

[57] 'Transfini et continu', in Œuvres..., p. 469, 472.

[58] *Ges. Math. Werke* III, pp. 339–340

[59] Hermann Weyl, *Das Kontinuum*, Leipzig, 1918. Reprint New York, Chelsea Publishing company, 1960, pp. 70–71.

[60] *Le nouvel esprit scientifique*, Paris, Félix Alcan, 1934. 9è édition, Paris, Presses Universitaires de France, 1966, p. 145.

recommends[61] giving up a theory where intuition serves as the *foundation* to the constitution of content and its chain-sequence.

It is important to understand that what one must give up is not intuition itself, but the foundational perspective. One of the most characteristic features of French epistemology is its anti-foundationalism. Cavaillès, for example, does not think that intuition plays no role in conceptual mathematics. Quite the contrary, it is to be found at the highest levels of abstraction. Rather, he thinks that intuition is not a founding basis and that it is neither external nor prior to concept. For him, intuition organizes itself in a systematic manner and by reference to an associated conceptual system. These two systems, intuitive and conceptual, are correlative and transform in parallel. The separation between the intuitive and the formal cannot be done in one step, nor in any clear-cut way. Intuition bears witness, at every stage, to the supposed 'naturalness' of a system of sequences conforming to specific rules. It is, in short, the mark of independence acquired by theories and methods, the sign of the objectivity of the concept. It is not the radiance from a consciousness of thought (the Cartesian *cogito*, the Kantian 'I think', the pure creative act of Brouwer, the intentionality of Husserl), but the effect of an effort of thought, which renders the formal intuitive.

So, intuition is not a subjective disposition. Or at least, what is of interest to French epistemology are intuition's objective traces, just as it only wants to examine the objective traces of experience and the objective traces of conceptualization. One is bound to recognize in that an original attempt to destroy the *myth of interiority*.

6 The dialectic of concept

6.1 The false problem of foundations

Anglo-Saxon pragmatism is often given credit for having eradicated the false problem that consists of wanting to give a rational account of knowledge. Wittgenstein observed that all the reasons that one could choose as the foundation for whatever it be, are in general less certain than those that we want to build upon them.[62] It must be recognized that French epistemology should be given credit for having repudiated the perspective of foundationalism and having understood that it is only a piece of knowledge that can justify another. There are no justifications for mathematics that are not themselves mathematics, teaches Cavaillès, closely followed by his successors. Put another way, mathematics contains its reasons within itself. This is in any case what is correctly meant by its 'autonomy'. To say that mathematics is autonomous is not to say that it is connected to nothing, not to a specific society, nor history, nor culture. It means that none of these

---

[61] *Les idéalités mathématiques*, p. 230.
[62] *Über Gewissheit*, §307.

links are useful in explaining the contents, which can only be explained between themselves, in relation to each other. 'The structure speaks of itself.'[63] To understand mathematics, you have to do some mathematics, either directly, or, as a philosopher, by way of its history. The historian goes over the mathematician's *way of working*, which constitutes a real remedy against the temptation of foundationalist prejudice. For French epistemology, cultivating historical research means the elimination of the problem of foundations in favour of an understanding of the progress of knowledge. And in the arena of history, the ground was readied for the putting in place of a new interpretative tool: the dialectic.

6.2 The internal development of concept

Concept is therefore not a free creation of the human mind (as certain mathematicians have affirmed, Cantor and Dedekind in particular). Concept, or expression, or meaning, is *the mathematical fact*. Mathematical theories develop from the formation of concepts. As we saw earlier, a concept lays down forms by the operation of idealization and 'lays itself down' by the operation of thematization. In the effective and objective operation of mathematical thought, concept is at once form and substance. Concept is born of concept and engenders new concepts.[64] Development is an internal dimension of the concept.

There is a strong analogy, often underlined by authors who knew Cavaillès, with the philosophy of Spinoza. Cavaillès himself, in the last part of his posthumous *La logique et la théorie de la science*, recalls that the true idea, in Spinoza's sense, leads to nothing else that is not a true idea. Nevertheless, in mathematics, the links are made across a complex network scattered with concepts connected to each other by organic links of different kinds. This 'organism' is not stable. It evolves constantly under the influence of local changes, which have repercussions on the configuration of the whole. The development of the concept is more important than the concept itself. With mathematics we are dealing with a 'conceptual progression'. The concept lives, and develops.

6.3 Explaining the primacy of progress: The dialectic of concepts

Cavaillès was looking for a theory of rational chain-sequences that would justify content, that is to say erase chance and arbitrariness from it, and explain progression, by which we do not mean an increase in volume but a perpetual revision of content through deepening. His answer was more or less this: it is in the perpetual development that the justification of contents is realized. *The very pursuit of development*, which integrates what came before into what comes next while modifying it, is a guarantee against the arbitrary and the contingent.

This development is 'material' in as much as, stimulated by the problems that crop up in the mathematician's work, it arises from the contents. But at the same

---

[63] Cavaillès, *Sur la logique et la théorie de la science*, Œuvres..., p. 506.

[64] This expression is literally due to Granger, *Formes, opérations, objets*, p. 345.

time it is conceptual, since the contents are methods and concepts, always 'intelligible', like Spinoza's ideas. To understand the development, it is not necessary to presuppose a driving act. On the other hand, it is useful to find the trajectories, the different 'moments' of the shifting of one concept into another: 'The idea of the idea manifests its generative power on the plane of limitless superposition which it defines without suffering harm'.[65] Development is a source of surprises without being a factor of contingency. Development is 'necessary' because it is inscribed in the 'internal bonds' between concepts, in their systematic organization. The progressive structure neutralizes contingency at source. The accidental and the accessory fade away in favour of rational connections.

Cavaillès called this material and rational progression a 'dialectic', employing a term much used at the time, notably by G. Bachelard, A. Lautman and F. Gonseth[66], and preserved, at least to begin with, in the works of Granger and Desanti. The dialectic of concepts, which was to replace the philosophy of consciousness, 'materialised the' autonomous and, so to speak, spontaneous progression between contents. The dialectic is a logic, but it is not a formal logic. It is a logic of content and development; we can speak of an 'internal dialectic'. This is required to reflect the development of knowledge in its two *inseparable* aspects, material and formal. It expresses the, so to speak, substantial link between the necessity and the unpredictability of mathematical development. It is a 'creative dialectic': it, and not the subject, brings about the creation of concepts. The last sentence Cavaillès wrote in *Sur la logique et la théorie de la science* is this: 'The generative necessity is not that of an activity, but of a dialectic'.

The dialectic seemed at the beginning of the twentieth century to be the most adequate method to the sciences, because every scientific principle is called upon to be specified, differentiated, revised. As the physician J. L. Destouches wrote, the dialectic was understood as a strategy to '*se mouvoir dans le mouvant*', to drive forward with the development. Although these authors did not recognize themselves much in Hegel's or Marx's dialectic, they could not entirely avoid the vocabulary and the new ideas these brought; G. Bachelard's *Philosophie du non* is witness to this. As for Cavaillès, he acknowledged that his own ideas were compatible with dialectical materialism, even if they were not a priori guided by it.[67]

But two things in particular were brought to the fore. 1) The dialectic was an improved replacement for logic, found much too rigid to fit the dynamism of thought. Thus Bachelard advised us to be wary of a concept that 'no-one has yet

---

[65] *Sur la logique et la théorie de la science, Œuvres...*, p. 514.

[66] In 1947, Gonseth and P. Bernays founded the revue *Dialectica*. The first two volumes of this revue contained a semantic explanation of the term 'dialectic'. One also finds there Gonseth's reaction to the work of Cavaillès, *Sur la logique et la théorie de la science*, which was about to be published.

[67] Letter of Cavaillès to the marxist mathematician Paul Labérenne, cited in Henri Mougin, Jean Cavaillès, *La Pensée*, n° 4, Juillet-Août-Septembre 1945, p. 79 : 'Although philosophically I am not oriented by the materialistic dialectic [...] I have said to you that I find myself led to results which are not exactly excluded by your attitude.'.

managed to dialectalize'.[68] 2) The dialectic introduced a dimension in which it was easy for the *possibility of progress* to find a place. In a way, the dialectic was seen less in the light of Marxism than as the shadow cast by Comte's 'positive spirit', which Léon Brunschvicg had injected into mathematics.

7 Conclusion

The reflective analysis of knowledge had shown the inevitable circularity that brings *the position* of the subject back to *the position* of the object, and vice versa. The circularity results from the replacement of the exteriority between these *positions* occupied, respectively, by the subject and object, concept and intuition, the abstract and the concrete, with an *internal* linking. It therefore is not useful to break this circularity, but instead we should look for a method to fit it. For a while, the intentional structure that Husserl aspired to had seemed a good candidate. Cavaillès, Granger and Desanti examined it. Without totally relinquishing it, they distanced themselves from it because it was rooted in consciousness. They chose the internal dialectic of concept because it avoided the difficulties of a subjective theory at the same time as it avoided the problems of a deductive-logical theory of knowledge. The dialectic installed an 'other' logic, or perhaps another meaning of the word 'logic', considering the circularity of thought not as a source of aporia, but as a normal, inevitable, fertile characteristic. Above all it allows the historical dimension to be put at the heart of this circularity. Thus is liberated all the dynamic potential of the continual and multiple back-and-forth between the opposing and correlative poles of the production of knowledge.

We can boil down all the correlations into just one of their number, which can be thought of as fundamental, that of subject and object. We can then attribute a *constituent* role to the subject, or a *constitutive* role to the object. Bachelard and Gonseth chose the first path. For his part, Cavaillès resolutely privileged the mathematical object, which he saw as a functional unit, capable of evolving in meaning, and as a structure (or concept). The dialectic of concept that he proposed avoids the need to resort to a transcendental or empirical subjectivity. But if the subject has been thrown out, its role persists: the *reflexive position* of the subject is taken on by the concept, as it appeared in expressions like 'the concept reveals itself', 'the concept transforms itself', 'mathematics occurs', 'it reflects on itself', 'it organizes itself into structures', etc. So it is legitimate to ask if this is not just a straightforward displacement, a transferral to object and concept of characteristics commonly recognized as belonging to the subject: autonomy, spontaneity, and dynamism.

---

[68] *Philosophie du non. Essai d'une philosophie du nouvel esprit scientifique*, Paris, Presses Universitaires de France, 1940, p. 134.

It is an altogether good question, and a pertinent one, that Granger and Desanti recognized, even if they did not tackle it head on. If they were categorical in choosing philosophy of concept, they nonetheless tried to find by what means, or within which limits, it is possible to eradicate all reference to subject.

By nuance, and to some extent by playing word games, Granger 'finally' settled on a form of transcendentalism that oscillated between Kant's doctrine and Husserl's. We might say of him the same thing *he* thinks of the Wittgenstein of the first part of *Philosophical Grammar*: he transposes the idea of the transcendental to an examination of the usage of language.[69] More exactly, he uses linguistic analysis to update the transcendental. Let us quickly say how. Granger opposes consciousness and concept as two modes of experience, the first centred on the subject, the second decentralized, organized on and open to a hierarchy of possible obviousness.[70] On the one hand, concept takes on the Kantian transcendental function: it allows us 'to establish the conditions of possibility for considering as objects the entities to which it relates'.[71] On the other hand, concept can itself be thematized (in the sense of the process of thematization discussed earlier) as a higher level object. Thus, the different actualizations of concept are 'categorial outlines' of the object. Therefore the progression of the transformation of the concept is projected onto the transcendental plane, of which the vanishing point is intuition itself; but here, it is Husserl's phenomenology that is entailed, that is to say a perspective that, despite its leaning towards the object, reintroduces the transcendental subject, in so far as it sets out the rules of knowledge and turns them into objects. In fact, Granger dismisses only the intimist aspect of phenomenology, which bogged down in perception and affection and knows nothing of science. Conscious experience is simply mistaken if it presents itself as a prototype of *all* experience and creates an illusion of stability and centricity, when scientific experience shows on the contrary that the subject is not the centre of the world. In the end, however decentralized objective thought might be, the transcendental subject reappears as a founder. Only, it is the subject of language, and not the subject of consciousness. An objectivized subject, so to speak. In fact language is, according to Granger, a 'store of forms, and the only thing responsible for their organization into a system'.[72] That is why it is necessary to root the forms of objectivity not in perception but in expression and to substitute for the aesthetic the semiotic, which sets up the 'general conditions of symbolic thought' and institutes the constituent categories of the object.

Desanti went another way, that of 'deconstructionalist philosophy'. If concepts were undeniably 'science's gift of a network', subject, history and work considered as entities were illusory, chimeras. Philosophy should undo them as units, break

---

[69] *Pour la connaissance philosophique*, pp. 235–239.

[70] *Pensée formelle et sciences de l'homme*, chapitre 6.

[71] Le transcendantal et le formel en mathématiques, in *Formes, Opérations, Objets*, chapitre 9, p. 150.

[72] On the idea of the mathematical concept 'natural', *Formes, Opérations, Objets*, chapitre 10.

the chimeras in order to approach mathematics as a cultural phenomenon.[73] In particular, it is necessary to decode the language and try to discover what expressions like 'the development of ideas', 'chain-sequence', 'conceptual necessity', 'the field of consciousness', and so on really mean. This decoding, which Desanti initiated without developing the work in a systematic way, led him, in principle, to the destruction of the subject and all phenomenological reasoning. In a way entirely original among supporters of historical epistemology, he further suggested the likely impossibility of comprehending the connection between historical contingence and necessary sequences. With Archimedes, Leibniz and Riemann on one side, quadratures, definite integrals and Riemann sums on the other, can we hope to do any better than *acknowledge* (empirically) the fact of the coexistence of these two distinct orders of reality? Desanti accepted what neither Cavaillès nor Granger was willing to do: to give up the principle of necessary reason.

---

[73] *Les idéalités mathématiques*, p. 11.

*Relative Consistency and Accessible Domains**

WILFRIED SIEG

... weil Nichts in der Mathematik
gefährlicher ist, als ohne genügenden
Beweis Existenzen anzunehmen ...[1]

Introduction

The goal of Hilbert's program – to give consistency proofs for analysis and set theory within finitist mathematics – is unattainable; the program is dead. The mathematical instrument, however, that Hilbert invented for attaining his programmatic aim is remarkably well: *proof theory* has obtained important results and pursues fascinating logical questions; its concepts and techniques are fundamental for the mechanical search and transformation of proofs; and I believe that it will contribute to the solution of classical mathematical problems.[2] Nevertheless, we may ask ourselves, whether the results of proof theory are significant for the foundational concerns that motivated Hilbert's program and, more generally, for a reflective examination of the nature of mathematics.

---

* This chapter was originally published in Synthese 84, 1990, pp. 259–297, an essay that in turn was a much-expanded version of *Relative Konsistenz* – written in German and published in (Börger, 1987). That collection of papers was dedicated to the memory of Dieter Rödding, my first logic teacher.

The text of the Synthese paper is essentially unchanged, except for the incorporation of some of the (still numerous) footnotes. In the meantime much illuminating historical research has been carried out and many significant mathematical results have been obtained. Some of these developments are reflected in four papers I have since published and that are most closely related to central issues in this essay; the references to those are found at the end of the references.

Translations in this chapter are my own, unless texts are taken explicitly from English editions. In the notes, some quotations that are not central to my arguments are given only in the original German.

[1] From Dedekind's letter to Lipschitz of July 27, 1876, published in (Dedekind, 1932), p. 477.

[2] Finally, there are real beginnings; see (Luckhardt, 1989).

The results I alluded to establish the consistency of *classical* theories relative to *constructive* ones and give in particular a constructive foundation to mathematical analysis. They have been obtained in the pursuit of a *reductive program* that provides a coherent scheme for metamathematical work and is best interpreted as a far-reaching generalization of Hilbert's program. For philosophers these definite mathematical results (should) present a profound challenge. To take it on means to explicate the reductionist point of constructive relative consistency proofs; the latter are to secure, after all, classical theories on the basis of more elementary, more evident ones. I take steps towards analysing the precise character of such implicitly epistemological reductions and thus towards answering the narrow part of the above question. But these steps get their direction from a particular view on the question's wider part.

As background for that view, I point to striking developments within mathematics, namely to the emergence of set-theoretic foundations, particularly for analysis, and to the rise of modern axiomatics with a distinctive structuralist perspective. These two developments overlap, and so do the problems related to them. Indeed, they came already to the fore in Dedekind's work and in the controversy surrounding it.[3] They were furthered by Hilbert's contributions to algebraic number theory and the foundations of geometry; the difficult issues connected with them prompted his foundational concerns during the late 1890s. Hilbert's program, though formulated only in the 1920s, can be traced to this earlier 'problematic'. I argue that it was meant to mediate between broad foundational conceptions and to address related, but quite specific methodological problems. An example of the latter is the use of 'abstract' (analytic) means in proofs of 'concrete' (number theoretic) results: the program – in its instrumentalist formulation – attempts to exploit the formalizability of mathematical theories for a systematic and philosophically decisive solution.

This instrumentalist aspect, as a matter of fact equivalent to the program's consistency formulation, has been overemphasized in the literature and leaves unaccounted-for critical features of Hilbert's thought. The historical part of this chapter brings into focus such neglected features and sets the stage for an analysis of proof-theoretic reductions as *structural* ones. The philosophical significance of relative consistency results is viewed in terms

---

[3] Dedekind played a significant role in the development of nineteenth-century mathematics. As far as our century is concerned I mention his influence on Hilbert and Emmy Noether, thus on Bourbaki's conceptions. An illuminating analysis of Dedekind's work is given in (Stein, 1988); the major influences on Bourbaki are documented in (Dieudonné, 1970). In (Zassenhaus, 1975) one finds on p. 448 the remark: '... we can see in Dedekind more than in any other single man or woman the founder of the conceptual method of mathematical theorization in our century. The new generation of mathematicians ... after the First World War realized in full detail Dedekind's self-confessed desire for conceptual clarity not only in the foundations of number theory, ring theory and algebra, but on a much broader front, in all mathematical disciplines.'

of the *objective underpinnings* of theories to which reductions are (to be) achieved.[4] The elements of *accessible domains* that provide such underpinnings have a unique build-up through basic operations from distinguished objects; the theories formulate principles that are evident – given an understanding of the build-up and a minimalist delimitation of the domain. But note that (i) the objects in accessible domains need not be constructive in any traditional sense: certain segments of the cumulative hierarchy will be seen to be accessible, and (ii) the restriction of logical principles used is not central: the theories of interest turn out to be such that the consistency of their classical versions is established easily relative to their intuitionistic versions (by finitist arguments).

Even in mathematical practice relative consistency proofs are prompted by epistemological concerns. One wants to guarantee the coherence of a complex (new) theory in terms of comprehended notions and does so frequently by devising suitable models. This general goal is pursued, e.g. when Euclidean models for non-Euclidean geometries are given. Proof-theoretic reductions have two special features: (i) they focus on the deductive apparatus of theories, and (ii) they are carried out within theories that have to measure up to restrictive epistemological principles. The latter are traditionally of a more or less narrow 'constructivist' character. In broadening the range of theories to 'quasiconstructive' ones and concentrating on one central feature, namely accessibility, we will be able to evaluate their (relative) epistemological merits. And in this way, it seems to me, we can gain a deepened understanding of what is *characteristic of and possibly problematic* in classical mathematics and of what is *characteristic of and taken for granted as convincing* in constructive mathematics.

In the current discussion, some do as if an exclusive alternative between platonism and constructivism had emerged from the sustained mathematical and philosophical work on foundations for mathematics; others do as if this work were deeply misguided and did not have any bearing on our understanding of mathematics. Both attitudes prevent us from using the insights (of pre-eminent mathematicians) that underly such work and the significant results that have been obtained. They also prevent us from turning attention to central tasks; namely, to understand the role of abstract structures in mathematical practice and the function of (restricted) accessibility notions in 'foundational' theories or 'methodical frames', to use Bernays's terminology. I attempt to give a perspective that includes traditional concerns, but that allows – most importantly – to ask questions transcending traditional boundaries. This perspective is deeply influenced by the writings of Paul Bernays.

---

[4] The reduced theories have to be mathematically significant. Indeed, the consistency program has been accompanied from its inception by work intended to show that the theories permit the formal development of substantial parts of mathematics.

1 Mathematical Reflections

These are concerned with mathematical analysis and theories in which its practice can be formally represented. So I start out by describing attempts to clarify the very object of analysis and thus, it was assumed, the role of analytic methods in number theory. These attempts came under the headings *arithmetization of analysis* and *axiomatic characterization of the real numbers*. I discuss two kinds of arithmetizations put forward by Dedekind and Kronecker, respectively. Dedekind proceeded axiomatically and sought to secure his characterization by a consistency proof relative to logic broadly conceived, whereas Kronecker insisted on a radical restriction of mathematical objects and methods. (Dedekind's arithmetization of analysis should perhaps be called *set-theoretic* and Kronecker's by contrast *strict*.) Hilbert's axiomatization of the real numbers grew directly out of Dedekind's and was the basis for two proposals to overcome at least for analysis the set-theoretic difficulties that had been discovered around the turn of the century. The second proposal, when suitably amended by the formalist conception of mathematics, led to Hilbert's program.

1.1 Consistent Sets

A systematic arithmetization is to achieve, Dirichlet demanded, that *any* theorem of algebra and higher analysis can be formulated as a theorem about natural numbers.[5] If that had been clearly so, Dirichlet's introduction of analytic methods to prove his famous theorem on arithmetic progressions would have been methodologically innocuous. But in using properties of 'continuous magnitudes' to prove facts concerning natural numbers, he pushed aside a traditional, partly epistemologically motivated boundary.[6] Dirichlet himself remarks: 'The method I employ seems to me to merit attention above all by the connection it establishes between the infinitesimal analysis and the higher arithmetic...'[7] In another paper that explores further uses of analytic methods in number theory he writes: '... I have been led to investigate a large number of questions concerning numbers from an entirely new point of view, that attaches itself to the principles of infinitesimal analysis and to the remarkable properties of a class of infinite series and infinite products...' The significance of these methodological innovations can be fathomed from remarks such as Kummer's, who compares them in his eulogy on Dirichlet to Descartes' 'applications of analysis to geometry', or Klein's, who stated that they gave 'direction to the entire further development of number theory'.

---

[5] That is reported in the preface to (Dedekind, 1888).

[6] I allude, of course, to Gauss's attitude; compare (Sieg, 1984), p. 162.

[7] (Dirichlet, 1838), p. 360, respectively (Dirichlet, 1839/40), p. 411.

The essays of Dedekind and Kronecker[8] seek an arithmetization satisfying Dirichlet's demand, but proceed in radically different ways. Kronecker admits as objects of analysis only natural numbers and constructs from them (in now well-known ways) integers and rationals. Even algebraic reals are introduced, since they can be isolated effectively as roots of algebraic equations. The general notion of irrational number, however, is rejected in consequence of two restrictive methodological requirements to which mathematical considerations have to conform: (i) concepts must be decidable, and (ii) existence proofs must be carried out in such a way that they present objects of the appropriate kind. For Kronecker there can be no infinite mathematical objects, and geometry is banned from analysis even as a motivating factor. Clearly, this procedure is strictly arithmetic, and Kronecker believes that following it analysis can be re-obtained. In (Kronecker, 1887) we read:

> 'I believe that we shall succeed in the future to 'arithmetize' the whole content of all these mathematical disciplines [including analysis and algebra]; i.e. to base it [the whole content] on the concept of number taken in its most narrow sense...'

Kronecker did prove, to his great pleasure, Dirichlet's theorem on arithmetic progressions satisfying his restrictive conditions.[9] But it is difficult for me to judge to what extent Kronecker pursued a program of developing (parts of) analysis systematically. In any event, such a program is not chimerical: from mathematical work during the last decade it has emerged that a good deal of analysis and algebra can indeed be done in conservative extensions of primitive recursive arithmetic.[10] Finally, let me mention that Kronecker begins the paper by hinting at his philosophical position – through quoting Gauss on the epistemologically special character of the laws for natural numbers; only these laws, in contrast to those of geometry, carry the complete conviction of their necessity and thus of their absolute truth.

Dedekind, a student of Gauss, emphasized already in his Habilitationsvortrag of 1854 a quite different and equally significant aspect of mathematical experience; namely, the introduction and use of new concepts to grasp composite phenomena that are being governed by the old notions only with great difficulty.[11] Referring

---

[8] Dedekind's relevant papers are the essays *Stetigkeit und irrationale Zahlen* (1872) and *Was sind und was sollen die Zahlen?* (1888); his letters to Lipschitz and Weber are also of considerable interest and were published in (Dedekind, 1932). As to Kronecker I refer to his *Über den Zahlbegriff* and Hensel's introduction to (Kronecker, 1901).

[9] (Kronecker, 1901), p. 11.

[10] It is most plausible that such work would be enriched by paying attention to Kronecker's. For references to the contemporary mathematical work see (Simpson, 1988). As PRA is certainly a part of number theory unproblematic even for Kronecker, this work can be seen as a partial realization of 'Kronecker's program' (and not, as it is done by Simpson, of Hilbert's).

[11] Dedekind mentions that Gauss approved of the 'Absicht' of his talk. Kneser reports in his *Leopold Kronecker*, Jahresbericht der DMV, 33, 1925, that Dedekind referred often to a remark of Gauss that (for a particular number theoretic problem) notions are more important than notations. In pointing to

to this earlier talk, Dedekind asserts in the preface to his (1888) that most of the great and fruitful advances in mathematics have been made in exactly this way. He gives, in contrast to Kronecker, a general definition of reals: cuts are explicitly motivated in geometric terms, and infinite sets of natural numbers are used as respectable mathematical objects. Kronecker's methodological restrictions are opposed by him, in particular the decidability of concepts; he believes that it is determined independently of our knowledge, whether an object does or does not fall under a concept. In this way Dedekind defends general features of his work in the foundations of analysis and in algebraic number theory.[12] But, the reader may ask, how does Dedekind secure the existence of mathematical objects? To answer this question I examine Dedekind's considerations for real and natural numbers.

The principles underlying the definition of cuts are for us set-theoretic ones, for Dedekind they belong to logic[13]: they allow — as Dedekind prefers to express it — the creation of new numbers, such that their system has 'the same completeness or ... the same continuity as the straight line'. Dedekind emphasizes in a letter to Lipschitz that the *stetige Vollständigkeit* (continuous completeness) is essential for a scientific foundation of the arithmetic of real numbers, as it relieves us of the necessity to assume in analysis *existences without sufficient proof*. Indeed, it provides the answer to his own rhetorical question:

> How shall we recognize the admissible existence assumptions and distinguish them from the countless inadmissible ones...? Is this to depend only on the success, on the accidental discovery of an internal contradiction?[14]

---

the 'Gaussian roots' of Dedekind's and Kronecker's so strikingly different positions, I want to emphasize already here that they can (and should) be viewed as complementary.

[12] Kronecker spurned Dedekind's algebraic conceptions. See, e.g., the note on p. 336 of his *Über einige Anwendungen der Modulsysteme*, Journal für Mathematik, 1886, and Dedekind's gentle rejoinder in, what else, a footnote of his (1888): 'but to enter into a discussion [of such restrictions] seems to be called for only when the distinguished mathematician will have published his reasons for the necessity or even just the advisability of these restrictions.' Kronecker expressed his views quite drastically in letters; for example in a letter to Lipschitz of August 7, 1883 he writes: 'Bei dieser Gelegenheit habe ich das lange gesuchte Fundament meiner ganzen Formentheorie gefunden, welches gewissermassen 'die Arithmetisierung der Algebra' – nach der ich ja das Streben meines mathematischen Lebens gerichtet habe – vollendet, und welches zugleich mir mit Evidenz zeigt, dass auch umgekehrt die Arithmetik dieser 'Association der Formen' nicht entbehren kann, dass sie ohne deren Hülfe nur auf Irrwege geräth oder sich Gedankengespinste macht, die wie die Dedekindschen, die wahre Natur der Sache mehr zu verhüllen als zu klären geeignet sind.' (Lipschitz, 1986), pp. 181–182.

[13] Why then 'arithmetization'? Dedekind views cuts as 'purely arithmetical phenomena'; see the preface to (1872) or (1888), where Dedekind talks directly about the 'rein arithmetische Erscheinung des Schnitts'. In the latter work he immediately goes on to pronounce arithmetic as a part of logic: 'By calling arithmetic (algebra, analysis) only a part of logic I express already that I consider the concept of number as completely independent of our ideas or intuitions of space and time, that I view it rather as an immediate outflow from the pure laws of thought.' (Dedekind, 1932), p. 335. – The next three references in this paragraph are to (Dedekind, 1932), namely, p. 321, p. 472, and p. 477, respectively.

[14] Letter to Lipschitz of July 27, 1876; in (Dedekind, 1932), p.477.

Dedekind is considering assumptions that concern the existence of individual real numbers; such assumptions are not needed, when we are investigating a complete system – ein denkbar vollständigstes Größen-Gebiet. By way of contrast, and in defense against the remark that all of his considerations are already contained in Euclid's *Elements*, he notices that such a complete system is not underlying the classical work. The definition of proportionality is applied only to those (incommensurable) magnitudes that occur already in Euclid's system and whose existence is evident for good reasons. And he argues in this letter of 1876 and later in the preface to *Was sind und was sollen die Zahlen?* that the algebraic reals form already a model of Euclid's presentation. For Euclid, Dedekind argues, that was sufficient, but it would not suffice, if arithmetic were to be founded on the very concept of number as proportionality of magnitudes.[15]

The question as to the existence of particular reals has thus been shifted to the question as to the existence of their complete system. If we interpret the essay on continuity in the light of considerations in *Was sind und was sollen die Zahlen?* and Dedekind's letter to Keferstein, we can describe Dedekind's procedure in a schematic way as follows. Both essays present first of all informal analyses of basic notions, namely of continuity by means of *cuts* (of points on the straight line and rationals, respectively) and of natural number by means of the components *system, distinguished object 1,* and *successor operation*. These analyses lead with compelling directness to the definitions of a complete, ordered field and of a simply infinite system. Then – in our terminology – models for these axiom systems are given. In *Stetigkeit und irrationale Zahlen* the system of all cuts of rationals is shown to be (topologically) complete and, after the introduction of the arithmetic operations, to satisfy the axioms for an ordered field. The parallel considerations for simply infinite systems in *Was sind und was sollen die Zahlen?* are carried out more explicitly. Dedekind gives in Section 66 of that essay his 'proof' of the existence of an infinite system. Such systems contain a simply infinite (sub-) system, as is shown in Section 72.

Dedekind believes to have given purely logical proofs for the existence of these systems and thus to have secured the consistency of the axiomatically characterized notions.[16] With respect to simply infinite systems he writes to Keferstein in a letter of February 27, 1890:

> After the essential nature of the simply infinite system, whose abstract type is the number sequence N, had been recognized in my analysis ..., the question arose: does such a system *exist* at all in the realm of our

---

[15] (1932), pp. 477–8, in particular top of p. 478.

[16] That such a proof is intended also in *Stetigkeit und irrationale Zahlen* is most strongly supported by the discussion in (Dedekind, 1888), p. 338. – The Fregean critique of Dedekind in section 139 of *Grundgesetze der Arithmetik*, vol. II, is quite misguided. For a deeper understanding of Dedekind's views on creation (Schaffung) of mathematical objects see also his letter of January 24, 1888 to H. Weber in (Dedekind, 1932), p. 489. That, incidentally, anticipates and resolves Benacerraf's dilemma in *What numbers could not be*.

ideas? Without a logical proof of existence it would always remain doubtful whether the notion of such a system might not perhaps contain internal contradictions. Hence the need for such a proof (articles 66 and 72 of my essay).[17]

I emphasize that Dedekind views these considerations not as specific for the foundational context of the essays analysed here, but rather as paradigmatic for a general mathematical procedure, when abstract, axiomatically characterized notions are to be introduced. That is unequivocally clear, e.g. from his discussions of ideals in (Dedekind, 1877a), where he draws direct parallels to the steps taken here.[18] The particular constructions leading to the general concept of real number provide an arithmetization of analysis: they proceed, as Dedekind believes, solely within logic and thus purely arithmetically (cf. footnote 13). Their specific logical character implies almost trivially that Dirichlet's demand is satisfied; any analytic statement can be viewed as (a complicated way of making) a statement concerning natural numbers. But Dedekind states, that it is nothing meritorious 'to actually carry out this tiresome re-writing (mühselige Umschreibung) and to insist on using and recognizing only the natural numbers'.

The very beginnings of the Hilbertian program can be traced back to these foundational problems in general and to Dedekind's proposed solution in particular. Hilbert turned his attention to them, as he recognized the devastating effect on Dedekind's essays of observations that Cantor communicated to him in letters, dated September 26 and October 2, 1897.[19] Cantor remarks there that he was led 'many years ago' to the necessity of distinguishing two kinds of totalities (multiplicities, systems); namely, *absolutely infinite* and *completed* ones. Multiplicities of the first kind are called *inconsistent* in his famous letter to Dedekind of July 28, 1899, and those of the second kind *consistent*. Only consistent multiplicities are viewed as sets, i.e. proper objects of set theory. This distinction is to avoid, and does so in a trivial way, the contradictions that arise from assuming that the multiplicity of all things (all cardinals, or all ordinals) forms a set.

In 1899 Hilbert writes *Über den Zahlbegriff*, his first paper addressing foundational issues of analysis. He intends – never too modest about aims – to rescue the set-theoretic arithmetization of analysis from the Cantorian difficulties. To this end he gives a categorical axiomatization of the real numbers following Dedekind's work in *Stetigkeit und irrationale Zahlen*. He claims that its consistency can be proved by a 'suitable modification of familiar methods'[20] and remarks that such a

---

[17] In (van Heijenoort, 1967), p. 101. The essay Dedekind refers to is (Dedekind, 1888).

[18] (Dedekind, 1877a), pp. 268–269; in particular the long footnote on p. 269.

[19] In particular in Section 66 of *Was sind und was sollen die Zahlen?*. That is clear from Cantor's response of November 15, 1899 to a letter of Hilbert's (presumably not preserved). Cantor's letter is published in (Purkert and Ilgauds, 1987), p. 154. See also remark A in Section 1.3.

[20] (Hilbert, 1900), p. 261. The German original is: 'Um die Widerspruchsfreiheit der aufgestellten Axiome zu beweisen, bedarf es nur einer geeigneten Modifikation bekannter Schlußmethoden.' (Bernays, 1935b) reports on pp. 198–199 in very similar words, but with a mysterious addition: 'Zur

proof constitutes 'the proof for the existence of the totality of real numbers or – in the terminology of G. Cantor – the proof of the fact that the system of real numbers is a consistent (completed) set'. In his subsequent Paris address Hilbert goes even further, claiming that the existence of Cantor's higher number classes and of the alephs can also be proved. Cantor, by contrast, insists in a letter to Dedekind, written on August 28, 1899, that even finite multiplicities cannot be proved to be consistent. The fact of their consistency is a simple, unprovable truth – 'the axiom of arithmetic'; and the fact of the consistency of those multiplicities that have an aleph as their cardinal number is in exactly the same way an axiom, the 'axiom of the extended transfinite arithmetic'.[21]

Hilbert recognized soon that his problem, even for the real numbers, was not as easily solved as he had thought. Bernays writes in his (1935b) on p. 199, 'When addressing the problem [of proving the above consistency claims] in greater detail, the considerable difficulties of this task emerged'. It is the realization, I assume, that distinctly new principles have to be accepted; principles that cannot be pushed into the background as 'logical' ones.[22] Dedekind's arithmetization of analysis has not been achieved without 'mixing in foreign conceptions', after all;[23] a rewriting, however tiresome, of analytic arguments in purely number-theoretic terms is seemingly not always possible.

1.2 Consistent Theories

In his address to the International Congress of Mathematicians, Heidelberg 1904, Hilbert examines again and systematically various attempts of providing foundations for analysis, in particular Cantor's. The critical attitude towards Cantor that was implicit in *Über den Zahlbegriff* is made explicit here. Hilbert accuses Cantor of not giving a rigorous criterion for distinguishing consistent from inconsistent multiplicities; he thinks that Cantor's conception on this point 'still leaves latitude for subjective judgment and therefore affords no objective certainty'. He suggests again that consistency proofs for suitable axiomatizations provide an appropriate

---

Durchführung des Nachweises gedachte Hilbert mit einer geeigneten Modifikation der in der Theorie der reellen Zahlen angewandten Methoden auszukommen.'

[21] (Cantor, 1932), p. 447–448.

[22] This general concern comes out in Husserl's notes on a lecture that Hilbert gave to the Göttingen Mathematical Society in 1901 and, in very similar terms, in (Hilbert, 1904), p. 266; Husserl's notes are quoted in full in Wang's *Reflections on Kurt Gödel*, Cambridge, 1987, p. 53.

[23] Dedekind points out emphatically, e.g. in the letter to Lipschitz (1932, p. 470) and in the introduction to (1888), that his constructions do not appeal anywhere to 'fremdartige Vorstellungen'; he has in mind appeals to geometric ones. – Bernays has again and again made the point that a 'restlose strikte Arithmetisierung' cannot be achieved. In (Bernays, 1941) one finds on p. 152 the remark: '... one can say – and that is certainly the essence of the finitist and intuitionist critique of the usual mathematical methods – that the arithmetization of geometry in analysis and set theory is not a complete one.' It is through the powerset of the set of natural numbers that our geometric conception of the continuum is connected to our elementary conception of number; e.g. in: *Bemerkungen zu Lorenzen's Stellungnahme in der Philosophie der Mathematik*, 1978.

remedy, but proposes a radically new method of giving such proofs: develop logic (still vaguely conceived) together with analysis in a common frame, so that proofs can be viewed as *finite* mathematical objects; then show that such formal proofs cannot lead to a contradiction. Here we have, seemingly in very rough outline, Hilbert's program as developed in the 1920s; but notice that the point of consistency proofs is still to guarantee the existence of sets, and that a reflection on the mathematical means admissible in such proofs is lacking completely. Before describing the later program, let me mention that this address and *Über den Zahlbegriff* are squarely directed against Kronecker. In his Heidelberg address Hilbert claims that he has refuted Kronecker's standpoint – by partially embracing it, as I hasten to add. I will explain below that this is by no means paradoxical. Indeed, a genuine methodological shift had been made; Bernays remarks that Hilbert started, clearly before giving this address, 'to do battle with Kronecker with his own weapons of finiteness by means of a modified conception of mathematics'.[24]

There are a number of general tendencies that influenced the Heidelberg address and the further development towards Hilbert's program. First of all, the *radicalization of the axiomatic method*; by that I mean the insight that the linguistic representation of a theory can be viewed as separable from its content or its intended interpretation. That was clear to Dedekind, was explicitly used by Wiener, and brought to perfection by Hilbert in his *Grundlagen der Geometrie*.[25] Secondly, the *instrumentalist view of (strong mathematical) theories*; the earliest explicit formulation I know of is due to Borel discussing the value of abstract, set-theoretic arguments from a Kroneckerian perspective.

> One may wonder what is the real value of these [set-theoretic] arguments that I do not regard as absolutely valid but that still lead ultimately to effective results. In fact, it seems that if they were completely devoid of value, they could not lead to anything ... This, I believe, would be too harsh. They have a value analogous to certain theories in mathematical physics, through which we do not claim to express reality but rather to have a guide that aids us, by analogy, in predicting new phenomena, which must then be verified.

Can one systematically explore, Borel asks, the sense of such arguments. His answer is this:

> It would require considerable research to learn what is the real and precise sense that can be attributed to arguments of this sort. Such research would be useless, or at least it would require more effort than it would be worth.

---

[24] The quotation is taken from a longer remark of Bernays in (Reid, 1970). It is preceded by: 'Under the influence of the discovery of the antinomies in set theory, Hilbert temporarily thought that Kronecker had probably been right there. *(That is, right in insisting on restricted methods.)* But soon he changed his mind. Now it became his goal, one might say, to do battle with ...'

[25] Dedekind describes, on p. 479 of (1932), such a separation before making the claim that the algebraic reals form a model of the Euclidean development. For a penetrating discussion of the general development see (Guillaume, 1985). Such a separation appears to us as banal, but it certainly was not around the turn of the century, as the Frege–Hilbert controversy amply illustrates.

> How these overly abstract arguments are related to the concrete becomes clear when the need is felt.[26]

To grapple with this problem clearly one has to use, thirdly, the *strict formalization of logic* that had been achieved by Frege (Peano, and Russell/Whitehead). That is a moment not yet appreciated in Hilbert's Heidelberg address, where one finds a discussion of logical consequence (Folgerung) quite uninformed by this crucial aspect of Frege's work. Hilbert succeeded to join these tendencies into a sharply focused program with a very special mathematical and philosophical perspective.

The *modified conception of mathematics* underlying the formulation of the program is characterized by Hilbert in the 1920s most pointedly and polemically: classical mathematics is a *formula game* that allows 'to express the whole thought content of mathematics in a uniform way'; its consistency has to be established within finitist mathematics, however. Finitist mathematics is taken to be a philosophically unproblematic part of number theory and, in addition, to coincide with the part of mathematics accepted by Kronecker and Brouwer.[27] Not every formula of this 'game' has a meaning but only those that correspond to finitist statements, i.e. universal sentences of the kind of Fermat's Theorem. For a precise description of the role of consistency proofs let **P** be a formal theory that allows the representation of classical mathematical practice and let **F** formulate the principles of finitist mathematics. The consistency of **P** is in **F** equivalent to the reflection principle

$$(\forall x)(\Pr(x, \text{'s'}) \Rightarrow s).$$

Pr is the finitistically formulated proof predicate for **P**, s a finitist statement, and 's' the corresponding formula in the language of **P**. A consistency proof in **F** was programmatically sought; it would show, because of the above equivalence, that the mere technical apparatus **P** can serve reliably as an instrument for the proof of finitist statements. After all, the consistency proof would allow to transform any **P**-derivation of 's' into a finitist proof of s (and thus give a quite systematic answer to Borel's question). Hilbert believed that consistency proofs would settle

---

[26] (Baire et al., 1905), p. 273. A striking, but different suggestion along these lines was made already in (Cantor, 1883), p. 173: 'If, as is assumed here [i.e. from a restrictive position], only the natural numbers are real and all others just relational forms, then it can be required that the proofs of theorems in analysis are checked as to their 'number-theoretic content' and that every gap that is discovered is filled according to the principles of arithmetic. The feasibility of such a supplementation is viewed as the true touchstone for the genuineness and complete rigor of those proofs. It is not to be denied that in this way the foundations of many theorems can be perfected and that also other methodological improvements in various branches of analysis can be effected. Adherence to the principles justified from this viewpoint, it is believed, secures against any kind of absurdities or mistakes.' This is in a way closer to Hilbert's belief that finitist statements must admit a finitist proof. That belief is implicitly alluded to in (Bernays, 1941), p. 151: 'The hope that the finitist standpoint (in its original sense) could suffice for all of proof theory was brought about by the fact that the proof-theoretic problems could be formulated from that point of view.'

[27] See remark B in Section 1.3. below.

foundational problems – once and for all and by purely mathematical means. Bernays judged in (1922a), p. 19:

> The great advantage of Hilbert's method is precisely this: the problems and difficulties that present themselves in the foundations of mathematics can be transferred from the epistemological-philosophical to the properly mathematical domain.

Because of Gödel's incompleteness theorems this advantage proved to be illusory, at least when finitist mathematics is contained in **P**:[28] for such **P**s the Second Incompleteness Theorem just states that their consistency cannot be established by means formalizable in **P**. The radical restriction of what was 'properly mathematical' had to be given up; a modification of the program was formulated and has been pursued successfully for parts of analysis.[29] The crucial tasks of this general reductive program are: (i) find an appropriate formal theory **P*** for a significant part of classical mathematical practice, (ii) formulate a 'corresponding' constructive theory **F***, and (iii) prove in **F*** the partial reflection principle for **P***, i.e.

$$Pr^*(d, \text{'s'}) => s$$

for each **P***-derivation d. Pr* is here the proof-predicate of **P*** and s an element of some class F of formulas. The provability of the partial-reflection principle implies the consistency of **P*** relative to **F***. (For the theories considered here, this result entails that **P*** is conservative over **F*** for all formulas in F.) Gödel and Gentzen's consistency proof of classical number theory relative to Heyting's formalization of intuitionistic number theory was the first contribution to the reductive program; as a matter of fact, their result made that program at all plausible.

I do not intend to sketch the development of proof theory and, consequently, I will comment only on some central results concerning theories for the mathematical continuum. Second-order arithmetic was taken by Hilbert and Bernays as the formal framework for analysis. The essential set-theoretic principles are the comprehension principle

$$(\exists X)(\forall y)(y \varepsilon X <=> S(y))$$

and forms of the axiom of choice

$$(\forall x)(\exists Y)S(x, Y) => (\exists Z)(\forall x)S(x, (Z)_x);$$

---

[28] And that is a more than plausible assumption for those **P**s Hilbert wanted to investigate and that contain elementary number theory. Consider the practice of finitist mathematics, for example in volume I of *Grundlagen der Mathematik*, the explicit remarks on p. 42 of that book, but also the analyses given by (Kreisel, 1965) and (Tait, 1981).

[29] Hilbert and Bernays, Gentzen, Lorenzen, Schütte, Kreisel, Feferman, Tait, and many other logicians and mathematicians have contributed; for detailed references to the literature see (Buchholz, et al., 1981) or (Sieg, 1985).

S is an arbitrary formula of the language and may in particular contain set quantifiers. These general principles are impredicative, as the sets X and Z whose existence is postulated are characterized by reference to all sets (of natural numbers). Subsystems of second-order arithmetic can be defined by restricting S to particular classes of formulas. The subsystems that have been proved consistent contain for example the comprehension principle for Π_1^1- and Δ_2^1-formulas; the latter have the shape $(\forall X)R$, respectively, are provably equivalent to formulas of the shape $(\forall X)(\exists Z)R$ and $(\exists Z)(\forall X)T$, where R and T are purely arithmetic.[30] These particular subsystems are of direct mathematical interest, as analysis can be formalized in them by (slightly) refining the presentation of Hilbert and Bernays in supplement IV of *Grundlagen der Mathematik II*. The proof-theoretic investigations have been accompanied by mathematical ones, showing that even weaker subsystems will do. Really surprising refinements have been obtained during the last fifteen years: all of classical analysis can be formalized in conservative extensions of elementary number theory, significant parts also of algebra already in conservative extensions of primitive recursive arithmetic.[31] These two complexes of results indicate corresponding complexes of problems for future development; namely, (1) to give constructive consistency proofs of stronger subsystems of analysis, first of all for the system with Π_2^1-comprehension, and (2) to find weaker, but mathematically still significant subsystems (whose consistency is easily seen from the finitist standpoint and) whose provably recursive functions are in complexity classes. These are mathematically and logically most fascinating problems.

1.3 Remarks

They are partly of historical, partly of systematic character and concern mostly the axiomatic method underlying Hilbert's program. Though they are intended to ease the transition to the philosophical reflections in the second part of this essay, they are also to defuse some of the widely held misconceptions of the program: e.g. its 'crude formalism' or its *ad hoc* character to serve as a 'weapon' against Brouwer's intuitionism.

A. PARADOXICAL BACKGROUND

The concern with consistent sets and the explicit use of Cantorian terminology in *Über den Zahlbegriff* show clearly that Hilbert was informed about the set-theoretic difficulties Cantor had found and communicated to Dedekind in the famous letter of July 28, 1899. The recently published earlier letters of Cantor's to Hilbert

---

[30] For details concerning the theories with versions of the axiom of choice see (Feferman and Sieg, 1981). The character of the consistency proofs is indicated below. Foundationally significant results are also described in (Feferman, 1988).

[31] For the discussion of these results and detailed references to the literature see (Simpson, 1988) and (Sieg, 1988).

I mentioned above throw light on this background. (They provide also surprising new information on the early history of the set-theoretic paradoxes and on the circumstances surrounding Cantor's letter to Dedekind.) There is no doubt that Hilbert was prompted by these difficulties to think seriously about foundational issues. After all, as I pointed out, he recognized the impact of Cantor's observations on Dedekind's logical foundations of arithmetic presented in *Was sind und was sollen die Zahlen?*. Here I just want to recall Dedekind's reaction to the Cantorian problems, reported in a letter of F. Bernstein to Emmy Noether and published in (Dedekind, 1932). Bernstein had visited Dedekind on Cantor's request in the spring of 1897. The express purpose was to find out what Dedekind thought about the paradox of the system of all things; Cantor had informed Dedekind about it already by letter in 1896. Bernstein reports: 'Dedekind had not arrived yet at a definite position and told me, that in his reflections he almost arrived at doubts, whether human thinking is completely rational.'[32] Strong words from a man as sober and clear headed as Dedekind.

B. ASSUMPTION

How is it possible to reconcile Hilbert's programmatic formalism with his deep trust in the correctness of classical mathematics? – Most easily, when the formal theories of central significance are complete or deductively closed, as the Hilbertians used to say. This completeness assumption is already found in *Über den Zahlbegriff*. Hilbert writes there that the set of real numbers should be thought of as '... a system of things, whose mutual relations are given by the above finite and closed system of axioms, and for which statements have validity, only if they can be deduced from those axioms by means of a finite number of logical inferences'. Hilbert talks about a non-formalized axiomatic theory. But if it is adequately represented by a formal theory **P**, then **P** must naturally be deductively closed. As a matter of fact, it was believed in the Hilbert school – until Gödel's incompleteness theorems became known – that the formalisms for elementary number theory and analysis were complete. For the purpose of obtaining a completeness proof Hilbert suggested in his Bologna address (1928) to reinterpret finitistically the familiar arguments for the categoricity of the Peano-axioms and of his axioms concerning the real numbers. The assumed completeness and the ensuing harmony of provability and truth help understand how Hilbert could take his radical formalist position, in order to simply bypass the epistemological problems

---

[32] (Dedekind, 1932), p. 449. Even six years later, in 1903, Dedekind still had such strong doubts that he did not allow a reprinting of his booklet. In 1911, he consented to a republication and wrote in the preface, 'Die Bedeutung und teilweise Berechtigung dieser Zweifel verkenne ich auch jetzt nicht. Aber mein Vertrauen in die innere Harmonie unserer Logik ist dadurch nicht erschüttert; ich glaube, daß eine strenge Untersuchung der Schöpferkraft des Geistes, aus bestimmten Elementen ein neues Bestimmtes, ihr System zu erschaffen, das notwendig von jedem dieser Elemente verschieden ist, gewiß dazu führen wird, die Grundlagen meiner Schrift einwandfrei zu gestalten.' (Dedekind, 1932), p.343.

associated with the classical infinite structures.[33] – The finitist mathematical basis was thought to be co-extensive with the part of arithmetic accepted by Kronecker and Brouwer. As to Kronecker, Hilbert mentions in his (1931): 'At about the same time [i.e. at the time of Dedekind's (1888)] ... Kronecker formulated most clearly a view, and illustrated it by numerous examples, that essentially coincides with our finitist standpoint.' The relation to intuitionism is discussed explicitly at a great number of places by Bernays; e.g. (Bernays, 1967), p. 502. A particularly concise formulation was given by Johann von Neumann in his *Formalistische Grundlegung der Mathematik*, Erkenntnis 2 (1931), pp. 116–7.

C. DOUBTS

Two mathematicians with quite different foundational views criticized Hilbert's formalism at exactly this point; i.e. they criticized the assumption that parts of mathematics can be represented (completely) by formal theories. The first of them was Brouwer, the second Zermelo.

Brouwer used in his development of analysis infinite proofs and treated them mathematically as well-founded trees. He wrote with respect to them: 'These *mental* mathematical proofs that in general contain infinitely many terms must not be confused with their linguistic accompaniments, which are finite and necessarily inadequate, hence do not belong to mathematics.'[34] He added that this remark contains his 'main argument against the claims of Hilbert's metamathematics'. The well-founded trees of Brouwer's can be viewed as inductively generated sets of sequences of natural numbers; that is the essential claim of the bar-theorem. In the case of the constructive ordinals the inductive generation proceeds by the following rules (on the right-hand side I indicate the graphic representation of the ordinal):

$$0 \varepsilon \mathbf{O}$$

$$\alpha \varepsilon \mathbf{O} \Rightarrow \alpha' \varepsilon \mathbf{O}$$

$$(\forall n) \alpha_n \varepsilon \mathbf{O} \Rightarrow \alpha := \sup \alpha_n \varepsilon \mathbf{O}$$

Notice, it is the bar-theorem together with the continuity principle that implies the fan-theorem and thus the properties of the intuitionistic continuum so peculiar from a classical point of view; e.g. the uniform continuity of all real-valued functions on the closed unit interval.

---

[33] Compare (Hilbert, 1929), pp. 14–15, (Bernays, 1930), pp. 59–60, and the discussion in Section 2.1.

[34] In (Brouwer, 1927), footnote 8, p. 460.

Also, Zermelo claimed that finite linguistic means are inadequate to capture the nature of mathematics and mathematical proof. In a brief note he argued: 'Complexes of signs are not, as some assume, the true subject matter of mathematics, but rather *conceptually ideal relations* between the elements of a conceptually posited *infinite manifold*. And our systems of signs are only *imperfect* and *auxiliary means* of our *finite* mind, changing from case to case, in order to master at least in stepwise approximation the infinite, that we cannot survey *directly* and *intuitively*.'[35] Zermelo suggested using an infinitary logic to overcome finitist restrictions. The concept of well-foundedness is fundamental for Zermelo's infinitary logic as well, but in an unrestricted set-theoretic framework. Zermelo's investigations of infinitary systems can be found in (Zermelo, 1935).

D. REDUCTION

Ironically, the constructive consistency proofs of impredicative theories, mentioned at the end of Section 1.2, use infinitary logical calculi; but the syntactic objects constituting them (namely, formulas and derivations) are treated in harmony with intuitionistic principles. The theories **F*** – in which the infinitary calculi are investigated and to which the impredicative theories are reduced – are extensions of intuitionistic number theory by definition and proof principles for constructive ordinals or other accessible i.d. [inductively defined] classes of natural numbers.[36] The above process of inductive generation for constructive ordinals can be expressed by an arithmetical formula $A(X,x)$. The two crucial principles are

(O1) $(\forall x)(A(\mathbf{O}, x) => \mathbf{O}(x))$, and

(O2) $(\forall x)(A(F, x) => F(x)) => (\forall x)(\mathbf{O}(x) => F(x))$.

The former expresses the closure principle for **O**, the latter the appropriate induction schema for any formula F. These principles and corresponding ones for other inductively defined classes are correct from an intuitionistic point of view; the theories **F*** are based on intuitionistic logic. Because of these facts we can claim that the consistency of the impredicative theories has been established relative to constructive theories.

---

[35] (Zermelo, 1931), p. 85.

[36] See (Buchholz et al., 1981) and for an informal introduction the second part of (Sieg, 1984). – Accessible or deterministic i.d. classes are distinguished by the fact that all their elements have unique construction trees. If the construction trees for all elements of an i.d. class are finite, we say that the class is given by a finitary inductive definition. For a detailed discussion of these notions see (Feferman and Sieg, 1981), pp. 22–23, and Feferman's paper *Finitary inductively presented logics* for Logic Colloquium '88.

2 Philosophical Reflections

The reductions of impredicative subsystems of analysis to intuitionistic theories of higher number classes or other distinguished inductively defined classes are certainly significant results; were not all impredicative definitions supposed to contain vicious circles? The question is, nevertheless, what has been achieved in a general, philosophical way. Gödel remarked that giving a constructive consistency proof for classical mathematics means 'to replace its axioms [i.e. those of classical mathematics] about abstract entities of an objective Platonic realm by insights about the given operations of our mind'.[37] This pregnant formulation gives a most dramatic philosophical meaning to such proofs; it seems to me to be mistaken, however, in its radical opposition of classical and constructive mathematics and even in the very characterization of their subject matters. I prefer to formulate the task of such proofs as follows: they are to relate two aspects of mathematical experience; namely, the impression that mathematics has to do with abstract objects arranged in structures that are independent of us, and the conviction that the principles for some structures are evident, because we can grasp the build-up of their elements. I will argue that this is indeed central to the mediating task of the (modified) Hilbert program. The starting point of my argument is a reanalysis of the reductive goals of the original program; that will lead to the notion of 'structural reduction' and to questions concerning its epistemological point.

2.1 Structural Reduction

The description of Hilbert's program in Section 1.2 brings out, appropriately, the goal of justifying the instrumentalist use of classical theories for the proof of true finitist statements; it captures also important features of Hilbert's approach in a natural way, for example his concern with 'Methodenreinheit' and the method of ideal elements. And yet, it truncates the program by leaving out essential and problematic considerations. Hilbert and Bernays both argue for a more direct mathematical significance of consistency proofs: such proofs are viewed as the last desideratum in justifying the existential supposition of infinite structures made by modern axiomatic theories.[38] It is clearly this concern that links the program to Hilbert's first foundational investigations and to Dedekind's attempted consistency proofs. Dedekind considers consistency proofs also as a last desideratum, but there seems to be a decisive difference as to the nature of theories: for him the theories (of natural and real numbers) are not just formal systems with some instrumentalist

---

[37] (Reid, 1970), p. 218; compare also Gödel's remarks in Hao Wang's *From Mathematics to Philosophy*, London, 1974, pp. 325–326.

[38] 'Existential supposition' is to correspond to the term 'existentielle Setzung' that is used by Hilbert and Bernays as a quasitechnical term. The problem pointed to is presented as a central one in *Grundlagen der Mathematik I*; see the summary of the discussion on p.19 of that work. As to the role of the reflection principle, compare the informative remarks on pp. 43–44.

use. On the contrary, they are contentually motivated, have a materially founded necessity, and mathematical efficacy. They play an important epistemological role by giving us a conceptual grasp of composite mathematical as well as physical phenomena; Dedekind claims, for example, that it is only the theory of real numbers that enables us 'to develop the conception of continuous space to a definite one'.[39]

None of these points are lost in the considerations of Hilbert and Bernays. The contentual motivation of axiom systems, for example, plays a crucial role for them, as is clear from the very first chapter of *Grundlagen der Mathematik I* where the relation between contentual and formal axiomatics ('inhaltliche', respectively 'formale Axiomatik') and its relevance for our knowledge is being discussed. 'Formal axiomatics,' they explain, 'requires contentual axiomatics as a necessary supplement; it is only the latter that guides us in the selection of formalisms and moreover provides directions for applying an already given formal theory to an objective domain'.[40] The basic conviction is that the contentual axiomatic theories are fully formalizable; formalisms, according to Hilbert (1928), provide 'a picture of the whole science'. Bernays (1930) discusses the completeness problem in detail and conjectures that elementary number theory is complete. Though there is 'a wide field of considerable problems', Bernays claims, 'this "problematic" is not an objection against the standpoint taken by us'. He continues, arguing as it were against the doubts of Brouwer's and Zermelo's:

> We only have to realize that the [syntactic] formalism of statements and proofs we use to represent our conceptions does not coincide with the [mathematical] formalism of the structure we intend in our thinking. The [syntactic] formalism suffices to formulate our ideas of infinite manifolds

---

[39] (Dedekind, 1932), p. 340. – The underlying general position is persuasively presented in (Dedekind, 1854). Dedekind viewed it as distinctive for the sciences (not just the natural ones) to strive for 'characteristic' and 'efficacious' basic notions; the latter are needed for the formulation of general truths. The truths themselves have, in turn, an effect on the formation of basic notions: they may have been too narrow or too wide, they may require a change so that they can 'extend their efficacy and range to a greater domain'. Dedekind continues, and that just cannot be adequately translated: 'Dieses Drehen und Wenden der Definitionen, den aufgefundenen Gesetzen und Wahrheiten zuliebe, in denen sie eine Rolle spielen, bildet die größte Kunst des Systematikers'. In mathematics we encounter the same phenomenon, e.g. when extending the definition of functions to greater number domains. In contrast to other sciences, however, mathematics does not leave any room for arbitrariness in how to extend definitions. Here the extensions follow with 'compelling necessity', if one applies the principle that 'laws, that emerged from the initial definitions and that are characteristic for the notions denoted by them, are viewed as generally valid; then these laws in turn become the source of the generalized definitions...' What a marvelous general description of his own later work in algebra (in particular the introduction of ideals) and in his foundational papers, the guiding idea of which is formulated clearly on p. 434 of this very essay.

[40] *Grundlagen der Mathematik I*, p. 2. – I translated by 'an objective domain' the phrase 'ein Gebiet der Tatsächlichkeit'. – Similar remarks can be found in earlier, pre-Gödel papers; see especially the comprehensive and deeply philosophical (Bernays, 1930).

and to draw the logical consequences from them, but in general it cannot combinatorially generate the manifold as it were out of itself.[41]

The close, but not too intimate connection between intended structure and syntactic formalism is to be exploited as the crucial means of reduction. This idea is captured in papers by Bernays through a mathematical image. (The papers are separated by almost fifty years; I emphasize this fact to point out that the remarks are not incidental, but touch the core of the strategy.) The first observation, from 1922, follows a discussion of Hilbert's *Grundlagen der Geometrie*.

> Thus the axiomatic treatment of geometry amounts to this: one abstracts from geometry, given as the science of spatial figures, the purely mathematical component of knowledge [Erkenntnis]; the latter is then investigated separately all by itself. The spatial relations are *projected* as it were into the sphere of the mathematically abstract, where the structure of their interconnection presents itself as an object of purely mathematical thinking and is subjected to a manner of investigation focused exclusively on logical connections.

What is said here for geometry is stated for arithmetic in (Bernays, 1922a) and for theories in general in (Bernays, 1970), where a sketch of Hilbert's program is supplemented by a clear formulation of the epistemological significance of such 'projections'.

> In taking the deductive structure of a formalized theory ... as an object of investigation the [contentual] theory is *projected* as it were into the number theoretic domain. The number theoretic structure thus obtained is in general essentially different from the structure intended by the [contentual] theory. But it [the number theoretic structure] can serve to recognize the consistency of the theory from a standpoint that is more elementary than the assumption of the intended structure.

Recalling that – according to Hilbert – the axiomatic method applies in identical ways to different domains, these projections have a uniform character. Thus Hilbert's program can be seen to seek a *uniform structural reduction*: intended structures are projected through their assumed complete formalizations into the properly mathematical domain (of Kronecker's and Brouwer's), i.e. finitist mathematics. The equivalence of consistency and satisfiability was claimed or at least conjectured;[42] consequently, it seemed that the existence of intended structures

---

[41] (Bernays, 1930), p. 59. The words in brackets were added by me to make the translation as clear as the German original. – The next longer quotations are taken from (Bernays, 1922b), p. 96, and (Bernays, 1970), p. 186.

[42] In (Bernays, 1930), p. 21, one finds the following phrase: 'It is for this reason necessary to prove for every axiomatic theory the *satisfiability*, i.e. the *consistency* of its axioms.' Compare also *Grundlagen der Mathematik I*, p. 19. – Minc and Friedman have shown that Gödel's completeness theorem for predicate logic can be established in a conservative extension of primitive recursive arithmetic. (That is an obvious improvement of Bernays' proof of the completeness theorem in elementary number theory.) The result is of considerable interest in this connection, as it justifies the equivalence of consistency and satisfiability from a finitist point of view – at least for formal first-order theories.

would be secured by the mathematical solution of the purely combinatorial consistency problem. The principles used in the solution were of course to be finitist; the epistemological gain of such reductions is described in *Grundlagen der Mathematik I*:

> Formal axiomatics, too, requires for the checking of deductions and the proof of consistency in any case certain evidences, but with the crucial difference [when compared to contentual axiomatics] that this evidence does not rest on a special epistemological relation to the particular domain, but rather is one and the same for any axiomatics; this evidence is the primitive manner of recognizing truths that is a prerequisite for any theoretical investigation whatsoever.[43]

This reconstruction of the intent of Hilbert's program is supported most explicitly by Bernays (1922a and 1930). Let me focus briefly on the earlier paper, not to report on all its detailed points, but rather to depict the structure of its argumentation. The problem faced by the program is seen in the following way. In providing a rigorous foundation for arithmetic (taken in a wide sense to include analysis and set theory) one proceeds axiomatically and starts out with the assumption of a system of objects satisfying certain structural conditions. But in the assumption of such a system 'lies something so-to-speak transcendental for mathematics, and the question arises, which principled position is to be taken [towards that assumption]'. Bernays considers two 'natural positions'. The first, attributed to Frege and Russell, attempts to prove the consistency of arithmetic by purely logical means; this attempt is judged to be a failure. The second position is seen in counterpoint to the logical foundations of arithmetic: 'As one does not succeed in establishing the logical necessity of the mathematical transcendental assumptions, one asks oneself, is it not possible to simply do without them'. Thus one attempts a constructive foundation, replacing existential assumptions by construction postulates; that is the second position and is associated with Kronecker, Poincaré, Brouwer, and Weyl. The methodological restrictions to which this position leads are viewed as unsatisfactory, as one is forced 'to give up the most successful, most elegant, and most proven methods only because one does not have a foundation for them from a particular standpoint'. Hilbert takes from these foundational positions what is 'positively fruitful': from the first, the strict formalization of mathematical reasoning; from the second, the emphasis on constructions. Hilbert does not want to give up the constructive tendency, but – on the contrary – emphasizes it in the strongest possible terms. The program, as described in Section 1.2, is taken as the tool for an alternative constructive foundation of all of classical mathematics.

It is not the case – as is so often claimed – that the difficult philosophical problems brought out by the axiomatic method and the associated structural view of

---

[43] *Grundlagen der Mathematik I*, p. 2. The parenthetical remark is mine. – Bernays uses the term 'primitive Erkenntnisweise' that I tried to capture by the somewhat unwieldy phrase 'primitive manner of recognizing truths'.

mathematics were not seen. They motivated the enterprise and were seen perfectly clearly; however, it was hoped, perhaps too naively, to either avoid them directly in a systematic-mathematical development (by presenting appropriate models) or to solve them in the case of 'fundamental' structures on the finitist basis. In any case, a so-to-speak absolute epistemological reduction was envisioned. These radical, philosophically motivated aspirations of Hilbert's program were blocked by Gödel's incompleteness theorems: according to the first theorem it is not possible, even in the case of natural numbers, to exclude systematically all contentual considerations concerning the intended structure; the second theorem implies that formal theories can be used at most as vehicles for partial structural reductions to strengthenings of the finitist basis. Bernays wrote in the epilogue to his (1930), p. 61:

> On the whole the situation is like this: Hilbert's proof theory – together with the discovery of the formalizability of mathematical theories – has opened a rich field of research, but the epistemological views that were taken for granted at its inception have become problematic.

At the inception of Hilbert's program, it seems, the epistemological views had not been dogmatically and unshakably fixed. As will be pointed out in the next section, Hilbert's original position had to be and was extended; in addition, to judge from (Bernays, 1922a), the focus on finitist mathematics was viewed as part of an 'Ansatz' to the solution of a problem. Having formulated the question as to a principled position towards the transcendental assumptions underlying the axiomatic foundations of arithmetic (see above), Bernays remarks, p. 11:

> Under this perspective[44] we are going to try, whether it is not possible to give a foundation to these transcendental assumptions in such a way that only primitive intuitive knowledge (primitive anschauliche Erkenntnisse) is used.

Viewing the philosophical position in this more experimental spirit, we can complement the metamathematical reductive program by a philosophical one that addresses two central issues: (i) what is the nature and the role of the reduced structures? and (ii) what is the special character of the theories to which they are reduced? As to the latter issue, our greater metamathematical experience allows us to point to perhaps significant general features.

2.2 Accessible Domains

The reductive program I described in Section 1.2. has been pursued successfully. I think there can be no reasonable doubt that (meta-) mathematically and, prima facie, also philosophically significant solutions have been obtained. As to the

---

[44] of taking into account the tendency of the exact sciences to use as far as possible only the most primitive 'Erkenntnismittel'. That does not mean, as Bernays emphasizes, to deny any other, stronger form of intuitive evidence.

mathematical results it can be observed:

- A considerable portion of classical mathematical practice, including all of analysis, can be carried out in a small corner of Cantor's paradise that is consistent relative to the constructive principles formalized in intuitionistic number theory. This is not trivial, if one bears in mind that in particular for analysis non-constructive principles seemed to be necessary.

The metamathematical results concerning the relative consistency of impredicative theories speak also for themselves.

- The constructive principles formalized in intuitionistic theories for i.d. classes[45] allow us to recognize the relative consistency of certain impredicative theories. This is again not trivial, if one takes into account that any impredicative principle, from a broad constructive point of view, seemed to contain vicious circles.[46]

These relative consistency results provide material for critical philosophical analysis. After all, they raise implicitly the traditional question: 'What is the (special) evidence of the mathematical principles used in (these) consistency proofs?' – The intuitionistic theories for i.d. classes formulate complex principles that are recognized by classical and constructivist mathematicians alike. On the one hand they are more elementary than the principles used in their set-theoretic justification, but on the other hand they cannot be given a direct (primitive) intuitive foundation. For a philosophical analysis that attempts to clarify extensions of the finitist standpoint and to explicate – relative to them – the epistemological significance of these particular results some clear and concrete tasks can be formulated.

At the very beginning of the development of Hilbert's program one finds an extension not of, but rather towards the finitist standpoint. Originally, Hilbert intended to make do with a mathematical basis that did not even include the 'Allgemeinbegriff der Ziffer': all mathematical knowledge (Erkenntnis) was to be reduced to primitive formal evidence.[47] This extremely restricted undertaking was given up quickly: how could the central goal of the program, consistency, be formulated within its framework? A 'finitist standpoint' that is to serve as the basis for Hilbert's investigations cannot be founded on the intuition of concretely given objects; it rather has to correspond to a standpoint, as Bernays explained,

---

[45] i.e. very special ones: higher number classes and, more generally, accessible i.d. classes.

[46] That point is clearly and forcefully made in Gödel's paper *Russell's Mathematical Logic*; see pp. 455–456 in the second edition of *Philosophy of Mathematics*, Benacerraf and Putnam (eds.), Cambridge, 1983.

[47] This is reported in (Bernays, 1946), p. 91; an example of the form of this quite primitive evidence can be found l.c. p. 89, but compare also (Bernays, 1961), p. 169. As to the historical point see (Bernays, 1967), p. 500: 'At the time of his Zurich lecture Hilbert tended to restrict the methods of proof-theoretic reasoning to the most primitive evidence. The apparent needs of proof theory induced him to adopt successively those suppositions that constitute what he then called the 'finite Einstellung'.'

'where one already *reflects* on the general characteristics of intuitive objects'.⁴⁸ A first task presents itself.

(I) Analyse this reflection for the natural numbers (and the elements of other accessible i.d. classes given by finitary inductive definitions) and investigate whether and how induction and recursion principles can be based on it.

Without attempting to summarize the extended (and subtle) discussion in (Bernays, 1930) I want to point to one feature that is crucial in it and important for my considerations. For Bernays the natural numbers (as ordinals) are the simplest formal objects; they are obtained by formal abstraction and are representable by concrete objects, numerals. This representation has a very special characteristic: the representing things contain ('enthalten') the essential properties of the represented things in such a way that relations between the latter objects obtain between the former and can be ascertained by considering those.⁴⁹ It is this special characteristic that has to be given up when extending the finitist standpoint: symbols are no longer carrying their meaning on their face, as they cannot exhibit their build-up.⁵⁰ For the consistency proofs mentioned in Remark D above one uses accessible i.d. classes of natural numbers; numerals

---

⁴⁸ From (Bernays, 1930), p. 40. The context is this: 'Diese Heranziehung der Vorstellung des Endlichen [used from the finitist standpoint] gehört freilich nicht mehr zu demjenigen, was von der anschaulichen Evidenz notwendig in das logische Schließen eingeht. Sie entspricht vielmehr einem Standpunkt, bei dem man bereits auf die allgemeinen Charakterzüge der anschaulichen Objekte *reflektiert*.' – This is a clearer and more promising starting point for an analysis than the one offered through Hilbert's own characterization in (van Heijenoort), p. 376. Important investigations, in addition to those of Bernays, have been contributed by i.a. Kreisel, Parsons, and Tait. See (Tait, 1981) and the references to the literature given there. However, in this systematic context (and also for a general discussion of feasibility) I should point out that weakenings of the finitist standpoint are of real interest; a penetrating investigation is carried out by R. Gandy in *Limitations of mathematical knowledge*, in: Logic Colloquium '80, Amsterdam 1982, pp. 129–146. Notice that the type-token problematic has to be faced already from weakened positions. – The crux of the additional problematic was compressed by Bernays into one sentence: 'Wollen wir ... die Ordnungszahlen als eindeutige Objekte, frei von allen unwesentlichen Zutaten haben, so müssen wir jeweils das bloße Schema der betreffenden Wiederholungsfigur als Objekt nehmen, was eine sehr hohe Abstraktion erfordert.' (Bernays, 1930, pp. 31–32.) It is for these formal objects that the 'Gedankenexperimente' are carried out, that play such an important role in *Grundlagen der Mathematik I* (p. 32) for characterizing finitist considerations.

⁴⁹ This is found on pp. 31–32, in particular in footnote 4 on that page. The general problematic is also discussed in (Bernays, 1935a), pp. 69–71. Compare the previous footnote to recognize that an isomorphic representation by a particular, physically realized object is not intended. The uniform character of the generation and the local structure of the schematic 'iteration figure' are important.

⁵⁰ Gödel described in his *Über eine bisher noch nicht benützte Erweiterung des finiten Standpunktes*, Dialectica 12 (1958), pp. 280–287, a standpoint that extends the finitist one and that is appropriate for the consistency proofs for number theory given by Gentzen and Gödel himself. The starting point of this proposal are considerations of Bernays – e.g. in his (1935b) and (1941) – concerning the question, 'In what way does intuitionism go beyond finitism?' Bernays's answer is 'Through its abstract notion of consequence.' And it is this abstract concept that is to be partially captured by the computable functionals of finite type. – Note that the specifically finitist character of mathematical objects requires, according to Gödel, that they are 'finite space–time configurations whose nature is irrelevant except

for the elements of such a class are now understood as denoting infinite objects, namely the unique construction trees associated effectively with the elements. So we have as a generalization of (I) – to begin with – the task:

> (II) Extend the reflection to constructive ordinals and the elements of other accessible i.d. classes and investigate whether and how induction and recursion principles can be based on it.

One delicate question has not been taken into account here. For the consistency proofs of strong impredicative theories the definition of i.d. classes has to be iterated uniformly; that means the branching in the well-founded construction trees is not only taken over natural numbers, but also over other already obtained i.d. classes. These trees are of much greater complexity. For example, it is no longer possible – as it is in the case of constructive ordinals – to generate effectively arbitrary finite subtrees; that has to be done now through procedures that are effective relative to already obtained number classes. Thus we have to modify (II).

> (III) Extend the reflection to uniformly iterated accessible i.d. classes, in particular to the higher constructive number classes.

Buchholz and Pohlers used in their investigations of theories for i.d. classes systems of ordinal notations. It is clearly in the tradition of Gentzen and Schütte to use for consistency proofs the principle of transfinite induction along suitable ordinals (represented through effective notation systems). But in parallel to proving the consistency of formal theories by such means, Gentzen wrote in a letter to Bernays of March 3, 1936, one has to pursue a complementary task, namely '... to carry out investigations with the goal of making the validity of transfinite induction constructively intelligible for higher and higher limit numbers'. It is only through such investigations that the philosophical point of consistency proofs can be made, namely, to secure a theory by reliable ('sichere') means. Gentzen's task is included in (III), since the systems of notations used in Buchholz and Pohlers' work are generated as accessible i.d. classes, and their well ordering is recognized through the proof principle for these i.d. classes. These systems of notations were quite complicated, but in their latest and conceptually best form they are given by clauses of the same character as those for higher number classes (Buchholz, 1990). I mentioned earlier the consistency problem for the subsystem of analysis with Π^1_2-comprehension; this is not only a mathematical problem, but also an open conceptual problem, as new 'constructive' objects are needed for a satisfactory 'constructive' solution.[51]

---

for equality and difference'. This seems to conflict with Bernays's analysis pointed to in the previous two footnotes.

[51] Work of Pohlers and his students to extend the method of local predicativity make it most likely that a close connection to set theory (in particular the study of large cardinals and the fine structure of the constructible universe **L**) is emerging.

The number classes provide special cases in which generating procedures allow us to grasp the intrinsic build-up of mathematical objects.[52] Such an understanding is a fundamental and objective source of our knowledge of mathematical principles for the structures or domains constituted by those objects: is it not the case that the definition and proof principles follow directly this comprehended build-up? Clearly, we have to complement an analysis of this source – as requested in (I)–(III) – by formulating (the reasons for the choice of) suitable deductive frames in which the mathematical principles are embedded. Thus there are substantial questions concerning the language, logic, and the exact formulation of schematic principles. But notice that for the concerns here these questions are not of primary importance. For example, the restriction to intuitionistic logic is rather insignificant: the double-negation translation used by Gödel and Gentzen to prove the consistency of classical relative to intuitionistic arithmetic can be extended to a variety of theories to yield relative consistency results. Indeed, Friedman showed for arithmetic, finite type theories, and Zermelo–Fraenkel set theory that the classical theories are Π_2^0-conservative over their intuitionistic version. Using Friedman's strikingly simple techniques Feferman and Sieg (1981) established such conservation results for some subsystems of analysis and also for the theories of iterated inductive definitions.[53] In the latter case it is the *further* restriction to accessible i.d. classes that is (technically difficult and) conceptually significant.

Disregarding the traditional constructive traits of the objects considered up to now we can extend the basic accessibility conception from i.d. classes of natural numbers to broader domains. A comprehensive framework for the 'inductive or rule-governed generation' of mathematical objects is given in (Aczel, 1977); it is indeed so general that it encompasses finitary i.d. classes, higher number classes, the set-theoretic model of Feferman's theory T_0 of explicit mathematics and of other constructive theories (like Martin–Löf's), but also segments of the cumulative hierarchy. Clearly, not all of Aczel's i.d. classes have the distinctive feature of accessible i.d. classes; those whose elements do have unique associated well-founded 'construction' trees are called deterministic and, here again, accessible. Segments of the cumulative hierarchy – that contain some ordinals (0, ω, or large cardinals) and are closed under the powerset, union, and replacement operations – are in this sense accessible: the uniquely determined transitive closure of their elements are 'construction' trees.[54] Here, as above, we have the task of

---

[52] By calling their build-up 'intrinsic' I point again to the parallelism between the generating procedure and the structure of the intended object; compare the case under discussion here, for example, with that of the computable functionals of finite type.

[53] The theory of arithmetic properties and ramified systems were shown to be Π_2^0-conservative over their intuitionistic versions in (Feferman and Sieg, 1981), pp. 57–59. – The generality of Friedman's techniques was brought out by Leivant in *Syntactic translations and provably recursive functions*, Journal of Symbolic Logic 50 (3), 1985, 682–688.

[54] Here is the basis for ε-induction and recursion. It seems to me that in this context the discussion and results concerning Fraenkel's Axiom of Restriction would be quite pertinent.

explicating (the difficulties in) our understanding of generation procedures. After all, for accessibility to have any cognitive significance such an understanding has to be assumed. The latter is in the present case relatively unproblematic, if we restrict attention to hereditarily finite sets; then we have an understanding of the combinatorial generation procedures and, in particular, of forming arbitrary subcollections. Indeed, ZF^-, i.e. Zermelo–Fraenkel set theory without the axiom of infinity, is equivalent to elementary number theory. The powerset operation is the critical generating principle; its strength when applied to infinite sets is highlighted by the fact that ZF without the powerset axiom is equivalent to second-order arithmetic.[55] But if we do assume an understanding of the set-theoretic generation procedure for a segment of the cumulative hierarchy, then it is indeed the case that the axioms of ZF^- together with a suitable axiom of infinity 'force themselves upon us as being true' – in Gödel's famous phrase; they just formulate the principles underlying the 'construction' of the objects in this segment of the hierarchy.[56] In summary, we have a wealth of accessible domains, and it seems that we can understand the pertinent mathematical principles quasiconstructively, as we grasp the build-up of the objects constituting such structures.

2.3 Contrasts

The 'ontological status' of mathematical objects has not been discussed. The reason is this: I agree with the subtle considerations of Bernays in the essay *Mathematische Existenz und Widerspruchsfreiheit* and suggest only one amendation, namely to distinguish the 'methodical frames' (methodische Rahmen) by having their objects constitute accessible domains. The contrast between 'platonist' and 'constructivist' tendencies is then not localized in the stark opposition formulated by Gödel; it comes to light rather in refined distinctions concerning the admissibility of operations, of their iteration, and of deductive principles. In this way, it

---

[55] Using powerset one obtains not the elements of a subclass from a given set, but rather all subclasses in one fell swoop. It is this utter generality that creates a difficulty even when the given set is that of the natural numbers; see the comprehensive discussion of Bernays's views in (Müller). The difficulty is very roughly this: in terms of the basic operations one does not have 'prior' access to all elements of the powerset, unless one chooses a second-order formulation of replacement. That would allow the joining of arbitrary subcollections, but 'arbitrary subcollection' has then to be understood in whatever sense the second-order variables are interpreted. – A focus on definable subsets leads to the ramified hierarchy, to Gödel's constructible sets, and to the consideration of subsystems of ZF. The investigations concerning subsystems of analysis can be turned into investigations of natural subsystems of set theory. That was done by G. Jäger. His work is presented in *Theories for admissible sets – A unifying approach to proof theory*; Bibliopolis, Naples, 1986.

[56] This reason for accepting the axioms of ZF seems to be (at least) consonant with Gödel's analysis in *What is Cantor's continuum problem?* and does not rest on the strong Platonism in the later supplement of the paper. The conceptual kernel of this analysis goes back to Zermelo's penetrating paper of 1930. A discussion of the rich literature on the 'iterative conception of set' is clearly not possible here. – Notice that the length of iteration is partly determined through the adopted axiom of infinity built into the base clause of the i.d. definition.

seems to me, methodical frames are not only distinguishable from each other, but also epistemologically differentiated from 'abstract' theories formulated within particular frames. I want to focus on this differentiation now and contrast the quasiconstructive aspect of mathematical experience I have been analysing to – what I suggest to call – its 'conceptional' aspect. The latter aspect is most important for mathematical practice and understanding, but also for the sophisticated uses of mathematics in physics; it is quite independent of methodical frames.

As a first step let us consider Dedekind's way of comprehending the accessible domain of natural numbers. The informal analysis underlying *Was sind und was sollen die Zahlen?* described in his letter to Keferstein, (Dedekind 1890), starts out with the question:

> What are the mutually independent fundamental properties of the sequence N, that is, those properties that are not derivable from one another but from which all others follow? And how should we divest these properties of their specifically arithmetic character so that they are subsumed under more general notions and under activities of the understanding *without* which no thinking is possible at all but *with* which a foundation is provided for the reliability and completeness of proofs and for the construction of consistent notions and definitions?

One is quickly led to infinite systems that contain a distinguished element 1 and are closed under a successor operation ϕ. Dedekind notes that such systems may contain non-standard 'intruders' and that their exclusion from **N** was for him 'one of the most difficult points' in his analysis; 'its mastery required lengthy reflection.'[57] The notion of chain allows him to give 'an unambiguous conceptual foundation to the distinction between the [standard] elements n and the [non-standard] t'. By means of this notion he captures the informal understanding that the natural numbers are just those objects that are obtained from 1 by finite iteration of ϕ, or rather the objects arising from any simply infinite system 'by entirely disregarding the special nature of its elements, and retaining only their distinguishability and considering exclusively those relations that obtain between them through the ordering mapping ϕ'.[58] He continues: 'Taking into account this freeing of the elements from every other content (abstraction), we can justifiably call the [natural] numbers a free creation of the human mind.' How startlingly close is this final view of natural numbers to that arrived at by Bernays through 'formal abstraction'!

For Dedekind the considerations (concerning the existence of infinite systems) guarantee that the notion 'simply infinite system' does not contain an internal

---

[57] The minimalist understanding is taken for granted when it is claimed that the induction principle is evident from the finitist point of view. Thus, even this seemingly most elementary explanation of induction leaves us with a certain 'impredicativity'. 'The same holds', Parsons rightly argues, 'for other domains of objects obtained by iteration of operations yielding new objects, beginning with certain initial ones. It seems that the impredicativity will lose its significance only from points of view that leave it mysterious why mathematical induction is evident.' (Parsons, 1983), pp.135–136.

[58] Section 73 of (Dedekind, 1888).

contradiction.[59] The 'purely logical' and presumably reliable foundation did not, of course, allow this goal to be reached. In Section 1.1 I emphasized methodological parallels between Dedekind's treatment of the natural and real numbers; here I want to bring out a most striking difference. We just saw that Dedekind's analysis of natural numbers is based on a clear understanding of their accessibility through the successor operation. This understanding allows the distinction between standard objects and 'intruders' and motivates directly the axioms for simply infinite systems. Given the build-up of the objects in their domains, it is quite obvious that any two simply infinite systems have to be isomorphic via a unique isomorphism. By way of contrast, consider the axioms for dense linear orderings without endpoints; their countable models are all isomorphic, but Cantor's back-and-forth argument for this fact exploits broad structural conditions and not the local build-up of objects. The last observation gives also the reason why these axioms do not have an 'intended model': it is the accessibility of objects via operations not just the categoricity of a theory that gives us such a model. Similar remarks apply to the reals, as the isomorphism between any two models of the axioms for complete, ordered fields is based on the topological completeness requirement, not any build-up of their elements. (That requirement guarantees the continuous extendibility of any isomorphism between their respective rationals.)

This point is perhaps brought out even more clearly by a classical theorem of Pontrjagin's, stating that connected, locally compact topological fields are either isomorphic to the reals, the complex numbers, or the quaternions. For this case Bourbaki's description, that the 'individuality' of the objects in the classical structures is induced by the superposition of structural conditions, is so wonderfully apt; having presented the principal structures (order, algebraic, topological) he continues:

> Farther along we come finally to the theories properly called particular. In these the elements of the sets under consideration, which, in the general structures have remained entirely indeterminate, obtain a more definitely characterized individuality. At this point we merge with the theories of classical mathematics, the analysis of functions of real or complex variable, differential geometry, algebraic geometry, theory of numbers. But they have no longer their former autonomy; they have become crossroads, where several more general mathematical structures meet and react upon one another.[60]

---

[59] What is so astonishing in every rereading of Dedekind's essay is the conceptual clarity, the elegance and generality of its mathematical development. As to the latter, it really contains the general method of making monotone inductive definitions explicit. The treatment of recursive definitions is easily extendible; that it has to be restricted, in effect, to accessible i.d. classes is noted. Compare (Feferman and Sieg, 1981), footnotes 2 and 4 on p. 75.

[60] p. 229 of N. Bourbaki, *The architecture of mathematics*, Math. Monthly (57), 1950, pp. 221–232. The natural numbers are not obtained at a crossroad.

Here we are dealing with abstract notions, distilled from mathematical practice for the purpose of comprehending complex connections, of making analogies precise, and to obtain a more profound understanding; it is in this way that the axiomatic method teaches us, as Bourbaki expressed it in Dedekind's spirit (l.c., p. 223),

> to look for the deep-lying reasons for such a discovery [that two, or several, quite distinct theories lend each other 'unexpected support'], to find the common ideas of these theories, buried under the accumulation of details properly belonging to each of them, to bring these ideas forward and to put them in their proper light.

Notions like group, field, topological space, differentiable manifold fall into this category, and (relative) consistency proofs have here indeed the task of establishing the consistency of abstract notions relative to accessible domains. In Bourbaki's enterprise one might see this as being done relative to (a segment of) the cumulative hierarchy. But note, this consideration cuts across traditional divisions, as it pertains not only to notions of classical mathematics, but also to some of constructive mathematics. A prime example of the latter is that of a choice sequence introduced by Brouwer into intuitionistic mathematics to capture the essence of the continuum; the consistency proof of the theory of choice sequences relative to the theory of (non-iterated) inductive definitions can be viewed as fulfilling exactly the above task.[61] The restriction of admissible operations (and deductive principles) can lead to the rejection of abstract notions; that comes most poignantly to the surface in the philosophical dispute between Kronecker and Dedekind, but also in Bishop's derisive view of Brouwer's choice sequences. Bishop is not only scornful of the 'metaphysical speculation' underlying the notion of choice sequence, but he also views the resulting mathematics as 'bizarre'. (Bishop, 1967), p. 6.

Concluding Remarks

The conceptional aspect of mathematical experience and its profound function in mathematics has been neglected almost completely in the logico-philosophical literature on the foundations of mathematics.[62] Abstract notions have been important for the internal development of mathematics, but also for sophisticated applications of mathematics in physics and other sciences to organize our experience of the world. It seems to me to be absolutely crucial to gain genuine

---

[61] Kreisel and Troelstra, *Formal systems for some branches of intuitionistic analysis*, Annals of Mathematical Logic 1(3), 1970, pp. 229–387. It is really just the theory of the second constructive number-class that is needed.

[62] The exception are papers of Bernays; there it is absolutely central.

insight into this dual role, if we want to bring into harmony, as we certainly should, philosophical reflections on mathematics with those on the sciences.

Results of mathematical logic do not give precise answers to large philosophical questions; but they can force us to think through philosophical positions. Broad philosophical considerations do not provide 'foundations' for mathematics; but they can bring us to raise mathematical problems. We shall advance our understanding of mathematics only if we continue to develop the dialectic of mathematical investigation and philosophical reflection; a dialectic that has to be informed by crucial features of the historical development of its subject. In Brecht's *Galileo Galilei* one finds the remark:

> A main reason for the poverty of the sciences is most often imagined wealth. It is not their aim to open a door to infinite wisdom, but rather to set bounds to infinite misunderstanding.[63]

What is said here for the sciences holds equally for mathematical logic and philosophy.

Acknowledgement

Notes of Hilbert's courses are kept at the *Nachlass* David Hilbert – Niedersächsische Staats- und Universitätsbibliothek Göttingen, Abteilung Handschriften und Seltene Drucke. Passages are quoted here with permission of the library.

---

[63] The German text is: 'Eine Hauptursache der Armut der Wissenschaften ist meist eingebildeter Reichtum. Es ist nicht ihr Ziel, der unendlichen Weisheit eine Tür zu öffnen, sondern eine Grenze zu setzen dem unendlichen Irrtum.' In its 'application' to philosophy it mirrors (for me) the views of the man who influenced so deeply Dedekind, Kronecker, Hilbert, ... ; they are reported in (Kummer, 1860), p. 340: 'Er [Dirichlet] pflegte von der Philosophie zu sagen, es sei ein wesentlicher Mangel derselben, dass sie keine ungelösten Probleme habe wie die Mathematik, dass sie sich also keiner bestimmten Grenze bewusst sei, innerhalb deren sie die Wahrheit wirklich erforscht habe und über welche hinaus sie sich vorläufig bescheiden müsse, nichts zu wissen. Je grössere Ansprüche auf Allwissenheit die Philosophie machte, desto weniger vollkommen klar erkannte Wahrheit glaubte er ihr zutrauen zu dürfen, da er aus eigener Erfahrung in dem Gebiete seiner Wissenschaft wusste, wie schwer die Erkenntnis der Wahrheit ist, und welche Mühe und Arbeit es kostet, dieselbe auch nur einen Schritt weiter zu führen.'

Coda

Modern mathematics as a cultural phenomenon

JEREMY GRAY

Introduction

One of the most significant changes mathematics has experienced in its long history is the transformation into 'modern mathematics'. As with all major changes, some historians look for, and find, much that did not change, while others accentuate significant differences from the immediately preceding period. The conflicting claims of continuity and disruption can both serve as guiding metaphors in the historical analysis of the period, and here they frame another question: where is history of modern mathematics to be placed? Is it, because of the nature of the mathematics, part of mathematics? Is it an island, just off the shore of the mainland of mathematics? Or is it part of the territory of history of science? Although the arrival of modern mathematics is a well-attested phenomenon, for obvious reasons most accounts of it are heavily internal. It is interesting, however, to discuss how the history of modern mathematics can be approached by proposing an analysis of modern mathematics as a particular case of Modernism, and this essay suggests some of the benefits that might come to the history of mathematics by adopting such an analysis.

Mathematics c. 1880

Historians of mathematics writing about the modern period have much to say, because, for an event so widely agreed to be momentous for mathematics, there is much that is un- or underresearched. But some things seem agreed – if still worthy of further, and better, research – and they point up the magnitude of the change about to take place. A brief look at the world of mathematics around 1880, before the transition to modern mathematics took place, would show that in Berlin, the indisputable centre of mathematics around 1880, the view was that the core discipline was analysis, indeed, complex analysis, around which were closely situated the fields of number theory and algebra. A hard-liner such as Kronecker explicitly placed geometry outside the core of mathematics[1], but nothing in Weierstrass's work suggests that he disputed that judgement. A defence of geometry amounting to an attack on this orthodoxy was only made by Felix Klein in 1895, when he felt able to do so after the death of both Weierstrass and Kronecker, and when his own position at Göttingen was becoming increasingly powerful; and even then it was wrapped up in a plea for intuition as an aid to discovery.[2] Part of Klein's position was his desire to promote what he saw as Riemannian ideas that had been largely neutralized by the heavyweights in Berlin, and his own credentials as Riemann's heir.

A similar generational shift was taking place in Paris, as young mathematicians such as Poincaré and Picard accepted Riemann's ideas in a way their mentor, Charles Hermite, had never been able to do.[3] Poincaré's own career exemplifies a tension in the French mathematical tradition that had largely been driven out in Germany: an attention to the needs, demands, and challenges of mathematical physics. A similar story might be told in Italy, where geometry was particularly well studied, and a mirror image would hold for Great Britain where, despite the life work of Cayley and Sylvester, the dominant tradition, at its best in Cambridge, put physics first and mathematics second.

There was little doubt in any of these places that analysis – largely, but by no means exclusively complex analysis – stood centre stage. The German enthusiasm for number theory was not so strongly matched elsewhere, and outside Germany there was more of an enthusiasm for geometry and mathematical physics. This can be seen as a consensus or as a division within the subject, depending on one's point of view, but it is a clear basis with which to contrast the changes that were to come.

The arrival of modern mathematics

Some of these changes form part of the subject of this volume. They are associated with such names as Hilbert and Frege, and with terms such as foundations, set theory, and logicism. Many accounts of this period in the literature lead towards 'modern algebra'; one might equally well arrive at topology, the quintessential new mathematical discipline of the twentieth-century. Even superficially, the interaction between logic, philosophy, and mathematics is truly remarkable; nothing in the above summary of the world of mathematics as it stood in 1880

---

[1] See (Boniface and Schappacher [2001]).
[2] (Klein [1895]).
[3] See (Gray [2000a]).

suggests that these three themes would come together in a way innovative for each of them and in a fruitful, indeed dramatic way for all three. The arrival of set-theoretic methods, the dramatic revival of abstract axiomatic thinking, the structural emphasis and stark conceptual abstractness of modern mathematics with its apparent disdain for applications is not only a matter of foundations, but embraces group theory, point-set topology, and the whole philosophy of mathematics. These changes were unexpected, massive and permanent.

Naturally and rightly, these changes have to be studied close up and in detail. When this is done, they reveal many interesting peculiarities specific perhaps to the branch of mathematics in question, or to the national setting, or in some other way. But historians who believe that these changes have some overall coherence, are somehow interrelated and not merely synchronous, have an opportunity to do more, and to seek to analyse that coherence and its significance. Those who take that opportunity naturally, and rightly, meet with opposition.

Continuity as opposed to change

To oppose such an account as the one just briefly outlined, which emphasizes novelty and change, at least three accounts emphasizing continuity can be advanced. The first is the most extreme claim. It argues that much of mathematics changed very little as a result of these developments, or at least changed no more than it had to, and that when the changes were all in place and a new orthodoxy prevailed, the familiar interconnections within mathematics, and the relations of mathematics to physics, could be seen to be more or less as they had been before. The vast empire of analysis, on this reply, along with much of algebraic and differential geometry show striking continuities with what went before, and it is hard to identify a topic that died out because modern mathematics came in.

The second claim is less extreme. It does not deny the substantial nature of various changes, but claims that they have roots deep in earlier developments (not only with Cauchy, but also Camille Jordan, for example, or Cantor's work on Fourier series representations). Any claim to special status for the late nineteenth-century changes is denied because, it is said, such putative changes dissolve on a close analysis of their origins into a long process of evolution.

The third response accepts the magnitude and timing of the changes in question, but replies that every so often there are major changes in the way mathematics gets done, and that the arrival of modern mathematics is just another one of these changes, important, of course, worth considering in detail, but not to be overstated. To cast this response in the form of rhetorical questions bracketing the supposed modernist arrival, it asks such questions as: Did not Cauchy initiate radical changes in real analysis? Is the modernism of twentieth-century mathematics the right framework for analysing the introduction of Hilbert-space methods and their subsequent employment in quantum mechanics?

Such debates are familiar in any historical account of any significant event. They will never go away, there is no reason why they should. Historians differ in their preferences, and, more interestingly, major events are surely always multifaceted. Conflicting analyses will always have much to commend them. The arrival of modern mathematics can never be fully understood, certainly not as a simple event, and the challenge presented by any analysis that stresses continuity requires a response rooted in an analysis of the place of history of mathematics within the panoply of historical enquiries.

First, I shall argue that the complexity of the changes brought in with modern mathematics, and specifically its relations with logic and philosophy, but also with physics, invites us to view that event as a cultural change in mathematics and to seek a cultural account of it. Then I shall argue that there is a framework ready to hand that fits these changes very well, which is the framework of modernism. Finally, I shall argue that the modernist framework does not enable us to decide once and for all between change or continuity, but enables us to structure a debate about the history of mathematics that recognizes both sorts of view.

Modernism

Modernism can be defined as an autonomous body of ideas, pursued with little outward reference, maintaining a complicated, rather than a naive, relationship with the day-to-day world and drawn to the formal aspects of the discipline. It is introspective to the point of anxiety; and is the *de facto* view of a coherent group of people, such as a professional, discipline-based group, who were profoundly serious in their intentions. As a philosophy (taking the term in its broadest sense) it is in sharp contrast to the immediately preceding one in each of its fields. Core examples are easy to point to: in painting, the cubists; in music, the serialism of Schoenberg; in literature the abandonment of narrative (Joyce's *Ulysses*) and the psychologization of character (Musil's *Der Mann ohne Eigenschaften*) – in poetry Mallarmé and later Pound and Eliot.

I shall suggest that the changes in mathematics around 1900 qualify it as a genuine modernism, then outline what seeing those developments in this way does for the history of mathematics, and finally I shall explore, rather more tentatively, the problems there are with Modernism as a tool and how they surface in the case of modern mathematics.[4]

---

[4] This essay concentrates on the intellectual side: modernism as an ideology. It is also important to take a more sociological perspective, looking at the related issues of modernity and modernization, but there is not space to do so here.

Mathematical Modernism

In arguing the case for mathematical modernism, it is simplest to concentrate on the case of geometry. There were two major trends in geometry in the nineteenth-century: projective geometry and non-Euclidean geometry.[5] Klein's view that all geometries could be seen as particular cases of projective geometry was set out in his Erlangen Program of 1872, where it languished until the 1890s, when it was republished, translated into several languages, and became more widely appreciated.[6] Birkhoff and Bennett (Birkhoff and Bennett [1988]) ascribed a 'great influence' to this 'classical document' (p. 173), writing that it was 'generally accepted as a major landmark in the mathematics of the nineteenth century' (p. 145).[7] Lately, however, historians have begun to differ over the importance of the Erlangen Program.[8] Despite the parade of major geometers and massive innovations in geometry that Birkhoff and Bennett attached to the Erlangen Program – a list that includes Lie, Cartan, and Weyl – only one innovation of the Program survived: the idea that every geometry has a group of allowable transformations attached to it, and the geometric properties of a space are those properties that are invariant under the action of the group. This is a profound idea, and it does nothing to diminish its importance if we observe that Poincaré seems to have come to it independently in 1880, at the start of his work on non-Euclidean geometry and Fuchsian functions.[9] But the idea that all geometries are in a hierarchy, with projective geometry as the most general, was not to survive. It is true that real and complex geometries with large transformation groups are described in a limited number of ways, but this is now regarded as a fact following from the classification of Lie groups. It is true that Cartan and Weyl expended a great deal of work in creating the theory of bundles with structure groups, and that this can be seen as a way of extending Klein's philosophy of groups to differential geometry, but note that, again, it only extends the simplest part of Klein's Program. The strikingly odd thing about the Erlangen Program, if it is taken as a claim about all geometry, is that it has nothing to say about almost all of Riemannian differential geometry, where, generally, the isometry group of the whole space is simply the trivial group.

Between 1872 and 1893 the Erlangen Program had rather little influence (this seems to have been one of Klein's reasons for wanting to republish it). But it can be usefully seen as indicating a crossroads, with one road pointing towards projective geometry and the other to metrical geometry. Projective geometry had

---

[5] For histories of non-Euclidean geometry, see (Bonola [1906]) and (Gray [1986]).

[6] (Klein Erlangen Program). (Klein [1871]) and (Klein [1873]) offer much more explicit accounts of non-Euclidean geometry from a projective standpoint.

[7] (Birkhoff and Bennett [1988]).

[8] (Hawkins [2000]).

[9] (Poincaré [1882]). His route to these discoveries is described in (Poincaré [1997]) and (Gray [2000a]).

become the hegemonic geometry by the 1870s. We still lack a good history of the subject, but it is clear from even a casual look at the work of Cremona and the Italian geometers, at Cayley and Sylvester, or at the German tradition, that a sort of middle age had set in. Birational geometry, first in Germany and then in Italy, had become the active research field; projective geometry became the essence of geometry for teaching purposes. But birational geometry may be thought of as a generalization of projective geometry; it enriches projective geometry, but does not challenge it. In particular, it leaves alone the idea that the fundamental geometric figures in the plane are the point and the line (with the obvious extension to planes and hyperplanes in projective spaces of higher dimension).

The projective tradition, as it might be called, is where Pasch was coming from in his famous *Vorlesungen über neuere Geometrie* [1882], and to which must be attached all explorations of geometry that take the straight line as a fundamental undefined concept, whether for mathematical or psychological purposes. There are at least four ways of thinking about straight lines in a plane. Two are projective, of which one is the heavily axiomatized version or versions produced by Pieri, Hilbert, Veblen and Young, Whitehead and others.[10] The aim here is almost to forget entirely what a line is, and indeed to allow the methods of projective geometry to be applied in areas where talk of lines seems on the face of it inappropriate. The other version is what one might call minimalist, or even the common sense way, and I believe it is what geometers used. It takes the least set of assumptions one can make about lines (any two points define a unique line, any two meet in a unique point – so the assumptions have been somewhat tidied up) allows only transformations that send lines to lines, and produces a theory, based on duality and the invariance of cross-ratio, that is synthetic and clear to the point of being transparent. Unlike many a branch of mathematics it has very few pre-requisites, but unlike some topics that can be produced by writing down rules, it is seemingly (unquestionably?) about something, even if that something is a strangely idealized version of the supposedly familiar world. The emphasis, however, as Nagel pointed out many years ago, is on the formal side.[11] This is because in plane projective geometry, duality puts points and lines on a par, something that intuition cannot do. Deprived of intuition, mathematicians perforce relied on formal arguments. A good way to see the projective-metrical split is to see that projective geometry was largely a syntactical theory, whereas metrical geometry was more semantic. Once projective geometry was rescued from the obscurities of Poncelet's original presentation[12], it was the clarity of its definitions and its rules for deduction that commended it. Metrical geometry is more obviously about something.

The other ways of thinking about a straight line are metrical and belong to differential geometry: first, the idea that a straight line is the shortest curve

---

[10] (Pieri [1899]), (Hilbert *Grundlagen*), (Veblen and Young *Projective Geometry*), (Whitehead [1906]).
[11] (Nagel [1939]).
[12] (Poncelet *Traité*).

joining any two of its points, and, second, that it is the straightest. A key episode in the metrical tradition in geometry in the period around 1900 was the final acceptance of non-Euclidean geometry, arising from the work of Riemann and Beltrami, as reinterpreted by Klein in projective terms and by Poincaré in metrical or differential geometric terms.[13] This metrical tradition embraces Helmholtz and all those who see rigid-body motions as the basic idea out of which geometry must be constructed (Lobachevskii, Poincaré, and others).[14] It is the non-Euclidean aspect of the metrical tradition that destroyed confidence in the truth of geometry. When in 1891 Poincaré pointed out that Euclidean and non-Euclidean geometry were relatively consistent he did something quite momentous.[15] Hitherto the consistency of Euclidean geometry had been taken for granted. A defence of the consistency of non-Euclidean geometry had always taken the form of giving it a full description in Euclidean or projective terms. For the first time, somebody observed that this was not really enough. The implication, hinted at by Poincaré but never explored, was that both Euclidean and non-Euclidean geometry might fall together. (Poincaré himself had a philosophy of rigid-body motions to rely on, but that was not to everyone's taste.)

The idea that geometry might not be true came as a powerful shock to mathematicians and other users of mathematics. It generated a large popular literature as well as a profusion of technical works, and a feature of this interest was not that space might be non-Euclidean, so much as that geometry might be. Once there was a plausible geometric description of space other than the traditional one, it followed that mathematics could in no simple way be true. At the very least, some empirical investigation was henceforth required. The simple dichotomy (Euclidean or non-Euclidean – for some reason Riemann's idea of a finite universe never caught on) disappeared with the advent of Einstein's general theory of relativity, but recall that Einstein himself brought in the idea of a non-Euclidean nature for physical space as early as 1909.[16]

Do we have then, Modernism in geometry? For an autonomous body of ideas, we have the formal axiomatic treatments of projective geometry offered by Pieri, Hilbert, and Veblen and Young. Less certainly, we have non-Euclidean geometry. Axiomatic projective geometry is autonomous in that, in the hands of Pieri, it explicitly renounced any claim to be about lines as commonly understood, and in Hilbert's version, it was denuded of any outward reference. Hilbert's *Grundlagen der Geometrie* was intended to show the merits of an entirely abstract way of defining objects. Non-Euclidean geometry happened, as a matter of fact, to interest only mathematicians, or rather to raise only mathematical questions about space, but the fact that it involves distance and measurement makes it harder to regard

---

[13] (Beltrami [1868]), (Poincaré [1882]).
[14] (Helmholtz [1870]), (?), (Poincaré [1997]).
[15] (Poincaré [1891]), better known when reprinted in 1902.
[16] See Stachel:1989.

as being autonomous. Mathematicians were by no means the only people to have such concerns. Rather, the autonomous aspect of non-Euclidean geometry is that it raised a deep philosophical question about the truth of mathematics. This was a question for mathematicians and philosophers.

It may therefore readily be granted that geometry around 1900 was pursued with little outward reference, that it maintained a complicated, rather than a naive, relationship with the day-to-day world, and that those who took it up were drawn to the formal aspects of the discipline. It was the *de facto* view of a small number of professional mathematicians, who were profoundly serious in their intentions and aimed to give, for the first time ever, as they saw it, valid foundations for ordinary geometry. Among these people were Italians such as Peano (see his (Peano [1889])) and Fano in his (Fano [1892])). Pieri took the same approach to any system of elementary geometry in (Pieri [1899]), and different geometries were simultaneously considered, with a view to suggesting how the study of axiom systems may be conducted, by Hilbert in his *Grundlagen*. Particularly in Hilbert's eyes this implied a philosophy of mathematics in sharp contrast to the immediately preceding one in the field (be that classical Euclidean geometry or, for that matter, Riemannian geometry).[17]

In many ways, it was highly introspective. From Riemann's hypotheses that lie at the foundations of geometry to the numerous investigations of the independence of axiom systems, geometrical research was full of enquiries into the nature of geometrical reasoning. It is less clear that this introspection reached the point of anxiety. Pasch, in his (Pasch *Geometrie*) certainly felt it was better to rebuild geometry from scratch than to continue with attempts to repair Euclid's *Elements*. The exchange between Hilbert and Frege[18], and the complaints of Peano and Padoa[19], capture a real concern. Hilbert, who was supremely confident of his ability in mathematics and largely immune from anxiety, put his confidence in syntactically defined systems that seemed to him to be consistent. Notoriously, he was even willing to licence axioms that asserted the existence of an essentially unique system of objects that obeys the axioms. His treatment of existence for axiomatically defined objects was decidedly lax, and struck his critics not so much as standing in need of repair as simply incoherent.

Geometry was far from being the only domain of mathematics that went through such changes. Algebraic number theory in Dedekind's extensive reworking is another example, and it generated a considerable extension of the concept of integer.[20] The whole topic of the definition of a number is a related issue, and Cantor showed that the clearest conceptual definitions of cardinal and ordinal number led unstoppably to new kinds of transfinite numbers. Taken together,

---

[17] See (Toepell [1986]).

[18] See (Frege [1980]).

[19] See (Padoa [1903]).

[20] See (Gray [1992]) and Avigad's essay in this volume.

the algebraic integers and the transfinite numbers exemplify the introduction of abstract sets as the foundational objects of mathematics.

Modern Analysis

Even mathematicians not known for their liking for transfinite mathematics contributed in characteristic ways to the modernizing, abstract tendency. A brief look at the advent of measure theory and point-set topology helps establish this point. Ferreirós (1996, 1999) has usefully pointed out that the naive concept of a set was developed under three distinct headings: as a foundational device, as part of a new theory of the infinite and the transfinite, and within several emerging branches of mathematics, including topology and the theory of integration. Here we can consider Lebesgue's theory of integral and measure. The primary meaning of the integral throughout the nineteenth-century was that it evaluated an area, area itself being taken as a primitive concept. That integration is the inverse operation to differentiation was established as a theorem, in this case the so-called fundamental theorem of the calculus. Multiple integrals stood for the evaluation of volumes, and so on. Specific integrands could be used for specific tasks, such as the evaluation of arc-length.

What drove the mathematicians onward was a deepening sense that integrals could, or should, be taken of a wide variety of discontinuous functions.[21] This was a natural matter to raise once a function could be compared with its Fourier-series representation. Once it was clear that a function and its representation do not take the same value at every point, it was natural to enquire, as Dirichlet, Cantor, and many others did, into what can be said about the ways in which they do not. The naive theory of Fourier-series displays the coefficients in a Fourier-series representation as integrals: is it always possible to do that, or are certain badly behaved functions going to cause significant amounts of trouble? It is such questions as these that led Cantor to begin to create his theory of transfinite sets and, a generation later, prompted mathematicians to break decisively the link between integral and area.

In the 1880s and 1890s the most significant and influential theory of the integral was Camille Jordan's. He took an arbitrary bounded set E in a Euclidean space, and supposed both that he could cover it with a set of disjoint regions of known area and that it could be approximated by disjoint regions of known area wholly contained within the set. Then the first set of regions has a known area that presumably overestimates the area of the set E, while the second set of regions provides an underestimate. If the first set of approximating regions can be

---

[21] There are a number of histories of the Lebesgue integral, and it will not be necessary to follow the developments in any detail, see (Hawkins [1970]).

successively refined, the limiting value of the regions that collectively cover the set is what Jordan called the outer content of the set E. Likewise, the limiting value of the underestimates is the inner content, and Jordan said that if the inner and outer content are equal then their common value is the measure of the set E.

For the theory to work, delicate topological and measure-theoretic questions have to be answered about the set E. These are particularly acute when E lies in a plane or higher-dimensional space, which is why the theory of how multiple and repeated integrals are related is so subtle, and was to be a productive topic for further investigation. Jordan's account, which became well known through his *Cours d'analyse* [1892], starts from a naive, intuitive definition of area, proceeds to confront, and generally solve, the problems that arise in extending the naive idea to difficult cases, and proposes on the way some general theorem about the measure of sets, such as that the measure of a union of disjoint sets should be the sum of the measures of the individual sets. The starting point is, one might say, given in nature; the task is to solve the problems that are also given in nature. For this reason, Lebesgue was later[22] to refer to Jordan as a traditionalistic innovator.

By contrast, the modern theory of measure and integral possesses all the characteristic features of mathematical modernism. The central idea is that of measurable sets, so, like Jordan's, the theory is an elaboration of naive set theory. But in Borel's presentation of 1895 (his *Leçons*) it is firmly axiomatic. As he put it (quoted in (Hawkins [1970]), p. 104) 'a definition of measure could only be useful if it had certain fundamental properties; we have posited these properties a priori and we have used them to define the class of sets which we regard as measurable ... new elements ... are introduced with the aid of their essential properties, that is to say those which are strictly indispensable for the reasoning that is to follow.' Borel did not connect his theory of the integral with contemporary theories of the integral, which may be one reason why Schoenflies was not much attracted to it in his 'Report on the Theory of Sets' [1900]. But Borel's ideas seem to have caught Lebesgue's sympathetic attention, for he took a similarly postulational approach to the definition of the area of surfaces in his paper on the Riemann integral, one of five that shortly formed the basis of his doctoral thesis. The thesis, entitled 'Intégrale, longueur, aire' (Lebesgue [1902]) was published in the Italian journal *Annali di matematica*, apparently because the older generation of French mathematicians thought that it was too abstract.

In his thesis, Lebesgue defined a non-negative measure on sets by some simple axioms, and gave theorems for computing the measure of a set that showed that it coincided with earlier notions due to Jordan and Borel. He then took the significant and original step of connecting his ideas of measurability with the definition of the integral. This led him to introduce the notion of a measurable function, and

---

[22] See his [1922], p. lxi, quoted in (Hawkins [1970]), p. 96.

to define the integral of a measurable function. It will be enough here to consider only the notion of measure. Lebesgue's axioms for a measure are that it should be defined and not identically zero on bounded sets E; it should be translation invariant ($m(E + a) = m(E)$); and it should be additive on pairwise disjoint sets (if the sets E_i, $i = 1, 2, \ldots$ are pairwise disjoint then $m(\bigcup_i E_i) = \sum_i m(E_i)$). Lebesgue then showed that if the measure of the interval [0, 1] is taken to be 1, then the measure of a set on his definition is less than or equal to the outer content of the set on Jordan's definition. He then showed that there is essentially only one definition of the measure of a set that meets the above four requirements, and that all Jordan-measurable sets are Lebesgue-measurable and have the same measure on each definition.

Lebesgue's theory of the integral rapidly turned out to have many advantages over the old Riemannian theory. Some of these advantages were discovered by Lebesgue and published in his thesis, others soon afterwards. Most famously, it is a good theory for taking limits, and, with some effort, it was shown to salvage as much as could be expected of the fundamental theorem of the calculus once more-or-less arbitrary (measurable) functions are admitted into the discussion. Of course, much of Lebesgue's success was a classic demonstration of good mathematics, rather than anything modernist. But Hawkins showed very clearly that Lebesgue's theory works because it is axiomatic and proceeds from an abstract definition of measure that produces, but does not start from, an idea of area. Integral, length, and area were given new definitions, as the title of the thesis suggests, based on the now more fundamental idea of measure that in essence only has to meet the requirements that any theory of the size of sets would have to meet.

At almost the same time, and as part of the same movement among French mathematicians, Maurice Fréchet produced a similarly abstract reformulation of a more technical problem area in mathematics that was to have a decisive influence on the creation of modern topology. Fréchet was interested in the calculus of variations, a topic in which the unknown is usually a function that is required to minimize a certain integral. Examples include finding geodesics in a space, surfaces of least area with given boundaries, and harmonic functions, but many problems in physics can be expressed in the language of the calculus of variations. However, reliable techniques for solving problems of this sort had proved hard to find. It was difficult to tell a minimal value from an extremal value, and therefore a stable from an unstable solution, and the complicated theory that existed was generally agreed to be in need of improvement. Hilbert had raised it as the last of his famous Paris problems, remarking that it was 'a branch of mathematics repeatedly mentioned in this lecture which, in spite of the considerable advance Weierstrass has recently given it, does not receive the general appreciation which, in my opinion, is its due'.

One technique much employed in contemporary calculus of variations was to find a sequence of successive approximations to the sought-for function. Fréchet proposed a simple analogy. Just as one might approximate π, say, by finding among all numbers a sequence of successive approximations to it, so one might situate the successive approximations to a sought-for function in a space of all plausibly relevant functions. There might indeed by many such spaces, depending on the properties of the functions involved, but a much graver difficulty facing Fréchet was that any such space was surely infinite, indeed infinite-dimensional, whatever that might mean. Undaunted, Fréchet proposed to show that in many contexts one could introduce a sense of distance in a space of functions and speak of a sequence of functions in this space as tending to a limit, exactly as a sequence of approximations to π tends to π. What were functions in one setting became mere points in a space in Fréchet's new view of things.

Fréchet succeeded in showing that there were many problems in the calculus of variations that could profitably be formulated his way. He found spaces of functions of various kinds and spaces of all curves of certain kinds could be made into metric spaces: spaces in which it made sense to speak of the distance between two points, and where a limiting process could be defined. Most importantly, he showed that these spaces could be complete (strictly, sequentially compact, but the details need not concern us) which means that if a sequence tends to a limit then the limiting value is also in the space.[23] Thus in Fréchet's vision, the concept of distance is greatly generalized away from any sense of distance in any Euclidean space, and subordinated to an approach that is axiomatic in spirit. Fréchet was clear that what was wanted was the ability to talk of distance and to take limits, in the abstract senses he had in mind, and one had this ability whenever a space satisfied certain axioms. Initially he talked of spaces of class D (or distance spaces) where it made sense to talk of distance. These spaces were later called metric spaces admitting a distance by Hausdorff. Then Fréchet introduced spaces of class V (for 'voisinage' or neighbourhood in French) where it made sense to talk of points being close, which he soon realized were equivalent to spaces of class D. Finally, Fréchet spoke of normal spaces of class V when it is possible to talk of limiting processes converging to an element of the space.

Examples such as these, drawn in the main from longer accounts given elsewhere by several writers, show that, some time around 1900, there were eminent mathematicians prepared to argue that mathematics was no longer based on the primitive acts of counting and measuring and that it was no longer any kind of idealized, abstracted, simplified science. It had cut itself off from such things. Its objects were defined independently of science, and its methods were naturally different. They establish that there is a real sense in which we may speak of Modernist mathematics. The question then becomes: so what?

---

[23] This need not be true: any sequence of rational numbers tending to π has a limit (π, of course) but π is not a rational number, so it is not an element in the set of rational numbers.

Mathematical modernism and changes in philosophy

If the transformation of mathematics was so sweeping, then an exclusive diet of tightly focused studies will inevitably fail to address the breadth of the transformation. But the alternative to a number of specialist studies is not to collapse into the arms of a generalization so sweeping that Picasso sits on the page with Einstein and Noether. Rather, we have an opportunity to compare the general features of this transformation of mathematics with changes in other cultural and intellectual activities may have around 1900. There are a number of interesting points of comparison. One is the struggle between the Leibnizians and the various strands of Kantians. Another is the use of language as a calculus. I shall briefly discuss each of these, and then move towards some more methodological points.

There is no dispute that there was a transformation of comparable magnitude in philosophy at this time, and many of those most active in those changes were either mathematicians by training or positively oriented towards mathematics. Among such people in England were Russell, Whitehead, and in due course Wittgenstein. In Germany, there were Husserl and Frege; in France and Italy Poincaré, Couturat, Peano and Enriques. On the other hand, the names of those active at the time but eclipsed reads like a list of discredited figures (whether fairly or not): Bradley in England, in Germany Lotze and various neo-Kantians, and in France Bergson and Brunschvicg.

Leibniz revived

The rallying cry for some of the leading movements in philosophy of mathematics was Leibniz. Anti-Kantians of various kinds spoke of themselves as followers of Leibniz, although, given the fragmentary state of Leibniz's collected work and the absence of significant systematic essays or books by Leibniz himself, it should not be asserted that the followers accurately followed the seventeenth-century philosopher after a gap of so many years. But then, it was the frequent cry of this or that neo-Kantian that only they, and not some other pretender, had accurately divined Kant. It would be very interesting to have a study of the Leibniz revival.[24] The Russell literature discusses the ways in which Leibniz was an influence on Russell[25], and I am sure that a sweeping study of Leibniz's influence would be rewarding.

Those who took up the Leibnizian cause, as they saw it, praised him for two different things. One was his belief that all truths are analytic, and the corollary that seemingly synthetic truths such as 'all mice are small' are rather to be understood

---

[24] It is regrettable that Peckhaus's book (Peckhaus [1997]), very informative though it is on the period from Leibniz to 1900, largely contents itself with Schröder.

[25] See, for example, (Griffin [1991])

as analytic truths that our finite minds have not properly understood. It is a limitation of our thinking that produces what seem to be contingent truths, and as we understand matters better such truths will become clearer until, with perfect understanding, we shall see that they could not be otherwise. This odd belief drove Russell and Couturat towards logicism, in the hope that at least for mathematics all truths could be proved to be analytic, which would be a significant step towards vindicating Leibniz's rationalism.

The second insight of Leibniz's was taken to be his emphasis on the importance of a proper language that would reduce philosophical disputes to matters of calculation. They agreed straight away that Leibniz had not created such a thing, nor had he had one at his disposal; his legacy in this respect was inadequate. But help was now at hand in the form of the mathematical logic of Boole, as improved by Jevons, Peirce, and Schröder. Thus Schröder advocated a pasigraphy of his own devising, and Russell found much to praise and adopt in the symbolism and stylized language of Peano. Indeed, as with Leibniz himself, the issue of a good artificial language for communication of all sorts was a related issue for several of these people.[26]

The situation with the Kantians is more complicated. As has already been remarked, it is not possible to give a simple, uniform characterisation of 'the' neo-Kantian position around 1900, because there were at least two such positions, the so-called Marburg school and the South-Western school. Then there were mathematicians and scientists who ventured into these waters flying some Kantian colours, most notably Helmholtz and Poincaré. There is also the striking fact that Hilbert's chosen philosophical ally was Leonard Nelson, a neo-Friesian, and therefore a Kantian of sorts.[27] Whereas the Leibnizians were concerned with restoring Leibniz's thought to a purported whole, the neo-Kantians were concerned to defend a stripped-out Kantianism, devoid of this or that feature that was taken to be discredited. Grounds for rejection were usually labelled as psychologistic, and while this alerts us to a notorious turf war of the period,[28] it is also the case that in rejecting the Kantian faculty of intuition the neo-Kantians made it difficult for them to produce a persuasive, coherent position.[29] Deprived of the resources they were forced to load everything onto logic, and this was to prove inadequate.

What caused the Kantians grief was, as is well known, non-Euclidean geometry. Any straight-forward reading of Kant's *Critique of Pure Reason* makes it clear that Kant believed that the intuitions that make knowledge of space possible delivered – as synthetic a priori truths – the theorems of Euclidean geometry. Once non-Euclidean geometry was accepted by mathematicians, anyone holding

---

[26] I explored some of these issues in (Gray [2002]).

[27] See the thorough account in (Peckhaus [1990]).

[28] See (Kusch *Psychologism*) for the turf war, and (Kohnke [1991]), for a largely sociological account of the earlier neo-Kantians.

[29] This is explained with admirable clarity in (Friedman [2000a]).

a Kantian position had a problem. What is to prevent the curious from giving in to temptation, and supposing that perhaps non-Euclidean geometry was true all along? A strict Kantian would object that we cannot know about space as such, and so we cannot be tempted in such a fashion; it might simply be that we cannot do otherwise than construe space as Euclidean. But common sense and, by the end of the nineteenth-century, popular fashion dictated that when two mathematical descriptions of space are available it should be possible to determine which is correct, and that it is unreasonable to suppose that we must merely choose one, rather than the other, in order to make space intelligible at all. If, as a species, we make such a choice, it would be a matter of human psychology, and Kantians wished to avoid that.

One way out was to deny that non-Euclidean geometry was given in intuition. It would then merely be a coherent logical system deprived of real existence, and therefore not a challenge to Kantian philosophy. Although I have not seen any philosopher in this period make this analogy, it would be exactly analogous to the points at infinity in projective geometry.[30] No-one ascribed real existence to them; they were not accessible, for example, in the way that points in the finite part of space were, but they were logically sound and geometry could be done validly with them. This denial was made problematic by Helmholtz, who argued in his famous paper [1870] that a non-Euclidean space would be as intelligible and intuitive as a Euclidean one. His grasp of Kantian niceties was contested by the orthodox, but much damage was done.

Leonhard Nelson

Nelson's views are particularly interesting, because they were intended to coexist with those of David Hilbert. They will also throw up a characteristic attitude to logic that distinguished the Kantians from the Leibnizians and, arguably, many mathematicians of the period, as will be discussed below. In his [1905] Nelson argued that mathematical axioms are synthetic (they are not matters of logic); they are not empirical because they are not about empirical objects (they are independent of experience), and so they are pure intuitions; and are therefore synthetic a priori. A synthetic judgement can be negated by contradicting other, accepted, synthetic judgments, but does not involve a self-contradiction. So the logical possibility of non-Euclidean geometry has nothing to do with the synthetic characteristic of geometric axioms, and nothing in Kantian philosophy prevents non-Euclidean geometry from being a logical possibility.

That did not mean, however, that non-Euclidean geometry was an intuition. It might be simply something that can be reasoned about. Nelson's view was

---

[30] Nelson, in his [1905], cited the example of spaces of various dimensions.

that the thinkable included everything that was not self-contradictory, and thus exceeded the domain of the intuitive. With this distinction in mind, he turned to the question of the origin of geometric axioms in intuition. The very possibility of non-Euclidean geometry shows that geometric axioms are not logical in nature, and this supports Kant's insight that logic and empirical investigations are not the only source of axioms, but that pure a priori intuition is another. This, Nelson said, quite wrongly, was a significant point that Helmholtz and Riemann were not the only ones to fail to recognize.

As for the relation of mathematics and logic, Nelson knew very well that Frege had proposed to reduce arithmetic to logic. His critique here was that for this to succeed it is necessary to have a criterion for recognizing when something is of a purely logical nature, such as Kant's distinction between analytic and synthetic judgements. But then it is easy to recognize the non-logical character of the premises of arithmetic, for the idea that every number has a successor is not at all a necessary idea but arises only in pure intuition. This showed, according to Nelson, that Frege's programme could not reduce arithmetic to logic, and of course it was only arithmetic, not geometry, that Frege sought to analyse in that way. It also shows rather clearly, the very different attitudes to logic that the Kantians and Frege (and the logicists) had.

Nelson now returned to the nature of non-Euclidean geometry. He distinguished mathematical existence sharply from logical existence (which he took to be freedom from contradiction) on the grounds that one rests on synthetic, the other merely analytic, hypothetical judgements. Logic provides only for statements of the form 'If X exists, then ...'. However, to guarantee the existence of the system of objects described by an axiom system a further existence axiom is required. This is because mathematical existence is a matter of pure intuition, not logic (nor, of course, is it an empirical matter). This led Nelson to consider Poincaré's views on the origins of geometric axioms. He then recognized that Poincaré had correctly dismissed the idea that geometrical ideas are empirical, and that Poincaré agreed with Kant that mathematical knowledge was apodictic without being reducible to logic. He recognized that 'this approach of the great French mathematician to the critical standpoint' was an advance, only to deplore all the more the 'total misunderstanding' that followed. This was Poincaré's assertion that a synthetic a priori judgement would impose itself with such force that a contrary proposition could not be conceived, and therefore non-Euclidean geometry could not exist. From this, Poincaré concluded that the axioms of geometry are conventions. In Nelson's view, by contrast, only an analytic judgement has a negation that cannot be conceived. And what could axioms be, if they are neither a posteriori *nor* a priori judgements?

Nelson replied to his own question that to distinguish a Euclidean from a non-Euclidean triangle in space could only be a task for someone equipped with the Kantian idea of pure intuition, for someone having only logic and experience to guide them could accept either Euclidean of non-Euclidean geometry, or indeed any geometry, as correct. An extension of Poincaré's conventionalist position would always be available and the question which geometry is valid would have no sense, and one could only ask which geometry is the most appropriate or convenient. However, the origin of geometry is not to be found in logic or experience. Geometry is a different form of knowledge, whose judgements are synthetic a priori. Poincaré, said Nelson, had failed to recognize this possibility, and believed that if a proposition was founded neither in logic nor in experience then it forfeited any claim to knowledge.

In conclusion, Nelson wrote, the apodictic nature of mathematics together with the possibility of non-Euclidean geometry is a reliable touchstone for any philosophy of mathematics. To hold only one of these views is to see it break on the rock of the other one. The way out, for Nelson and his followers, was to regard non-Euclidean geometry as thinkable but not a synthetic a priori judgement such as Euclidean geometry rests upon.

Nelson thus emerges as an ultraorthodox Kantian. Kant had argued that the way we automatically make sense of the world and the objects in it, although we have no access to objects in themselves, and so we cannot picture or represent them, and although sensory experience is totally unstructured, is transcendentally. We structure or constitute these objects via the a priori structures that are the pure forms of intuition – space and time – that Nelson was also prepared to invoke. Transcendental logic was Kant's term for the rules that make it possible that thoughts can apply to objects, as opposed to other thoughts (the subject matter of general, or, one might say, traditional logic). It works by means of certain a priori concepts so it is resolutely anti-empiricist.

Ernst Cassirer and the neo-Kantians

The neo-Kantians were equally anti-empiricist, but they dropped the idea of a faculty of pure intuition as being suspiciously psychological. As Friedman very nicely shows, (Friedman [2000a]), they had therefore to explain how thought could apply to objects by means of a priori concepts which could not be spatio-temporal. How could this 'pure logic' relate to logic as understood by the Leibnizians, and also to mathematics. Here the neo-Kantians divided, and I shall consider only the Marburg school (Natorp and Cassirer, specifically). It will be convenient to set out their views on number before returning to the analysis of space and geometry.

For Natorp, number belonged to pure logic, not to intuition, and not to logic as, for example Couturat understood it. The 'one' that forms the unit of counting was not to be defined by some formula, because no such formula can explain how counting is possible. Cassirer, in his 'Kant und die moderne Mathematik' (Cassirer [1907]) was of much the same opinion: critical philosophy began where logistic stopped. The term 'critical philosophy' was the neo-Kantian's way of identifying the key to their general approach. They sought to analyse the conditions that made science possible (not, of course, to add to science itself). This inevitably took them into historical analyses of the development of science, mostly physics, but also mathematics, because Kant's own route, the understanding of human cognition, seemed fraught to them. So they analysed scientific reasoning, critically as they said, in order to find the philosophical grounds for its success. Hypothesis and theory were their focus, and because science was in flux around 1900 it was scientific method that they focused on most of all – hypothesis and theory formation, rather than this or that theory or hypothesis.

Mathematics too was in flux. Cassirer and Natorp found themselves largely in agreement with the Russell of *Principles of Mathematics* and ultimately with Dedekind that numbers are a particular type of serial order. 'The essence of the 'numbers' is completely expressed in their positions', wrote Cassirer.[31] The succession and distinctness of the elements is purely conceptual, and arithmetic is not, as Kant had proposed, the science of pure time, except insofar as time is removed from the fundamental idea and only 'order in progression remains', an approach Cassirer noted that had already been taken by Hamilton. In the same historically motivated spirit, Cassirer could accept that the formal side of logic now embraced the theory of relations, which again he seems to have taken from Russell's *Principles of Mathematics*. Historical enquiry went only so deep – but the point was not to write history but to obtain from the historical record enough evidence of scientific, or in this case, mathematical method to ground a critique.

Cassirer's historical survey of nineteenth-century geometry in (Cassirer *Substance and Function*) comes close to being uncritical, even endorsing at one point Russell's views in *The Foundations of Geometry* that projective geometry was the most fundamental geometry, a view that Russell had abandoned, first after heavy criticism from Poincaré and then as he moved away from Kant and towards Leibniz. Cassirer also endorsed Klein's more modest, because not philosophical, view in the *Erlangen Program*, as somehow bringing change into the orbit of geometry while retaining the Platonic ideal of geometry as the study of the changeless and eternal.

The critical analysis comes with Cassirer's turn to metageometry. It emerges that Cassirer too had moved towards Leibniz: not only does logic embrace the study of relations, but Hilbert's formulation of geometry is a 'pure doctrine of relations'; points, lines, and planes are defined by Hilbert's axioms much as numbers are by Dedekind's. Cassirer argued that this approach surpasses the empiricism of Pasch in many ways, for example, it yields geometries of many dimensions. These naturally exceed the capability of spatial intuition, so spatial intuition is not the means by which we construct geometry. Such geometries are made of abstract, hypothetical truths accessible to the intellect. On the other hand, some sort of experience is clearly involved in our fastening on to Euclidean three-dimensional geometry as the correct geometry of space. What experience enables us to do is pass judgement on theories, but even then it is experience shaped into a possibly

---

[31] In (Cassirer *Substance and Function*) p. 39.

rival theory (there are no raw facts for Cassirer). As a result, neither experience nor experiment can decide whether a piece of physics or of geometry should be given up when there is a conflict; to that extent Poincaré's conventionalism is upheld. Indeed, because Euclidean geometry is the simplest (and, in Cassirer's opinion the only possible infinitesimal geometry) it will be preferred. On the other hand, any logically consistent system of relations is capable of being given a meaning, so non-Euclidean geometry could become meaningful. And it might even be, Cassirer concluded by saying, that one day some firmly established observations could appear that could not be accommodated by modifications to our scientific theory of nature, and then we might agree to modify the 'form of space'. 'It is only the pure system of conditions, which mathematics erects, that is absolutely valid ... the pure concept ... is prepared and fitted for all conceivable changes in the empirical character of perception'.[32]

It is hard to avoid a feeling that Cassirer was facing in too many directions at once. Poincaré was adamant that geometric conventionalism meant that no experiment could decide the issue of space. Cassirer feels the force of this, but holds the door open for some, unspecified, future change. Now, it can be argued the general theory of relativity indeed makes geometric conventionalism untenable, as Friedman does in his *Foundations of Space-Time Theories*[33] thus vindicating Cassirer's caution, but Cassirer did not identify a weak point in Poincaré's analysis. Rather, deprived of a priori intuitions as a means of structuring experience, Cassirer has nothing but logic to provide us with geometry. That is an overgenerous source of supply, so experience (theorized to some degree) helps us out. The methodological secret of science, as identified by critical philosophy, is the way it adjudicates between these various theories.

Continuity, change, and Modernism

An intellectual position that asserts a complete break with the past and that nothing can any longer be as it was, can be that of certain protagonists at the time, be they artists, writers, mathematicians, or critics. It is very unlikely to be the position of people writing well after the event, because there is a great element of propaganda in such claims that sober reflection cannot sustain. One can see that ideologies were in play at the time of mathematical modernism that it would be interesting to trace, and ideology as a force in shaping mathematics is a theme that has been insufficiently pursued.

The term 'ideology' is defined variously. I use it here in the sense of 'integrated assertions, theories, and aims that constitute a sociopolitical progamme'[34] where

---

[32] (Cassirer *Substance and Function*), p. III.

[33] (Friedman *Foundations of Space-Time Theories*).

[34] This is definition 2c in *Webster's*.

the 'sociopolitical progamme' is to be understood as one aimed at mathematics. So an ideology is a normative position, and to espouse an ideology for mathematics is to believe that at least an appreciable area of mathematics should be done in a particular way. There is a sense that some new method should be of general application, and will precede most specific problems. Thus the belief that geometric problems should be done by algebra was part of the ideology of the enthusiasts for Cartesian co-ordinate geometry, and the contrary view that geometry is really about lengths, angles and similar concepts, and algebra is alien, is part of the ideology of synthetic geometry.

Modernism in mathematics can be an ideology without there being tub-thumping manifestos. Descartes launched his new vision of geometry in his *La Geometrie* in a variety of rhetorical ways that declared its ideological character, without forgetting that in mathematics results speak more clearly than words. The sweep of his programme, its apparent embrace of an entire class of problems most of which had never been contemplated before, the way it generalized and dispatched Pappus's problem, and the careful way it delineated to what it was applicable and what was beyond its scope all displayed its programmatic nature. Descartes' nonchalant refusal to show how the transition from geometry to algebra works in the simplest cases and his occasional remark about how powerful the new methods are, are rhetorical devices intended to force the reader to accept the work on his terms. By marvelling at it, the reader will be inspired to do likewise.

The modernist ideology in mathematics preached renunciation from the world, in the sense that one did not do geometry or analysis, to take the two topics discussed above, by taking as given what is presented by idealized common-sense. That tells you what a straight line is by accepting what is generally agreed and refining it into a definition. From the feeling that every rational person can recognize a straight line when they see one, the old argument moved to a belief that all such people share the concept, which, if need be, can be articulated. The modernist argument preferred to define straight lines only as part of a system of definitions for, as it might be, plane or higher-dimensional geometry, and it did this not by telling you what a straight line is, but by telling you what you could say about it. Whatever met the definition was a straight line, even if they might look very strange. There was no attempt to show that the new, implicit, definitions somehow captured the essence of the real object, because the real object was only incidentally what it was about. The same was true, mutatis mutandis, of area. Measure theory was an axiomatized theory from the start in the belief that specifying the allowed deductions about size would give the requisite degree of control of the concept in a way that idealizing it could not. Indeed, there were those commenting on the empiricist position in geometry who pointed out that experience could not get started because there were *no* points, lines and planes in our experience.

Axiomatics was a key component of mathematical modernism. With it came a belief that mathematics could and should be about abstract systems capable of many interpretations. Cassirer picked up this point from a source that deserves to be brought back to the attention of historians of mathematics: the essay on *Grundlagen der Geometrie* by Josef Wellstein in (Wellstein Weber and Wellstein). He rightly praised the essay for its 'very instructive examples and explanation'. There Wellstein showed how mathematicians now commonly used an abstract system of geometry to prove results about whole families of shapes, by conceiving of a space the points of which were conics, or spheres, or whatever it might be, in some other space. In the same way, although there was no space to discuss it above, pure, abstract measure theory did not require that one consider only the limited stock of measures associated with the nineteenth-century calculus. Perhaps the best simple example of the freedom this brought was the Radon's work of 1913.[35]

The new ideology advocated the axiomatic formulation of theories and their study in the abstract as free creations. It was not part of the ideology that mathematics should be studied purely for its own sake, or that all systems of axioms were equally deserving of attention. The relation of the new mathematics to the world was complicated but vital, and while examples can be given of mathematics 'for its own sake' it should also be noted that throughout the period when he was, by force of his example, the leading moderniser, Hilbert was actively involved in the study of physics. He advocated an abstract approach, but his interest was sincere and deep, as Corry has shown in his (Corry [1997]) and (Corry [2004]).[36]

Mathematicians do not often draw attention to the errors of their predecessors. They prefer to put in place a new way of proceeding. But, as discussed, when it came to geometry they could not do that. The profusion of new geometries, some of them theoretically applicable, formulated in ways that were neither Euclidean, synthetic, nor Cartesian, made it impossible to avoid the impression of a revolution. Some mathematicians drew out the philosophical implications of non-Euclidean geometry as they saw them, others made a point of saying nothing. In the main, those who wrote about the significance of the novel methods saw them as marking an abrupt change, and while historians might want to stress certain continuities, it seems impossible to deny that mathematics was changed permanently by the arrival of abstract axiomatics and multiple interpretations. Some other branches of mathematics, complex analysis perhaps, or differential geometry, might seem to have continued to operate as if unchanged by the upheaval in the foundations, just as cars can be given new types of engine, but even they adapted to the new ideology. Even if it is conceded (too generously, as I hope to show elsewhere) that these disciplines were unaffected, this would not show that there were not massive changes elsewhere. The claim for continuity in the presence of change may be granted in this case, but not if it implies that the continuities eclipse the changes.

---

[35] Hawkins, (Hawkins [1970]) p. 179 writes that 'Radon [1913] initiated the study and application of integrals based upon measures more abstract than those of Lebesgues'.

[36] See also his essay in this volume.

What of the claim that these changes were indeed significant, but were merely the result of work done much earlier: by Cauchy in analysis and, one might suppose, Bolyai and Lobachevskii in geometry? Or, if the wretched reception of the work of Bolyai and Lobachevskii is recalled, may we not ascribe much of the change to the work of Riemann? This second question brings a methodological division to the surface that I alluded to at the start of this chapter. We may dismiss the claims of Cauchy to have pre-empted modernism, on the grounds that the shift from an incoherent or at best naive theory of the real numbers to any theory advocated in 1900, with the consequent arrival of naive set theory, is too big a shift to be treated as a mere consequence. But the example of Riemann is quite different. It cannot be claimed that his vision of geometry was properly appreciated for many years after his death. It was essential for Beltrami, but the study of spaces of higher dimension than two, and of spaces of non-constant curvature, was scarcely taken up until the twentieth-century. Indeed, it can be argued that it took Einstein's general theory of relativity to push mathematicians to take up where Riemann had left off.[37]

One could say that Riemann is too early to count as a modernist. Or that, on the contrary, the delayed reception of his most 'modernist' work shows precisely how deserving of the label it is (which, on philosophical grounds, one could surely argue that it was). Or, one could try to argue that the strong intellectual affinities his work has with 'mathematical modernism' and his position in the middle of the nineteenth-century imperils the concept of 'mathematical modernism'. Either it must be stretched to accommodate him, in which case it is not only too long a period but its claim to the modernist label is unfounded on chronological grounds, or he is denied membership, and the intellectual coherence of mathematical modernism evaporates.

I believe a useful answer is given by tackling this last claim head-on. Riemann in his approach to geometry certainly did hold a view of mathematics in its relation to the world that was abstract and deliberately open to many interpretations. But it was not axiomatic, and, as noted, it was not taken up until much later. Riemann is a typical precursor of a movement, the sort of person who is described unthinkingly as 'ahead of his time'. I would not want to try and discover a worthwhile meaning in the idea of being ahead of one's time, but a precursor is a precursor, a challenge to the accepted chronology for those who like their history tidy and fear that too much disorder in an account hints at an intellectual incoherence. Whatever it might mean to understand a piece of the past, the historian's arguments should make a reasonable amount of sense. I believe that the wrong way to deal with Riemann here would be to accept the terms of the alternatives just sketched. Rather, the historian should note that many different

---

[37] There are examples of work on Riemannian geometry in the spirit of Riemann in the period 1866–1914, quite a lot of it by Bianchi, and they are worthy of proper study, but misunderstandings of the concept of the curvature of space abound.

interpretations all have something to commend them, and that that is the best we can hope for.

Mathematical modernism is offered as a useful characterization of a major change in mathematics. As such, it captures enough of what the original actors saw as its significance, its novelty, and the nature of the transformation for it to be a guide to the historian. But no one analysis can capture all of an event of the size of that transformation. The past is untidy because it was alive. Riemann is not a refutation of the case for mathematical modernism, but he is an eloquent example of what mathematical modernism does not fully encompass. Once that is recognized, one could make a sharper than usual case for his exceptional nature as a result. Or one could ask for a more detailed study of how his ideas made their way into mainstream mathematics. Tappenden, for example, argues that Frege is better understood by seeing that his appreciation of Riemann is part of his hostility to Weierstrassian analysts.

The third objection to the case for mathematical modernism concedes its importance, but argues that the changes it purports to capture are better understood as one of a sequence of major changes mathematics periodically undergoes. None of these changes should be exaggerated (claimed to be revolutions, for example) and this one has no justified claim on the label modernist, other than the rough chronological coincidence. Indeed, there have been a number of such changes, and most likely each of them has raised the level of abstraction in mathematics. One droll example is Cauchy's rigorization of the calculus, which so annoyed the military cadets at the École Polytechnique who were its first audience that they stamped their feet in protest and had to be disciplined – as was Cauchy.[38]

The possibility of seeing modern mathematics as a species of modernism rests on the significant analogy that exists between it and contemporary developments in painting, literature, or music. Painters, writers, composers know they have to adapt their skills to the market-place that, generally, sustains them, although some make their living in other ways, perhaps as instructors, teachers, or performers. They also know they have to distinguish themselves from their contemporaries by their mastery of technique and by some element of originality or independence of mind. On occasion, and for the leaders in the field, the inner requirement of originality drives them to innovate in an unusual way that nonetheless has enough structure to be adoptable by other original artists, and a style is born.

To be brief about it, the striking feature of the innovations of Schoenberg or Picasso is surely that a novel grammar was offered for the composition of new works, one that, in each case, moved sharply away from existing canons of representation and placed a greater emphasis on the formal side of music or painting. There was a consequent shock that spread through the cultural milieus as this novelty was confronted, which signals the magnitude of the changes they were introducing and establishes very clearly (not that it needs to be said today)

---

[38] Cauchy's teaching, and the varied responses it drew, is discussed at length in (Belhoste *Cauchy*) Chapter 5.

that theirs was not the ordinary, year-on-year innovation that sustains the artistic worlds. In these cases, the inner necessity that prompted these artists to take such risks with their careers, and that brought followers to them, was allowed to flourish by a heightened degree of independence and autonomy for creative artists by comparison with previous generations. It was possible to proclaim a new style on the grounds of artistic sensibility and with little regard for customers' and patrons' tastes.

How this played out is well known. Picasso became a cultural icon, but everyone who writes about Schoenberg has to confront the fact that no twelve-tone composer wrote crowd-pulling music. The point to insist on, however, is that their innovations appealed to and attracted significant numbers of other artists at the time, and that it was barely possible after their interventions to paint or compose as if they had never existed. One can also look back and see that there were precursors, and that Modernism is a sprawling thing. From late Cézanne to early Picasso is a thought-provoking journey; probably no painter ever worked without thinking of painting as putting paint on a flat canvas. Schoenberg aimed at a theory of harmony to match that of Bach. Major changes, to be sure, but part of a series of such in the history of art or music, and with continuities too.

Mathematical modernism likewise surpasses the requirement of originality imposed on any ambitious mathematician of the late nineteenth and early twentieth centuries. It offers a style that others could, and did, adopt and adapt. It emphasized the formal aspects of mathematics, and allowed for complicated ways in which mathematics can relate to the world. Mathematical modernisers such as Hilbert could do so with some chance of success because of a heightened degree of autonomy from, in their case, physicists and engineers, but they did so from a powerful sense of the inner necessity of their work. Finally, the shocked response to their achievements is visible in the extensive discussion of its implications, the controversies it generated, the wholesale shift in the philosophy of mathematics that resulted. It was not necessary that every mathematician thereafter work in the new way: for many purely mathematical purposes, not to speak of practical ones, the well known (Halmos *Naive Set Theory*) is all that a working mathematician need know in order to do excellent work in many areas of real analysis.

The large literature on cultural modernisms hardly suggests a straight jacket.[39] There is no party line, rather, a vigorous debate. The issue of precursors has already been mentioned, and there are others: national considerations, for example the anomalous case of America; East European literature as opposed to Western European; other artistic styles, as, for example, Stravinsky's. The existence of so many versions of Modernism, and so many historical accounts, should be seen as an opportunity for historians of mathematics. Most likely, some of these debates will not resonate in the mathematical arena – it will be hard, I suspect, to find many strongly anti-bourgeois mathematicians – and I doubt

---

[39] I have not seen a wholesale attempt to discredit the concept of Modernism, such attempts exist for the Enlightenment, and one supposes a deconstruction of 'Modernism' would not be without point.

if there was much cross-fertilization.[40] The prospects for historians of mathematics are similar. Indeed, America is an anomalous case. One eastern European country, Poland, staked its entire mathematical future on the most modernist disciplines of modern mathematics: logic, topology, and their implications for real analysis. Mehrtens, in his innovative book (Mehrtens *Moderne Sprache Mathematik*) has already explored some of the implications for mathematical modernism as he understands it in the German context, and he certainly does find Nazi mathematicians and those who sympathized with the Nazis.

Conclusion

Where then would an account of the history of mathematics that accepted mathematical modernism be placed? Again, a plurality of answers is possible, and even advisable. A book such as (Moore [1982]) needs its rich detail in mathematics and mathematical logic, and, however regrettable it is, many historians of science would have to wrench themselves around considerably to appreciate it. Peckhaus's book (Peckhaus [1997]) looks at a bigger picture, tracing a journey from Leibniz to Schröder, where it stops, but it is no criticism to say that it probably appeals to much the same audience as Moore's. However, one may also acknowledge that the whole question of what is logic strikes at the heart of the intense nineteenth-century debate about psychologism.[41] Is logic normative, or is it an account of how we reason? It was not a casual gesture when Boole called his most famous book *The Laws of Thought*.[42] There are massive currents at work here, some to do with the nineteenth-century psychologization of Kant. Issues to do with language run right through philosophical positions held at the time: can one step outside language or is one, as it were, trapped within it? Is a largely syntactical account of reasoning, such as Leibniz may have wanted, an aid to thought, an account of thought, or a way of avoiding philosophical questions?

It is well known that the world of linguistics went through its own 'modernization', brought about by the work of Ferdinand de Saussure. I have been unable to trace a direct influence between mathematics and linguistics, in either direction. Ferdinand had a brother, René, who was a mathematician, but I have found no evidence that the one learned from the other. They did meet, however, in the world of artificial languages, Esperanto, latine sine inflexione, and the like. It is a striking fact that among those most involved in the move for an international language were a number of linguists (such as Jespersen, a structuralist in

---

[40] Hausdorff might qualify because of his upbeat Nietzscheanism, and it would be interesting to know if Hilbert, or Emil Artin to pick a more likely candidate, listened to Schoenberg.

[41] For which, see Martin Kusch's (Kusch *Psychologism*).

[42] (Boole [1854]).

Saussure's tradition) and mathematicians (Peano most notably, but also Schröder and Couturat; even Frege was not opposed). There is a marked overlap between the drive of Peano and some of his followers for an unambiguous language of mathematics and an unambiguous language for scientific exchange[43], and it is particularly interesting that Leibniz was a rallying cry here too.

Many of these questions came round when the search for the logical foundations of mathematics was most intense. I think many aspects of this congerie of topics could be explored much further. I wish to offer one observation, which concerns the implications for logic. Questions of the finite versus the infinite, and of the capacity of the human mind to grasp the infinite, are well known. Distinctions of a somewhat Kantian kind were made between the mathematical and the logical by some, Kronecker for example: the mathematical involves a construction of some kind. Others, who retreated from the paradoxes of intuitive set theory, did so in the name of some form of constructivism, which ultimately invokes human understanding (Emile Borel, for example).[44] Those who attempted to drive the paradoxes away, such as Zermelo, found that they were forced to produce axioms, the truth of which was not as obvious as they would have liked. Could these axioms be ascribed to logic? If so, did their awkward nature signal that they were not innate, or could they be innate but not, actually, correct?[45]

The more mathematicians sought to reform logic and mathematics simultaneously, the more their logic ceased to resemble laws of thought. Hilbert's finitistic reasoning is a concession to how humans reason, as Hallett pointed out, but I think a careful analysis of the issue would show that the question of psychology has dropped out of sight entirely.[46] The Modernist foundations of mathematics ultimately dispensed with the idea that the subject matter of logic was the correct rules of reason – those that would be followed by any undamaged mind. A part of logic does consider such rules, but it seemed ever more obvious that the logic needed to create genuine mathematics is not a candidate for even an idealized description of the way people think. Not only geometry, not only the conception of number, but eventually any simple-minded association of logic with correct thinking was made anew.

It is now clear that if the story of mathematics and logic necessarily goes through topics in philosophy, questions about the nature of language, and the history of psychology, then no-one can write it all. There will be several authors, with several, necessarily partial, views. And within that chorus, there will be places for the history of mathematics within the history of culture.

---

[43] Described in (Gray [2002]).

[44] See (Moore [1982]).

[45] A possibility admitted for geometry by Voss, who wrote 'It is indeed the task of mathematics to show that the space underlying our intuition is not necessarily to be described as Euclidean geometry teaches', (Voss, Wesen), p. 91.

[46] See (Hallett [1994]).

References

Abel N. 1827/1828, Recherches sur les fonctions elliptiques *Journal für die Reine und Angewandte Mathematik* pp. 101–181, pp. 160–191 (= *Oeuvres* I pp. 263–388).
Abrusci, V. M., 1980, 'Proof', 'Theory', and 'Foundations' in Hilbert's mathematical work from 1885 to 1900; in: Dalla Chiara (ed.), *Italian Studies in the Philosophy of Science*, Dordrecht 1980, 453–491.
Aczel, P., 1977, An introduction to inductive definitions; in: *Handbook of Mathematical Logic*, J. Barwise (ed.), Amsterdam, 739–782.
Ajdukiewicz, K., 1921, *Z metodologii nauk dedukcyjnych*, Lwow. Translated by J. Giedymin as *From the Methodology of the Deductive Sciences* in *Studia Logica*, 19, 1966, pp. 9–46.
Aguilar, M., S. Gitler, et al. 2002, *Algebraic Topology from a Homotopical Viewpoint*. New York, Springer-Verlag.
Ahlfors L. 1953, Development of the Theory of Conformal Mapping and Riemann Surfaces through a Century in *Contributions to the theory of Riemann Surfaces*, Annals of Mathematics Studies 30 Princeton: Princeton University Press pp. 3–13.
Ahlfors L. 1966, *Complex Analysis* (2nd edn) New York: McGraw-Hill.
Alexandroff, P. & Heinz H. 1935, *Topologie*, Julius Springer, Berlin. Reprinted 1965, New York: Chelsea Publishing.
Alexandroff, P. 1925a, 'Über stetige Abbildungen kompakter Räume', *Proceedings of the Section on Sciences, KNAW* 28, 997–99.
Alexandroff, P. 1925b, 'Zur Begründung der n-dimensionalen mengentheoretischen Topologie', *Mathematische Annalen* 94, 296–308.
Alexandroff, P. 1926, 'Über stetige Abbildungen kompakter Räume', *Mathematische Annalen* 96, 555–71.
Alexandroff, P. 1927, 'Review of Vietoris 1927', *Jahrbuch über die Fortschritte der Mathematik* 53, 552.
Alexandroff, P. 1928, 'Zusammenhangszahlen abgeschlossener Mengen', *Nachrichten, Gesellschaft der Wissenschaften zu Göttingen* pp. 323–29.
Alexandroff, P. 1932, *Einfachste Grundbegriffe der Topologie*, Julius Springer, Berlin. Translated as *Elementary Concepts of Topology*. New York: Dover, 1962.
Alexandroff, P. 1969, 'Die Topologie in und um Holland in den Jahren 1920-1930', *Nieu Archief voor Wiskunde* **XXVII**, 109–27.
Alexandroff, P. 1981, *In memory of Emmy Noether*, Marcel Dekker, New York, pp. 99–114. This 1935 eulogy at the Moscow Mathematical Society is also in N. Jacobson ed. *Emmy Noether Collected Papers*, Springer Verlag, 1983, 1–11; and Dick 1970, 153–80.
Archibald, T., 1991, Riemann and the Theory of Nobili's Rings, *Centaurus* 34, 247–71.
Aristotle, 1984, 'Metaphysics.' In *The Complete Works of Aristotle*. Ed. J. Barnes, 2 vols.. Princeton: University Press, II: 1552–1722.
Aspray, W. and Kitcher, P. (eds.), 1988, *History and Philosophy of Modern Mathematics*, Minnesota Studies in the Philosophy of Science, Volume XI, Minneapolis.

Atiyah, M., 2002, Mathematics in the 20th Century. *Contemporary Trends in Algebraic Geometry and Algebraic Topology*. S.-S. Chern, L. Fu and R. Hain. Singapore, World Scientific. 5, 1–21.

Avigad, J., forthcoming Mathematical method and proof. To appear in *Synthese*.

Awodey, S., and Carus, A.W., 2001, Carnap, Completeness, and Categoricity: the Gabelbarkeitssatz of 1928, *Erkenntnis*, 54, 145–172.

Awodey, S., and Reck, E., 2002, Completeness and Categoricity, Part I: 20th century metalogic and 20th century metalogic, *History and Philosophy of Logic*, 23, 1–30.

Awodey, S., and Reck, E., 2003, Completeness and Categoricity, Part II: 19th century axiomatics and 21st century semantics, *History and Philosophy of Logic*, [complete]

Bach, C., Tarski's, 1936, account of logical consequence, *Modern Logic*, vol. 7, no.2, pp. 109–130.

Bachelard, G., 1934, Le nouvel esprit scientifique, Paris, Félix Alcan,. 9th edition, Paris, Presses Universitaires de France, 1966.

Bachelard, G., 1940, *Philosophie du non. Essai d'une philosophie du nouvel esprit scientifique*, Paris, Presses Universitaires de France.

Bachelard, G., 1945, 'La philosophie scientifique de Léon Brunschvicg', *Revue de métaphysique et de morale*, , n° 1–2, reprinted in dans *Léon Brunschvicg. L'œuvre. L'homme*, Paris, A. Colin, 1945.

Bachmann P., 1872, *Die Lehre von der Kreistheilung und ihre Beziehungen zur Zahlentheorie* Leipzig: Teubner.

Bacon, F., *Instauratio magna* or *Great Restoration*, in *The Works of F. Bacon* (London, Longman, 1858–1874).

Baird, D., 2004, *Thing knowledge: a philosophy of scientific instruments*. Berkeley, University of California Press.

Baire, R., Borel, E., Hadamard, J., Lebesgue, H., 1905, Cinq lettres sur la théorie des ensembles; Bulletin de la Société Mathématique de France 33, 261–273. (Translated in Moore's book *Zermelo's Axiom of Choice*, New York, Heidelberg, Berlin 1982, 311–320.)

Baldus, R., 1924, *Formalismus und Intuitionismus in der Mathematik*, Braun, Karlsruhe.

Baumann, J.J. 1868, ed., *Die Lehren von Zeit, Raum und Mathematik*, Berlin.

Bays, T., 2001, On Tarski on Models, *The Journal of Symbolic Logic*, 66, pp. 1701–1726.

Beaney, M., 1996, *Frege: Making Sense*, London: Duckworth.

Beaney, M., 1997, 'Introduction' to Frege FR, pp. 1–46.

Beaney, M., 2003a, 'Russell and Frege', in N. Griffin, ed., *The Cambridge Companion to Bertrand Russell*, Cambridge: Cambridge University Press, pp. 128–70.

Beaney, M., 2003b, 'Analysis', *Stanford Encyclopedia of Philosophy*; http://plato.stanford.edu/entries/analysis/.

Beaney, M., 2005, 'Sinn, Bedeutung and the Paradox of Analysis', in Beaney and Reck 2005, Vol. IV, pp. 288–310.

Beaney, M. and Reck, E.H., 2005, eds., *Gottlob Frege: Critical Assessments*, 4 vols., London: Routledge.

Becker, O. 1927/28, 'Das Symbolische in der Mathematik.' *Blätter für deutsche Philosophie* 1:329–348.

Belhoste, B., 1991, *Augustin-Louis Cauchy, a Biography*, Springer-Verlag, New York.

Bell, E.T., 1937, *Men of Mathematics*, Gollancz, London.

Beller, Mara, 1999, *Quantum Dialogue: The Making of a Revolution*. Chicago: University Press.

Bellotti, L., 2002, Tarski on Logical Notions, *Synthese*, [complete]

Beltrami, E., 1868 Saggio di interpretazione della geometria non Euclidea, *Giornale di Matematiche*, VI, 284–312.
Benacerraf, P., 1973, Mathematical Truth, *Journal of Philosophy* 70, reprinted in Benacerraf and Putnam 1983.
Benacerraf, P. and H. Putnam, 1983, *Philosophy of Mathematics: Selected readings*, Cambridge University Press.
Benis Sinaceur, H., 1995, 'Formes et concepts', in *La connaissance philosophique. Essais sur l'œuvre de G. G. Granger*, Paris, Presses Universitaires de France.
Benis Sinaceur, H., 1987, 'Structure et concept dans l'épistémologie de Jean Cavaillès', *Revue d'Histoire des Sciences*, XL-1, pp. 5–30, 117–129.
Benis Sinaceur, H., 1994, *Jean Cavaillès. Philosophie mathématique*, Paris, Presses Universitaires de France.
Benis Sinaceur, H., 1997, Les 'enfants naturels' de Descartes, in *Actes du colloque Descartes* (Louvain-la-Neuve, 21–22 Juin 1996), ed. P. Radelet-de Grave, J.-F. Stoffel, Brepols, pp. 205–221.
Benis Sinaceur, H., 2000, Address at the Princeton University Bicentennial Conference on problems of mathematics, by A. Tarski, *Bulletin of Symbolic Logic* 6, pp. 1–44.
Benis Sinaceur, H., 2001, Alfred Tarski: Semantic Shift, heuristic shift in metamathematics, *Synthese* 126, 49–65.
Benke, H. and K. Kopfermann (eds), 1966, *Festschrift zur Gedächtnisfeier für Karl Weierstraß, 1815–1965* Köln: Westdeutscher Verlag.
Bernays, P., 1922a, Über Hilberts Gedanken zur Grundlegung der Arithmetik; *Jahresberichte DMV* 31, 10–19.
Bernays, P., 1922b, A Hilberts Bedeutung für die Philosophie der Mathematik; Die Naturwissenschaften 4, 93–99. Translated in Mancosu 1998, pp. 189–197.
Bernays, P., 1930, Die Philosophie der Mathematik und die Hilbertsche Beweistheorie *Blätter für Deutsche Philosophie* 4, pp. 326–367; in: (Bernays, 1976), 17–61. Translated in Mancosu 1998, pp. 234–265.
Bernays, P., 1935a, Sur le platonisme dans les mathématiques, *L'Enseignement Mathématique* 34. English version in Benacerraf and Putnam 1983. German in (Bernays, 1976), 62–78.
Bernays, P., 1935b, Hilberts Untersuchungen über die Grundlagen der Arithmetik; in: (Hilbert, 1935), 196–216.
Bernays, P., 1941, Sur les questions méthodologiques actuelles de la théorie hilbertienne de la démonstration; in: *Les entretiens de Zurich sur les fondements et la méthode des sciences mathématiques*, F. Gonseth (ed.), 1938, Zurich 1941, 144–152.
Bernays, P., 1946, Gesichtspunkte zum Problem der Evidenz; in: (Bernays, 1976), pp. 85–91.
Bernays, P., 1950, Mathematische Existenz und Widerspruchsfreiheit; in: (Bernays, 1976), 92–106.
Bernays, P., 1961, Zur Rolle der Sprache in erkenntnistheoretischer Hinsicht. In Bernays, *Abhandlungen zur Philosophie der Mathematik*, Darmstadt : Wissenschaftliche Buchgesellschaft, 1976.
Bernays, P., 1967, Hilbert, David; in: *Encyclopedia of Philosophy*, P. Edwards (ed.), New York and London, 1967, 496–504.
Bernays, P., 1970, Die Schematische Korrespondenz und die Idealisierten Strukturen; in: (Bernays, 1976), 176–188.
Bernays, P., 1976, *Abhandlungen zur Philosophie der Mathematik*; Wissenschaftliche Buchgesellschaft, Darmstadt.

Betti E., 1885 *Lehrbuch der Potentialtheorie und ihrer Anwendungen auf Elektrostatik und Magnetismus.* (German translation by W. F. Meyer) Stuttgart: Kohlhammer.
Biermann O., 1887 *Theorie der Analytischen Funktionen* Liepzig.
Birkhoff, G. and Bennett, M.K., 1988, Felix Klein and His 'Erlangen Program', in *History and Philosophy of Modern Mathematics*, W. Aspray and P. Kitcher (eds.) Minnesota Studies in the Philosophy of Science, vol. XI, pp. 145–176.
Bishop, E., 1985, *Constructive Analysis*, Berlin, Springer.
Bishop, E., 1967, *Foundations of constructive analysis*, New York, McGraw-Hill.
Blok, W.J., Pigozzi, D., 1988, Alfred Tarski's work on general metamathematics, *Journal of Symbolic Logic*, 53, pp. 36–50.
Bôcher, M., 1904, The fundamental conceptions and methods of mathematics, *Bulletin of the American Mathematical Society*, X, pp. 115–135.
Boltzmann, L., 1902, Model, in L. Bolzmann, *Theoretical Physics and Philosophical Problems. Selected Writings*, ed. by B. McGuinness, Reidel, Dordrecht, 1974, pp. 213–220.
Boniface, J. and Schappacher, N., 2001, Sur le concept de nombre en mathématique', Cours inédit de Leopold Kronecker à Berlin (1891), *Revue d'histoire des Mathématiques*, 7.2, 207–276.
Bonola, R., 1906, *La geometria non-Euclidea*, Zanichelli, Bologna, tr. H.S. Carslaw, as *History of non-Euclidean geometry*, preface by F. Enriques, Open Court, Chicago, 1912.
Boole, G., 1854, *An Investigation of the Laws of Thought, on which are founded the mathematical theories of logic and probabilities*, Walton and Maberly: London.
Borevich, A. I. and I. R. Shafarevich. , *Number theory*. Translated from the 1964 Russian first edition by Newcomb Greenleaf. Academic Press, New York, 1966.
Börger, E. (ed.), 1987, *Computation Theory and Logic*; Springer Lecture Notes in Computer Science 270, Berlin, Heidelberg, New York.
Bos, H.J.M., 1984, Arguments on motivation in the rise and decline of a mathematical theory: the "construction of equations", 1637–ca.1750,*Archive for history of exact sciences* 30, 331–380.
Bottazzini U. and Tazzioli R., 1995, '*Naturphilosophie* and its Role in Riemann's Mathematics' *Revue d'histoire des Mathématiques* 1 pp. 3–38.
Bottazzini U., 1986, *The Higher Calculus: A History of Real and Complex Analysis from Euler to Weierstrass* W. van Egmund (trans.) Berlin: Springer.
Bottazzini, U., 1994, 'Three Traditions in Complex Analysis: Cauchy, Riemann and Weierstrass' in I. Grattan-Guinness (ed.) *Companion Encyclopedia of the History and Philosophy of the Mathematical Sciences* vol I London: Routledge pp. 419–431.
Bourbaki, N., 1948, L'architecture des mathématiques, *Les grands courants de la pensée mathématique*, Paris, Cahiers du Sud. English version in the *American Mathematical Monthly* 57 (1950), reprinted in Ewald 1996, vol.2.
Brieskorn, E: 'Wer in aller Welt glaubt denn noch an nothwendige Übereinstimmung zwischen Leben und Denken, Mensch und Werk!' Manuscript of a talk given in Bonn, 8 September 1997. (Unpublished.)
Brill A. and Nöther M. 1894, 'Die Entwickelung der Theorie der algebraischen Functionen in älterer und neuerer Zeit' *Jahresbericht der Deutschen Mathematiker-Vereinigung* 3 pp. 107–566.

Brouwer, L.E.J. 1911, 'Über Abbildungen von Mannigfaltigkeiten', *Mathematische Annalen* 71, 97–115. Reprinted as 1911D in Arendt Heyting ed. *Brouwer, Collected Works*, North Holland, 1975, vol. 2, pp. 454–74.
Brouwer, L. 1912a 'Beweis der Invarianz der geschlossen Kurve', *Mathematische Annalen* 72, 422–25. Reprinted as 1912L in Arendt Heyting ed. *Brouwer, Collected Works*, North Holland, 1975, vol. 1, pp. 523–26.
Brouwer, L.E.J., 1912b *Intuitionisme en formalisme*. Groningen, Noordhoff. Traslation in the *Bulletin of the American Mathematical Society* 20 (1913), 81–96; reprinted in Benacerraf & Putnam, eds., *Philosophy of Mathematics: Selected readings* (Cambridge UP 1983)
Brouwer, L.E.J., 1927, Über Definitionsbereiche von Funktionen; *Mathematische Annalen* 97, 60–75; translated in: van Heijenoort, 446–463.
Brouwer, L.E.J., 1929, 'Mathematik, Wissenschaft und Sprache', *Collected Works* I, A. Heyting ed., North-Holland, 1975, pp. 416–428.
Brouwer, L.E.J., 1948, 'Consciousness, Philosophy, and Mathematics', *Proceedings of the tenth International Congress of Philosophy*, Amsterdam, 1948, 1, Fasc. 2, North-Holland. Reprint in *Collected Works* I, pp. 480–494.
Brouwer, L.E.J., 1976, *Collected Works*, vol. II, ed. H. Freudenthal, Amsterdam: North Holland.
Brunschvicg, L., 1912, *Les Étapes de la philosophie mathématique*, Paris, Alcan, p. 562–577.
Brunschvicg, L., 1924, 'L'idée critique et le système kantien', *Revue de métaphysique et de morale*, n° 2, reprinted in Brunschwicg, *Écrits philosophiques* I, Paris, Presses Universitaires de France, 1951, pp. 206–270.
Buchholz, W., 1990, Proof theory of iterated inductive definitions revisited; to appear in: Archive for Mathematical Logic.
Buchholz, W., Feferman, S., Pohlers, W. Sieg, W. 1981, *Iterated Inductive Definitions and Subsystems of Analysis: Recent Proof-theoretical Studies*; Springer Lecture Notes in Mathematics 897, Berlin, Heidelberg, New York.
Burgess, J. and G. Rosen, 1997, *A Subject with No Object: Strategies for nominalistic interpretation of mathematics*, Oxford University Press.
Burkhardt, H., 1897, *Einführung in die Theorie der analytischen Funktionen einer complexen Veränderlichen* Leipzig: Verlag Von Vliet.
Burkhardt, H., 1906/1913 *Einführung in die Theorie der analytischen Funktionen einer complexen Veränderlichen* (2nd ed.) Leipzig: Verlag Von Vliet.
Byers, N., 1996, The life and times of Emmy Noether, *in* H. Newman & T. Ypsilantis, eds, *History of original ideas and basic discoveries in particle physics*, Plenum Press, New York, pp. 945–64.
Cantor, G., 1879 'Über unendliche lineare Punktmannigfaltigkeiten, Nr. 1.' In: Georg Cantor, *Gesammelte Abhandlungen mathematischen und philosophischen Inhalts*. Berlin: Springer, 1932, pp. 139–144.
Cantor, G., 1883, *Grundlagen einer allgemeinen Mannichfaltigkeitslehre*, Leipzig: Teubner. Also (without preface) in *Mathematische Annalen* 21. References to *Abhandlungen mathematischen und philosophischen Inhalts* (Berlin: Springer, 1932), 165–208. English version in Ewald 1996, vol. 2, 878–919.
Cantor, G., 1932, *Gesammelte Abhandlungen mathematischen und philosophischen Inhalts*; E. Zermelo (ed.); Berlin.
Carnap, R., 1927, Eigentliche und Uneigentliche Begriffe, *Symposion* I, pp. 355–374.
Carnap, R., 1929, *Abriss der Logistik*, Springer Vienna.

Carnap, R., 1930, Bericht über Untersuchungen zur allgemeinen Axiomatik, *Erkenntnis* 1, pp. 303–310.
Carnap, R., 1934a, Die Antinomien und die Unvollständigkeit der Mathematik, *Monatshefte für Mathematik und Physik*, 41, 263–284.
Carnap, R., 1934b, *Logische Syntax der Sprache*, Springer, Vienna.
Carnap, R., 2000, 1928, *Untersuchungen zur allgemeinen Axiomatik*, T. Bonk and J. Mosterin. Eds., Wissenschaftliche Buchgeselllschaft, Darmstadt.
Carnap, R., Bachmann, F., 1936, Über Extremalaxiome, *Erkenntnis* 6, 166–188.
Casorati, F. 1868, *Teorica delle funzioni di variabili complexi* Pavia.
Cassirer, E., 1907, 'Kant und die moderne Mathematik' *Kantssudien* 12, 1–49.
Cassirer, E., 1923, *Substance and Function*, W.C. and M.C. Swabey, trans., Open Court, Dover reprint, New York, 1953, (German original 1910).
Cassirer, E. 1922, *Philosophie der symbolischen Formen I. Die Sprache*. Berlin. Further editions Darmstadt: Wissenschaftliche Buchgesellschaft ^{2}1953, ... ^{10}1994. English translation by R. Manheim, Oxford 1954.
Cavaillès, J., 1937, 'Réflexions sur le fondement des mathématiques', Travaux du IXe Congrès international de philosophie, t. VI, Paris, Hermann, reprinted in *Œuvres complètes de philosophie des sciences*, Paris, Hermann, 1994, pp. 577–580.
Cavaillès, J., 1938a, *Méthode axiomatique et formalisme. Essai sur le problème du fondement des mathématiques*, Paris, Hermann, reprinted in *Œuvres complètes de philosophie des sciences*, pp. 177, 179.
Cavaillès, J., 1938b, *Remarques sur la formation de la théorie abstraite des ensembles, étude historique et critique*, Paris, Hermann, reprint in *Œuvres complètes de philosophie des sciences*, pp. 221–374.
Cavaillès, J., 1946, 'La pensée mathématique' (en collaboration avec Albert Lautman), Bulletin de la Société française de philosophie, 40, n° 1, pp. 1–39, reprinted in *Œuvres complétes de philosophie des sciences*, pp. 593–630.
Cavaillès, J., 1947a, 'Transfini et continu', Paris, Hermann, reprinted in *Œuvres complètes de philosophie des sciences*, pp. 451–472.
Cavaillès, J., 1947b, *Sur la logique et la théorie de la science*, Paris, Presses Universitaires de France, reprinted in *Œuvres complètes de Philosophie des Sciences*, pp. 473–560.
Chevalley, C., 1993, Niels Bohr's words and the atlantis of Kantianism. In *Niels Bohr and Contemporary Philosophy*, ed. Henry Faye, Jan; Folse. Dordrecht etc.: Kluwer pp. 33–55.
Chevalley, C., 1995, Philosophy and the birth of quantum mechanics. In *Physics, Philosophy, and the Scientific Community*, Essays in the Philosophy and History of the Natural Sciences and Mathematics in Honor of Robert S. Cohen, ed. K. Gavroglu; J. Stachel. Dordrecht etc.: Kluwer pp. 11–37.
Chevalley, C., 1951 *Introduction to the Theory of Algebraic Functions of One Variable* New York: American Mathematical Society, Reinhart and Wilson.
Chrystal G., 1886 *Algebra* (vols I and II) Edinburgh.
Clebsch A. and Lindemann F., 1876, *Vorlesungen Über Geometrie* (lectures by Clebsch transcribed and edited by Lindemann) (vol. 1) Leipzig.
Clebsch, A. and Gordon, P., 1866, *Theorie der Abelschen Funktionen* Leipzig: Teubner.
Coleman, R.; Korté, Herbert, 2001, Hermann Weyl: Mathematician, physicist, philosopher. In *(Scholz* 2001*)*. pp. 161–388.
Corcoran, J., 2003, The Absence of Multiple Universes of Discourse in the 1936 Tarski Consequence-definition Paper, unpublished typescript, 9 pages.

Corfield, D., 2003, *Towards a Philosophy of Real Mathematics*, Cambridge University Press.

Corry, L., 1996, *Modern Algebra and the Rise of Mathematical Structures*, Birkhäuser, Basel.

Corry, L., 1997 David Hilbert and the Axiomatization of Physics (1894–1905), *Archive for history of exact sciences* 51, 83–198.

Corry, L., 1999, 'From Mie's Electromagnetic Theory of Matter to Hilbert's Unified Foundations of Physics', *Studies in History and Philosophy of Modern Physics* 30 B (2), 159–183.

Corry, L., 2001, 'Mathematical Structures from Hilbert to Bourbaki: The Evolution of an Image of Mathematics', in A. Dahan and U. Bottazzini (eds.) *Changing Images of Mathematics in History. From the French Revolution to the new Millenium*, London: Harwood Academic Publishers, 167–186.

Corry, L., 2003, *Modern Algebra and the Rise of Mathematical Structures*, Basel and Boston, Birkhäuser, 2nd revised edn (1st edn - 1996).

Corry, L., 2004, *David Hilbert and the Axiomatization of Physics (1898–1918). From Grundlagen der Geometrie to Grundlagen der Physik*. Dordrecht etc.: Kluwer.

Courant R., 1950, *Dirichlet's Principle, Conformal Mapping and Minimal Surfaces* New York: Wiley interscience.

Couturat, L., 1905, *Principes des mathématiques* Alcan, Paris.

Currie, G., 1982, *Frege: An Introduction to his Philosophy* Totowa N.J.: Barnes and Noble.

Curtis, M. L., 1956, 'The Covering Homotopy Theorem.' *Proceedings of the American Mathematical Society* 7: 682–684.

Czelakowski, J., and Malinowski, G., 1985, Key notions of Tarski's methodology of deductive systems, *Studia Logica* 44, 4, pp. 321–351.

van Dalen, D., 1984, 'Four letters from Edmund Husserl to Hermann Weyl.' *Husserl Studies* 1:1–12.

van Dalen, D., 2000, Brouwer and Weyl on proof theory and philosophy of mathematics, in V. F. Heudricks, S.A. Pedersen, K.F. Jørgensen (eds.), *Proof Theory: history and Philosophical Significance* Kluwer Academic Publishers, 2000, pp. 117–152.

Dedekind, R., 1854. Über die Einführung neuer Funktionen in der Mathematik. Delivered as a *Hablitationsvorlesung* in Göttingen on June 30, Reprinted in (Dedekind, 1968), Volume 3, Chapter LX, pp. 428–438. Translated by William Ewald as 'On the introduction of new functions in mathematics' in (Ewald, 1996), Volume 2, pages 754–762.

Dedekind, R., 1857. Abriß einer Theorie der höheren Kongruenzen in bezug auf einen reellen Primzahl-Modulus. *Journal für reine und angewandte Mathematik*, 54:1–26, Reprinted in (Dedekind, 1968), Volume 1, Chapter V, 40–67.

Dedekind, R., 1871. Über die Komposition der binären quadratischen Formen. Supplement X to the second edition of (Dirichlet, 1863), pp. 423. Parts reprinted in (Dedekind, 1968), Volume 3, Chapter XLVII, pp. 223–261 and (Dedekind, 1964), pp. 223–261. Translated by Jeremy Avigad, with introductory notes, as 'On the composition of binary forms,' Technical report CMU-PHIL-162, Carnegie Mellon University. A short excerpt is also translated, with introductory notes, by William Ewald, in (Ewald, 1996), volume 2, pages 762–765.

Dedekind, R., 1872. *Stetigkeit und irrationale Zahlen*. Braunschweig, Vieweg, Reprinted in (Dedekind, 1968), Volume 3, Chapter L, pages 315–334. Translated by Wooster Beman as 'Continuity and irrational numbers' in *Essays on the theory of numbers*, Open Court, Chicago, 1901; reprinted by Dover, New York, 1963. The Beman translation is reprinted, with corrections by William Ewald, in (Ewald, 1996), volume 2, pp. 765–779.

Dedekind, R., 1873, Anzeige von P. Bachmann: *Die Lehre von der Kreisteilung und ihre Beziehungen zur Zahlentheorie*, in *Gesammelte mathematische Werke*, vol. 2 (New York, Chelsea, 1969), 408–20.

Dedekind, R., 1876a B. Riemanns Lebenslauf, in Riemann's *Werke*, 541–58.

Dedekind, R., 1876b. Letter to R. Lipschitz, October 6, Reprinted in (Dedekind, 1968), Volume 3, Chapter LXV, pp. 468–474.

Dedekind, R., 1877a. *Sur la théorie des nombres entiers algébriques*. Gauthier-Villars, Paris, Also *Bulletin des sciences mathématiques et astronomiques (1)*, 11 (1876) 278–288, (2), 1 (1877) 17–41, 69–92, 144–164, 207–248; parts reprinted in (Dedekind, 1968), Volume 3, Chapter XLVII, pages 263–296 and (Dedekind, 1984), pages 263–296. Translated as *Theory of algebraic integers* with an editorial introduction by John Stillwell, Cambridge University Press, Cambridge, 1996.

Dedekind, R., 1877b 'Schreiben an Herrn Borchardt über die Theorie der elliptischen Modulfunctionen' *Journal für Reine und Angewante Mathematik* 83 pp. 265–292 (=Werke I pp. 174–201).

Dedekind, R., 1878, Über den Zusammenhang zwischen der Theorie der Ideale und der Theorie der höheren Kongruenzen. Abhandlungen der Königlichen Gesellschaft der Wissenschaften zu Göttingen, Vol. 23, pages 1–23. Reprinted in (Dedekind, 1968), Volume 1, Chapter XV, pages 202–237.

Dedekind, R., 1879, Über die Theorie der ganzen algebraischen Zahlen. Supplement X to the third edition of (Dirichlet, 1863), pages 515–530. Parts reprinted in (Dedekind, 1968), Volume 3, Chapter XLIX, pages 297–213 and (Dedekind, 1964), pages 297–213.

Dedekind, R., 1888. *Was sind und was sollen die Zahlen?* Vieweg, Braunschweig Vieweg, A later edition reprinted in (Dedekind, 1968), Volume 3, Chapter LI, pages 335–391. The second edition, with a new preface, was published in 1893, and is translated by Wooster Beman as 'The nature and meaning of numbers' in *Essays on the theory of numbers*, Open Court, Chicago, 1901; reprinted by Dover, New York, 1963. The Beman translation is reprinted, with corrections by William Ewald, in (Ewald, 1996), volume 2, pages 787–833.

Dedekind, R., 1890, Letter to Keferstein; in: van Heijenoort, 98–103.

Dedekind, R., 1894, Über die Theorie der ganzen algebraischen Zahlen. Supplement XI to the fourth edition of (Dirichlet, 1863), pages 434–657. Reprinted in (Dedekind, 1968), Volume 3, Chapter XLVI, pages 1–222, and (Dedekind, 1964), pages 1–222.

Dedekind, R., 1895, Über die Begründung der Idealtheorie. *Nachrichten von der Königlichen Gesellschaft der Wissenschaften zu Göttingen*, Mathem.-phys. Klasse, pages 106–113. Reprinted in (Dedekind, 1968), Volume 2, Chapter XXV, pages 50–58.

Dedekind, R., 1900, 'Über die von drei Moduln erzeugte Dualgruppe', *Mathematische Annalen* 53, 371–403.

Dedekind, R., 1930–1932, *Gesammelte mathematische Werke*, Friedr. Vieweg & Sohn, Braunschweig. Three volumes.

Dedekind, R., 1964. *Über die Theorie der ganzen algebraischen Zahlen*. F. Vieweg, Braunschweig, Excerpts on ideal theory from (Dedekind, 1968), with a foreword by B. van der Waerden.

Dedekind, R., 1968. *Gesammelte mathematische Werke*. Edited by Robert Fricke, Emmy Noether and Öystein Ore. Chelsea Publishing Co., New York. Volumes I–III. Reprinting of the original edition, published by F. Vieweg & Sohn, Braunschweig, 1930–1932. Online at http://gdz.sub.uni-goettingen.de/en/.

Dedekind, R., 1996, *Theory of Algebraic Integers*, Cambridge University. Originally published in French 1877.
Dedekind, R., and Weber H., 1882, 'Theorie der algebraischen Funktionen einer Veränderlichen' *Journal für Reine und Angewante Mathematik* 92 p. 181–290 (dated 1880) (page reference to the reprinting in Dedekind's *Werke* vol. I p. 238–349).
Dehn, M. & P. Heegaard 1907, Analysis situs, *in* 'Encyklopadie der mathematischen Wissenschaften III A B 3', B.G. Teubner, Leipzig, pp. 153–220.
Demopoulos, W. 1994, 'Frege and the Rigorization of Analysis' *Journal of Philosophical Logic* 23 pp. 225–245, page references to the reprinting in Demopoulos [1995].
Deppert, W. et al. (eds.), 1988, *Exact Sciences and Their Philosophical Foundations*. Frankfurt/Main: Peter Lang. Weyl, Kiel Kongress 198.
Desanti, J. T., 1968, *Les idéalités mathématiques*, Paris, Le Seuil.
Descartes, R., 1641, *Meditationes de prima philosophia: méditations métaphysiques*, Paris, Vrin, 1953.
Detlefsen, M., 1986, *Hilbert's Program: An Essay on Mathematical Instrumentalism*, Dordrecht, Reidel.
Detlefsen, M., 1993, 'Hilbert's Work on the Foundations of Geometry in Relation to his Work on the Foundations of Arithmetic', *Acta Analytica* 8 (11), 27–39.
Dieudonné, J., 1962, 'Les méthodes axiomatiques modernes et les fondements des mathématiques', in F. Le Lionnais (ed.) *Les grands Courants de la Pensée Mathématique* (Second, enlarged edn), Paris, Blanchard, 443–555.
Dieudonné J., 1970, The work of Nicholas Bourbarki; American Mathematical Monthly 77, 134–145.
Dieudonné, J., 1984, 'Emmy Noether and algebraic topology', *J. Pure Appl. Algebra* 31, 5–6.
Dieudonné, J., 1989, *A History of Algebraic and Differential Topology 1900–1960*. Basel, Birkhäuser.
Dieudonné, J., 1994, Une brève histoire de la topologie. *Development of Mathematics 1900–1950*. J.-P. Pier. Basel, Birkhäuser: 35–156.
Dirichlet, G. P. Lejeune, 1837, Beweis des Satzes, dass jede unbegrenzte arithmetische Progression, deren erstes Glied und Differenz ganze Zahlen ohne gemeinschaftlichen Faktor sind, unendlich viele Primzahlen enthält; reprinted in (Dirichlet, 1889), 313–342.
Dirichlet, G. P. Lejeune, 1838, Sur l'usage des séries infinies dans la théorie des nombres; reprinted in (Dirichlet, 1889), 357–374.
Dirichlet, G. P. Lejeune, 1839/40, Recherches sur diverses applications de l'analyse infinitésimale à la théorie des nombres; reprinted in (Dirichlet, 1889), 411–496.
Dirichlet, G. P. Lejeune, 1863 *Vorlesungen über Zahlentheorie*. Vieweg, Braunschweig, Edited by R Dedekind. Subsequent editions in 1871, 1879, 1894, with 'supplements' by R Dedekind. The first edition is translated by John Stillwell, with introductory notes, as *Lectures on number theory*, American Mathematical Society, Providence, RI, 1999.
Dirichlet, G. P. Lejeune, 1889, *Gesammelte Werke I*, edited by L. Kronecker, Berlin, 1889.
Dirichlet, G. P. Lejeune, 1897, *Gesammelte Werke II*, edited by L. Kronecker and continued by L. Fuchs, Berlin 1897. (Both volumes were reprinted by Chelsea in one volume, New York, 1969.)
Dugac, P., 1976, *Dedekind et les fondements des mathématiques*, Paris, Vrin.
Dühring E., 1873, *Kritische Geschichte der Allgemeinen Prinzipien der Mechanik* Berlin.
Dummett M., 1991, *Frege: Philosophy of Mathematics* Cambridge Mass.: Harvard University Press.

Dwyer, W. G. and J. Spalinski, 1995, Homotopy Theories and Model Categories. *Handbook of Algebraic Topology*. I. M. James. Amsterdam, Elsevier: 73–126.

Dwyer, W. G., P. S. Hirschhorn, *et al.* 2004, Homotopy Limit Functors on Model Categories and Homotopical Categories. 2004.

Earman, J., 1989, *World-Enough and Space-Time*, Cambridge, MIT Press.

Eckmann, B., 1942, 'Zur Homotopietheorie gefaserter Raüme.' *Commentarii Mathematici Helvetici* 14: 141–192.

Eckmann, B., 1999, 'Birth of Fibre Spaces, and Homotopy.' *Expositiones Mathematicae* 17: 23–34.

Edwards, H. M., 1974, *Riemann's zeta function*. Dover Publications Inc., Mineola, NY, 2001. Reprint of the original, Academic Press, New York.

Edwards, H. M., 1980, The genesis of ideal theory. *Archive for History of Exact Sciences*, 23, 321–378.

Edwards, H. M., 1983, Dedekind's invention of ideals. *Bull. London Math. Soc.*, 15, 8–17. Reprinted in E. Phillips (ed.) *Studies in the History of Mathematics* MAA monographs 26 pp. 8–20.

Edwards, H. M., 1984, *Galois theory*. Springer-Verlag, New York.

Edwards, H. M., 1988, Kronecker's place in history. In (Aspray and Kitcher, 1988), pp. 139–144.

Edwards, H. M., 1989, Kronecker's views on the foundations of mathematics. In D. E. Rowe and J. McCleary, editors, *The History of Modern Mathematics*, pages 67–77. Academic Press, 1989.

Edwards, H. M., 1990, *Divisor theory*. Birkhäuser Boston Inc., Boston, MA

Edwards, H. M., 1992, Mathematical ideas, ideals, and ideology. *Math. Intelligencer*, 14:6–19.

Edwards, H. M., 1994, *Advanced calculus: a differential forms approach*. Birkhäuser Boston Inc., Boston, MA, Corrected reprint of the 1969 original.

Edwards, H. M., 1995, *Linear algebra*. Birkhäuser Boston Inc., Boston, MA.

Edwards, H. M., 1996, *Fermat's last theorem: a genetic introduction to algebraic number theory*. Springer-Verlag, New York, Corrected reprint of the 1977 original.

Edwards, H. M., 2005, *Essays in constructive mathematics*. Springer-Verlag.

Ehresmann, C. and J. Feldbau, 1941, 'Sur les propriétés d'homotopie des espaces fibrés.' *Comptes Rendus de l'Académie des Sciences de Paris* 212: 945–948.

Eichhorn, E: 'In memoriam Felix Hausdorff (1868–1942).' In: Eichhorn, Eugen and Thiele, Ernst-Jochen: *Vorlesungen zum Gedenken an Felix Hausdorff*. Berlin: Heldermann, 1994.

Eilenberg, S. & S. Mac Lane, 1942, Appendix A: On homology groups of infinite complexes and compacta, *in* 'Algebraic Topology', American Mathematical Society, pp. 344–49.

Einstein, A., 1912a, The Speed of Light and the Statics of the Gravitational Field, in *The Collected Papers of Albert Einstein vol. 4, The Swiss Years: Writings, 1912–1914*, pp. 95–106.

Einstein, A., 1912b, On the Theory of the Static Gravitational Field, in *The Collected Papers of Albert Einstein vol. 4, The Swiss Years: Writings, 1912–1914*, pp. 107–120.

Einstein, A., 1916, Relativity; the special and general theory, tr R.W. Lawson, 1920, in *The Collected Papers of Albert Einstein vol. 6, The Berlin Years: Writings, 1914–1917*, pp. 247–420.

Einstein, A., 2002, *The Collected Papers of Albert Einstein, Volume 7: The Berlin Years: Writings, 1918–1921*. Ed. M. Janssen *et al.*, Princeton University Press.

Enneper A., 1876, *Elliptische Funktionen: Theorie und Geschichte* Halle.

Epple, M., 1999, *Die Entstehung der Knotentheorie. Kontexte und Konstruktionen einer modernen mathematischen Theorie*, Wiesbaden, Vieweg Verlag.

Epple, M., 2004, Knot Invariants in Vienna and Princeton during the 1920s: Epistemic Configurations of Mathematical Research, *Science in Context*, 17.1/2, 131–164.

Epple, M. et al., 'Zum Begriff des topologischen Raumes.' In: (Hausdorff, *Werke* II, 2002, 675–744.

Epple, M., 'From Quaternions to Cosmology: Spaces of Constant Curvature, ca. 1873–1925.' In: Li Tatsien (Ed.), *Proceedings of the International Congress of Mathematicians, Beijing 2002*, vol. III, Beijing: Higher Education Press, 2002, pp. 935–945.

Etchemendy, J., 1988, Tarski on Truth and Logical Consequence, *Journal of Symbolic Logic*, 53, 51–79.

Etchemendy, J., 1990, *The Concept of Logical Consequence*, Harvard, Cambridge.

Etchemendy, J., 1999, Reflections on Consequence, unpublished typescript available on line in the author's web page.

Euler, L. *Vollständige Anleitung zur Algebra*. Kays. Akademie der Wissenschaften, St. Petersberg, 1770. Reproduced as Volume 1 of Euler's *Opera Omnia. Series Prima: Opera Mathematica*, Societas Scientiarum Naturalium Helveticae, Geneva, 1911. Translated into French by Bernoulli, with additions by Lagrange. The French edition was, in turn, translated into English as *Elements of algebra* by Rev. John Hewlett, (reprinted by Springer, Berlin, 1972, with an introduction by C.Truesdell).

Ewald, W., (ed.) 1996, *From Kant to Hilbert. A source Book in the Foundations of Mathematics*, 2 Vols., Oxford, Clarendon Press.

Fano, G., 1892, Sui postulati fondamentali della geometria proiettiva *Giornale di Matematiche*, 30, 106–131.

Feferman, S., 1977, Theories of finite type related to mathematical practice; in: J. Barwise (ed.), *Handbook of Mathematical Logic*, Amsterdam, 913–971.

Feferman, S., 1988, Hilbert's Program relativized: proof-theoretical and foundational reductions, Journal of Symbolic Logic, 53 (2), 364–384.

Feferman, S., 1998, The Logic of Mathematical Discovery versus the Logical Structure of Mathematics, reprinted in *In the Light of Logic*, Oxford University Press, 77–93.

Feferman, S., 2000, Relationships between constructive, predicative and classical systems of analysis. In *Hendricks e.a.* 2000. pp. 221–236.

Feferman, S., 2002, Tarski, set theory and the conceptual analysis of semantical notions, [complete].

Feferman, S., und Sieg, W. 1981, Iterated inductive definitions and subsystems of analysis; in: (Buchholz, et al., 1981), 16–77

Ferreirós, J., 1996, Traditional Logic and the Early History of Sets, 1854–1908. *Archive for History of Exact Sciences* 50: 5–71.

Ferreirós, J., 1999, *Labyrinth of Thought. A history of set theory and its role in modern mathematics*. Basel/ Boston: Birkhäuser.

Ferreirós, J., 2000, La multidimensional obra de Riemann: estudio introductorio, in Ferreirós, ed., *Riemanniana Selecta*, Madrid, CSIC (Colección Clásicos del Pensamiento), pp. ix–clvii.

Ferreirós, J., 2003, Kant, Gauss y el problema del espacio, in J. Ferreirós and A. Durán, eds., *Matemáticas y matemáticos* (Publicaciones de la Universidad de Sevilla), 105–33.

Ferreirós, J., 2004, The Motives Behind Cantor's Set Theory: Physical, biological and philosophical questions, *Science in Context* 17, n° 1/2, 1–35.

Ferreirós, J., 2005, Certezas e hipótesis: enfoques históricos y naturalistas sobre el conocimiento matemático, in A. Estany (ed.), *Filosofía de las ciencias naturales, sociales y matemáticas* (Madrid, Trotta).

Ferreirós, J., 2005, Notice on Dedekind and Peano, *in* I.Grattan-Guinness, ed., 'Landmark Writings in Western Mathematics 1640–1940', Elsevier, pp. 613–626.

Ferreirós J., 2006 'Ο θεος ἀριθμητίζει: The Rise of Pure Mathematics as arithmetic with Gauss' in C. Goldstein, N. Schappacher, J. Schwermer (eds.), *The Shaping of Arithmetic: Number Theory after Carl Friedrich Gauss's* Disquistiones Arithmeticae, Springer, Berlin.

Ferreiros, J., 2007, 'Hilbert, Logicism, and Mathematical Existence'. Forthcoming.

Ferreirós, J., forthcoming. The Crisis in the Foundations of Mathematics, in T. Gowers, ed. (with the collaboration of J. Barrow-Green), *Princeton Companion to Mathematics*, Princeton University Press, in press.

Fisher, G., 1981, The infinite and infinitesimal quantities of du Bois-Reymond and their reception, *Archive for history of exact sciences* 24, 101–163.

Fox, R. H., 1943, 'On Fibre Spaces. I.' *Bulletin of the American Mathematical Society* 49: 555–557.

Fox, R. H., 1945, 'On Topologies for Function Spaces.' *Bulletin of the American Mathematical Society* 51: 429–432.

Fraenkel 1928, *Einleitung in die Mengenlehre*, 3rd edn, Springer, Berlin.

Fraenkel, A. Y. Bar-Hillel and A Levy, 1973, *Foundations of Set Theory*, Amsterdam, North-Holland.

Frege, G., 1877 [*RTh*] Review of Thomae [1876a], in *Collected Papers*, Frege 1984.

Frege, G., 1879 [B],[1] *Begriffsschrift, A Formula Language Modeled on that of Arithmetic, for Pure Thought*, Stephan Bauer-Mengelberg trans. In *From Frege to Gödel*, Heijenoort 1967.

Frege, G., 1880/81 [*BLC*], Boole's logical Calculus and the Concept-script, in Frege 1969, *NS*, pp. 9–52; tr. in *PW*, Frege 1979, pp. 9–4.

Frege, G., 1882 [*BLF*], Boole's logical Formula-language and my Concept-script, in Frege 1969, *NS*, pp. 53–9; tr. in *PW*, Frege 1979, pp. 4–52.

Frege, G., 1884 [*GL*], *Die Grundlagen der Arithmetik*, Breslau: W. Koebner; tr. as Frege *FA*, 1953; Introd., §§1–4, 4–69, 87–91, 104–9 tr. M. Beaney in Frege 1997, *FR*, pp. 84–129.

Frege, G., 1890? [*DRC*], Draft towards of a review of Cantor's *Gesammelte Abhandlungen zur Lehre vom Transfiniten*, in Frege 1979.

Frege, G., 1891 [*FC*], Function and Concept, in *Collected Papers*, Frege 1984.

Frege, G., 1892a [*RC*], Review of Cantor, G. *Zur Lehre vom Transfiniten: Gesammelte Abhandlungen aus der Zeitschrift für Philosophie und philosophische Kritik*, in *Collected Papers*, Frege 1984.

Frege, G., 1892b [*CO*], On Concept and Object, *Vierteljahrsschrift für wissenschaftliche Philosophie* 16, pp. 192–205; in *Collected Papers*, Frege 1984; tr. P.T. Geach in Frege 1952, *TPW*, pp. 4–55; Frege 1997, *FR*, pp. 181—93.

Frege, G., 1892? [*OCN*], On The Concept of Number, in *PW*, Frege 1979.

Frege, G., 1893 [*GG*] *Grundgesetze der Arithmetik, begriffsschriftlich abgeleitet*, vol I, Jena: H. Pohle. Repr. together with vol II, Hildesheim: Georg Olms, 1962; repr. 1998 with corrigenda by C. Thiel. Most of Preface, Introd., and §§1–7, 26–9, 32–3 of Vol. I tr. M. Beaney in Frege 1997, *FR*, pp. 194–223.

Frege, G., 1894 [*RH*], Review of Husserl, E. *Philosophy of Arithmetic*, in *Collected Papers*, Frege 1984.

---

[1] In the case of Frege, like in that of Russell, references are sometimes given by means of the keys that are usual among specialists. In our bibliography the key is given within square brackets, next to the year of publication – e.g. [*GG*] for the first volume of Frege's *Grundgesetze* (1893).

Frege, G., 1899? [LDM], Logical Defects in Mathematics, in PW, Frege 1979 (ca. 1899–1903).

Frege, G., 1903a [GGII] *Grundgesetze der Arithmetik, begriffsschriftlich abgeleitet*, vol II, Jena: H. Pohle. Repr. together with vol I, Hildesheim: Georg Olms, 1962; repr. 1998 with corrigenda by C. Thiel. §§56–67, 86–137, 139–4, 14–7, and 'Nachwort' of Vol. II tr. P.T. Geach and M. Black in Frege 1952, *TPW*, pp. 159–24, §§55–67, 138–4, and 'Nachwort' of Vol. II tr. P.T. Geach repr. in Frege 1997, *FR*, pp. 258–89.

Frege, G., 1903b [FGI], On the Foundations of Geometry: First Series, in Frege *KS*, pp. 262–72, tr. in Frege *CP*, pp. 273–84.

Frege, G., 1906a [FGII], On the Foundations of Geometry: Second Series, in Frege *KS*, pp. 281–323, tr. in Frege *CP*, pp. 293–34.

Frege, G., 1906b [RT], Reply to Mr. Thomae's Holiday Causene (1906), in Frege *KS*, pp. 324–8, tr. in Frege *CP*, pp. 34–5

Frege, G., 1914 [LM], Logic in Mathematics, in Frege *NS*, pp. 219–70, tr. in Frege *PW*, pp. 203–50, extract in Frege *FR*, pp. 308–18.

Frege, G., 1952 [TPW], *Translations from the Philosophical Writings of Gottlob Frege*, tr. and ed. Peter Geach and Max Black, Oxford: Blackwell, 1st edn 1952, 2nd edn 1960, 3rd edn 1980.

Frege, G., 1953 [FA], *The Foundations of Arithmetic*, tr. of Frege 1884 by J.L. Austin, with German text, Oxford: Blackwell, 2nd edn 1953.

Frege, G., 1964 [BLA], *Basic Laws of Arithmetic: Exposition of the System*, Furth, M. trans Berkeley: University of California Press 1964 (1893) (Gz II appendix: 1903).

Frege, G, 1967 [KS], *Kleine Schriften*, ed. I. Angelelli, Hildesheim: Georg Olms; tr. as Frege *CP*, 1984.

Frege, G., 1969 [NS], *Nachgelassene Schriften* (second, expanded edition), Hermes H. Kambartel F. Kaulbach F. (eds.) Hamburg: Felix Meiner Verlag, tr. as Frege *PW*, 1979; 2nd edn 1983.

Frege, G., 1971, *On the Foundations of Geometry and Formal Theories of Arithmetic*, Kluge, E.H. (ed.) New Haven: Yale University Press.

Frege, G., 1972, *Conceptual Notation and Related Articles* Bynum T. (ed.) Oxford: Oxford University Press.

Frege, G., 1976 [WB], *Wissenschaftlicher Briefwechsel* Gabriel, G. Hermes, H. Kambartel, F. Thiel, C. and Veraart, A. (eds.) Hamburge: Felix Meiner Verlag; abr. And tr. as Frege *PMC*, 1980.

Frege, G., 1979 [PW], *Posthumous Writings*, tr. of Frege 1969, *NS*, tr. P. Long and R. White, Oxford: Basil Blackwell.

Frege, G., 1980 [PMC] *Philosophical and Mathematical Correspondence*, tr. of Frege 1976, *WB*, abridged from the German edition by McGuinness, B. Translated by Kaal, H. Chicago: University of Chicago Press.

Frege, G., 1984 [CP], *Collected Papers on Mathematics, Logic and Philosophy*, tr. of Frege 1967, *KS*, McGuinness B. (ed.) Oxford: Basil Blackwell.

Frege, G., 1997 [FR], *The Frege Reader*, ed. with an introd. by M. Beancy, Oxford: Blackwell.

Freudenthal, H., 1975, 'Riemann' in *Dictionary of Scientific Biography* vol. 11 New York, pp. 447–56.

Freudenthal, H., Ed. 1976, *L.E.J. Brouwer-Collected Works*. Amsterdam, North Holland.

Friedman, M., 1983, *Foundations of Space-Time Theories: Relativistic Physics and Philosophy of Science* Princeton University Press.

Friedman, M., 2000a *A Parting of the Ways* Open Court.

Friedman, M., 2000b 'Geometry, Construction and Intuition in Kant and his Successors' *Between Logic and Intuition* Sher, G. and Tieszen R. (eds.) Cambridge: Cambridge University Press pp. 186–218.

Friedman 2002, Geometry as a Branch of Physics: Background and Context for Einstein's 'Geometry and Experience,' in D. B. Malament, ed., *Reading Natural Philosophy (Essays Dedicated to Howard Stein on His 70th Birthday)*, Open Court Press, 2002.

Frobenius, F. & L. Stickelberger, 1879, 'Ueber Gruppen von vertauschbaren Elementen', *Journal für die reine und angewandte Mathematik* 86, 217–62.

Frostman O., 1966, 'Aus dem Briefwechsel von G. Mittag-Leffler' in H. Benke and K. Kopfermann (eds) [1966] p. 53–56.

Gabriel, G., 2001, 'Existenz- und Zahlaussage. Herbart und Frege', in A. Hoeschen and L. Schneider, eds., *Herbarts Kultursystem*, Würzburg: Königshausen and Neumann, pp. 149–62; tr. as 'Existential and Number Statements: Herbart and Frege' in Beaney and Reck 2005, Vol. I, pp. 109–23.

Galison, P. L., 1997, *Image and logic: a material culture of microphysics*. Chicago, University of Chicago Press.

Gauss, C. F., *Werke*, 12 vols., Hildesheim, Olms, 1973 (reprint of the original edition Göttingen, Dieterich, 1863–1929).

Gauss, C.F., 1801, *Disquisitiones Arithmeticae*. G. Fleischer, Leipzig, Translated with a preface by Arthur A. Clarke, Yale University Press, New Haven, 1966, and republished by Springer Verlag, New York, 1986. reprinted as Gauss' *Werke* vol. I.

Gauss, C.F., 1828, Disquisitiones generales circa superficies curvas, *Werke*, vol. IV, 217–58.

Gauss, C.F., 1831, Selbstanzeige der Theoria residuorum biquadraticorum, commentatio secunda, *Werke*, vol. II, 169–78.

Gauss, C.F., 1832, Theoria residorum biquadraticum, commentatio secunda. *Commentationes Societatis Regiae Scientiarum Gottingensis Recintiores*, 7:89–148, Reprinted in Gauss' *Werke*, volume 2, Königliche Gesellschaft der Wissenschaft, 1876, pp. 93–148.

Gauss, C.F., 1849, Beiträge zur Theorie der algebraischen Gleichungen, *Werke*, vol. III, 71–102.

George, A. and D. Velleman, 2001, *Philosophies of Mathematics*, Oxford, Blackwell.

Giaquinto, M., 2002, *The Search for Certainty*, Oxford University Press.

Gispert, H., 1991, La France mathématique, *Cahiers d'histoire et de philosophie des sciences*. 34, 13–180.

Gödel, K., 1929, Über die Vollständigkeit des Logikkalküls (doctoral dissertation, University of Vienna), reprinted in K. Gödel, 1986, pp. 60–100.

Gödel, K., 1930a, Vortrag über die Vollständigkeit des Logikkalküls, in K. Gödel, 1995, pp. 16–28.

Gödel, K., 1930b, Die Vollständigkeit der Axiome des logischen Funktionenkalküls, *Monatshefte für Mathematik und Physik*, 37, pp. 349–360, reprinted in K. Gödel, 1986, pp. 102–123.

Gödel, K., 1931, Über formal unentscheidbare Sätze der *Principia Mathematica* und verwandter Systeme. I, *Monatshefte für Mathematik und Physik*, 38, 173–198. Translated and reprinted in Kurt Gödel, 1986, 144–195.

Gödel, K., 1932, Eine Eigenschaft der Realiserungen des Aussagenkalküls, *Ergenbnisse eines mathematischen Kolloquiums* 2, 20–21, reprinted in K. Gödel, 1986, pp. 238–241.

Godel, K., 1964, What is Cantor's Continuum Problem?, in Kurt Gödel, *Collected Works*, vol. 2, Oxford University Press, 254–270.

Gödel, K., 1986, *Collected Works I*, eds. S. Feferman *et al.*, Oxford University Press, Oxford.
Gödel, K., 1995, *Collected Works III*, eds. S. Feferman *et al.*, Oxford University Press, Oxford.
Goldfarb, W., 1979, Logic in the twenties: the nature of the quantifier, *Journal of Symbolic Logic* 44, pp. 351–368.
Goldman, J. R. *The queen of mathematics: a historically motivated guide to number theory.* A K Peters Ltd., Wellesley, MA, 1998.
Gomez-Torrente, M. 1996, Tarski on logical consequence, *Notre Dame Journal of Formal Logic*, 37, pp. 125–151.
Gomez-Torrente, M. 1999, Logical truth and Tarskian logical truth, *Synthese*, 117, pp. 375–408.
Gomez-Torrente, M., 1998, On a fallacy attributed to Tarski, *History and Philosophy of Logic*, 19, 227–234.
Gonseth, F., 1939, *Philosophie mathématique*, Paris, Hermann.
Goursat E., 1900, 'Sur La Définition Générale des Functions Analytiques, d'après Cauchy' *Transactions of the American Mathematical Society* 1 1 pp. 14–16.
Granger G. G., 1960, *Pensée formelle et sciences de l'homme*, Paris, Aubier.
Granger G. G., 1986, 'Pour une épistémologie du travail scientifique', *La philosophie des sciences aujourd'hui*, sous la direction de Jean Hamburger, Paris, Gauthier-Villars.
Granger G. G., 1988, *Pour la connaissance philosophique*, Paris, Odile Jacob.
Granger G. G., 1994, *Formes, Opérations, Objets*, Paris, Vrin.
Gray, J. D., S. A. Morris, 1978, When is a Function that Satisfies the Cauchy-Riemann Equations Analytic? *American Mathematical Monthly*, Vol. 85, No. 4 (Apr., 1978), pp. 246–256.
Gray, J.J., 1989, *Ideas of Space, Euclidean, non-Euclidean and relativistic*, Oxford University Press.
Gray, J.J., 1992, A nineteenth-century revolution in mathematical ontology, in *Revolutions in mathematics*, 226–248, ed D. Gillies, Oxford University Press.
Gray, J.J., 2000a *Linear differential equations and group theory from Riemann to Poincaré*, 2nd edn, Birkhäuser, Boston and Basel.
Gray, J.J., 2000b, *The Hilbert Challenge*, Oxford University Press.
Gray, J.J., 2002, Languages for mathematics and the language of mathematics in a world of nations, in *Mathematics Unbound: The Evolution of an International Mathematical Community, 1800–1945*, pp. 201–228, K.H. Parshall, A.C. Rice (eds) American and London Mathematical Societies, 23, Providence, Rhode Island.
Griffin, N., 1991, *Russell's idealist apprenticeship*, Oxford: Clarendon.
Griffiths, H., 1976, *Surfaces* Cambridge: Cambridge University Press.
Gronau D., 1997, 'Gottlob Frege, A Pioneer in Iteration Theory' in *Iteration Theory (ECIT 94) Proceedings of the European Conference on Iteration Theory, Opava* Grazer mathematische Berichte #334 p. 105–119.
Grünbaum B., 1976, 'Lectures in Lost Mathematics' unpublished mimeographed notes.
Guillaume, M., 1985, Axiomatik und Logik; in: *Geschichte der Mathematik, 1700–1900*, J. Dieudonné (ed.), Braunschweig, 748–882.
Guillaume, M., 1994, La logique mathématique en sa jeunesse, in J.-P. Pier, *Development of Mathematics 1900–1950*, Birkhäuser Verlag, Basel, Boston, Berlin, pp. 185–367.
Guinand A., 1979, 'The Umbral Method: A survey of Elementary Mnemonic and Manipulative Uses' *American Mathematical Monthly* 86 3 pp. 187–195.

Hale, B. and Wright, C, 2001, *Reason's Nearest Kin: Essays Towards a Neo-Fregean Philosophy of Mathematics*, Oxford: Clarendon Press.

Hallett M., 1984, *Cantorian Set Theory and Limitation of Size* Oxford: Oxford University Press.

Hallett, M., 1994, Hilbert's Axiomatic Method and the Laws of Thought, pp. 158–198 in *Mathematics and Mind*, A. George, ed., Oxford University Press, New York and Oxford.

Hallett, M., 2003, 'Foundations of Mathematics', in T. Baldwin, ed., *The Cambridge History of Philosophy 1870–1945*, Cambridge: Cambridge University Press, pp. 128–56.

Hallett M. and U. Majer (eds.), 2004, *David Hilbert's Lectures on the Foundations of Geometry, 1891–1902*, Berlin-Heidelberg-New York, Springer Verlag.

Halmos, P. 1960, *Naive Set Theory* Van Nostrand, Princeton.

Hamel, G., 1903, "Ueber die Geometrieen, in denen die Geraden die Kuerzesten sind." *Mathematische Annalen* 57 (1903), 231–264.

Hatcher, A., 2002, *Algebraic Topology*. Cambridge, Cambridge University Press.

Hattendorff, K. ed., *Partielle Differentialgleichungen und deren Anwendung auf physikalische Fragen*, Brunswick, Vieweg, 1869 (3rd edn. 1881).

Haubrich, R. *Zur Entstehung der algebraischen Zahlentheorie Richard Dedekinds*. PhD thesis, Göttingen, 1992.

Hausdorff, F., 1891 'Zur Theorie der astronomischen Strahlenbrechung.' *Berichte über die Verhandlungeen der Königlich-Sächsischen Gesellschaft der Wissenschaften zu Leipzig, Mathematisch-physikalische Classe* 43, pp. 481–566.

Hausdorff, F., 1893 'Zur Theorie der astronomischen Strahlenbrechung II, III.' *Berichte über die Verhandlungeen der Königlich-Sächsischen Gesellschaft der Wissenschaften zu Leipzig, Mathematisch-physikalische Classe* 45, pp. 120–162, 758–804.

Hausdorff, F., 1897/2004, *Sant' Ilario – Gedanken aus der Landschaft Zarathustras*. Leipzig: C.G. Naumann. Also in: (Hausdorff, *Werke* VII, 2004).

Hausdorff, F., 1895 'Über die Absorption des Lichtes in der Atmosphäre.' *Berichte über die Verhandlungeen der Königlich-Sächsischen Gesellschaft der Wissenschaften zu Leipzig, Mathematisch-physikalische Classe* 47, pp. 401–482.

Hausdorff, F. 1898/2004, *Das Chaos in kosmischer Auslese – Ein erkenntniskritischer Versuch*. Leipzig: C.G. Naumann. Also in: (Hausdorff, *Werke* VII, 2004).

Hausdorff, F., 1903 'Das Raumproblem.' *Ostwalds Annalen der Naturphilosophie* 3, pp. 1–23.

Hausdorff, F., 1914, *Grundzuge der Mengenlehre*, Von Veit, Leipzig.

Hausdorff, F., 1914a, *Grundzüge der Mengenlehre*. Leipzig: Veit & Co., 1914. Also in: (Hausdorff, *Werke* II, 2002, 93–576).

Hausdorff, F., 1914b, 'Bemerkungen über den Inhalt von Punktmengen.' *Mathematische Annalen* 75 (1914), pp. 428–433. Also in: (Hausdorff, *Werke* IV, 2001, 5–10).

Hausdorff, F., 1919 'Dimension und äußeres Maß.' *Mathematische Annalen* 77, pp. 157–179. Also in: (Hausdorff, *Werke* IV, 2001, 21–43).

Hausdorff, F., 2001. *Gesammelte Werke einschließlich der unter dem Pseudonym Paul Mongré erschienenen philosophischen und literarischen Schriften*, ed E. Brieskorn e.a.. Berlin etc.: Springer. [9 volumes planned, published: II (2002), IV (2001), V (2005), VII (2004)].

Hausdorff, 2002, *Gesammelte Werke*, vol. II, eds. E. Brieskorn *et al.* Berlin : Springer.

Hausdorff, 2004, *Gesammelte Werke*, vol. VII, ed. W. Stegmaier. Berlin : Springer.

Hawkins, T., 1920. Lebegue's Theory of Integration, Chelsea, New York.
van Heijenoort J. (ed.), 1967, *From Frege to Gödel: A Source Book in Mathematical Logic*, Harvard University Press. Reprinted 2002.
Heidegger, M., 1928, *Sein und Zeit*. Tübingen: Neomarius.
Hellman, G., 1989, *Mathematics without Numbers*, Oxford University Press.
Helmholtz, H. von, 1870, 'Über den Ursprung und die Bedeutung der geometrischen Axiome: Vortrag, gehalten im Docentenverein zu Heidelberg im Jahre 1870.' In: Helmholtz, Hermann, v.: *Populäre Wissenschaftliche Vorträge*. Vol. III, Braunschweig: Vieweg und Sohn, 1876, pp. 21–54, tr. M.F. Lowe, in *Hermann von Helmholtz, Epistemological Writings*, P. Hertz and M. Schlick, eds, Boston Studies in the physics of science, 37, Reidel, Dordrecht and Boston, 1977.
Helmholtz, H, v., 1878 'Die Thatsachen in der Wahrnehmung: Rektoratsrede vom 3. August 1878.' In: Helmholtz, Hermann von: *Vorträge und Reden*. Vol. 2, Braunschweig: Vieweg und Sohn, 1884; pp. 217–271.
Helmholtz, H von, 1887, Zählen und Messen erkenntnistheoretisch betrachtet. In *Philosophische Aufsätze Eduard Zeller zu seinem fünfzigjährigen Doktorjubiläum gewidmet*. Leipzig: Fues pp. 17–52. In *Wissenschaftliche Abhandlungen* III (1895), 356–391. English in (Helmholtz 1977, 72–114).
Helmholtz, H von, 1921, *Schriften zur Erkenntnistheorie*. Ed. P. Hertz, M. Schlick. Berlin: Springer.
Helmholtz, H von, 1977, *Epistemological Writings*. English translation of (Helmholtz 1921) ed. R.S. Cohen, Y. Elkana. Dordrecht/Boston: Reidel.
Hendry, J, 1984, *The Bohr-Pauli Dialogue and the Creation of Quantum Mechanics*. Dordrecht: Reidel.
Herbart, J. F., *Die Psychologie als Wissenschaft, neu gegründet auf Erfahrung, Metaphysik und Mathematik*, vol. I (1824) in *Sämtliche Werke* (Aalen, Scientia, 1964), vol. 5, 177–402; vol. II (1825) in *Sämtliche Werke*, vol. 6, 1–339.
Herreman, A, 2000, *La topologie et ses signes: Éléments pour une histoire sémiotique des mathématiques*, L'Harmattan, Paris.
Hertz, H., 1956, *The Principles of Mechanics*, Dover.
Hess, K., 2002, 'Model Categories in Algebraic Topology.' *Applied Categorical Structures* 10: 195–220.
Hesseling, D, 2003, *Gnomes in the Fog. The Reception of Brouwer's Intuitionism in the 1920s*. Basel: Birkhäuser.
Hilbert, D., 1897, *Bericht über die Theorie der algebraischen Zahlkörper*, in *Abhandlungen*, vol. 1. English translation as *The Theory of Algebraic Number Fields*, Springer-Verlag, Berlin, 1998.
Hilbert, D., 1899, *Grundlagen der Geometrie (Festschrift zur Feier der Enthüllung des Gauss-Weber-Denkmals in Göttingen)*, Leipzig, Teubner. 6th edn. Berlin, Springer, 1930.
Hilbert, D., 1900, 'Über den Zahlbegriff', *Jahresbericht DMV* 8, 180–184. (English translation in Ewald (ed.) 1996, 1089–1095.)
Hilbert, D., 1901, 'Über das Dirichletsche Prinzip' *Jahresbericht der Deutschen Mathematiker-Vereinigung* 8 pp. 184–8.
Hilbert, D., 1902, 'Sur les problémes futurs des mathématiques', 2e *Congrès international des mathématiciens*, Paris, 1900 Gauthier-Villars, pp. 58–114.
Hilbert, D., 1904, 'Über das Dirichletsche Prinzip' *Mathematische Annalen* 59 pp. 161–86.
Hilbert, D. 1904, Über die Grundlagen der Logik und Arithmetik; in: *Grundlagen der Geometrie*, 5th edn, Leipzig and Berlin, 1922, 243–258; translated in: van Heijenoort, 129–138.

Hilbert, D., 1905, *Logische Principien des mathematischen Denkens* (Manuscript of a Lecture Course, SS 1905. Annotated by E. Hellinger. Bibliothek des Mathematischen Seminars der Universität Göttingen.

Hilbert, D., 1916, 'Die Grundlagen der Physik (Erste Mitteilung)', *Nachrichten von der Königlichen Gesellschaft der Wissenschaften zu Göttingen, Mathematische-Physikalische Klasse* (1916), 395–407.

Hilbert, D. 1916–17, *Die Grundlagen der Physik, II* (Manuscript of a Lecture Course, WS 1916–17, annotated by R. Bär. *Nachlass* David Hilbert).

Hilbert, D., 1917, 'Die Grundlagen der Physik (Zweite Mitteilung)', *Nachrichten von der Königlichen Gesellschaft der Wissenschaften zu Göttingen, Mathematische-Physikalische Klasse* (1917), 53–76.

Hilbert, D., 1918, 'Axiomatisches Denken', *Mathematische Annalen*, 18, reprinted in Hilbert, D., *Gesammelte Abhandlugen* III, Berlin, Springer 1935, pp. 146–156. Translated in Edward 1996, pp. 1105–15.

Hilbert, D., 1919–20, *Natur und Mathematisches Erkennen: Vorlesungen, gehalten 1919–1920 in Göttingen. Nach der Ausarbeitung von Paul Bernays* (Edited and with an English introduction by David E. Rowe), Basel, Birkhäuser (1992).

Hilbert, D., 1921, Adolf Hurwitz, *Math. Annalen* 1921, 83, 161–62. Also in Hurwitz's *Mathematische Werke* (Basel, Birkhäuser, 1963).

Hilbert, D., 1922, 'Neubegründung der Mathematik. Erste Mitteilung', *Abhandlungen aus dem mathematischen Seminar der Hamburgischen Universität* 1, 157–177. (English translation in W. Ewald (ed.) 1996, 1134–1148; also in Mancosu 1998, pp. 198–214).

Hilbert, D., 1922–23, *Wissen und mathematisches Denken* (Manuscript of a Lecture Course, WS 1922–23, annotated by W. Ackermann. Edited and published in Göttingen (1988) by C.F. Bödinger).

Hilbert, D., 1926, Über das Unendliche, *Mathematische Annalen* 95, 161–190.

Hilbert, D. 1928, Die Grundlagen der Mathematik; *Abhandlungen aus dem mathematischen Seminar der Hamburgischen Universität* 6 (1/2), 65–85.

Hilbert, D. 1929, Probleme der Grundlegung der Mathematik; in: K. Reidemeister (ed.), *Hilbert*, Berlin, 9–19; appeared first in: *Math. Annalen* 102(1), 1–9.

Hilbert, D., 1930, 'Naturerkennen und Logik', *Die Naturwissenschaften* 9, 59–63. (English translation in W. Ewald (ed.) 1996, 1157–1165.)

Hilbert, D. 1931, Die Grundlegung der elementaren Zahlenlehre; *Math. Annalen* 104(4), 485–494.

Hilbert, D., 1932–1935, *Gesammelte Abhandlungen*, 3 vols., Berlin, Springer.

Hilbert, D., 1971, *Foundations of Geometry*, 10th English edition, translation of the second German edition by L. Unger.

Hilbert, D., and Ackermann, W., 1928, *Grundzüge der theoretischen Logik*, Springer, Berlin.

Hilbert, D; von Neumann, J; Northeim L, 1928, 'Über die Grundlagen der Quantenmechanik.' *Mathematische Annalen* 98:1–30.

Hilbert, D. & S. Cohn-Vossen, 1932, *Anschauliche Geometrie*, Julius Springer, Berlin. Translated to English as *Geometry and the Imagination*, New York: Chelsea Publishing, 1952.

Hilbert, D. and P. Bernays, 1934/39, *Grundlagen der Mathematik*, Berlin, Springer, 2 vols.

Hirsch, G., 1978, Topologie. *Abrégé d'histoire des mathématiques*. J. Dieudonné. Paris, Hermann. II: 211–266.

Hocking, J. & G. Young, 1961, *Topology*, Addison-Wesley.

Hodges, W., 1985/6, Truth in a Structure, *Proceedings sof the Aristotelian Society*, 86, pp. 135–151.
Hopf, H., 1928, 'Eine Verallgemeinerung der Euler-Poincaréschen Formel', *Nachrichten, Gesellschaft der Wissenschaften zu Göttingen* pp. 127–36.
Hopf, H., 1929, 'Über die algebraische Anzahl von Fixpunkten', *Mathematische Zeitschrift* 29, 493–524.
Hopf, H., 1966, 'Eine abschnitt aus der Entwicklung der Topologie', *Jahresberichte der DMV* 68, 182 (96)–192 (106). The same lecture is printed as 'Einige Persönliche Erinnerungen aus der Vorgeschichte der heutigen Topologie', *Colloque de Topologie Tenu a Bruxelles Centre Belge de Recherches Mathématique*, 9–20. Paris: Gauthier Villars, 1966.
Hovey, M., 1999, *Model Categories*. Providence, AMS.
Humboldt, W. von, 1836, *Über die Kawi-Sprache auf der Insel Java (Einleitung)*. Berlin. In *Gesammelte Schriften VII* (1901), 1–344, als 'Über die Verschiedenheit des menschlichen Sprachbaues und ihren Einluß auf die Entwicklung des Menschengeschlechts'.
Huntington, E. V., 1904, Sets of independent postulates for the algebra of logic, *Transactions of the American Mathematical Society*, 5, pp. 288–309.
Huntington, E. V., 1906–1907, The fundamental laws of addition and multiplication in elementary algebra, *Annals of Mathematics*, second series, 8, pp. 1–44.
Huntington, E.V., 1911, The fundamental propositions of algebra, in J.W.A. Young, *Monographs on Topics of Modern Mathematics*, Longmans, Green and Co., New York.
Huntington, E. V., 1913, A set of postulates for abstract geometry, *Mathematische Annalen*, 73, 522–599.
Hurewicz, W., 1955, 'On the Concept of Fiber Space.' *Proceedings of the National Academy of Sciences U.S.A.* 41: 956–961.
Hurewicz, W., and N. E. Steenrod 1941, 'Homotopy Relations in Fiber Spaces.' *Proceedings of the National Academy of Sciences U.S.A.* 27: 60–64.
Husserl E., 1891, Philosophie der Arithmetik. Psychologische und Logische Untersuchungen vol 1 Halle: C. Pfeffer.
Husserl. E., 1900, *Logische Untersuchungen*, I, Halle, Niemeyer.
Husserl, E., 1929, *Formale und transzendentale Logik*, Halle, Niemeyer.
Illich, I. 1973, *Tools for Conviviality*. New York: Harper and Row. French, Paris: Seuil 1973, German, Hamburg: Rowohlt 1980, etc.
Jacobi C.G.J., 1829, *Fundamenta Nova Theoriae Functiorum Ellipticarum* Frankfurt.
Jacobson, N., ed., 1983, *E. Noether: Gesammelte Abhandlungen*, Springer Verlag.
James, I. M., 2001, Combinatorial Topology versus Point-Set Topology. *Handbook of the History of General Topology*. C. E. Aull and R. Lower. Dordrecht, Kluwer. 3: 809–834.
Jungnickel, C. and McCormmach R., 1986 *Intellectual Mastery of Nature: Theoretical Physics from Ohm to Einstein* vol. 1 Chicago: University of Chicago Press.
Kant, I., 1787, *Kritik der reinen Vernunft*, Riga, Hartknoch, 2nd edn. (references to this edition, noted B). *CPR, Critique of Pure Reason* (1781, 1787), tr. P. Guyer and A.W. Wood, Cambridge: Cambridge University Press, 1997.
Kaufmann, F., 1930, *Das Unendliche in der Mathematik und seine Ausschaltung*, Duticke, Vienna.
Kellogg O., 1929, *Foundations of Potential Theory* New York: Dover (1959 reprint).
Kerry B., 1889, 'Anschauung und ihr Psychische Verarbeitung' (5th article) in *Vierteljahrsschrift für Wissenschaftliche Philosophie* pp. 71–124.
Keyser, C., 1918a, Doctrinal functions, *Journal of Philosophy*, 15, 262–267.

Keyser, C.J., 1918b, Concerning the number of possible interpretations of any axiom system of postulates, *Bulletin of the American Mathematical Society*, 24, pp. 391–93.
Keyser, C., 1922, *Mathematical Philosophy. A Study of Fate and Freedom*, Dutton & Co., New York.
Killing, W.: *Einführung in die Grundlagen der Geometrie*. 2 vols., Paderborn, F. Schöningh, 1893–1898.
Kitcher. P., 1981, 'Mathematical Rigor – Who Needs It?' *Noûs* 15 pp. 469–93.
Kitcher P., 1984, *The Nature of Mathematical Knowledge* Oxford: Oxford University Press.
Kitcher, P., 1986, 'Frege, Dedekind, and the Philosophy of Mathematics' in L. Haaperanta and J. Hintikka (eds.) *Frege Synthesized* Dordrecht: D. Reidel pp. 299–344.
Kleene, S.C., 1952, *Introduction to Metamathematics*, North Holland.
Klein C.F., 1871, Über die sogenannte Nicht Euklidische Geometrie, I, *Mathematische Annalen*, 4, in *Gesammelte Mathematische Abhandlungen* I, (no. XVI) 254–305.
Klein C.F., 1872, *Vergleichende Betrachtungen über neuere geometrische Forschungen* (Erlanger Programm) 1st pub. Deichert, Erlangen, in *Gesammelte Mathematische Abhandlungen* I (no. XXVII) 460–497.
Klein, C.F., 1873, Über die sogenannte Nicht Eukidische Geometrie, II, *Mathematische Annalen* 6, in *Gesammelte Mathematische Abhandlungen* I (no. XVIII) 311–343.
Klein, C.F., 1884/1893, *On Riemann's Theory of Algebraic Functions and Their Integrals* Cambridge: McMillan and Bowes 1893 (page references to Dover reprint, German original published 1884).
Klein, C.F., 1895, Ueber Arithmetisirung der Mathematik. Gött. Nachr. Geschäftl. Mitt. 82–91. English translation by I. Maddison appears in Ewald [1996] p. 965–971 page references to this translation.
Klein, C.F., 1897, Riemann und seine Bedeutung für die Entwicklung der modernen Mathematik, *Jahresbericht der Deutschen Mathematiker-Vereinigung* 4, 71–87.
Klein, C.F., 1926, *Vorlesungen über die Entwicklung der Mathematik im 19. Jahrhundert*. 2 vols. Berlin, Springer, 1926/27. Reprint 1979.
Kline, M., 1972, *Mathematical Thought from Ancient to Modern Times*. Oxford U. P.
Kneale, W. and M. Kneale, 1962, *The Development of Logic*, Oxford University Press, Oxford.
Kohnke, K.C., *The Rise of neo-Kantianism*, CUP, 1991, translated by R.J. Hollingdale (German original, *Entstehung und Aufstieg des Neukantianismus*, 1986).
Korselt, A., 1913, Was ist Mathematik?, *Archiv der Mathematik und Physik*, (series 3), 21, pp. 371–373.
Kossak, E., 1872, *Die Elemente der Arithmetik* Berlin: Nicolai'sche Verlagsbuchhandlung.
Kratzsch, I., 1979, 'Material zu Leben und Wirken Freges aus dem Besitz der Universitäts- bibliothek Jena' in *Begriffsschrift – Jenaer Frege Konferenz*, Fr. Schiller Universität Jena pp. 534–46.
Kreisel, G., 1958, Hilbert's programme; *Dialectica* 12, 346–372; reprinted in: Benacerraf, and Putnam, 1983, 207–238.
Kreisel, G., 1965, Mathematical Logic; in: *Lectures on modern mathematics*, vol. III, T.L. Saaty (ed.), New York, 95–195.
Kreisel, G., 1968, Survey of Proof Theory; *Journal of Symbolic Logic* 33, 321–388.
Kreiser L., 1983, 'Nachschrift einer Vorlesung und Protokolle mathematischer Vorträge Freges' Supplement to G. Frege *Nachgelassene Schriften* H. Hermes, F. Kambartel and F. Kaulbach eds. Hamburg: Felix Meiner Verlag.

Kreiser, L., 1984, 'G. Frege Grundlagen der Arithmetik – Werk und Geschichte' in G. Wechsung (ed.) *Frege conference 1984* Berlin.

Kreiser, L., 1995, 'Die Hörer Freges und sein Briefpartner Alwin Korselt' in *Wittgenstein Studies*, Diskette 1/1995.

Kreiser, L., 2001, *Gottlob Frege: Leben – Werk – Zeit*, Hamburg: Felix Meiner.

Kriz, I., 2001, A Brief Introduction to the Work of J. Peter May, on the occasion of his 60th birthday. *Homotopy Methods in Algebraic Topology*. J.P.C. Greenlees *et al*. Providence, American Mathematical Society: xix–xlii.

Kronecker, L. 1870 Auseinandersetzung einiger Eigenschaften der Klassenzahl idealer complexer Zahlen. *Koniglich Akademie der Wissenschaft Berlin, Monatsbericht*, pages 881–889, Reprinted in (Kronecker, 1968), vol. I, pages 273–282.

Kronecker, L. 1882 *Grundzüge einer arithmetischen Theorie der algebraischen Grössen*. Riemer, Berlin, Also published in *Journal für reine und angewandte Mathematik*, volume 92, 1882, pages 1–122, and (Kronecker, 1968), volume II, pages 237–387.

Kronecker, L., 1886, Über einige Anwendungen der Modulsysteme; Crelles Journal für die reine und angewandte Mathematik 99, 329–371, also: Werke III, 145–208.

Kronecker, L., 1887 Über den Zahlbegriff. In *Philosophische Aufsätze, Eduard Zeller zu seinem fünfzigjährigen Doctorjubiläum gewidmet*, pages 261–274. Fues, Leipzig, Reprinted in (Kronecker, 1968), volume IIIa, pages 249–274. Translated as 'On the concept of number' by William Ewald in (Ewald, 1996), volume 2, pages 947–955.

Kronecker, L., 1887a Ein Fundamentalsatz der allgemeinen Arithmetik. *Journal für die reine und angewandte Mathematik*, 100:490–510.

Kronecker, L., 1895–1930 *Werke*. Edited by K. Hensel, Teubner, Leipzig, 5 volumes. Reprinted by the Chelsea Publishing Co., New York, 1968.

Kronecker, L., 1901 *Vorlesungen über Zahlentheorie*. Teubner, Leipzig, Edited by Kurt Hensel. Republished by Springer, Berlin, 1978.

Krull, W., 1935, *Idealtheorie*, Julius Springer, Berlin.

Kummer, E.E. 1846, Zur Theorie der complexen Zahlen. *Koniglich Akademie der Wissenschaft Berlin, Monatsbericht*, pages 87–97, Also in *Journal für die reine und angewandte Mathematik* 35:319–326, 1847, and in Kummer's *Collected Papers*, edited by André Weil, Springer-Verlag, Berlin, 1975, volume 1, 203–210.

Kummer, E.E., 1860, Gedächtnisrede auf Gustav Peter Lejeune Dirichlet; reprinted in (Dirichlet, 1897), 311–344.

Kuperberg, K. Ed., 1995, *Collected Works of Witold Hurewicz*. Providence, American Mathematical Society.

Kusch, M., 1995, *Psychologism: a case study in the sociology of philosophical knowledge*. London. Routledge.

Latakos, I. 1970 Falsification and the methodology of scientific research programme. In Imre Lakatos and Alan Musgrave, editors, *Criticism and the growth of knowledge*, pages 91–196. Cambridge University Press, 1976.

Lakatos, I. 1976, *Proofs and Refutations*, Cambridge University Press.

Lakatos, I., 1967, A Renaissance of Empiricism in the Recent Philosophy of Mathematics?, in *Philosophical Papers*, vol. 2, Cambridge University Press, 1978, 24–42.

Lang, S., 1993, *Algebra*, Addison-Wesley.

Langford, C. H., 1927a, Some theorems on deducibility, *Annals of Mathematics*, series 2, 28, pp. 16–40.

Langford, C. H., 1927b, Some theorems on deducibility, *Annals of Mathematics*, series 2, 28, pp. 459–471.

Laugwitz D., 1992, 'Das letzte Ziel ist immer die Darstellung einer Funktion': Grundlagen der Analysis bei Weierstraß 1886 historische Wurzeln und Parallelen' *Historia Mathematica* 19 pp. 341–355.

Laugwitz, D., 1999, *Bernhard Riemann, 1826–1866. Turning Points in the Conception of Mathematics*, Basel, Birkhäuser (original German edn. 1996).

Lawvere, F. W., and R. Rosebrugh, 2003, *Sets for Mathematics*, Cambridge University Press.

Lefschetz, S., 1930, *Topology*, American Mathematical Society, Providence.

Lefschetz, S., 1942, *Algebraic Topology*, American Mathematical Society, Providence.

Lefschetz, S., 1949, *Introduction to Topology*, Princeton University Press.

Levine, J., 2002, 'Analysis and Decomposition in Frege and Russell', *Philosophical Quarterly* 52, pp. 195–216; repr. in Beaney and Reck 2005, Vol. IV, pp. 392–414.

Lewis, C. I., Langford, C. H., 1932, *Symbolic Logic*, The Century Company, Toronto.

Lie S., and Scheffers G., 1896, *Geometrie der Berührungstransformationen* Leipzig (1956 reprint: Chelsea, New York).

Liebmann, O.: *Zur Analysis der Wirklichkeit: Eine Erörterung der Grundprobleme der Philosophie*. 2nd enlarged edition, Strassburg: Trübner, 1880; first edition, Strassburg: Trübner, 1876.

Lipschitz, R., 1986, *Briefwechsel mit Cantor, Dedekind, Helmholtz, Kronecker, Weierstrass*; Vieweg, Braunschweig.

Lobatchevskii, N.I., 1899, *Zwei geometrische Abhandlungen*, tr. F. Engel, Teubner, Leipzig.

Luckhardt, H., 1989, Herbrand-Analysen zweier Beweise des Satzes von Roth: polynomiale Anzahlschranken; *Journal of Symbolic Logic* 54(1), 234–263.

Lützen, J., 1990, *Joseph Liouville, 1809–1882: Master of Pure and Applied Mathematics*, Springer-Verlag, New York.

Mac Lane, S., 1938, 'A lattice formulation for transcendence degrees and p-bases', *Duke Mathematical Journal* 4, 455–68.

Mac Lane, S., 1978, 'Origins of the cohomology of groups', *Enseignement Mathématique* 24, 1–29.

Mac Lane, S., 1986, 'Topology becomes algebraic with Vietoris and Noether', *J. Pure Appl. Algebra* 39, 305–307.

Mac Lane, S., 1988, 'Group extensions for 45 years', *Mathematical Intelligencer* 10, 29–35.

Mac Lane, S., 1998, *Categories for the Working Mathematician*, 2nd edn, Springer-Verlag.

MacFarlane, J., 2002, 'Frege, Kant, and the Logic in Logicism', *Philosophical Review* 111, pp. 25–65; repr. in Beaney and Reck 2005, Vol. I, pp. 71–108.

Mackey, G., 1988, 'Hermann Weyl and the application of group theory to quantum mechanics.' In (Deppert 1988, 131–160).

Mackey, G., 1993, 'The mathematical papers.' In (Wigner 1992ff., A I, 241–290).

Maddy, P., 1990, *Realism in Mathematics*, Oxford University Press.

Maddy P., 1997, *Naturalism in Mathematics* Oxford: Oxford University Press

Maddy, P. Three forms of naturalism. In Stewart Shapiro, editor, *Oxford handbook of philosophy of logic and mathematics*. Oxford University Press, Oxford, 2005.

Majer, U., 1988, 'Zu einer bemerkenswerten Differenz zwischen Brouwer und Weyl.' In (Deppert 1988, 543–551).

Mancosu, P., ed., 1998, *From Brouwer to Hilbert. The Debate on the Foundations of Mathematics in the 1920s*, Oxford University Press.

Mancosu, P., 2001, Mathematical Explanation: Problems and prospects, *Topoi* 20 (2001), no. 1, 97–117.

Mancosu, P., 2005, 'Harvard 1940–41: Tarski, Carnap and Quine on a finitistic language of mathematics for science', *History and Philosophy of Logic*. 26: 327–357.

Mancosu, P.; Ryckman, T., 2002. 'The correspondence between O. Becker and H. Weyl.' *Philosophia Mathematica* 10:130–202.

Mancosu, P., Zach, R., and Badesa, C., forthcoming, The development of mathematical logic from Russell to Tarski: 1900–1935, to appear in *Handbook of the History of Logic*, L. Haaparanta ed., Oxford University Press.

Mancosu, P., K.F. Jorgensen and S.A. Pedersen (eds.),2005, *Visualization, Explanation and Reasoning Styles in Mathematics* Berlin, Springer(Synthese Library, vol.327).

Maxwell, J. C., 1873. *A Treatise on Electricity and Magnetism*, Oxford, Clarendon Press.

Mazur, B., 1997 Conjecture, *Synthese* 111, 197–210.

McLarty, C., 1990, 'The uses and abuses of the history of topos theory', *British Journal for the Philosophy of Science* 41, 351–75.

McLarty, C., 1992: *Elementary Categories, Elementary Toposes*, Oxford University Press.

Mehrtens, H., 1990 *Moderne Sprache Mathematik. Eine Geschichte des Streits um die Grundlagen der Disziplin und des Subjekts formaler Systeme*, Frankfurt am Main: Suhrkamp.

Minkowski, H., 1905. Peter Gustav Lejeune-Dirichlet und seine Bedeutung für die heutige Mathematik, *Jahresbericht der D. M. V.* 14, 149–63.

Minkowski, H., 1896. *Geometrie der Zahlen*. Leipzig, B.G. Teubner.

Mongré, P. (= Hausdorff, F.), 1898, *Das Chaos in kosmischer Auslese: Ein erkenntniskritischer Versuch*. Leipzig: C. G. Naumann, Also in: (Hausdorff, Werke VII, 2004, 499–807).

Mongré, Paul (= Hausdorff, F.) 1897, *Sant' Ilario: Gedanken aus der Landschaft Zarathustras*. Leipzig: C. G. Naumann. Also in: (Hausdorff, Werke VII, 2004, 87–473).

Monna, A., 1972, 'The Concept of Function in the 19th and 20th Centuries in Particular with Regard to the Discussions between Baire, Borel, and Lebesgue' *Archive for the History of Exact Sciences*. 9: 57–84.

Monna, A., 1975, *Dirichlet's Principle: A Mathematical Comedy of Errors and its Influence on the Development of Analysis* Utrecht: Oosthoek, Scheltema & Holkema.

Moore, G.H., 1982, *Zermelo's Axiom of Choice: Its Origins, Development and Influence* Springer: New York.

Moore, G.H., 1988, The emergence of first-order logic; in: (Aspray and Kitcher), 95–138.

Müller, G.H., 1981, Framing mathematics; *Epistemologia* 4(1), 253–286.

Nagel, E., 1939, The formation of modern concepts of formal logic in the development of geometry *Osiris*, 7, 142–224.

Neuenschwander E., 1997 *Riemanns Einführung in die Funktionentheorie. Eine quellenkritische Edition seiner Vorlesungen mit einer Bibliographie zur Wirkungsgeschichte der Riemannschen Funktionentheorie* Göttingen: Vandenhoeck and Ruprecht.

Neumann, C., 1877 *Untersuchungen über das Logarithmische und Newton'sche Potential* Leipzig: Teubner.

von Neumann, J., 1925, Eine axiomatisierung der Mengenlehre, *Journal für die reine und aangewandte Mathematik*, 154, pp. 219–240. Translation in van Heijenoort, 1967, pp. 393–413.

Nicholson, J., 1993, The development and understanding of the concept of quotient group. *Historia Mathematica*, 20:68–88.

Noack, H., 1936, *Symbol und Existenz in der Wissenschaft*. Halle/Saale: Niemeyer.

Noether, E., 1921, 'Idealtheorie in Ringbereichen', *Mathematische Annalen* 83, 2466. In (Jacobson 1983), 354–96.

Noether, E., 1924, 'Abstract of a talk "Abstrakter Aufbau der Idealtheorie im algebraischen Zahlkörper"', *Jahresbericht DMV* 33, 104. In (Jacobson 1983), 482.

Noether, E., 1925, 'Abstract of a talk "Gruppencharaktere und Idealtheorie"', *Jahresbericht DMV* 34, 144. In (Jacobson 1983), 484.

Noether, E., 1926, 'Abstract', *Jahresbericht DMV* 34, 104.

Noether, E., 1927, 'Abstrakter Aufbau der Idealtheorie in algebraischen Zahl- und Funktionenkörpern', *Mathematische Annalen* 96, 26–91. I cite this from (Jacobson 1983). 493–528.

Noether, E., 1929, 'Hyperkomplexe Grössen und Darstellungstheorie', *Mathematische Zeitschrift* 30, 641–92. I cite this from (Jacobson 1983), 563–614.

Noether, E., & J. Cavaillès, eds 1937, *Briefwechsel Cantor-Dedekind*, Hermann, Paris.

Norton Wise, M., 1981, German Concepts of Force, Energy, and the Electromagnetic Ether: 1845–1880, in G. N. Cantor and M. J. S. Hodge, eds. *Conceptions of Ether* (Cambridge University Press), 269–307.

Nowak, G., 1989, Riemann's *Habilitationsvortrag* and the Synthetic *a priori* Status of Geometry, in D. Rowe and J. McLeary, eds. *The History of Modern Mathematics* (Boston/London, Academic Press), vol. 1, 17–46.

Ore, Ö., 1935, 'On the foundations of abstract algebra, I', *Annals of Mathematics* 36, 406–37.

Padoa, A., 1901, Essai d'une théorie algébrique des nombres entiers, precede d'une introduction logique à une théorie deductive quelconque, *Bibliothèque du Congrès international de philosophie*, Paris 1900, A. Colin, Paris, 1901, pp. 249–256. Partially translated in van Heijenoort 1967, pp. 118–123.

Padoa, A., 1902a, Un nouveau système de définitions pour la géométrie euclidienne *Compte Rendu du Deuxième Congrès international des mathematicians*, Paris, Gauthiers-Villars.

Padoa, A., 1902b, Un nouveau système irréductible de postulats pour l'algèbre, *Compte rendu du Deuxième congrès international des mathematicians tenu à Paris du 6 au 12 août 1900*, Gauthiers-Villars, Paris, pp. 249–256.

Padoa, A., 1903, Le Problème no. 2 de M. David Hilbert, *L'Enseignement Mathématique*, 5, 85–91.

Parshall, K.H. and Rowe, D.E., 1994, *The Emergence of the American Mathematical Research Community; J.J. Sylvester, Felix Klein, and E.H. Moore*, American and London Mathematical Societies, 8, Providence, Rhode Island.

Parshall, K.H., 2004, Defining a mathematical research school: the case of algebra at the University of Chicago, 1892–1945, *Historia Mathematica*, 31, 263–278.

Parsons, C.D., 1983, The impredicativity of induction; in: *How many questions? – Essays in honor of Sidney Morgenbesser*, Cauman, Levi, Parsons, Schwartz (eds.), Indianapolis, 132–153.

Pasch, M., 1882, *Vorlesungen übber neuere Geometrie*, Teubner, Leipzig.

Peano, G., 1889, *I principii di geometria logicamente espositi*, Turin, rep. in *Opere scelte*, 2, Rome, 1958, 56–91.

Peano, G., 1894, Sui fondamenti della Geometria, *Rivista di matematiche* 4, 51–90.

Peckhaus, V., 1997, *Logik, Mathesis universalis und allgemeine Wissenschaft: Leibniz und die Wiederentdeckung der formal Logik im 19. Jahrhundert* Akademie Verlag, Berlin.

Pieri, M., 1895, Sui principi che reggiono la geometria di posizione, *Atti Accademia Torino*, 30, 54–108.

Pieri, M., 1899, I Principii della Geometria di Posizione, composti in sistema logico deduttivo, *Memorie della Reale Accademia delle Scienze di Torino* (2) 48, 1–62.

Pieri, M., 1901, Sur la géométrie envisagée comme un système purement logique, *Bibliothèque du Congrès international de philosophie*, Paris 1900, A. Colin, Paris, 1901, pp. 367–404.

Pincherle S., 1880, 'Saggio di una introduzione alla teoria delle funzione analitiche secondo i principi del prof. C. Weierstrass' *Giornale di Mathematiche* 18 pp. 178–254, 314–57.

Poincaré H., 1882, Théorie des Groupes Fuchsiens, *Acta Mathematica*. 1, 1 62, in *Oeuvres*, II, 108–168.

Poincaré, H., 1891, Les Géométries non-euclidiennes, *Revue générale des Sciences pures et appliqués*, 2, 769–774, in Poiucaré 1902, pp. 63–76.

Poincaré, H., 1895, 'Analysis situs', *Journal de l'école Polytechnique* 1, 1–121. Reprinted in *Oeuvres* vol. 6, 193–288.

Poincaré, H., 1899–1904, 'A series of five "compléments à l'analysis situs"'. Extending and correcting Poincaré 1895. Collected in *Oeuvres* vol. 6, 289–498.

Poincaré, H., 1901, 'Sur les propriétés arithmétiques des courbes algébriques', *Journal des Mathématiques* 7, 161–233. Reprinted in *Oeuvres* V, 483–550.

Poincaré, H., 1902 *La science et l'hypothèse*. Paris: Flammarion.

Poincaré, H., 1906 *La valeur de la science*. Paris: Flammarion.

Poincaré, H., 1908a *Science et méthode*. Paris: Flammarion.

Poincaré, H., 1908b, L'avenir des mathématiques, *in Atti del IV Congresso Internazionale dei Matematici*, Accademia dei Lincei, pp. 167–182. Translated in heavily edited form as 'The future of mathematics' in G. Halstead ed. 1921: *The Foundations of Science*. New York: The Science Press, 369–82.

Poincaré, H., 1916-1956, *Oeuvres de Henri Poincaré*, Gauthier-Villars. In 11 volumes.

Poincaré, H., 1983, Analyse de ses travaux scientifiques, *in* F.Browder, ed., *The mathematical heritage of Henri Poincaré*, Vol. 2, American Mathematical Society, pp. 257–357. A description of his own work written in 1901 for Mittag Lefler. Published in *Acta Mathematica* 38 (1921) 36–135.

Poincaré, H., 1997, *Three Supplementary Essays on the Discovery of Fuchsian Functions* (J.J. Gray and Scott Walter, eds. with an introductory essay) Akademie Verlag, Berlin and Blanchard, Paris.

Polanyi, M., 1958, *Personal Knowledge: Towards a Post-Critical Philosophy*. Chicago, Chicago University Press.

Polanyi, M., 1960–61, Science: Academic and Industrial. *Jounal of the Institute of Metals*, 89: 401–406.

Poncelet, J.V., 1822, *Traité des propriétés projectives des figures*, Paris, Gauthier-Villars.

Pont, J.-C., 1974, *La topologie algébrique des origines à Poincaré*, Presses Universitaires de France, Paris.

Potter, M., 2000: *Reason's Nearest Kin: Philosophies of Arithmetic from Kant to Carnap*, Oxford University Press.

Purkert, W., 2002, 'Grundzüge der Mengenlehre. Historische Einführung' in W. Purkert et al. (eds.) *Felix Hausdorff, Gesammelte Werke*, Vol. 2, Berlin, Springer Verlag, 1–89.

Purkert, W. and H. J. Ilgauds, 1987. *Georg Cantor 1845–1918*. Basel/Boston: Birkhäuser.

Putnam, H., 1994, Philosophy of Mathematics: Why nothing works (orig. 1979), in his *Words and Life* (Harvard University Press), 499–512.

Quillen, D., 1967, *Homotopical Algebra*. New York, Springer-Verlag.

Quillen, D., 1969, 'Rational Homotopy Theory.' *Annals of Mathematics* 90: 205–295.

Quine, W.V.O., 1951, Two Dogmas of Empiricism, *Philosophical Review* (Jan. 1951), reprinted as chap. 2 of *From a Logical Point of View* (Harvard University Press, 1953).
Quine, W.V.O., 1960, *Word and Object* Cambridge: MIT Press.
Quine, W.V.O., Goodman, N., 1940, Elimination of extra-logical postulates, *The Journal of Symbolic Logic*, 5, pp. 104–109.
Ramsey, F.P., 1931, *The Foundations of Mathematics*, London: Routledge.
Ravenel, D., 1986, *Complex Cobordism and Stable Homotopy Groups of Spheres*, Academic Press.
Ravenel, D., 1987, 'Localization and Periodicity in Homotopy Theory.' *Homotopy Theory Proceedings Durham (1985)*, LMS 117, Cambridge, Cambridge University Press: 175–194.
Ray, G., 1996, Logical consequence: a defence of Tarski, *The Journal of Philosophical Logic*, 25, 617–677.
Reck, E.H. and Beaney, M., 2005 'Frege's Philosophy of Mathematics', introduction to Beaney and Reck 2005, Vol. III, pp. 1–12.
Reck, E.H., 2003, 'Dedekind's Structuralism: An Interpretation and Partial Defense', *Synthese* 137, pp. 369–419.
Reck, E.H., 2005, 'Frege's Natural Numbers: Motivations and Modifications', in Beaney and Reck 2005, Vol. III, pp. 270–301.
Reed, D., 2005, *Figures of thought: mathematics and mathematical texts*. Routledge, London.
Reid, C., 1970, *Hilbert*; Springer-Verlag, Berlin, Heidelberg, New York.
Renn, J.; Sauer, T., 1999, Heuristics and mathematical representation in Einstein's search for a gravitational field equation. In *The Expanding Worlds of Relativity*, ed. Hubert Goenner, Jürgen Renn, Jim Ritter, Tilman Sauer. Basel: Birkhäuser pp. 87–126.
Riemann B., 1851, *Grundlagen für eine allgemeine Theorie der Funktionen einer veränderlichen complexen Grösse* Inaugural dissertation Göttingen (=*Werke*, p. 3–45).
Riemann B., 1854/1868, 'Ueber die Hypothesen, welche der Geometrie zu Grunde Liegen' (Habilitation lecture, first published in posthumously in *Abhandlungen der Koenigliche Gessellschaft d. Wissenschaft zu Göttingen* XIII (=*Werke* p. 272–287) English translation by W. Clifford (revised by Ewald) in Ewald [1996] p. 652–661 page references to Ewald translation.
Riemann B., 1854/1868a, 'Ueber die Darstellbarkeit einer Function durch eine Trigonometrische Reihe' (Habilitation paper, first published posthumously in *Abhandlungen der Koenigliche Gessellschaft d. Wissenschaft zu Göttingen* XIII (=*Werke* p. 272–287).
Riemann B., 1857a, 'Theorie der Abel'schen Funktionen' *Journal für Reine und Angewandte Mathematik* (Crelle's Journal) 54 pp. 101–55 (=*Werke* pp. 88–153).
Riemann B., 1857b, 'Beiträge zur Theorie der durch die Gauss'sche Riehe F(α, β, γ, x) darstellbaren Functionen' *Abhandlungen der Königlichen Gesellschaft der Wissenschaften zu Göttingen* 6 (=*Werke* pp. 67–83).
Riemann B., 1859, 'Über die Anzahl die Primzahlen unter einer gegebener Grösse' in Werke, pp. 145–155.
Riemann, B., 1865, 'Ueber das Verschwinden der Theta – Functionen' *Journal für Reine und Angewandte Mathematik* (Crelle's Journal) 65 (=*Werke* p. 212–224).
Reimann, B. 1876, Fragmente philosophischen Inhalts (I. Zur Psychologie und Metaphysik; II. Erkenntnisstheoretisches; III. Naturphilosophie), in Werke, 509–38.
Riemann, B., *Werke: Gesammelte mathematische Werke und wissenschaftlicher Nachlass* (1st edn. 1876, 2nd 1892), revised and expanded by R. Narasimhan, New York, Springer,

1991. References to this edition respect the pagination of 1892 (Leipzig, Teubner; reprinted in New York, Dover, 1953).

Roch G., 1863, 'Ueber Functionen complexer Grössen' *Zeitschrift für Mathematik und Physik* 8 p. 12–26 p. 183–203.

Roch G., 1865, 'Ueber Functionen complexer Grössen' *Zeitschrift für Mathematik und Physik* 10 p. 169–94.

Rosenfeld, B. A., 1965, The analytic principle of continuity. *American Mathematical Monthly*, to appear. A translation, by Abe Shenitzer, of the first four sections of the original Russian article, which apeared in the Russian journal *Historico-Mathematical Investigations*, issue XVI, Nauka, Moscow, pages 273–294.

de Rouilhan, P., 1998, Tarski et l'universalité de la logique, in F. Nef and D. Vernant, *Le Formalisme en Question. Le Tournant des Années 30*, Vrin, Paris, pp. 85–102.

Rowe, D., 1989, 'Klein, Hilbert, and the Göttingen Mathematical Tradition' *Osiris* 2nd Series 5 pp. 186–213.

Rowe, D., 2000, 'Episodes in the Berlin – Göttingen Rivalry 1870–1930' *Mathematical Intelligencer* 22.1 pp. 60–69.

Rowe, D., 1994 [-a-] 'The Philosophical Views of Klein and Hilbert' in C. Sasaki *et al.*, The intersection of history and mathematics, Basel: Birkhäuser.

Russ, S., 2004, *The Mathematical Works of Bernhard Bolzano*, Oxford University Press.

Russell, B., 1903 [POM],[2] *The Principles of Mathematics* (2nd edn 1937), with a new introd. by John G. Slater, London: Routledge, 1992.

Russell, B., 1905 [OD], 'On Denoting', *Mind* 14, pp. 479–93; repr. in Russell 1956, pp. 41–56; Russell 1973, pp. 103–19.

Russell, B., 1907 [RMDP], 'The Regressive Method of Discovering the Premises of Mathematics', in Russell 1973, pp. 272–83.

Russell, B., 1913 [TK], *Tlteory of Knowledge: The 1913 Manuscript*, ed. Elizabeth Ramsden Eames in collaboration with Kenneth Blackweii, in Russell CP, Vol. 7.

Russell, B., 1918 [PLA], 'The Philosophy of Logical Atomism', in Russell 1956, pp. 175–281; orig. in *Monist* 28 and 29.

Russell, B., 1919 [IMP], *Introduction to Mathematical Philosophy*, with a new introd. by John G. Slater, London: Routledge, 1993; orig. publ. London: George Allen and Unwin.

Russell, B., 1945 [HWP], *History of Western Philosophy*, London: George Allen and Unwin, 2nd edn 1961.

Russell, B., 1956 [LK], *Logic and Knowledge: Essays 1901–1950*, ed. R.C. Marsh, London: George Allen and Unwin.

Russell, B., 1959 [MPD], *My Philosophical Development*, London: Unwin Paperbacks, 1985; orig. publ. London: George Allen and Unwin, 1959.

Russell, B., 1973 [EA], *Essays in Analysis*, ed. Douglas Lackey, London: George Allen and Unwin.

Russell, B., 1983 CP, *The Collected Papers of Bertrand Russell*, 28 vols., London: Routtedge.

Sagüillo, J., 1997, Logical consequence revisited, *The Bulletin of Symbolic Logic*, 3, pp. 216–241.

---

[2] In the case of Russell, like in that of Frege, references are sometimes given by means of the keys that are usual among specialists. In our bibliography the key is given within square brackets, next to the year of publication – e.g. [PM] for the work of Whitehead and Russell (1910).

Sarkaria, K., 1999, The topological work of Henri Poicaré, *in* I. James, ed., 'History of Topology', Elsevier, pp. 123–68.

Scanlan, M., 1991, Who were the American postulate theorists?, *Journal of Symbolic Logic*, 56, pp. 981–1002.

Scanlan, M., 2003, American postulate theorists and Alfred Tarski, *History and Philosophy of Logic*, 24, pp. 307–325.

Schaeffer H. (ed.), [1877] *Erinnerungsblätter der Mathematischen Gessellschaft zu Jena* Jena.

Schappacher, N., 2007? 'What is Arithmetization?' forthcoming in *Two Hundred Years of Number Theory after Carl Friedrich Gauss' Disquisitiones Arithmeticae* Berlin: Springer.

Schappacher, N & R Schoof, 1996, 'Beppo Levi and the arithmetic of elliptic curves', *Mathematical Intelligencer* 18, 57–69.

Schering, E., 1866, Zum Gedächtniss an B. Riemann, in E. Schering, *Gesammelte mathematische Werke* (Berlin, Mayer & Müller, 1909), 367–83; reproduced in Riemann's *Werke*, 828-47.

Schlimm, D. *Axiomatics as engine for discovery driving in mathematics and science*. PhD thesis, Carnegie Mellon University, 2005.

Schlömilch O., 1862, *Handbuch der Algebraischen Analysis* Jena: Frommann.

Schlömilch O. 1868, *Übungsbuch der Höheren Analysis* Leipzig: Teubner.

Schlick, M., *Raum und Zeit in der gegenwärtigen Physik: Zur Einführung in das Verständnis der allgemeinen Relativitätstheorie*. First edition, Berlin: Springer, 1917; 3rd edn, Berlin: Springer, 1920.

Schmidt, A., 1962, *Der Begriff der Natur in der Lehre von Karl Marx*. Frankfurt/Main: Europäische Verlagsanstalt.

Schoenflies, A., 1900, 'Die Entwicklung der Lehre von den Punktmannigfaltigkeiten', *Jahresberichte DMV* 82, 1–250.

Schoenflies, A., 1913, *Entwickelung der Mengenlehre und ihrer Anwendungen*, B.G Teubner, Leipzig.

Scholz, E., 1980, *Geschichte des Mannigfaltigkeitsbegriff von Riemann bis Poincaré*, Basel, Birkhäuser.

Scholz, E., 1982a, Herbart's influence on Bernhard Riemann. *Historia Mathematica* 9: 413–40.

Scholz, E., 1982b, Riemanns frühe Notizen zum Mannigfaltigkeitsbegriff und zu den Grundlagen der Geometrie, *Archive for Hist. of Exact Sciences* 27, 213–32.

Scholz, E., 1996, 'Logische Ordnungen im Chaos: Hausdorffs frühe Beiträge zur Mengenlehre.' In: Brieskorn, Egbert (Ed.): *Felix Hausdorff zum Gedächtnis: Aspekte seines Werkes*. Wiesbaden: Vieweg, pp. 107–134.

Scholz, E., 1999, The Concept of Manifold, 1850–1950, in I. M. James, ed. *History of Topology* (Elsevier, 1999), 25–64.

Scholz, E., 2001 Bernhard Riemanns Auseinandersetzung mit der Herbartschen Philosophie. A. Hoeschen and L. Schneider, eds., *Herbarts Kultursystem* (Würzburg: Königshausen & Neumann).

Scholz, E. (ed.), 2001 *Hermann Weyl's* Raum - Zeit - Materie *and a General Introduction to His Scientific Work*. Basel etc.: Birkhäuser.

Scholz, E., 2004a, 'The changing concept of matter in H. Weyl's thought, 1918–1930.' Preprint Wuppertal, to appear in J. Lützen (ed.), *The interaction between Mathematics, Physics and Philosophy from 1850 to 1940*, Dordrecht: Kluwer. [http://arxiv.org/math.HO/0409576].

Scholz, E, 2004b, 'The introduction of groups into quantum theory.' Preprint Wuppertal, to appear in *Historia Mathematica* [http://arxiv.org/math.HO/0409571].

Scholz, E, 2005, 'Philosophy as a Cultural Resource and Medium of Reflection for Hermann Weyl.' *Révue de Synthèse* 126. [http://arxiv.org/math.HO/0409596]

Scholz H. and Bachmann, F., 1935, 'Der Wissenschaftliche Nachlass von Gottlob Frege' *Actes du Congrès International de Philosophie Scientifique VII: Histoire de la Logique ou de la Philosophie Scientifique* Paris: Sorbonne p. 24–30.

Schwartz H., 1870, 'Über einen Grenzübergang durch altirnierendes Verfahren' *Vierteljahrsschrift der Naturforschenden Gesellschaft in Zürich* 15 1870 pp. 272–286. (=*Abhandlungen* 2 pp. 133–143).

Schwarzschild, K., 'Über das zulässige Krümmungsmaß des Raumes.' *Vierteljahresschrift der Astrononomischen Gesellschaft* 35 (1900), pp. 337–347.

Seifert, H. & W. Threlfall, 1934, *Lehrbuch der Topologie*, Teubner, Leipzig. Translated as *A Textbook of Topology*. New York: Academic Press, 1980.

Selick, P., 1997, *Introduction to Homotopy Theory*. Providence, American Mathematical Society.

Shapiro, S., 1997, *Philosophy of Mathematics: Structure and ontology*, Oxford University Press.

Shapiro, S., 2000, *Thinking about Mathematics*, Oxford University Press.

Shapiro, S., 2005, *Oxford Handbook of Philosophy of Mathematics and Logic*, Oxford University Press.

Sher, G., 1991, *The Bounds of Logic*, MIT Press, Cambridge.

Sher, G., 1996, Did Tarski commit Tarski's fallacy?, *Journal of Symbolic Logic*, 61, 653–686.

Sichau, C., *Die Viskositätsexperimente von J.C. Maxwell und O.E. Meyer: Eine wissenschaftshistorische Studie über die Entstehung, Messung und Verwendung einer physikalischen Größe*. Berlin: Logos, 2002.

Sieg, W., 1984, Foundations for Analysis and Proof Theory; *Synthese* 60 (2), 159–200.

Sieg, W. 1985, Reductions of Theories for Analysis; in: *Foundations of Logic and Linguistics*, G. Dorn and P. Weingartner (eds.) New York and London, 199–230.

Sieg, W. 1988, Hilbert's Program Sixty Years Later; *Journal of Symbolic Logic*, 53(2), 338–348.

Sieg, W., 1999, 'Hilbert's Programs: 1917–1922', *Bulletin of Symbolic Logic* 5(1), 1–44.

Sieg, W., 2002, Beyond Hilbert's Reach?; in: *Reading Natural Philosophy*, D.B. Malament (ed.), Open Court, Chicago, 363–405.

Sieg, W. and D. Schlimm 2005, Dedekind's Analysis of Number: systems and axioms; *Synthese*, 147, 121–170.

Sieg, W., 2005, (with Mark Ravaglia) David Hilbert and Paul Bernays, *Grundlagen der Mathematik*; in: *Landmark Writings in Western Mathematics: Case Studies, 1640–1940*, I. Grattan-Guinness (ed.), 19 pp.

Simons, P., 1987, 'Frege's Theory of Real Numbers' reprinted in Demopoulos [1995] pp. 358–85.

Springer G., 1957, *Introduction to Riemann Surfaces* New York: Addison-Wesley.

Simpson, S.G., 1988, Partial Realizations of Hilbert's Program; *Journal of Symbolic Logic*, 53(2), 349–363.

References

Skolem, T., 1923, Einige Bemerkungen zur axiomatischen Begründung der Mengenlehre, in his *Selected Works in Logic* (Oslo, Universitetsforlaget, 1970). English translation in van Heijenoort 1967, 290–301.

Sluga, H.D., 1980, *Gottlob Frege*, London ; Boston: Routledge and Kegan Paul.

Smorynski, C., 1977, The incompleteness theorems; in: *Handbook of Mathematical Logic*, J. Barwise (ed.), Amsterdam, 821–866.

Soames, S., 1999, *Understanding Truth*, Oxford University Press, New York.

Spanier, E. H., 1966, *Algebraic Topology*. New York, Springer-Verlag.

Speiser, A., 1927, Naturphilosophische Untersuchungen von Euler und Riemann, *Journal für reine und angew. Math.* 157, 105–14.

Stachel, J., 1989, The Rigidly Rotating Disc as the 'Missing Link' in the History of General Relativity, in *Einstein and the History of General Relativity*, D. Howard and J. Stachel (eds) Birkhäuser, Boston and Basel, pp. 48–62.

Stahl H., 1896, *Theorie der Abel'schen functionen* Leipzig: Teubner.

Stahl H., 1899, *Elliptischen functionen. Vorlesungen von Bernhard Riemann* Leipzig: Teubner.

von Staudt, C.G.C., 1847, *Geometrie der Lage*, Nürnberg.

Stegmaier, W, 2002, 'Ein Mathematiker in der Landschaft Zarathustras. Felix Hausdorf als Philosoph.' *Nietzsche Studien* 31:195–240. [Similarly in (Hausdorff 2001ff., VII, 1–83)].

Stein, H., 1988, Logos, Logic, Logistiké: Some philosophical remarks on the 19th century transformation of mathematics; in: (Aspray and Kitcher), 238–259.

Tait, W.W., 1981, Finitism; *Journal of Philosophy* 78, 524–546.

Tait, W. W., 1997, 'Frege versus Cantor and Dedekind: On the Concept of Number', in W. W. Tait, ed., *Early Analytic Philosophy: Frege, Russell, Wittgenstein*, Chicago: Open Court, pp. 213–48; repr. in Beaney and Reck 2005, Vol. III, pp. 115–56.

Tappenden J., 1995a, 'Geometry and Generality in Frege's Philosophy of Arithmetic' *Synthèse* 102 pp. 319–361.

Tappenden, J, 1995b, 'Extending Knowledge and 'Fruitful Concepts': Fregean Themes in the Foundations of Mathematics', *Nous* 29, pp. 427–67; repr. in Beaney and Reck 2005, Vol. III, pp. 67–114.

Tappenden J., 2005, 'The Mathematical Background of the Caesar Problem' Forthcoming in a *Dialectica* special issue on the Julius Caesar problem.

Tappenden J., forthcoming *Philosophy and Mathematical Practice: Frege in his Mathematical Context* to appear with Oxford University Press.

Tappenden J., forthcoming a, Explanation and Mathematical Proof: why do Elliptic Functions have Two Periods? in preparation.

Tarski, A., 1924a, Sur quelques théorèmes qui èquivalent à l'axiome du choix, *Fundamenta Mathematicae*, vol. 5, pp. 147–154, reprinted in Tarski 1986, vol I, pp. 41–48.

Tarski, A., 1924b, Sur les ensembles finis, *Fundamenta Mathematicae*, vol. 6, pp. 45–95, reprinted in Tarski 1986, vol I, pp. 67–117.

Tarski, A., 1929a, Les fondements de la géométrie des corps, *Annales de la Societé Polonaise de Mathématiques*, pp. 29–33, reprinted in Tarski 1986, vol I, pp. 227–231. English translation with substantial modifications in Tarski 1983, pp. 24–29.

Tarski, A., 1929b, Geschichtliche Entwicklung und gegenwärtiger Zustand der Gleichmässigkeitstheorie und der Kardinalzahlarithmetik, *Annales de la Societé Polonaise de Mathématiques*, pp. 48–54, reprinted in Tarski 1986, vol I, pp. 235–241.

Tarski, A., 1930a, Über einige fundamentale Begriffe der Metamathematik, *Sprawozdania z posiedzen Towarzystwa Naukowego Warszwskiego*, wydzial III, 23, 22–29; English translation with revisions in Tarski, 1983, 30–37. Reprinted in Tarski 1986, vol. I.

Tarski, A., 1930b, Fundamentale Begriffe der Methodologie der deduktiven Wissenschaften, I, *Monatshefte für Mathematik und Physik*, 37, 361–404; reprinted in Tarski 1986, vol. I. English translation with revisions in Tarski, 1983, 60–109.

Tarski, A., 1931, Sur les ensembles définissables de nombres réels, I., *Fundamenta Mathematicae*, 17, 210–239; reprinted in Tarski 1986, vol. I. English translation with revisions in Tarski, 1983, 110–142.

Tarski, A., 1932, Der Wahrheitsbegriff in der Sprachen der deduktiven Disziplinen, *Anzeigen der Akademie der Wissenschaften in Wien*, 69, 23–25. Reprinted in Tarski 1986, vol. I.

Tarski, A., 1933a, Einige Betrachtungen über die Begriffe der ω-Widerspruchsfreiheit und der ω-Vollständigkeit, *Monatshefte für Mathematik und Physik*, 40, 97–112; reprinted in Tarski 1986, vol. I. English translation with revisions in Tarski, 1983, 279–295.

Tarski. A., 1933b, *Pojecie prawdy w jezykach nauck dedukcyjnych* (The concept of truth in the languages of deductives sciences), Prace Towarzystwa Naukowego Warszawskiego, wydzial III, no. 34.

Tarski, A., 1935a, Der Wahreitsbegriff in den formalisierten Sprachen, *Studia Philosophica* (Lemberg), 1, 261–405. Reprinted in Tarski 1986, vol. II. English translation in Tarski 1983, 152–278.

Tarski, A., 1935b, Einige methodologische Untersuchungen über die Definierbarkeit der Begriffe, *Erkenntnis*, 5, pp. 80–100. English translation with revisions in Tarski 1956, 296–319. Reprinted in Tarski 1986, vol. I.

Tarski, A., 1935c, Grundzüge des Systemenkalküls. Erster Teil, *Fundamenta Mathematicae*, 25, 503–526; reprinted in Tarski 1986, vol. II. English translation with revisions in Tarski, 1983, 342–383.

Tarski, A., 1935d, Zur Grundlegung der Boole'schen Algebra, I, *Fundamenta Mathematicae*, vol. 24, pp. 177–198; reprinted in Tarski 1986, vol. II, pp. 3–24. English translation with revisions in Tarski, 1983, pp. 320–341.

Tarski, A., 1936a, Über den Begriff der logischen Folgerung, *Actes du Congrès International de Philosophie Scientifique* 7, Actualités Scientifiques et Industrielles, Herman, Paris, pp. 1–11; reprinted in Tarski 1986, vol. II. English translation with revisions in Tarski, 1956, 409–420.

Tarski, A., 1936b, O pojeciu wynikania logicznego, *Przeglad Filozoficzny*, 39, pp. 58–68. English translation in Tarski 2002.

Tarski, A., 1936c, Grundzüge des Systemenkalküls. Zweiter Teil, *Fundamenta Mathematicae*, 26, 283–301; reprinted in Tarski 1986, vol. II, pp. 225–243. English translation with revisions in Tarski, 1983, pp. 364–383.

Tarski, A., 1937a, *Einführung in die mathematische Logik und die Methodologie der Mathematik*, Springer, Vienna.

Tarski, A., 1937b, Sur la méthode deductive, *Travaux du IXe Congrès International de Philosophie*, Tome 6, Actualités Scientifiques et Industrielles, Herman, Paris, 95–103; reprinted in Tarski 1986, vol. II.

Tarski, A., 1938a, Einige Bemerkungen zur Axiomatik der Boole'schen Algebra, *Sprawozdania z posiedzen Towarzystwa Naukowego Warszwskiego*, wydzial III, 31, 33–35; reprinted in Tarski 1986, vol. II, pp. 353–355.

Tarski, A., 1938b, Ein Beitrag zur Axiomatik der Abelschen Gruppen, *Fundamenta Mathematicae*, 30, pp. 253–256; reprinted in Tarski 1986, vol. II, pp. 447–450.

Tarski, A., 1939a, New Investigations on the completeness of deductive theories, *Journal of Symbolic Logic*, 4, p. 176.

Tarski, A., 1939b, On undecidable statements in enlarged systems of logic and the concept of truth, *The Journal of Symbolic Logic*, 4, pp. 105–112.
Tarski, A., 1940, On the Completeness and Categoricity of Deductive Systems, unpublished typescript, Alfred Tarski Papers, Carton 15, Bancroft Library, U.C. Berkeley.
Tarski, A., 1948, *A Decision Method for Elementary Algebra and Geometry*, (second, rev. edn; first edition 1948), University of California Press, Berkeley.
Tarski, A., 1983, *Logic, Semantics, Metamathematics*, Oxford University Press, Oxford. Second edition. (First edition, 1956).
Tarski, A., 1986, *Collected Papers*, S. Givant and R. McKenzie eds., vols. I–IV, Birkhäuser, Basel.
Tarski, A., 1986/1966, What are logical notions?, *History and Philosophy of Logic*, 7, pp. 143–154.
Tarski, A., 1995, Some current problems in metamathematics, *History and Philosophy of Logic* 16, 159–168.
Tarski, A., 2002, On the concepts of following logically, *History and Philosophy of Logic*, 23, 155–196. [A translation of 1936b]
Tarski, A., Lindenbaum, A., 1926, Communication sur les recherches de la théorie des ensembles, *Sprawozdania z posiedzen Towarzystwa Naukowego Warszwskiego*, wydzial III, 19, 299–330. Reprinted in Tarski 1986, vol. 2, pp. 173–204.
Tarski, A., Lindenbaum, A., 1936, Über die Beschränktheit der Ausdrücksmittel deduktiver Theorien, *Ergebnisse eines mathematischen Kolloquiums*, 7, pp. 15–22. English Translation in Tarski 1983, pp. 384–392.
Tarski, A., Lukasiewicz, J., 1930, Untersuchungen über den Aussagenkalkül, *Sprawozdania z posiedzen Towarzystwa Naukowego Warszwskiego*, wydzial III, 23, 30–50; English translation with revisions in Tarski, 1983, 38–59.
Tarski, A., Mostowski, R. Robinson, 1953,*Undecidable Theories*, North-Holland, Amsterdam.
Temple, G., 1981, *100 Years of Mathematics* Berlin: Springer.
Thomae J., 1876, *Sammlung von Formeln, welche bei Anwendung der Elliptischen und Rosenhainschen Funktionen Gebraucht Werden* Halle: Louis Nebert.
Toda, H., 1962, *Composition methods in homotopy groups of spheres*. Princeton, N.J., Princeton University Press.
Toda, H., 1982, 'Fifty Years of Homotopy Theory.' *Iwanami-Sugaku* 34: 520–582.
Toepell, M.-M., 1986, *Über die Enstehung von David Hilberts 'Grundlagen der Geometrie'*, Vandenhoeck and Ruprecht, Göttingen.
Torretti, R., 1978, *Philosophy of Geometry from Riemann to Poincaré*, Dordrecht, Reidel. (2nd edition 1984).
Torretti, R., 1990, *Creative Understanding: Philosophical Reflections on Physics*, University of Chicago Press.
Tymoczko, T., 1998, *New Directions in the Philosophy of Mathematics*, Basel, Boston, Birkhäuser. Revised and expanded edition in Princeton University Press, 1998.
Ueberweg, F., 1882, *System der Logik und Geschichte der logischen Lehren*, 5th edn., Bonn, A. Marcus.
Vanden Eynde, R, 1992, 'Historical evolution of the concept of homotopic paths', *Archive for History of Exact Sciences* 29, 127–88.
Vanden Eynde, R., 1999, 'Development of the concept of homotopy', *History of Topology*, I.M. James, ed., Amsterdam: North Holland, 65–102.
Vaught, R., 1974, Model theory before 1945, in L. Henkin et al. (eds.), *Proceedings of the Tarski Symposium*, Providence, AMS, pp. 152–172.

Veblen, O., 1904, A system of axioms for geometry, *Transactions of the American Mathematical Society*, 5, pp. 343–384.

Veblen, O., 1905, *Princeton lectures 'On the foundations of geometry*, Chicago University Press.

Veblen, Oswald, 1922, *Analysis Situs*, American Mathematical Society, New York.

Veblen, O. and Young, J.W., 1910, *Projective Geometry*, Ginn and Company, Boston.

Veraart, A., 1976 'Geschichte des Wissenschaftlichen Nachlassen Gottlob Freges und seiner Edition: Mit einem Katalog des ursprünglichen Bestands der nachgelassenen Schriften Freges' in Schirn M. ed. *Studien zu Frege/Studies on Frege vol I* Stuttgart-Bad Cannstatt: Fromann-Holzboog 49–106.

Vercelloni, L., 1988, *Filosofia delle Strutture*, La Nuova Italia, Firenze.

Vick, J., 1994, *Homology Theory: An Introduction to Algebraic Topology*, Springer-Verlag.

Vietoris, L., 1926a, 'Ueber den höheren Zusammenhang von kompakten Räumen', *Proceedings of the Section on Sciences, KNAW* 29, 1008–13.

Vietoris, L., 1926b, 'Ueber stetige Abbildungen einer Kugelfläche', *Proceedings of the Section on Sciences, KNAW* 29, 443–53.

Vietoris, L., 1927, 'Ueber den höheren Zusammenhang von kompakten Räumen', *Mathematische Annalen* 97, 454–72.

Volkov, S., *Antisemitismus als symbolischer Code*. 2nd edn, München: Beck, 2000.

Voss, A., 1913, *Über das Wesen der Mathematik* Teubner, Leipzig and Berlin.

Vossler, K., 1925, *Sprache und Wissenschaft*. Heidelberg: Winter.

van der Waerden, Bartel L., *Moderne Algebra* 1930, 2nd edn 1937, Springer Verlag, Berlin.

Wagner S., 1992, 'Logicism' in M. Detlefsen (ed.) *Proofs and Knowledge in Mathematics* Routledge, London pp. 65–110.

Warwick, A., 2003, *Masters of theory: Cambridge and the Rise of Mathematical Physics*, University of Chicago Press.

Webb, J., 1995, Tracking contradictions in geometry: the idea of model from Kant to Hilbert, in J. Hintikka, ed., *Essays on the development of the foundations of mathematics*, Kluwer, pp. 1–20

Weber, H., 1899, *Lehrbuch der Algebra*, 2nd edn, F. Vieweg und Sohn, Braunschweig.

Weierstrass K., 1870, 'Über das sogennante Dirichlet'sche Prinzip' read in 1870 (= Werke 2 [1895] pp. 49–54).

Weierstrass K., 1875, 'Letter to H. Schwartz' *Werke* 2 p. 235.

Weierstrass, K., 1886/1988 *Ausgewählte Kapitel aus der Funktionenlehre. Vorlesung, gehalten in Berlin. Mit der akademischen Antrittsrede, Berlin 1857 und drei weiterenn Originalarbeiten von K. Weierstrass aus den Jahren 1870 bis 1880/6* edited with commentary by R. Siegmund-Schultze Teubner – *Archiv zur Mathematik*, Bd. 9 Liepzig: Teubner.

Weiner, J., 1984 'The Philosopher Behind the Last Logicist' in C. Wright (ed.) *Frege: Tradition and Influence* Oxford: Blackwell.

Weiner, J. 1990, *Frege in Perspective* Ithaca NY: Cornell University Press.

Weiner, J., 2000, *Frege* Oxford: Oxford University Press.

Weiner, J., 2001, 'Theory and Elucidation: The End of the Age of Innocence', in J. Floyd and S. Shieh, eds., *Future Pasts: The Analytic Tradition in Twentieth-Century Philosophy*, Oxford University Press, pp. 43–65.

Weiner, J., 2005, 'On Fregean Elucidation', in Beaney and Reck 2005, Vol. IV, pp. 197–214.

Wellstein, J., 1905, *Enzyklopädie der Elementar-Mathematik. Ein Handbuch fur Lehrer und Studierende* Weber, H. and Wellstein, J (eds) vol 2., *Elemente der Geometrie*.

Weyl, H., 1910, Über die Definitionen der mathematischen Grundbegriffe, *Mathematisch-naturwissenschaftliche Blätter*, 7, 93–95, 109–113. Reprinted in Weyl (1968) I, 298–304.

Weyl, H., 1913, *Die Idee der riemannschen Fläche*, B.G. Teubner, Leipzig, Berlin.

Weyl, H., 1918, *Das Kontinuum: Kritische Untersuchungen über die Grundlagen der Analysis*, Leipzig, Veit. Reprinted in New York, Chelsea, 1954. English translation, *The Continuum: A Critical Examination of the Foundation of Analysis* by Stephen Pollard & Thomas Bole, Dover Publications, New York, 1994.

Weyl, H., 1919, *Raum–Zeit–Materie*. Berlin, Springer, 5th edn. 1923.

Weyl, H., 1919a, Preface and Annotations to R. Riemann, *Über die Hypothesen, welche der Geometrie zu Grunde liegen* (new edn., Berlin, Springer, 1919).

Weyl, H., 1921, 'Über die neue Grundlagenkrise der Mathematik.' *Mathematische Zeitschrift* 10:39–79. In (Weyl 1968, II, 143–180) [41]. English in (?, 86–121).[3]

Weyl, H., 1923, 'Análisis situs combinatorio', *Revista Matematica Hispano-Americana* 5, 209–18, 241–48, 278–79. Cited this from (?) vol. 2, 390–415.

Weyl, H., 1924, 'Was ist Materie?' *Die Naturwissenschaften* 12:561–568, 585–593, 604–611. Reprint Berlin: Springer 1924. Darmstadt: Wissenschaftliche Buchgesellschaft 1977. In (Weyl 1968, II, 486–510) [66].

Weyl, H., 1925–27, 'The Current Epistemological Situation in the Mathematics', *Symposion* 1, 1–32. Reprinted in Paolo Mancosu (ed.) (1998), *From Brouwer to Hilbert*, Oxford, Oxford University Press, 123–142.

Weyl, H., 1927a, *Philosophie der Mathematik und Naturwissenschaft*, Handbuch der Philosophie, Abt. 2A. . München: Oldenbourg. Weitere Auflagen ^{2}1949, ^{3}1966. English with comments and appendices (Weyl 1949b).

Weyl, H., 1927b, 'Quantenmechanik und Gruppentheorie.' *Zeitschrift für Physik* 46:1–46. In (Weyl 1968, III, 90–135) [93] [75].

Weyl, H., 1928, Diskussionsbemerkungen zu dem zweiten Hilbertschen Vortrag über die Grundlagen der Mathematik, *Abhand. Math. Seminar Hamburg. Universität* 6, 86–88. English version in van Heijenoort (1967), 480–84.

Weyl, H., 1931, 'Geometrie und Physik.' *Die Naturwissenschaften* 19:49–58. In (Weyl 1968, III, 336–345) [93].

Weyl, H., 1939, 'Invariants', *Duke Mathematical Journal* 5, 489–502. Also in (?) vol. 3, 670–83.

Weyl, H., 1940 *Algebraic theory of numbers*. Princeton University Press, Princeton, NJ, Reprint of the 1940 original, Princeton Paperbacks.

Weyl, H., 1948, 'Wissenschaft als symbolische Konstruktion des Menschen.' *Eranos-Jahrbuch* pp. 375–431. In (Weyl 1968, IV, 289–345) [142].

Weyl, H., Ms 1949a, 'Entwicklungslinien der Mathematik seit 1900. Problems and methods of 20th century mathematics. Gastvorlesung nach 1950, englisch.' *Nachlaß Weyl*, Zürich: ETH Bibliothek Hs 91a:72.

Weyl, H., 1949b, *Philosophy of Mathematics and Natural Science*. 2nd edn. 1950. Princeton: University Press.

Weyl, H., 1951, A Half Century of Mathematics, *American Math. Monthly* 58, 523–553, and *Gesammelte Abhandlungen*, vol. 4, 464–49, Springer-Verlag, New York 1968.

---

[3] Numbers in square brackets [YY] refer to publication list of (Weyl 1968).

Weyl, H., 1953, 'Über den Symbolismus der Mathematik und mathematischen Physik.' *Studium generale* 6:219–228. In (Weyl 1968, IV, 527–536) [156].
Weyl, H., 1954, 'Erkenntnis und Besinnung (Ein Lebensrückblick).' *Studia Philosophica*. In (Weyl 1968, IV, 631–649) [166].
Weyl, H., 1968, *Gesammelte Abhandlungen*, 4 vols. ed. K. Chandrasekharan. Berlin etc.: Springer.
White, S. D., 2004. Axiomatics, methodology, and Dedekind's theory of ideals. Master's thesis, Carnegie Mellon University.
Whitehead, A. N., 1906, *The Axioms of projective Geometry*, Cambridge University Press Tract nr. 4, Cambridge.
Whitehead, A. N., 1907, Introduction logique à la géométrie, *Revue de Metaphysique et de Morale*, pp. 34–39.
Whitehead, A.N. and Russell, B., 1910 [PM], *Principia Mathematica*, 3 vots., Cambridge: Cambridge University Press, 1910–13.
Whitehead A.N. and Russell, B., 1925, *Principia Mathematica* (2nd edn), 3 vols., Cambridge: Cambridge University Press, 1925–27.
Whitehead, G. W., 1978, *Elements of Homotopy Theory*. New York, Springer-Verlag.
Whitehead, G. W., 1983, 'Fifty Years of Homotopy Theory.' *Bulletin of the American Mathematical Society (New Series)* 8(1): 1–29.
Wiener, H., 1891, Über Grundlagen und Aufbau der Geometrie; Jahresberichte der DMV 1, 45–48.
Wigner, Eugene, 1992ff, *Collected Works*. Part A: The Scientific Papers, Part B: Historical, Philosophical, and Socio-Political Papers. Berlin etc.: Springer.
Wilson, Mark, 2005, 'Ghost World: A Context for Frege's Context Principle', in Beaney and Reck 2005, Vol. III, pp. 157–75.
Wittgenstein, Ludwig, 1922, *Tractatus logico-philosophicus*. London etc: Routledge and Kegan. English and revised German edition; first published in Annalen der Naturphilosophie (1921). In *Schriften I*, Frankfurt/Main:Suhrkamp 1960.
Wittgenstein, L., 1970, *Über Gewissheit*, G. E. M. Anscombe und G. H. von Wright. Frankfurt am Main: Suhrkamp Verlag.
Wolenski, J., 1989, *Logic and Philosophy in the Lvov-Warsaw School*, Kluwer, Dordrecht.
Wussing, H., *The genesis of the abstract group concept*. MIT Press, Cambridge, MA, 1984. Translation of *Die Genesis des abstrakten Gruppenbegriffes*, VEB Deutscher Verlag der Wissenschaften, Berlin, 1969.
Young, J.W., 1911, *Lectures on the fundamental concepts of algebra and geometry*, The MacMillan Company, NY.
Youskevich A., 1976, 'The Concept of a Function up to the Middle of the Nineteenth Century' *Archive for the History of Exact Sciences* 7 pp. 37–85.
Zach, R., 1999, 'Completeness before Post: Bernays, Hilbert, and the Development of Propositional Logic', *Bulletin of Symbolic Logic* 5 (3), 331–365.
Zassenhaus, H.J., 1975, On the Minkowski-Hilbert dialogue on mathematization; Canadian Math. Bulletin 18, 443–461.
Zermelo, E., 1908, 'Untersuchungen über die Grundlagen der Mengenlehre', *Math. Ann.* 65, 261–281.
Zermelo, E. 1929, Über den Begriff der Definitheit in der Axiomatik, *Fundamenta Mathematicae*, 14, pp. 339–344.
Zermelo, E. 1930, Über Grenzzahlen und Mengenbereiche: neue Untersuchungen über die Grundlagen der Mengenlehre, *Fundamenta Mathematicae*, 16, pp. 29–37.

Zermelo, E., 1931 Über Stufen der Quantifikation und die Logik des Unendichen; Jahresberichte DMV 41, 85–88.

Zermelo, E., 1935 Grundlagen einer allgemeinen Theorie der mathematischen Satzsysteme; Fund. Math. 25, 136–146.

Zisman, M., 1999, Fibre Bundles, Fibre Maps. *History of Topology*. I. M. James. Amsterdam, Elsevier: 605–629.

Name Index

Abbe, E. 110, 123–124
Abel, N.H. 124
Ahlfors, L.V. 79n32, 110, 117, 121n49
Ajdukiewicz, K. 215
Alexander, J.W. 203
Alexandroff, P. 39, 187, 188, 192, 195, 202–5, 207
Archimedes 327, 337
Aristotle 14, 308
Aspray and Kitcher ix, 4, 12, 28, 37, 47–48
Avenarius, R. 268
Avigad, J 189

Bachelard, G. 311, 321, 334
Bachmann, P. 124
Bacon, F. 2
Baer, R. 196, 269
Baire, R. 121
Baumann, J.J. 55
Bäumler, A. 291
Bays, T. 209, 218, 234
Becker, O. 300–1
Bell, E.T. 22, 301
Beltrami, E. 377
Benacerraf, P. 9, 30, 40
Bergson, H. 383
Bernays, P. 7, 8, 14, 65n21, 95, 138, 325, 326n41, 372n64
Bianchi, L. 392n37
Biermann, O. 126–127
Birkhoff and Bennett 375
Bishop 66, 367
Bôcher, M. 213
Bohr, N. 299, 302
Boltzmann, L. 210n2
Bolyai, J. 90, 392
Bolzan, B. 316, 326
Bonola, R. 375n5
Borel, E. 121, 348, 380, 396
Borevich and Shafarevich 175
Born, M. 292, 298–9
Borsuk, K. 252
Bos, H.J.M. 27

Bottazzini, U. 69n8, 82, 89, 116n37, 121n49
Bourbaki, N. 4, 30, 39, 136, 317, 366
Bradley, F.H. 383
Brecht, B. 368
Brentano, F. 319
Brewster, D. 83
Brill, A. 109–110, 119
Brouwer, L.E.J. 187–8, 202–5, 207, 239, 246–9, 291, 295, 305, 313, 318, 325, 328, 331, 332, 349, 353–4, 358, 367
Bruns, H. 267
Brunschvicg, L. viii, 42, 311–2, 314, 322–3, 335, 383
Buchholz et al 350 n29, 354 n36
Buchwald, J.Z. 36
Burkhardt, H. 110

Canguilhelm, G. 313
Cantor, G. 3, 7, 15, 41, 70, 91, 95, 126, 174, 189, 265, 271, 278, 295, 333, 346–7, 384 n19, 349 n26, 351, 373, 379
Carnap, R. 210n2, 230n21, 236
Cartan, É 375
Casorati, F. 120
Cassirer, E. 305n31, 387–9, 391
Cauchy, A.L. 22, 27, 36, 92, 115–116, 246, 316, 373, 393
Cavaillès, J. viii, 1, 17, 42, 96, 312, 314, 317–33, 337
Cayley, A. 372, 376
Cech, E. 250
Cézanne, P. 394
Chevalley, C. 118n21
Chrystal, G. 91n53, 119n43
Clebsch, R.F.A. 114, 123
Clebsch-Gordan 124
Clifford, W.K. 87n48, 282
Cohen, I.B., 83n42
Comte, A. 335
Corcoran, J. 224
Corfield, D. 17n24
Corry, L. 160–1, 210n2, 391

Name Index

Couturat, L. 321–2, 383–4, 396
Curtis, C.W. 255

Dalen, D. von 202
Dedekind, R. 1, 3, 15n22, 19, 37, 39, 48, 54–59, 66, 80, 83, 90–92, 95, 99, 102, 107, 108n21, 111–114, 122, 131, all of *Avigad*, 189, 193, 194, 199–201, 207, 282, 316, 333, 340, 340n3, 342 ff, 352, 352n32, 355, 356n39, 365, 378, 388
Dedekind-Weber 101, 108n21
Dehn, M. 144, 247
Deleuze, G. 313
Demokritos 293
Desanti, J.T. viii, 42, 313–5, 317, 319, 321–2, 322n33, 323–4, 326–7, 329–30, 334–7
Desargues, G. 327
Descartes, R. 2, 3, 26, 133–4, 293, 390
Destouches, J.L. 334
Detlefsen, M. 136n2
Dieudonné, J. 136–8, 188, 252
Dirichlet, P.G.L. 67, 92, 117, 166, 342, 348 n63, 379
Dugac, P. 160
Dummett, M. 102
Dwyer, W.G. 256

Eckehardt, M. 306
Eckmann, B. 252, 254
Eddington, A. 301
Edwards, H.M. 112, 161, 162, 163–4, 166n11, 168, 175n16, 178, 181
Ehresmann, C. 252, 254
Eilenberg, S. 256n22
Einstein, A. 67, 81n35, 84–85, 90, 137, 149, 151, 298, 302, 377, 392
Eisenstein, G. 109n25
Enneper, A. 104n15, 124
Enriques, F. 383
Epple, M. 20n26, 283n44
Etchemendy, J. 40, 209, 211, 218, 236,
Euclid 14, 95
Euler, L. 72, 82, 84, 91, 163
Ewald, W. 4, 65n22, 160

Fano, G. 378
Faraday, M. 84
Fechner, G.T. 268
Feferman, S. 8, 13, 28, 30, 351n30, 354n36
Feldbau 252, 254
Ferreirós 13, 16n23, 37, 70n12, 88n49, 160–1, 379

Fichte 291, 306
Fisher 28
Fitzgerald 288
Floyd, J., 8n10
Foucault 313, 314
Fraenkel, A. 8
Fréchet, M. 250n14, 381–2
Frege, G. Chapter 2; vii, 3–6, 26, 37–8, 71, 97–99, 103, 105, 107–108, 114–116, 118, 120n45, 121, 123–124, 126–129, 133–4, 161, 316, 326, 328, 358, 378, 383, 396, 393
Freudenthal, H. 115
Friedman, M. 90, 387, 389
Fries, J. 72n17
Frobenius and Stickelberger 192, 210

Gassendi, P. 293
Gauss, C. F. 1, 14, 72–73, 80, 87–93, 95, 105, 112, 139, 167, 168, 316, 321n29, 343
Gentzen, G 350, 362–63
Giaquinto, M. 6n6, 235
Gieseking, H. 191
Gispert, H. 23
Gödel, K. 6, 7, 13, 95, 142, 210n2, 233–4, 350, 355, 361n50, 364, 364n56
Goethe, J.W. 268, 305n31
Gomez-Torrente, M. 209, 211–2, 216, 218, 219–220, 223, 225,
Gonseth, E. 1, 95, 302, 334
Granger, G.G. viii, 42, 313–5, 319–24, 326–9, 334–7
Grassmann, H.G. 26, 316
Gray, J.J. 13, 160–1
Grothendieck, A. 195

Hale, R., 47
Hallett, M. 62n17, 396
Halley, E. 244
Halmos, P. 394
Hamel, G. 144
Hattendorff, K. 90n53
Haubrich, R. 160
Hausdorff, F. viii, 17, 20, 26, 41, 148, 263–7, 269n17, 271–2, 274, 278–84, 286, 288–9, 295–6, 382
Hawkins, T. 381, 391n35
Heegard, P. 247
Hegel, G.W.F. 334
Heidegger, M. 304
Heijenoort, J. van 160
Heisenberg, W. 292, 298–9, 302
Hellman, G. 11n15

Name Index

Helmholtz, H. von 89, 277n35, 305, 377, 384–6
Hensel, K. 179
Herbart, J.F. 55n9, 72, 73n21, 77–77, 82
Herreman, A. 190
Hertz, H. 36, 74, 139, 210n2
Hessenberg, G. 33
Heyting, A. 350
Hilbert, D. vii, 1, 3n3, 4–7, 13–4, 20, 26, 33, 37–8, 41, 58–62, 68–71, 91, 115, 119, all of *Corry*, 173, 187, 213, 215, 266, 279, 280, 282, 283, 291–2, 294, 300, 303, 313, 316, 317n13, 322–5, 328–9, 331, 339, 346, 347–8, 350, 351–2, 353, 356–6, 360, 376–8, 385, 391
Hirsch, G. 239, 240
Hopf, H. 40, 202, 205–6, 239, 246, 248–9, 252
Humboldt, H. von 305
Huntington, E.V. 213–4
Hurewicz, W. 202, 239, 246, 249–55
Hurwitz, A. 70, 183–4
Husserl, E. 43, 126–127, 134, 291, 300, 318–9, 323–4, 326n44, 330, 332, 336, 385
Huygens, C. 293, 306

Illich, I. 307–8

Jacobi, C.G.J. 124
Jaspers, K. 304
Jespersen, O. 395
Jevons, W.S. 384
Jordan, C. 246, 373, 379–80
Jordan, P. 292
Joyce, J. 374

Kant, I. vii, 14, 37, 48–52, 64, 66, 73, 75, 78, 95–96, 154, 269, 311, 322–4, 330, 331, 336
Kaufmann, E. 210n2
Keferstein, H. 345, 365
Keller, G. 268
Kellogg, O. 117n39
Kepler, J. 244, 293
Keyser, C.J. 214
Killing, W. 282
Kitcher, P. 1, 5, 11n16, 14n21, 16, 18, 28, 30, 47–48, 95, 102–104
Klein, C.F. 19, 23, 32-3, 70, 90, 94, 106, 108n21, 109, 140, 246, 372, 376, 388
Kline, M. 21
Kohlrausch, F.W.G. 81, 82n38

Korselt, A. 214
Kossak, E. 126–127
Kreiser, L. 56
Krömer, R 191
Kronecker, L. 1, 3, 23, 106n18, 109, 114, 162, 166, 168–9, 171–2, 178–9, 182–4, 325, 328, 331, 342–3, 344 n12, 348, 349, 353, 358, 372, 396
Krull, W. 194–5
Kuhn, T.S. 25
Kummer, E.E. 23, 109, 162, 164–5, 168–9, 170, 172, 176, 184, 326
Kurosh, A.G. 105, 196,

Lagrange, J.L. 246
Lakatos, I. 1, 5, 13, 15–8, 26–8, 95, 184
Lang, S. 198
Langford, C.H. 215, 221–2,
Laugwitz, D. 67n3, 73n21, 74n23, 82
Lautman, A. 317, 334
Lawvere, F.W. 8n9
Lebesgue, H. 121, 380
Lefschetz, S. 205, 209
Leibniz, G.W. 26, 292, 327, 337, 383, 388, 395–6
Lesniewski, S. 217n10
Levi-Civita, T. 297
Lewis-Langford 215
Lie, S. 108n23, 109, 113, 140, 375
Liebmann, O 268–9
Lindemann, F. 109,
Lindenbaum, A. 222, 234
Lindenbaum-Tarski 225
Lipschitz, R.O.S 112, 344
Lobachevskii, N. I. 90, 392
Lorentz, H.A. 288
Lorenzen, P. 347n23, 350n29
Lorenzo, De, P. 1
Lotze, R.H. 72, 383
Lützen, J. 24

Mac Lane, S. 188, 190, 191, 195, 209,
Mach, E. 291
Maddy, P. 5, 9, 11n16, 16, 20
Mancosu, P. 13, 19n25, 300n21
Marx, K. 297, 334
Maxwell, J.C. 84, 154, 267,
Medicus, F. 291
Mehrtens, H. 265n7, 268n16, 395
Menger, K. 202
Michelson, A.A. 154
Mie, G. 149
Mill, J.S. 14n21
Minkowski., H. 81, 282
Mises, von, R. 82n39

Name Index

Mittag-Leffle, G. 113–114
Mongré, P 41, 264–5, 267, 269–71, 273, 275–80, 284–5, 287
Moore, G.E. 60
Moore, G.H. 13n19, 395
Mostowski, A. 236
Moulton, F.R. 33
Mumma, J. 162
Musil, R. 374

Nagel, E. 376
Natorp, P. 387–8
Nelson, L. 134, 384–7
Neumann, C.F. 119n43, 120
Newton, I. 67, 72, 82–91, 154, 293
Nicholson, J. 160
Nietzsche, F. 268–70, 272
Noack, H. 305
Noether, E vii, 39,110n26, 168, 187, 321n29, 352
Noether, M. 109–110, 119
Northeim 300
Nowak, G. 67n3, 74n23

Ore, O 190, 192, 194, 195, 210n2

Padoa, A. 212–3, 378
Parshall and Rowe 24
Parshall, K.H. 24–5
Parsons, C. 361n48, 365n57,
Pasch, M. 212, 316, 376, 378, 388
Paulsen, F. 267–8
Peano, G. 3, 70, 91, 212–3, 378, 383, 396
Peckhaus, V. 395
Peirce, C.S. 384
Piaget, J. 14n21, 77, 95
Picard, E. 252, 372
Picasso, P. 393–4
Pieri, M. 212–3
Pincherle, S. 116n37
Plato 14, 95
Poggendorff, J.C. 82n38
Poincaré, H. 1, 38, 41, 68, 89, 95, 119n43, 188–192, 207, 209–211, 239, 246–50, 267, 285, 287–8, 291–2, 303, 325, 328, 331, 358, 372, 375, 377, 383–4, 386–7, 389
Polanyi, M. 241–2
Pólya, G. 17, 27
Poncelet, J.V. 376n12
Potter, M. 14n21
Prym, F. 110n26
Puiseux, P.H. 246
Putnam, H. 9, 10n12, 17, 40

Quillen, D. 39–40, 246, 257
Quine, W.V.O. 9, 17, 31, 77, 95, 100–101, 159, 236

Radon, J.K.A. 391
Ravenal, D. 244
Ray, G. 209, 218n12
Reck, E.H. 59n11, 60n12, 160–1
Reed, D. 161
Ricci-Curbastro, G. 297
Riemann, G.B.F. 1, 3, 13, 19–20, 37–8, 41, 67–81, 85–96, 98–99, 100, 107, 110–112, 115–116, 118n41, 119, 120n45, 121–122, 127, 129, 151, 166, 246, 316, 327, 337, 377–8, 386, 392
Ritter, A. 91
Robinson, A. 27, 236
Roch, G. 110n26
Rosebrugh, R. 8
Russ, S. 26
Russell, B. 4–5, 7, 13, 37, 59–62, 66, 70, 100–103, 214, 303, 348, 383–4, 388

Sánchez Ron, J.M. 67
Saussure, F. de 395
Saussure, R. de 397
Scheffers, G. 109,
Schering, E.C.J. 83, 87, 123–124
Schlick, M. 41, 267, 285–8
Schlimm, D. 160, 162
Schlömilch, O. 104n14, 195
Schmalfuss, C. 91
Schoenberg, A. 393–4
Schoenflies, A.M. 380
Scholz, E. 71, 78n31, 88n49, 91n58, 92, 192
Schopenhauer, A. 268
Schröder, E. 127, 383, 396
Schrödinger, E. 292, 298, 302
Schröter, H.E. 291
Schwarz, H.A. 119n43, 120
Schwarzschild, K. 282
Seifert and Threlfall 204
Seifert, K.J.H. 252
Selick, P. 258
Serre, J.-P. 246, 254–5, 257
Shapiro, S. 5, 9, 10n13, 10n14, 11n16, 159
Sheffer, H. 101
Sher, G. 209, 211
Sieg, W. 160
Skolem, T.A. 6
Soames, S. 211–2

Name Index

Spalanski, j. 256
Spanier, E.H. 243, 248n10
Speiser, A. 82,
Spinoza, B. 296, 306, 333–4
Springer, T. 108n21
Stachel, J. 377n16
Stahl, H. 110–111, 113
Staudt, C.G.C. von 32–4
Steenrod, N. 252–4, 256n22
Stein, H. 160
Stillwell, J. 163
Stolz, O. 32
Stone, M.H. 299
Stravinsky, I. 394
Sylveste, J.J. 372, 376

Tait, W. 350n28, 350n29, 361n48
Tappenden, J. 19n25, 51n3, 393
Tarski, A. vii, 40, all of *Mancosu*, 313, 323
Tarski-Lindenbaum 223, 224
Tazzioli, R. 82, 89
Thomae, J. 110n26, 124, 126
Thomson, J.J. 36
Toda, H. 246n5
Toretti, R. 72n18, 85, 285n46
Tymoczko, T. 17n24

Ulam, S. 202
Urysohn, P.S. 202

van den Eynde, R. 246n6, 247n7
van der Waerden, B.L. 7, 39, 193, 195, 201
Veblen and Young 376–7
Veblen, O. 191, 204, 205, 217n10, 222, 223n14, 224–5
Vellman, D. 10n14
Vercelloni, L. 210n2
Veronese, G. 140, 143
Vietoris, L. 39, 188, 191, 202–5

Voevodski, V. 240
von Neumann, J. 210n2, 299
Voss, A. 109, 123, 396n45
Vossler, K. 305

Wagner, R. 268
Warschauer, A. 267
Warwick, A. 25, 32
Weber, W 67n2, 72, 80n34, 81, 82n39, 83n42, 88n50, 210
Weierstrass, C. 23, 92, 95, 99–100, 102–103, 105–107, 109, 110–6, 12, 127–129, 130, 372
conjectures Weil 35
Weiner, J. 54n8, 103–4
Wellstein, J. 391
Weyl, H. viii, 1, 3n4,6, 8, 15, 17, 20, 26, 38, 41, 68, 81n35, 90, 94–95, 136n2, 175, 187, 191, 205, 207, 210n2, all of *Scholz*, 291–310, 331, 348, 375
Whitehead, A.N. 13n18, 59, 214, 376, 383
Whitehead, G 243, 246n5, 248n10
Whitney, H. 252n18
Wiener, L 140
Wilson, M. 57n11
Wise, Norton M. 84
Wittgenstein, L. 60, 73, 305, 319n24, 322, 336, 383
Wright, C., 47
Wundt, W. 268
Wussing, H. 168,

Young, J.W. 214n5

Zermelo, E. 7, 144–5, 210n2, 353ff, 364n56
Zisman, M. 252

Subject Index

Abelian functions 113, 119, 120, 124
Abelian integrals 125, 129
Algebra 7, 15, 19, 20, 23, 24, 40, 110n26, McLarty passim, 323, 329n51, 330, 342, 351, 372, 390
Algebraic geometry 101n7, 110, 206, 366
Algebraic integer 39, 163n5, 169–72, 174, 175n16, 179, 379
Algebraic topology 40, 187–8, 206
Analysis 6, 7, 16, 23–24, 59, 79, 92, 140, 226, 234, 263, 266, 284, 289, 303, 320, 339–344, 346–351, 355, 360, 366, 372–373, 379–382, 390–392, 394–395
  constructive analysis 6, 29, 95, 349–351, 353, 367
  differential equations 19, 22, 24, 69, 82–83, 85, 109n, 115, 152, 269, 274
  non-standard analysis 27
Analytic philosophy (and philosophical analysis) 1, 9, 37, 42, Beaney passim, 132, 162, 186, 292, 313, 319, 360
Analytic, analyticity 29–30, 48, 49–52, 66, 74n, 228n, 230n, 318, 383–84, 386
Anschauung 48, 93, 94, 137–8, 141, 143, 148, 154–5, 284n45, 286n47
Applied mathematics, applications 20, 23, 24, 36, 148, 267
Apriorism, a priori 13, 14n21, 18, 38, 48, 72, 74, 95, 141, 154–5, 284, 311, 312, 314–15, 325, 326,328, 334, 380, 384–87, 389
Architecture (of mathematics) 2–3, 282, 285, 316, 317, 330, 366n60
Arithmetic 6, 9, 14, 29, 47–50, 52, 58–60, 62, 70, 91, 95, 103, 105, 125, 127, 138–147, 150, 153,156, 177, 316, 317, 344–47, 350–4, 357–8, 363–64, 386, 388
Arithmetisation of analysis 92, 99–101, 105–106, 122, 342, 346–347
Axioms 64–5, 75, Corry passim, 194n14, 195, 212, 213, 215–7, 219, 221, 222–3, 234, 240, 256–7, 266, 276–77, 280–82, 284, 289, 311–13, 316–17, 319–21, 323, 327, 330, 340, 342, 345–48, 351–52, 355–59, 364, 366, 373, 376–8, 380–82, 385–86, 390–92, 396
axiom of Archimedes 33
axiom of choice 6, 7, 13, 14, 16, 29, 196n, 350–51
axiom of infinity 6, 7, 13, 14, 217–18, 226, 228, 232, 364
axiom of parallels 13, 139, 142, 152
axioms of a model category 256–57
completeness axiom 13, 16, 143, 144
Frege's axiom V (basic law V) 56–8, 63, 65
Hilbert's axioms 71n, 142, 342, 346, 366, 388
power set axiom 7, 364
Zermelo-Fraenkel axioms 29, 280, 363n, 364
Axiomatization, axiomatics 137, 144–9, 160, 167, 185

Begriffsschrift 52
Boole's logic 52

Calculus 22, 36, 59n, 69, 379, 391, 393
  of variations 113, 239, 246, 381–82
Cantor-Hume principle 60n15
categorical 217n10, 222–3, 225, 227, 235, 250–1, 336, 346
Category theory 8, 9, 39–40, 256, 260
Cauchy-Riemann equations 69, 109n25, 115–7
Clifford-Klein space forms 282, 283n44
Completeness 141–2, 144, 222, 226–7, 228, 229–31, 350, 352, 356–57
Complex analysis (see Function theory) 26, 38, 67, 103–4, 108–122, 205, 372, 391
Conjecture: 7, 17, 34–36, 185, 244, 248, 356–57
  Cantor's continuum hypothesis 29, 145, 364n

conjectures of Taniyama-Shimura 35
conjectures of Weil 35
Riemann hypothesis 28
Consistency 215, 339–342, 347–351, 355–358, 360–62, 367–368, 377
Constructivism (constructive math) 6–7, 29, 95, 341, 349–351, 353, 367
Continuity 27, 36, 51, 77–78, 79, 99n, 101, 103, 104–05, 131, 140, 143, 176, 200–01, 205, 278, 283, 289, 344, 345, 353
Continuum 29, 70, 75, 79, 96, 270–72, 274, 276, 278, 283, 297, 331, 347n, 350, 353, 367
Conventionalism 152, 386, 389
cyclotomic integers 164–5, 168, 170–1

Dedekind's *Was sind*... 54, 56, 343n, 345, 346n, 352, 365
Dialectics, dialectical 297, 302–303, 314, 320n, 326n, 332–335, 368
Dirichlet principle 92, 107, 117–120, 130
Disquisitiones Arithmeticae 112, 167
Duhem-Quine thesis 31

Ecole Normale 23, 24, 311
Ecole Polytechnique 23, 24
Electromagnetism 148, 154
Empiricism 9, 39, 41, 74, 155, 266–67, 279, 284, 288, 289, 368
Epistemic configurations 20
Epistemology 9–11, 31, 39, 67, 72–3, 76–7, 84, 88, 95, 98, 114, 184, 240, 266, 273, 280, 281, 288, 311–14, 317, 319, 324, 326, 330–33, 337
Erlangen Program 32, 375, 388
Euclid's *Elements* 28, 33, 49, 345, 378

Field theory 72, 144, 160, 161n, 162, 163n, 169–71, 178, 181, 194n, 329n, 345, 366
  physical theory of fields 84, 87, 137, 149, 151–52, 298, 301, 306
Finitism 38, 134, 136, 138, 142, 153, Sieg passim, 396
Formalism 5–6, 9, 17, 31, 38, 62–5, 135–6, 137, 148, 153, 155, 292, 312, 316, 351–53, 356–57
Foundations (foundational studies) 5–8, 47, 48, 53, 58, 63n19, 66, 69–70, 100, 263, 280, 282, 285, 287, 292, 294–96, 303, 332–33, 340–41, 347, 349–50, 358–59, 367–68, 372, 373, 378, 388–89, 396

Function (see also Complex analysis) 24, 27, 36, 56, 68, 69, 71, 89, 92, 98, 101, 104, 107, 111, 131, 166, 173, 178, 183, 185–86, 204–05, 274, 298, 306, 356n, 366, 379, 380–81, 388
  Function space 250–51, 254, 382
  propositional function 214, Mancosu passim
  recursive function 341, 351
Function theory (see Complex analysis) 15, 68, 69, 72, 82, 92–93, 112–3, 117, 125, 138, 166–67, Tappenden passim, 239, 246
Functor 204–06

Gaussian integers 39, 164, 171
General Relativity Theory, GRT 38, 41, 87, 137, 147, 149–155, 279, 285–88, 293, 294, 297–302, 305
Geometry:
  differential 38, 80, 82, 85, 88–90, 91, 94, 252, 366, 373, 375, 376–377, 391
  Euclidean 13, 33, 49, 50–51, 60, 65, 75, 88, 119, 138–140, 142, 147–48, 150–52, 154–55, 210n2, 228, 276–78, 282–83, 284, 287–88, 331, 348n, 377, 384, 386–87, 388–89, 396n
  non-Archimedean 71, 143, 316
  non-Euclidean 29, 36, 95, 100n, 101n, 152, 210n, 277–278, 281–283, 287, 311, 316, 341, 375, 377–378, 384–387, 389, 391–392
  projective 32–34, 108n, 140, 142, 280, 375–377, 385, 388
Group theory 40, 64, McLarty passim, 217, 231, 251, 299, 323, 373, 375
  fundamental group 207, 247, 249–251
  homology groups 188, 190–91, 202–04, 207–08, 249
  homotopy groups 243–57
  Lie group 190, 253, 295, 375
Grundgesetze 50, 52, 54, 56–7, 60, 100, 125, 126, 129n64, 130
Grundlagen (Cantor) 70
Grundlagen (Frege) 50, 52, 53, 54–6, 60, 64, 71, 104, 108, 125–8
Grundlagen der Geometrie (GdG, Hilbert) 32–33, 38–39, 41, 64, 71, 140–148, 151, 266, 279, 282, 316, 331, 348, 350n28, 351–52, 356–58, 361, 377, 378, 391

440

Subject Index

Hilbert programme 38–39, 43, 135–36, 138, 142, 153, 303, 339–40, 355, Sieg passim
Homology 188–91, 197, 200, 202–4, 206, 239, 243, 245, 247–51, 256, 258–59
Homomorphism 188, 192–207
Homotopy theory 40, 207–8, Marquis passim
Hypothesis (hypothetical conception, quasi-empirical) 13–16, 37, 68, 74–5, 326

Ideals, ideal theory 15, 39, 113, 132n, Avigad passim, 193–94, 196, 199, 326, 346
Idealism 6, 84, 269, 273, 286, 291–292, 293, 303, 305, 313, 328
Image of mathematics 3, 39, 135, 148, 161
Integral 379–381
Intuition (see also *Anschauung*) 23, 92, 105, 107–9, 117–9, 122–23, 129–130, 137, 138, 141–48, 150, 154, 280, 282–84, 298, 312, 316, 318, 322, 324–25, 328, 330–32
Intuitionism 5–6, 8, 70, 73–4, 92–4, 318, 341, 347, 350–51, 353–55, 361, 363, 367, 384–88

Language 383–84, 395–96
Logic, mathematical 15, 19, 20, 30, 40, 368
Logical consequence 40, Mancosu *passim*
Logicism 5–6, 47, 52, 321, 372, 384

Manifold 3, 69, 70, 74–6, 80, 86–9, 91–2, 93, 174n14, 188, 197, 246–9, 271, 301
Mapping 57, 69, 73, 116, 160, McLarty passim, 246, 248–250, Marquis passim, 271–72, 365
Materialism 84, 87, 273, 288n, 293, 297n, 334
Mathematical physics 16, 24, 144, 301, 348, 372
Metamathematics 3n, 6, 7, 234, Mancosu passim, 284, 323, 325, Sieg passim
Metaphysics 2, 9, 39, 75, 78, 84–85, 90, 106, 108n, 113, 114n, 159, 161, 185, 266–267, 268, 272–273, 275, 279, 284–286, 288–289, 293–295, 296, 298, 300, 301, 306, 307–308, 311, 367

Michelson's experiment 154
Model 40, 209–11, 216, 219, 223–4, 226–30, 233, 317, 323, 331, 341, 345, 355, 358, 363, 366
Modernism 43, 371, 374–75, 377, 380, 383, 389–90, 392–395
Module 192–3, 196, 198

Naturalism 5, 11, 14n, 18–20
Neo-humanism 23, 24
Numbers 3, 14, 55, 61, 63–4, 66, 127, 128, 150, 160, 178, 323, 325, 326, 331
  natural 6, 15, 28, 30, 55, 57, 58–59, 128, 146, 150, 343
  real 100–01, 128–29, 176, 344, 345, 366
  complex 105, 114, 127, 163–65, 169, 178
Number theory 15, 19, 23, 28, 29, Avigad passim, 340, 342–44, 349–352, 356–57, 360–61, 364, 372, 378

Object, mathematical 6, 7, 9, 11, 31, 40, 49, 63–64, 116, 148, 167–68, 172–74, 185, 197, 199, 241, 258–59, 317, 325, 326, 331, 335, 341, 343, 344–46, 348, 355, 357, 363, 364
  logical object 58
Ontology 9, 11–12, 17, 161, 217n10, 246, 299, 303, 307

Paradoxes (antinomies) 71, 78–79, 96, 101, 174n, 348n, 352, 396
  Banach-Tarski paradox 264
  Hausdorff paradox 263–64
  Russell's paradox 56n, 62, 145
Peano arithmetic 14, 15, 60, 212, 233, 352
Phenomenology 42, 43, 291, 296, 300, 318–319, 324, 326, 327, 336–337
Platonism 5, 7, 9, 17, 43, 95, 341, 364
Postulational math 7, 79
Practice, mathematical 1, 5–6, 10–12, 15–20, 25–26, 38, 42, 59, 65, 68, 97, 101n, 102–04, 115, 123n, 130, 134, 136, 147, 159–60, 184, 233, 241–42, 291–92, 303, 305–06, 307, 314–15, 318–20, 341, 349–50, 365, 367
Predicativism (impredicative) 5, 7, 8, 29, 303, 351, 355, 360, 362n
Proof theory 8, 40, 43, 313, 328, 339, 350, 359
Proof 18–19, 30

Pure mathematics 24, 28, 36, 63, 76, 88, 91, 93, 94n, 106, 117, 138, 140–41, 148, 241
Purity of method 19, 112, 167, 183

Quantum theory 42, 292–94, 296–303, 305

Rationalism 328, 384
Rationality 9, 12, 18
Realism 9, 42, 77, 266–67, 268, 273, 274, 279, 291–92, 301–06, 315, 322, 325, 326
Ring theory 39, 164–65, 169–71, 176, 188, 193–94, 196–97, 199, 244–45

Set theory 9, 11, 12, 14, 19, 20, 24, 28–31, 40, 41, 76, 91n56, 95, 145, 150, 210n2, 217n9, 222n14, 235, 263, 266, 272, 279–80, 294–95, 317, 320–21, 339, 346–47, 358, 363, 372, 380, 392, 394, 396
Space problem 80, 89–90, 281
Space-time 79, 81, 297, 361n, 389
Structuralism 5, 11–12, 30, 42, 43, 59n, 204

Technology 241–3, 249, 257, 259–60
Theorems:
  Abel's theorem 115
  Brouwer's fixed point 188, 248
  Dedekind's Prague theorem 183
  Dirichlet's on arithmetical progressions 342–43
  Gödel's incompleteness theorems 227n, 350, 352, 359
  Kronecker's 'Fundamentalsatz' 178
  Fundamental theorem of algebra 88n, 178
  Fundamental theorem of calculus 379, 381
  HLP theorem 252–53
  Homorphism and isomorphism theorems 188, 192–96
  Löwenheim-Skolem 212n, 234–35
  theorem of Desargues, 32–35, 316, 327
  theorem of Fermat, 28–29, 35, 349
  theorem of Pappus, 4, 33–34, 316
  theorem of unique factorisation 163–65, 174–76
  Zermelo's well ordering theorem 145, 196n
  Wilson's theorem 112, 167
Topology (*analysis situs*) 39, 88, 92, 206, McLarty passim, 239, 247, Marquis passim, 320, 323
Truth 9, 211, 214, 218–19, 230, 232–6, 266, 289, 316–18, 343, 347, 352, 358, 377, 378, 383–84, 388
Types, theory of 217–9, 222, 224, 228, 232–3, 235

Vienna circle 285, 286
Visualization 18–19, 116, 323, 325

Zermelo-Fraenkel (ZF) 8, 29, 363–64